D3P1595:

Handbook of
ELECTRONIC
TEST EQUIPMENT

PRENTICE-HALL SERIES IN ELECTRONIC TECHNOLOGY

DR. IRVING L. KOSOW, editor

CHARLES M. THOMSON, JOSEPH J. GERSHON, JOSEPH A. LABOK, consulting editors

PRENTICE-HALL INTERNATIONAL, INC., *London*
PRENTICE-HALL OF AUSTRALIA, PTY. LTD., *Sydney*
PRENTICE-HALL OF CANADA, LTD., *Toronto*
PRENTICE-HALL OF INDIA PRIVATE LTD., *New Delhi*
PRENTICE-HALL OF JAPAN, INC., *Tokyo*

Handbook of ELECTRONIC TEST EQUIPMENT

JOHN D. LENK

Prentice-Hall, Inc., Englewood Cliffs, N.J.

© 1971 by
PRENTICE-HALL, INC.
Englewood Cliffs, N.J.

All rights reserved. No part of this
book may be reproduced in any form
or by any means without permission
in writing from the publisher.

Current printing (last digit):
10 9 8 7 6 5 4 3 2

13-377366-3
Library of Congress Catalog Card Number: 78-135753
Printed in the United States of America

Dedicated to my wife, Irene,

whose idea it was to write this book

PREFACE

All electronic technicians and engineers use some form of test equipment. Most technicians use many types of test equipment in their daily work. It is therefore essential that technicians know what test instruments are available, what they do, and how they operate. This handbook fills that specific need.

No single book could list all items of test equipment available in today's electronics field, much less describe such equipment. However, this handbook describes both the purpose and operational principles related to *all types* of test equipment. By concentrating on generalized circuits, at the block diagram (or simplified schematic) level, the reader will find that once the information in this handbook has been digested the *specific circuit* of any electronic test instrument can be understood. For example, an entire appendix is devoted to digital logic circuits. If the readers study and understand this information, they will be able to read and interpret the logic diagrams found in much of today's electronic test equipment.

This book, therefore, serves as a refresher for the experienced technician and a basic text for the student technician.

The most commonly used test instruments in both shop and laboratory are meters, oscilloscopes, and signal generators. For that reason, the *basic operating procedures* for these three instruments are included in applicable chapters. These operating procedures are typical and can be used as a supplement to the procedures found in the instruction manual for the equipment. The operating procedures included here are especially useful when no instruction manual is available.

If, after digesting these basic procedures and the instruction manual data, the readers want further information on applications of test equipment, their attention is invited to the author's Handbook of Oscilloscopes, Handbook of Electronic Meters, and Handbook of Practical Electronic Tests and Measurements, all published by Prentice-Hall, Inc.

All test equipment should be tested and calibrated periodically against known standards, since the instruments can shift in accuracy because of damage or normal aging. Test equipment should also be checked out thoroughly when first placed in use. If the equipment is new, it can be checked against manufacturers specifications. If the equipment is used and no specifications are available, the technician can establish a reference point for future periodic calibrations.

Laboratory test equipment must be checked against precision standards available in the laboratory or must be sent out to calibration laboratories. A separate chapter is included on laboratory standards and calibration procedures. These data permit the technician to perform the calibration or to understand the basis for outside calibration.

A special problem is encountered with shop test equipment. Most shops are not equipped with precision standards, and it is expensive and time-consuming to send equipment out for calibration. Therefore, means must be devised to check test equipment with available standards. Procedures have been included to permit the calibration and testing of shop equipment against commonly available standards and to make maximum use of these standards. If the test reveals that a particular test instrument is not up to standard, the manufacturer's data can be consulted to find such information as location of calibration controls, test limits, and so on.

The author has received much help from various organizations in writing this book. He wishes to give special thanks to the following: B & K Manufacturing, Fairchild, General Radio, Hewlett-Packard, Radio Corporation of America, Sencore, Simpson Electric Co., Tektronix Inc., and Triplett Electrical Instrument Co.

J. D. LENK

Los Angeles

CONTENTS

INTRODUCTION

1. ANALOG METERS—1

1-1.	Meter Basics	1
1-2.	D'Arsonval Movement and the Basic VOM	2
1-3.	Basic Alternating-Current Meters	10
1-4.	Laboratory Analog Meters	15
1-5.	Meter Scales and Ranges	22
1-6.	Meter Protection Circuits	31
1-7.	Meter Accuracy	33
1-8.	Basic Meter-Operating Procedures	36
1-9.	Basic Ohmmeter (Resistance) Measurements	37
1-10.	Basic Voltmeter (Voltage) Measurements	39
1-11.	Basic Ammeter (Current) Measurements	41
1-12.	Basic Decibel Measurements	42
1-13.	Ohmmeter Test and Calibration	42
1-14.	Voltmeter Test and Calibration	43
1-15.	Frequency Test and Calibration	47
1-16.	Ammeter Test and Calibration	49

2. DIGITAL AND DIFFERENTIAL METERS—51

2-1.	Introduction to Digital Circuits	51
2-2.	Basic Digital Meter	52
2-3.	Ramp-Type Digital Meter	53
2-4.	Integrating-Type Digital Meter	55
2-5.	Integrating/Potentiometric-Type Digital Meter	58
2-6.	Dual-Slope Integration-Type Digital Meter	59
2-7.	Basic Differential Meter	64

2-8. Differential Meter Circuits 66
2-9. Common-Mode Signal Problems 69
2-10. Measurement of Resistance with a Differential Voltmeter 71

3. BRIDGE-TYPE TEST EQUIPMENT—74

3-1. Wheatstone Bridges 74
3-2. Alternating-Current Bridges 77
3-3. Universal Bridges 78
3-4. Standard Capacitance and Inductance Bridges 83
3-5. Automatic Bridges 86
3-6. Q Meters 87
3-7. RX Meters 88
3-8. Admittance Meters 89
3-9. Vector Impedance Meters 91
3-10. Low-Current Meters 92

4. SIGNAL GENERATORS—95

4-1. Signal Generator Basics 95
4-2. Radic-Frequency Signal Generators 96
4-3. Sweep and Swept-Frequency Generators 106
4-4. Marker Generator and Time Base Generator 116
4-5. Audio Generators and Function Generators 122
4-6. Frequency Synthesizers 130
4-7. Pusle Generators and Digital Delay Generators 139
4-8. Frequency-Modulated Stereo Generator 144
4-9. Color Television Signal Generators 150

5. OSCILLOSCOPES AND RECORDERS—163

5-1. Oscilloscope Cathode-Ray Tube 164
5-2. Oscilloscope Beam Deflection System 165
5-3. Basic Frequency Measurement with an Oscilloscope 166
5-4. Basic Voltage Measurement with an Oscilloscope 167
5-5. Basic Oscilloscope Circuits 167
5-6. Oscilloscope Operating Controls and Indicators 173
5-7. Storage Oscilloscope 182
5-8. Sampling Oscilloscope 184
5-9. Vectorscope 189
5-10. Curve Tracers 195
5-11. Recorders 198
5-12. Oscilloscope Operating Precautions 200

5-13.	Placing an Oscilloscope in Operation	202
5-14.	Recording an Oscilloscope Trace	204
5-15.	Measuring Voltage and Current with an Oscilloscope	205
5-16.	Measuring Time and Frequency with an Oscilloscope	220
5-17.	Measuring Phase with an Oscilloscope	227

6. ELECTRONIC COUNTERS—232

6-1.	Basic Electronic Counter	232
6-2.	Counter Circuit Elements	233
6-3.	Totalizing Operation	234
6-4.	Frequency Measurement Operation	235
6-5.	Period Measurement Operation	236
6-6.	Time-Interval Measurement Operation	238
6-7.	Ratio Measurement Operation	239
6-8.	Counter/Readout and Divider Circuit Operation	239
6-9.	Counter Display Problems	246
6-10.	Counter Accuracy	247

7. AMPLIFIERS—250

7-1.	Amplifiers as Test Equipment	250
7-2.	Types and Functions of Amplifiers	250
7-3.	Typical Amplifier Characteristics and Test Methods	261
7-4.	Operating Precautions for Amplifiers	266
7-5.	Differential and Operational Amplifiers in Test Applications	267

8. FREQUENCY AND TIME STANDARDS—278

8-1.	Timekeeping Methods	279
8-2.	Time and Frequency Broadcasts	281
8-3.	Time and Frequency Standard Equipment	289
8-4.	Practical Frequency Calibration Techniques	300
8-5.	Practical Time Calibration Techniques	307
8-6.	Practical Phase Comparison Techniques	311
8-7.	Definitions for Terms Used in Frequency and Time Standard Instruments	312

9. PROBES AND TRANSDUCERS—317

9-1.	The Basic Probe	317
9-2.	Low-Capacitance Probes	318

xii   Contents

9-3.   Resistance-Type Voltage-Divider Probes 319
9-4.   Capacitance-Type Voltage-Divider Probes 320
9-5.   Radio-Frequency Probes 320
9-6.   Demodulator Probes 322
9-7.   Special Purpose Probes 323
9-8.   Probe Compensation and Calibration 324
9-9.   Probe Operating Techniques 326
9-10. Transducers 327
9-11. Resistance-Changing Transducers 328
9-12. Self-Generating Transducers 330
9-13. Inductance- and Capacitance-Changing Transducers 335

10. WAVE ANALYSIS INSTRUMENTS—337

10-1. Distortion Measurement 337
10-2. Modulation Measurement 343
10-3. Wave Analyzers 344
10-4. Spectrum Analyzers 349
10-5. Time Domain Reflectometers 358

11. MICROWAVE TEST EQUIPMENT—364

11-1. Microwave Transmission Lines 364
11-2. Microwave Resonant Circuits 366
11-3. Transfer of Energy in Microwave Circuits 370
11-4. Microwave Signal Sources 372
11-5. Modulation, Attenuation, and Level Control for
       Microwave Signal Sources 374
11-6. Microwave Power Measurements 378
11-7. Microwave Measurement Devices 382
11-8. Fixed-Frequency Versus Swept-Frequency Measurements 387
11-9. Microwave Impedance Measurements 387
11-10. Microwave Attenuation Measurements 393
11-11. Microwave Frequency Measurements 394

APPENDIX Digital Logic Circuits   401

INDEX   453

INTRODUCTION

A thorough study of this handbook will make the reader familiar with the *basic* principles and operating procedures for all types of electronic test equipment. It is assumed that the reader will take the time to become equally familiar with the principles and operating controls for any *particular test equipment* he will use. Such information is contained in the instruction manual for the particular equipment. It is absolutely essential that the operators become thoroughly familiar with their particular test instruments. No amount of textbook instruction will make the operator an expert in operating test equipment; it takes actual practice.

It is strongly recommended that readers establish a *routine operating procedure or sequence of operation* for each item of test equipment in the shop or laboratory. This will save time and familiarize the readers with the capabilities and limitations of their particular equipment, thus eliminating false conclusions based on unknown operating conditions.

SAFETY PRECAUTIONS

In addition to a routine operating procedure, certain precautions must be observed during operation of any electronic test equipment. Many of these precautions are the same for all types of test equipment; other precautions are unique to special test instrument such as meters, oscilloscopes, signal generators, and so on. Some of the precautions are designed to prevent damage to the test equipment or the circuit under test; others are to prevent injury to the operator. Where applicable, special safety precautions are included throughout the various chapters of this handbook.

The following *general safety precautions* should be studied thoroughly and then compared to any specific precautions called for in the equipment instruction manuals and in the related chapters of this handbook.

1. Many test instruments are housed in metal cases. These cases are connected to the ground of the internal circuit. For proper operation, the ground terminal of the instrument should always be connected to the ground of the equipment under test. Make certain that the chassis of the equipment under test is not connected to either side of the a-c line, or to any potential above ground. If there is any doubt, connect the equipment under test to the power line through an *isolution transformer*.

2. Remember that there is always danger in testing electrical equipment that operates at hazardous voltages. Therefore, the operator should familiarize himself thoroughly with the equipment under test before working on it, bearing in mind that high voltage may appear at unexpected points in defective equipment.

3. It is good practice to remove power *before connecting* test leads to high-voltage points. (High-voltage probes are often provided with alligator clips.) It is preferable to make all test connections with the power removed. If this is impractical, be especially careful to avoid accidental contact with equipment and objects that can procide a *ground*. Working with one hand in your pocket and standing on a properly insulated floor lessens the danger of shock.

4. Capacitors may store a charge large enough to be hazardous. Therefore, discharge filter capacitors *before* attaching the test leads.

5. Remember that leads with broken insulation offer the additional hazard of high voltages appearing at *exposed* points doing the leads. Check test leads for frayed or broken insulation before working with them.

6. To lessen the danger of accidental shock, disconnect test leads *immediately after* the test is completed.

7. Remember that the risk of severe shock is only one of the possible hazards. Even a minor shock can place the operator in danger of more serious risks, such as a bad fall or contact with a source of higher voltage.

8. The experienced operator guards continuously against injury and does not work on hazardous circuits unless another person is available to assist in case of accident.

9. Even if you have had considerable experience with test equipment, always study the instruction manual of any instrument with which you are not familiar.

10. Use only shielded leads and probes. Never allow your fingers to slip down to the metal probe tip when the probe is in contact with a "hot" circuit.

11. Avoid vibration and mechanical shock. Most electronic test equipment is delicate.

12. Study the circuit under test *before making any test connections*. Try to match the capabilities of the instrument to the circuit under test.

1 ANALOG METERS

Most meters used in both shop and laboratory work are *analog meters;* that is, they use rectifiers, amplifiers, and other circuits to generate a *current proportional to the quantity being measured.* This current, in turn, drives a meter movement. Even the digital meters described in Chapter 2 are of the analog type, but they differ from the conventional analog meter in the manner of readout. Although both shop and laboratory meters use the same basic analog principles, laboratory-type meters include many circuit refinements to improve their accuracy and stability. This chapter describes the basic principles of analog meters and shows how these basic principles are adapted to both shop and laboratory use.

1-1. METER BASICS

The main purpose of any meter used in electronics is to check on circuits and components—to find what voltage is available, how much current is flowing, and so on. The simplest and most common instrument that will measure the three basic electrical values (voltage, current, and resistance) is the *Volt-ohm-milliammeter,* or VOM. There are dozens of VOMs available in all price ranges. As the price goes up, accuracy is increased, more scales or functions are added, and the scales are given greater range.

The first improvement on the VOM was the *vacuum-tube voltmeter,* or VTVM. Today the VTVM is generally being replaced by the *electronic voltmeter,* such as the transistorized meter, field-effect meter, and so on. The sensitivity of the VTVM or the electronic voltmeter is much greater than that of the VOM since they both contain a vacuum-tube or transistor *amplifier* between the meter movement and the input. The VTVM and electronic meter have another advantage over the VOM in that they present a *high impedance* to the circuit or component being measured. Thus they draw little or no current from the circuit and have little effect on circuit operation.

1-2. D'ARSONVAL MOVEMENT AND THE BASIC VOM

Except for digital meters (Chapter 2) all meter circuits are designed around the basic meter movement. Virtually all nondigital meters use some form of the D'Arsonval meter movement shown in Figure 1-1.

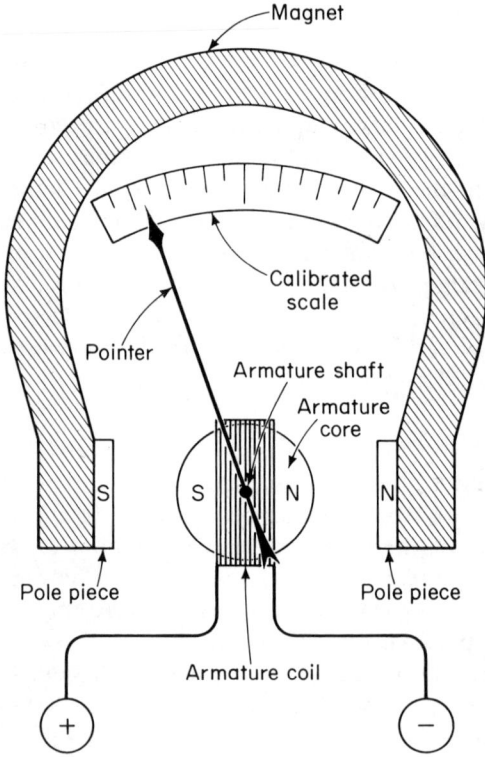

Fig. 1-1. The basic D'Arsonval meter movement.

This movement is also known as the *moving-coil galvanometer*. Early D'Arsonval movements had a core made of soft iron. A coil of very fine wire was wound on an aluminum bobbin formed around the core. The iron core is now usually omitted from the movement. The coil and aluminum form function somewhat like an armature mounted on a shaft seated in jewel bearings in order to be free to turn (rotate). Springs on each end of the shaft act as current leads to the coil and help steady the movement.

The coil is placed between the poles of a U-shaped permanent magnet. One end of a pointer is fastened to the armature shaft. As the shaft rotates, the other end of the pointer moves over a calibrated scale. Current through the armature coil sets up a magnetic field that reacts with the permanent magnet's field to rotate the coil with respect to the magnet. When current passes through the coil, its magnetic field is such that the poles repel and, since the permanent magnet cannot move, the coil rotates on its shaft. Current through the coil makes the coil turn a proportional amount. Thus the basic meter movement is an analog device.

The amount of travel of the pointer attached to the coil is related directly to the amount of current flowing through the movement. The meter scale is then related to some particular current. For example, if 1 mA is required to rotate the coil and pointer across the full scale, a half-scale reading will be equal to 0.5 mA, a quarter-scale reading will be equal to 0.25 mA, and so on.

Usually, maximum rotation of the armature (full-scale reading) is completed in less than a half-turn in the clockwise direction. The complete assembly is enclosed in a glass-faced case that protects it from dust and air currents. This enclosed meter movement can be used *by itself* as a very sensitive *ammeter*. However, the movement is usually part of another instrument, such as a VOM, or of a panel connected to an external circuit. In the case of a laboratory or shop meter, there is a resistor network to extend the range of the basic movement (as an ammeter), or to convert the basic movement to a voltmeter.

1-2.1. Basic Ammeter

By itself, the basic D'Arsonval movement forms an *ammeter* (*am*pere *meter*). A true ammeter measures current in amperes. In electronics, current is more often measured in milliamperes or microamperes. Most movements used in electronic meters will produce full-scale deflection when 50 μA are passed through them.

A *shunt* must be connected across the meter movement if it is desired to measure currents greater than the full-scale range of the basic meter. The shunt can be a precision resistor, a bar of metal, or a simple piece of wire. Electronic meter shunts are usually precision resistors that may be selected by means of a switch. Panel meters for heavy industrial work use metal bar shunts. Shunt resistance is only a fraction of the movement resistance. Current divides itself between the meter and shunt, with most of the current flowing through the shunt. Shunts must be precisely calibrated to match the meter movement.

Figures 1-2 and 1-3 show the two typical milliammeter range-selection circuits for VOMs. In Figure 1-2, individual shunts are selected by the range-scale selector. In Figure 1-3, the shunts are cut in or out of the circuit by the

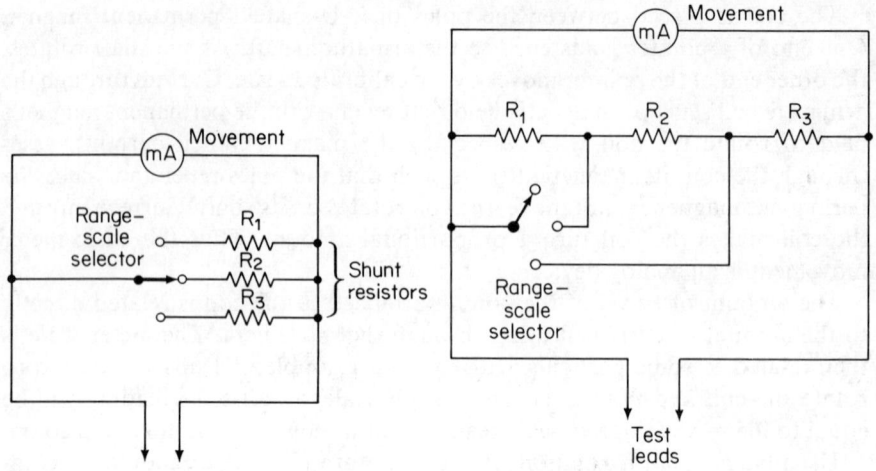

Fig. 1-2. Typical milliammeter range-selector circuit (individual shunt method).

Fig. 1-3. Typical milliammeter range-selector circuit (series shunt method).

selector. If the selector is in position 1, all three shunts are across the meter movement, giving the least shunting effect (most current through the movement). In position 2, resistor R_1 is shorted out of the circuit, with resistors R_2 and R_3 shunted across the movement, increasing the meter's current range. In position 3, only R_3 is shunted across the movement, and the meter reads maximum current.

1-2.2. Calculating Shunt Values

The shunt resistance required to convert a basic meter movement to an ammeter capable of measuring a given current range can be calculated by either of two ways.

With method 1, use the following equation:

$$\text{Shunt resistance } (\Omega) = \frac{R_m}{N-1}$$

where R_m is the internal resistance of the meter movement in ohms
N is the multiplication factor by which the scale factor is to be increased

For example, assume that a meter has a 50-μA full-scale movement (with 50 equal divisions on the scale) and an internal resistance of 300 Ω and that it is desired to convert the meter to measure 0 to 100 mA. (Each of the 50 scale divisions will then represent 2 mA.)

1. Find the multiplication factor (or N factor) by which the movement is to be increased: N equals the desired scale factor (0.1 A or 100 mA) divided by movement current (0.00005 A), or 2000.
2. Use the equation to find shunt resistance:

$$\text{Shunt resistance} = \frac{300}{2000 - 1} = 0.1508 \, \Omega$$

With method 2, follow these steps:

1. Find the voltage required for full-scale deflection of the meter movement: $E = IR$, or $(3 - 10^2) \times (5 \times 10^{-5}) = 0.015$ V.
2. Subtract the meter movement full-scale current (0.00005 A) from the desired current (0.1 A) to find the current that must flow through the shunt: $(0.1 - 0.00005 = 0.09995$ A).
3. Using Ohm's law, find the shunt resistance: $R = E/I$, or $0.015/0.09995 = 0.1508 \, \Omega$.

1-2.3. Calculating Meter Movement Internal Resistance

To find shunt values, it is necessary to know the full-scale deflection current and the resistance of the basic meter movement. Usually, the full-scale deflection current is indicated on the scale face. However, the internal resistance is usually not marked.

The internal resistance of a meter movement can be found by using the test circuits of Figures 1-4 and 1-5. These same test circuits can be used to find the internal resistance of a VOM on its various direct-current ranges.

To use the circuit of Figure 1-4, disconnect R_2 from the circuit by opening switch S_1. Adjust R_1 until the meter movement is at full-scale deflection. Close switch S_1 and adjust R_2 until the meter movement is at exactly one-half

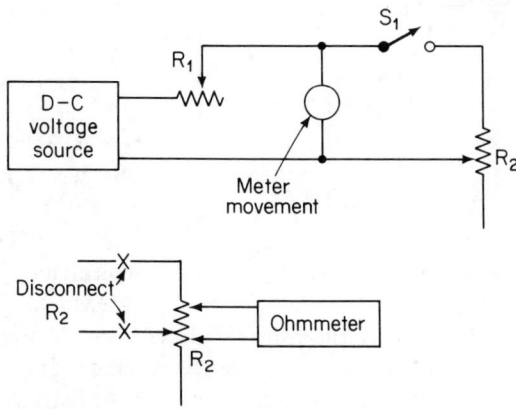

Fig. 1-4. Finding internal resistance of meter movement (half-scale method).

6 Analog Meters

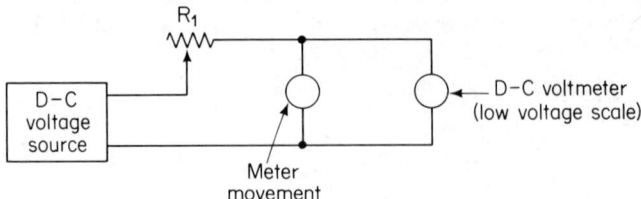

Fig. 1-5. Finding internal resistance of meter movement (voltage-current method).

scale deflection. Disconnect R_2 from the circuit and measure the R_2 resistance value with an ohmmeter. The R_2 resistance value is equal to the internal resistance of the meter movement.

To use the circuit of Figure 1-5, adjust R_1 until the meter movement is at some exact current value. Calculate the internal resistance, using Ohm's law ($R = E/I$). Note that the voltage drop across a typical movement will be less than 1 V. Therefore, it will probably be necessary to use the lowest scale of the voltmeter.

Caution: Never connect an ohmmeter directly across the meter movement. This will damage (and probably burn out) the instrument.

1-2.4. Basic Voltmeter

When the basic D'Arsonval movement is connected in series with resistors, a voltmeter is formed. The series resistance is known as a *multiplier*, since the resistance multiplies the range of the basic meter movement.

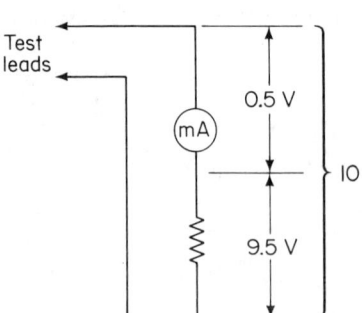

Fig. 1-6. Basic voltmeter circuit.

The basic voltmeter circuit is shown in Figure 1-6. As shown, the voltage divides itself across the meter movement and the series resistance. If an 0.5-V full-scale deflection meter movement were used and it were desired to measure a full scale of 10 V, the series resistor would have to drop 9.5 V. If a 100 V full scale were desired, the series resistance would have to drop 99.5 V, and so on.

Figures 1-7 and 1-8 show the two typical voltmeter range-selection circuits for VOMs. In Figure 1-7, individual multipliers are selected by the range-scale selector. In Figure 1-8, the multipliers are cut in or out of the circuit by the selector. If the selector is in position 1, only resistor R_1 is in the circuit, giving the least voltage drop (meter will read the

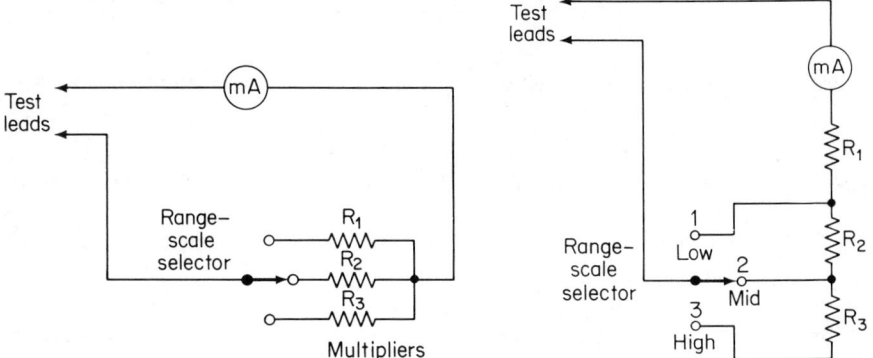

Fig. 1-7. Typical voltmeter range-selector circuit (individual multiplier method).

Fig. 1-8. Typical voltmeter range-selector circuit (series multiplier method).

lowest voltage). In position 2, both R_1 and R_2 are in the circuit, giving the meter a higher voltage range. In position 3, all three resistors drop the voltage, permitting the meter to read maximum voltage.

The term *ohms per volt* (Ω/V) is used to describe commercial VOMs. Ohms per volt is a measure of a VOM's sensitivity and represents the number of ohms required to extend the range by 1 V. For example, if the meter movement requires 1 mA for full-scale deflection, then 1000 Ω (including the movement's internal resistance) are needed for each volt that could be measured if the movement were used as a voltmeter. If the movement requires only 100 μA for full-scale deflection, then 10,000 Ω/V are needed. Thus the more sensitive the meter movement (those requiring the least current), the higher the ohms-per-volt requirement.

Voltmeters with a high ohms-per-volt rating put less load on the circuit being measured and have a less disturbing effect on the circuit. For example, assume that a 1-V drop across a 1000-Ω circuit is to be measured with both a 100-Ω/V meter and a 20,000-Ω/V meter. A 1-V drop across a 1000-Ω circuit will produce a 1-mA current flow. With the 1000-Ω/V meter across the circuit, the 1-mA current will divided itself between the meter and the circuit. This will cut the circuit's normal current in half. With a 20,000-Ω/V meter across the same circuit one-twentieth of the current will pass through the meter and nineteen-twentieths will remain in the circuit.

1-2.5. Calculating Multiplier Values

The multiplier resistance required to convert a basic meter movement to a voltmeter capable of measuring a given voltage range can be calculated by using the following equation:

$$R_x = \frac{R_m(V_2 - V_1)}{V_1}$$

where R_x is the multiplier resistance in ohms (in series with the meter movement)

R_m is the internal resistance of the meter movement
V_1 is the voltage required for full-scale deflection of the meter movement
V_2 is the voltage desired for full-scale deflection (maximum voltage range desired)

For example, assume that a meter has a 50-μA full-scale movement (with 50 equal divisions on the scale), an internal resistance of 300 Ω (see Section 1-2.3), and that it is desired to convert the meter movement to measure 0–100 V. (Each of the 50 scale divisions will then represent 2 V.)

1. Find the voltage required for full-scale deflection of the meter movement: $E = IR$, or $(3 \times 10^2) \times (5 \times 10^{-5}) = 0.015$ V. This is voltage V_1.
2. Use the equation given above to find the value of R_x:

$$R_x = \frac{300(100 - 0.015)}{0.015} = 1{,}999{,}700 \; \Omega$$

3. To verify this multiplier resistance, add the meter internal resistance (300 Ω) to it and then divide by the full-scale voltage obtained with the multiplier (100) to find the ohms-per-volt rating:

$$1{,}999{,}700 + 300 = 2{,}000{,}000$$

$$\frac{2{,}000{,}000}{100} = 20{,}000 \; \Omega/V$$

4. Then divide the full-scale current (50 μA) into one to find an ohms-per-volt rating of 20,000. Both ohms-per-volt ratings should match.

1-2.6. Basic Ohmmeter

An *ohmmeter* (or resistance-measuring device) is formed when a basic meter movement is connected in series with a resistance and a power source (such as a battery). The basic ohmmeter arrangment is shown in Figure 1-9. Here a 3-V battery is connected to a meter movement with a full-scale reading of 5 mA. The current-limiting resistor R has a value (600 Ω less meter resistance) such that exactly 5 mA will flow in the circuit when the test leads are clipped together.

When there is no connection across the test leads, the current will be zero. The meter's pointer will rest at the "infinity" mark (∞) on the scale. When the two leads are shorted, the meter will move to its full 5-mA reading, which, on the scale, will indicate that there are 0 Ω at the test leads.

If a 600-Ω resistance were connected across the leads, the total resistance would be 1200 Ω, and the meter would drop to one-half in full-scale reading, or 2.5 mA.

If the battery voltage and limiting resistor R would remain constant, the pointer would always move to 2.5 mA whenever 600 Ω were connected across the test leads. The 2.5-mA point on the meter scale could be marked "600 Ω."

With a 2400-Ω resistance across the leads, the total resistance would be 3000 Ω, and the pointer would drop to a 1-mA reading, since $I = E/R$, or $3/3000 = 0.001$ A (1 mA).

Again, if the battery voltage and the limiting resistor remain constant, the meter will always read 1 mA when a resistance of 2400 Ω is placed across the test leads. Therefore, the 1-mA point on the meter scale could be marked "2400 Ω."

The ohmmeter arrangment of Figure 1-9 would then be capable of measuring 600 Ω and 2400 Ω. Any number of resistance values could be plotted on the scale, provided resistances of known value were placed across the leads.

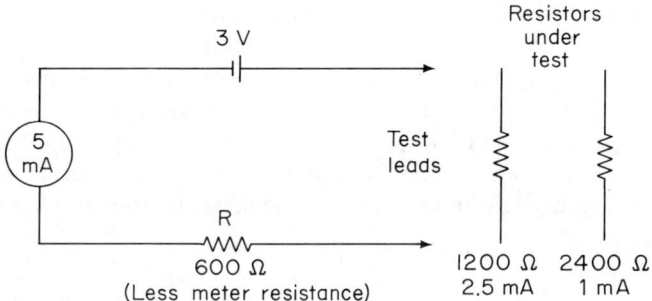

Fig. 1-9. Basic ohmmeter circuit.

The scale of a commercial VOM will have its own markings or calibration. As discussed later in this chapter, the ohmmeter scale is printed on the meter face along with the voltage and current scales. However, the ohmmeter scale is quite different from the other scales in two respects. The zero point is at the right-hand side (usually), and the maximum resistance (usually marked "infinity") is at the left-hand side. Also, the scale is not linear (lower-resistance divisions are wider, and higher-resistance divisions are narrower).

The ohmmeter circuit of a typical VOM is shown in Figure 1-10. Here the ohmmeter has two range scales that can be selected by means of a switch. In the "high" position, a series multiplier (similar to that of a voltmeter) is connected to the circuit and drops the voltage by a corresponding amount.

10   Analog Meters

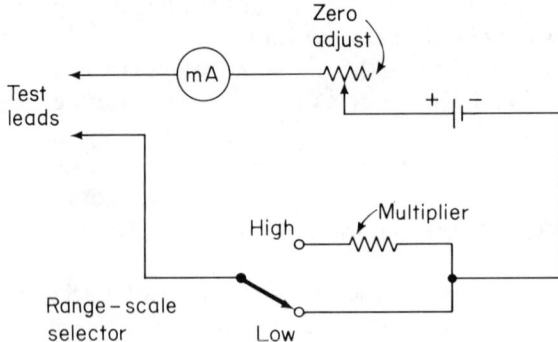

Fig. 1-10. Ohmmeter circuit of a typical VOM.

This reduces current flow through the entire circuit, usually by a ratio of 10:1, 100:1, or 1000:1, so that the ohmmeter scale represents 10, 100, or 1000 times the indicated amount. In the case of the ohmmeter circuit (Figure 1-9) just described, the "600-Ω" point (half scale) would represent 6000, 60,000, or 600,000 Ω.

No matter which scale is used, the meter and battery are in series with a variable resistor that allows the circuit to be "zeroed." As a battery ages, its output drops. Also, it is possible that with extended age or extreme temperature the resistance values (or meter movement itself) could change in value. Any of the conditions would make the ohmmeter scale inaccurate. The variable resistor (usually marked "zero adjust" or "zero") is included in a commercial VOM circuit. In use, the leads are shorted together, and the variable resistor is adjusted until the meter is at zero (at the right-hand side of the ohmmeter scale). When the leads are opened, the meter then drops back to "infinity" or "open" (left-hand side), and the meter is ready to read resistance accurately.

1-2.7. Basic Galvanometer

The basic D'Arsonval meter movement described thus far can be used as a galvanometer. However, the term galvanometer has come to mean a meter where the *zero of the scale is at the center*, with negative current reading to the left and positive current reading to the right. (See Figure 1-11.) Generally, a galvanometer is used to read *proportional* positive or negative changes in circuits rather than the actual unit value of current. The main use for such a meter is in *bridge circuits*, such as those described in Chapter 3.

1-3. BASIC ALTERNATING-CURRENT METERS

Most a-c meters are similar to d-c meters in that they are analog current-measuring devices. However, since ac reverses direction during each

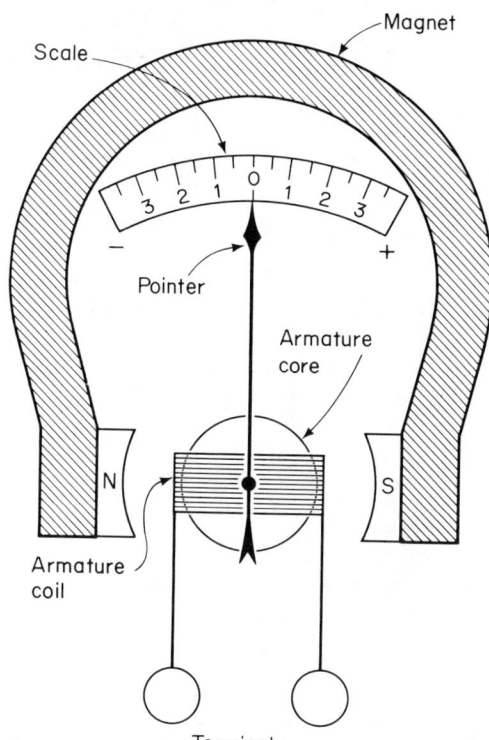

Fig. 1-11. The basic zero-center galvanometer movement.

cycle, the basic D'Arsonval movement cannot be connected directly to ac. Instead, the meter movement is connected to the a-c voltage through a *rectifier*. Both half-wave and full-wave rectifiers are used. However, the full-wave bridge rectifier of Figure 1-12 is most efficient, since a direct current will flow through the meter movement on both half-cycles. The remainder of the a-c meter circuit can be identical to that of a d-c meter.

Such an arrangement will work well with alternating currents of low frequency, but it presents a problem as frequency increases. This is because the movement and multiplier resistances may load the circuit being tested. A *radio-frequency probe* can be connected ahead of the basic meter circuit to overcome this condition. (Meter probes are discussed in Chapter 9.)

Alternating-current meter scales also present a problem that does not occur on d-c scales. As shown in Figure 1-13, there are four ways to measure an a-c voltage. We can measure the average, RMS or effective peak, or peak-to-peak voltage.

The *peak voltage* is measured from the crest of one half-cycle, while *peak-to-peak* is measured from the crests of both half-cycles. However, the direct current to the meter movement will be less than the peak alternating current, since the voltage and current drop to zero on each half-cycle.

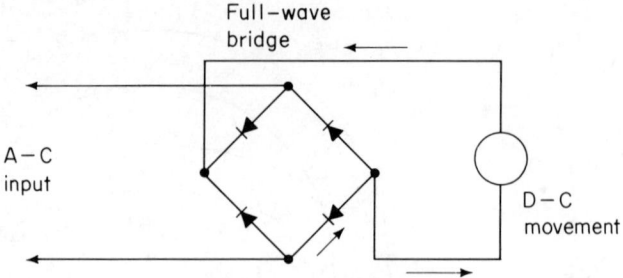

Fig. 1-12. Basic a-c meter circuit with full-wave bridge.

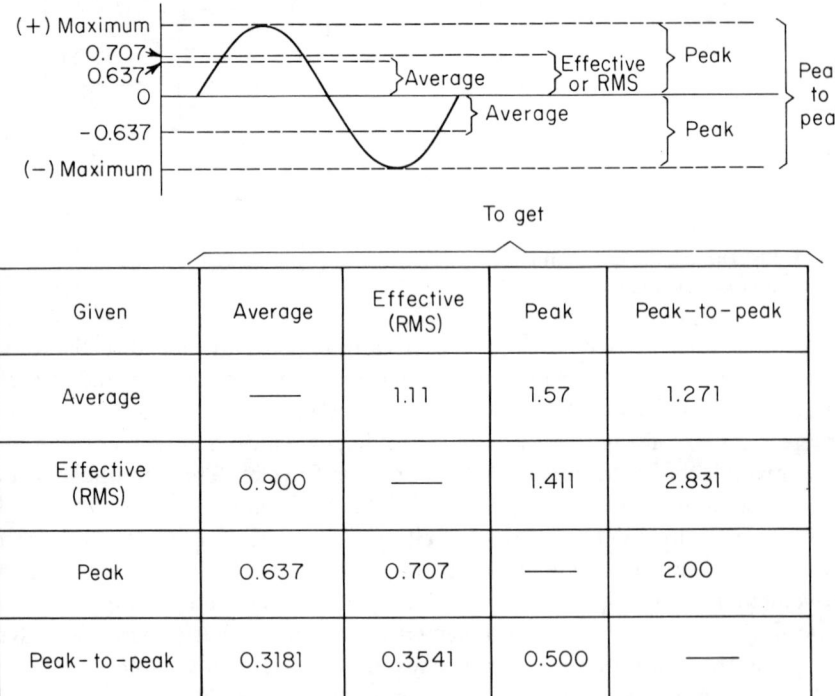

Given	Average	Effective (RMS)	Peak	Peak-to-peak
Average	—	1.11	1.57	1.271
Effective (RMS)	0.900	—	1.411	2.831
Peak	0.637	0.707	—	2.00
Peak-to-peak	0.3181	0.3541	0.500	—

Fig. 1-13. Relationship of average, effective or RMS, peak and peak-to-peak values for alternating-current sine waves.

With a full-wave bridge rectifier, the current or voltage will be 0.637 of the peak value (a half-wave rectifier will deliver 0.318 of the peak value). This is known as *average* value, and some meters are so calibrated. Most meters have RMS, or root–mean–square, scales. In an RMS meter, the scale indicates 0.707 of the peak value (assuming the usual full-wave rectifier is used). This value is the *effective* value of an alternating current.

A direct current flowing through a resistor produces heat. So does an alternating current. The effective value of an alternating current or voltage is that value which will produce the same amount of heat in a resistor as direct current or voltage of the same numerical value. The term "RMS" is used because it represents the square root of the average of the squares of all instantaneous values in a *perfect sine wave*. Since nearly pure sine waves are frequently measured, this mathematical representation is of particular importance, and it is important to know that the effective value of a sine wave (its heat-producing equivalent of dc) is 0.707 of the peak value.

1-3.1. Current Probes and Clip-on Meters

One type of meter that is unique to a-c measurements is the clip-on meter or current probe shown in Figure 1-14. Alternating currents

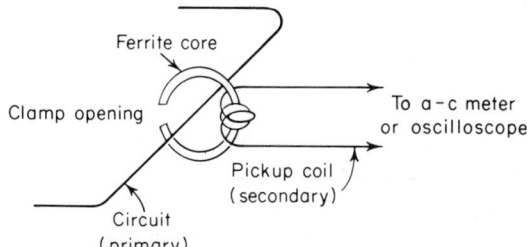

Fig. 1-14. Basic clip-on meter or current probe circuit.

set up alternating fields around a wire or conductor. These currents can be picked up by a coil of wire around the conductor, stepped up through a transformer, and measured by a voltmeter. A clip-on meter, complete with built-in coil, transformer, and meter movement, is particularly useful where conductors are carrying heavy currents and where it is not convenient to open the circuit to insert an ammeter. Clip-on meters are most often used for heavy industrial work.

Current probes are similar to clip-on meters, except that probes are generally used in conjunction with an amplifier to measure small currents. Most electronic laboratories use current probes rather than clip-on meters. A typical probe clips around the wire carrying the current to be measured and, in effect, makes the wire the one-turn primary of a transformer formed by ferrite cores and a many-turn secondary within the probe. The signal induced in the secondary is amplified and can be applied to any suitable a-c voltmeter for measurement. In commercial units, the amplifier constants are chosen so that 1 mA in the wire being measured produces 1 mV at the amplifier output. In this way, current can be read directly on the voltmeter.

1-3.2. Thermocouple Meters

The thermocouple meter measures dc, ac, and even rf. Figure 1-15 shows the basic circuit used in thermocouple meters. When bars of two

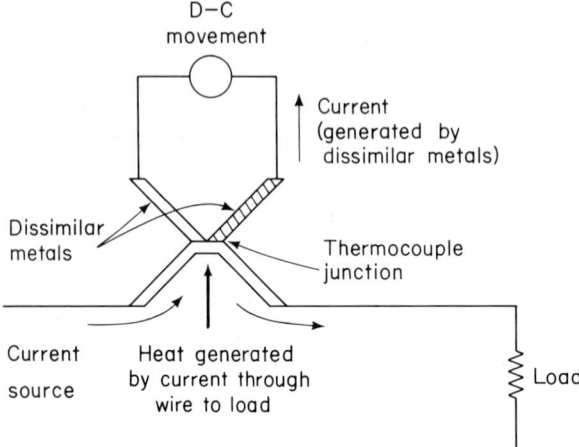

Fig. 1-15. Basic thermocouple meter circuit.

dissimilar metals are connected at one end and heat is applied to the connected ends, a d-c voltage is developed across the open ends of the two dissimilar metals. This voltage is directly proportional to the temperature of the wires in the heated junction. The generation of d-c voltage by heating the junction is called *thermoelectric* action, and the device is called a *thermocouple*.

Any two bars of dissimilar metals will produce a voltage across their open ends when their junction is heated. However, certain combinations of metals will produce various specific voltages for each degree of temperature difference. Commonly used metal combinations are copper-constantan, iron-constantan, chromel-constantan, chromel-alumel, and platinum-rhodium. Tables are available that show the voltages produced by each of the various metal combinations at sepcific temperatures.

An electric current passing through a wire or conductor will produce heat in that wire in *proportion to the square* of the current. Therefore, if a current is passed through the junction of a thermocouple, heat will be generated in the wires and a voltage will be developed at the open ends. If a calibrated meter movement is connected to the free ends of the thermocouple, the generated voltage can be measured. Of course, the meter scale must be calibrated to relate the reading to the amount of current passing through the thermocouple and heating the junction, rather than the voltage produced by the thermocouple.

The direction of current in the thermocouple has no effect on the heating of the junction, so the instrument can measure dc, ac, or rf. When measuring very low currents or very high frequencies (some thermocouple systems operate up to 50 MHz), the thermocouple junction is usually sealed in a vacuum similar to the filament of a vacuum tube. This gives the greatest amount of heat for a minimum amount of current.

1-4. LABORATORY ANALOG METERS

Laboratory analog meters operate by producing a current proportional to the quantity being measured, as do basic VOMs. However, laboratory meters include many circuit refinements to improve their accuracy and stability. Also, there are special-purpose analog meter circuits that are unique to laboratory instruments.

1-4.1. Basic Vacuum-Tube Voltmeter and Electronic Meter Measurements

The basic VTVM and electronic meter circuit is shown in Figure 1-16. The amplifier can be either vacuum-tube (for VTVM) or tran-

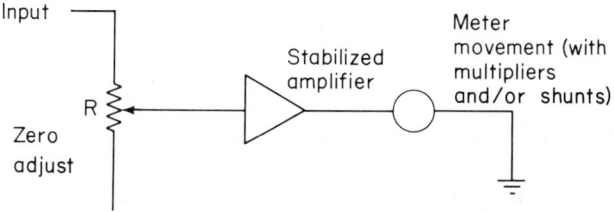

Fig. 1-16. Basic electronic voltmeter circuit.

sistor (for electronic meter) and is usually direct-coupled. Field-effect transistors are often used for electronic meter amplifiers.

When the basic circuit is used as a voltmeter, resistance R has a large value (usually several megohms) and is connected in parallel across the voltage circuit being measured. Because of the high resistance, very little current is drawn from the circuit, and operation of the circuit is not disturbed. The voltage across resistance R raises the voltage at the amplifier input from the zero level. This causes the meter at the amplifier output to indicate a corresponding voltage.

When the basic circuit is used as an ammeter, resistance R has a small value (a few ohms or less) and is connected in series with the circuit being measured. Because of the low resistance, there is little change in the total circuit current, and operation of the circuit is not disturbed. Current flow through resistance R causes a voltage to be developed across R, which raises the amplifier input level from zero and causes the meter to indicate a corresponding (current) reading.

One of the most common circuits used in VTVMs and electronic meters is shown in Figure 1-17. This is essentially a differential amplifier, in which

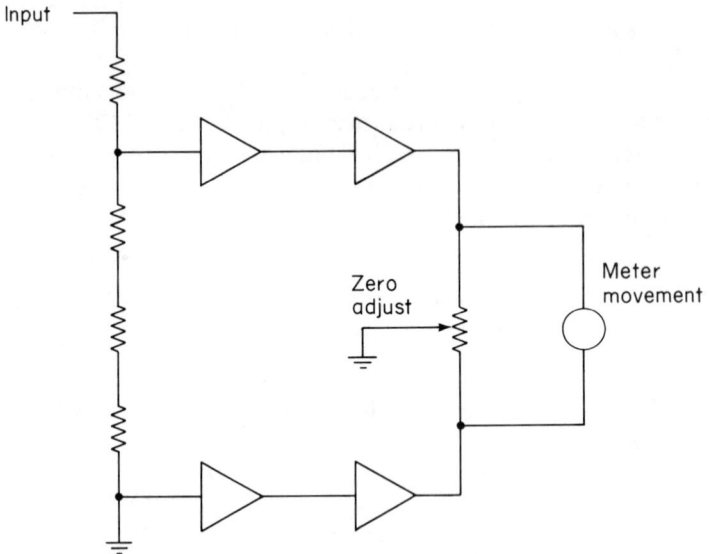

Fig. 1-17. Basic electronic voltmeter with differential amplifier.

the voltage to be measured is applied to one input and the other input is grounded. The zero-set resistance is adjusted so that the meter reads zero when no input voltage is applied. When the voltage to be measured is applied across the input resistance, the circuit is unbalanced, and the meter indicates the proportional unbalance as a corresponding voltage reading. One of the reasons for using the differential amplifier circuit is to minimize drift due to power supply changes.

The amplifier in Figures 1-16 and 1-17 performs three basic functions. First, the effective sensitivity of the meter movement is increased. An amplifier changes the measured quantity to a current of sufficient amplitude to deflect the meter movement. Thus a few microvolts that would not show up on a typical VOM could be amplified to several volts to deflect any meter movement.

The second function of the amplifier is to increase the *input impedance* of the meter so that the instrument draws little current from the circuit under test. A typical VTVM amplifier will provide about 11-MΩ input impedance, while a field-effect transistor meter amplifier can provide up to about 100-MΩ impedance.

The third amplifier function is to limit the maximum current applied to the meter movement. Therefore, there is little danger when unexpected overloads occur that could burn out the meter movement.

1-4.2. Drift Problems in Electronic Meters

One of the problems common to any direct-coupled amplifier is drift due to power supply change. The amplifier cannot tell the difference between power supply change and change in the voltage being measured. This is especially aggravated when a VTVM or electronic meter is used to measure small voltages.

A common technique for eliminating drift is to convert the d-c voltage being measured to an a-c voltage (Figure 1-18). This is done by alternately

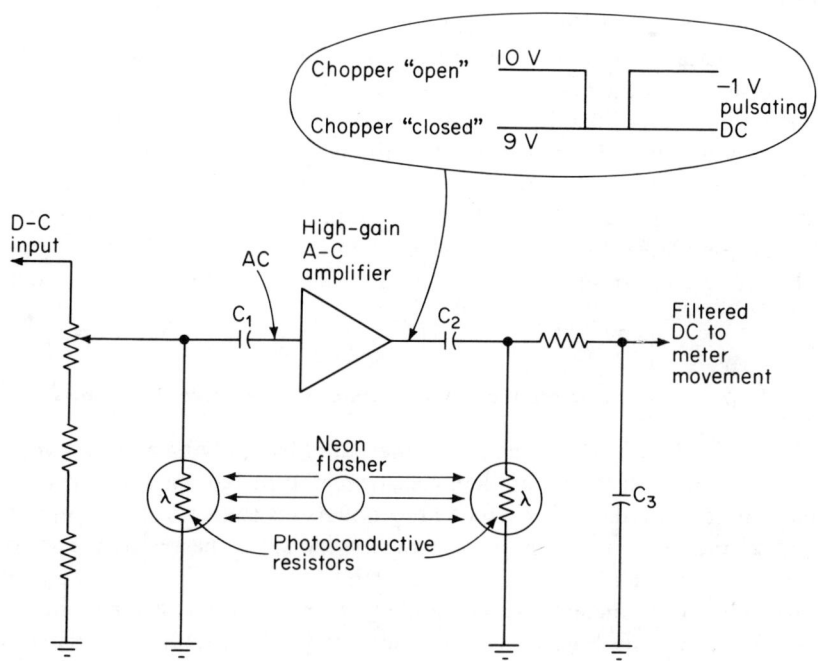

Fig. 1-18. Basic circuit for eliminating drift in electronic meters.

applying and removing the d-c input voltage through a "chopper," amplifying the "chopped" signal in an a-c amplifier, and then synchronously rectifying the amplifier output back to a d-c voltage for application on a meter movement. Overall d-c feedback ensures accuracy of the d-c gain. (The a-c amplifier will not be affected by changes in power supply voltage.) Therefore, d-c drift is limited to a value set by the input chopper.

An electromechanical chopper can be used, but some form of electronic chopper (such as the photoconductive chopper, also shown in Figure 1-18) is used more often. In a typical photoconductive chopper, the d-c input voltage is converted to a comparable a-c voltage by periodically illuminating a group of photoconductive resistors. This results in a low-noise, high-impedance chopper action.

The photoconductive resistors are illuminated by a flasher, such as a neon lamp driven by a 60-Hz line source. The input and output signals are synchronized since they are both driven by the same source. When the photoconductive resistors are illuminated they provide low resistance and act as a short circuit. Without light, the resistors provide very high resistance and appear as an open circuit.

With light, the input to the amplifier is grounded (zero input). One side of capacitor C_2 is also grounded. The opposite side of C_2 is connected to the collector (or plate) of the amplifier output circuit. Capacitor C_2 charges up to the collector (or plate) value. When the flasher light is removed on alternate half-cycles of the 60-Hz driving source, the d-c voltage to be measured is applied to the amplifier input. This causes the output collector (or plate) to drop by an amount proportional to the input signal. Capacitor C_2 then discharges into the meter circuit by a corresponding amount. For example, assume that C_2 is charged to 10 V with no input and that the amplifier output collector (or plate) drops to 9 V when the input is applied (chopper open). Under these conditions, the output is then ideally a square wave euqal to -1 V.

1-4.3. Resistance Measurements in Electronic Meters

In a VOM, resistance is measured by applying a known voltage to the unknown resistance and then measuring the current passing through the circuit. When voltage and current are known, resistance can be computed. In actual practice, computation is unnecessary since the resistance scale of the meter is precalculated.

Most electronic meters use a modified procedure for resistance measurement. As shown in Figure 1-19, the current in the circuit depends on the series combination of the unknown resistor R_x and the internal resistor R_i. This means that both the voltage and current in the external circuit will change according to the value of the unknown. The resistance scales of the meter are calibrated for the measurement of the unknown resistance.

If R_x were infinite, the meter would read the full battery voltage. Full-scale deflection would correspond to a resistance of infinity. If R_x were zero (short circuit), the meter would read zero. The mid-scale range occurs when R_x equals R_i.

Laboratory Analog Meters   19

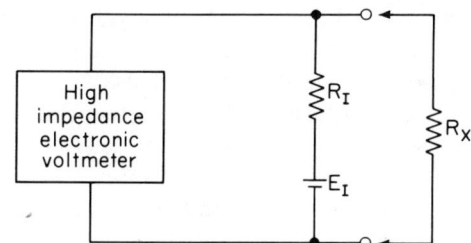

Fig. 1-19. Basic resistance measurement procedures in electronic meters.

The resistance R_i included as part of the ohmmeter circuit provides a convnient means of changing the range of the instrument. When values of low resistance are being measured, the resistance of the ohmmeter leads (included in the total resistance measurement) can add considerable error. To overcome this condition, the circuit can be altered to that shown in Figure

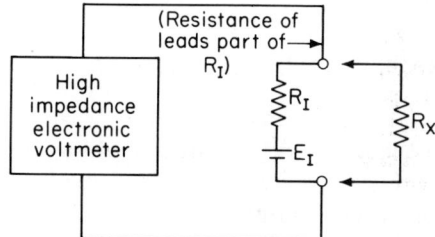

Fig. 1-20. Improved resistance measurement procedure for electronic voltmeters.

1-20. Here the resistance of the current-carrying leads is calibrated as part of R_i, and the resistance in the voltmeter leads is small compared with the high input impedance of the metering circuit.

1-4.4. Resistance Measurements with External Power Source

An *external power power source* is often used in laboratory work when it is necessary to measure very high or very low resistances.

For *high resistances*, a high voltage is applied to the unknown, and the current is measured on a sensitive current meter (or microammeter), as shown in Figure 1-21. This method is generally preferable to the use of an electronic meter for high-resistance measurement. High-resistance measurements can be distrubed by the high-impedance of the measuring electronic voltmeter when this impedance is close to the resistance being measured. Many laboratory meters account for this by adjusting the value of R_i on the high-resis-

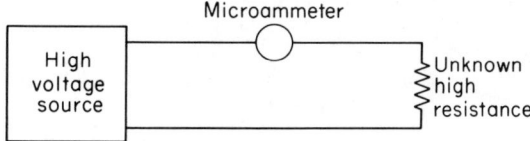

Fig. 1-21. High resistance measurement with external power source.

tance ranges to compensate for the voltmeter input impedance. For example, on a 100-MΩ scale the value of R_i is actually 200 MΩ. The parallel combination of the 200-MΩ R_i and the 200-MΩ input impedance of the meter gives an effective internal impedance of 100 MΩ.

To measure *very low resistances*, such as those found in short lengths of wire or in relay contacts, a constant-current source may be used to supply a fixed amount of current through the unknown resistance. The voltage drop across the resistance being measured is then indicated by a sensitive voltmeter (or microvoltmeter), as shown in Figure 1-22. Resistance measurements as low as 1 μΩ can be made by this technique.

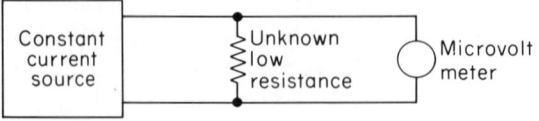

Fig. 1-22. Low resistance measurement with external power source.

1-4.5. Alternating-Current Measurements in Electronic Meters

Electronic meters for measuring a-c voltages also use an amplifier with the meter movement but add a rectifier circuit to convert the ac to dc. Most shop-type meters are RMS-reading instruments. This is also true of laboratory meters, although it is possible to use average, peak, or peak-to-peak reading meters for special applications.

Although a meter may be RMS-*reading*, it is usually average-or peak-*responding;* that is, the scale reads RMS values, but the meter movement operates on an average or peak value.

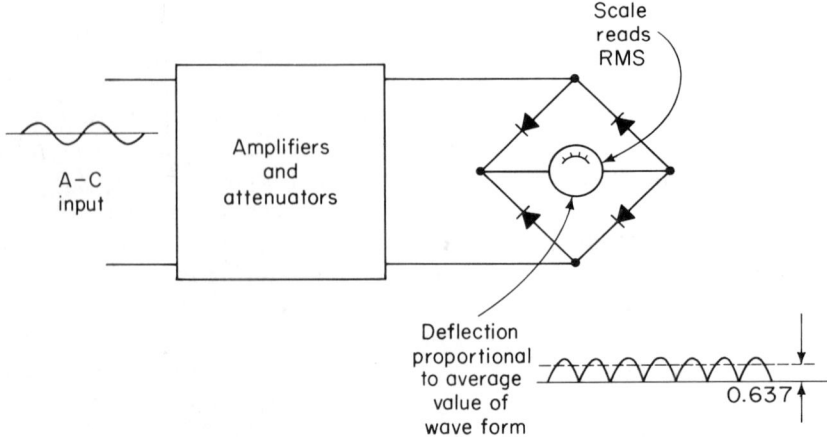

Fig. 1-23. Basic average-responding (RMS-reading) a-c voltmeter (Hewlett-Packard).

Figure 1-23 shows the basic circuit of an average-responding meter. Here the a-c signal is amplified in a gain-stabilized a-c amplifier and then is rectified by the diodes. The resulting current pulses drive the meter. The meter deflection is proportional to the average value of the wave form being measured (even though the scale may or may not read RMS).

The *peak-responding* voltmeter shown in Figure 1-24 places the rectifier

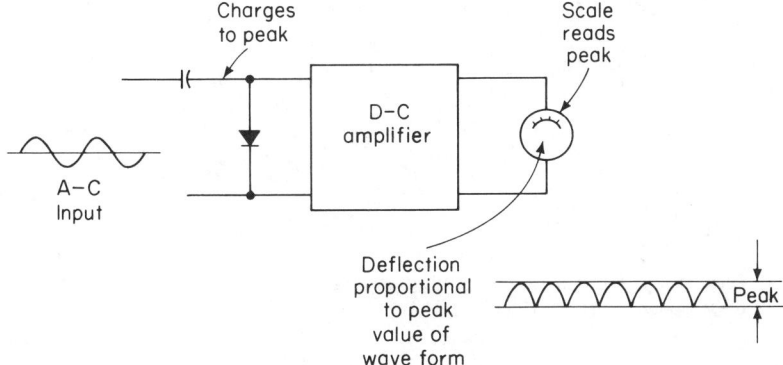

Fig. 1-24. Basic peak-responding a-c voltmeter (Hewlett-Packard).

in the input circuit, where it charges the small input capacitor to the peak value of the input signal. This voltage is passed to a d-c amplifier that drives the meter. Meter deflection is proportional to the peak amplitude of the input wave form. The meter scale can be calibrated in RMS or peak voltage, as required.

If highly compex wave forms (non-sine waves) are to be measured, a true RMS-*responding* voltmeter should be used. Such a circuit is shown in Figure 1-25. Here the complex wave form is used to heat the junction of thermocouples.

One of the major problems in this technique is the nonlinear behavior, as well as low response and possible burnout, of the thermocouple. Nonlinear behavior complicates calibration of the indicating meter. This difficulty can be overcome by the use of two thermocouples mounted in the *same thermal environment*. Nonlinear effects in the measuring thermocouple are offset by similar nonlinear operations of the second thermocouple.

As shown in the circuit of Figure 1-25, developed by Hewlett-Packard, the amplified input signal is applied to the measuring thermocouple, and a d-c feedback voltage is fed to the balancing thermocouple. The d-c voltage is obtained from the voltage output *difference* between the thermocouples. The circuit can be considered as a feedback-control system that matches the heating power of the d-c feedback voltage to the input waveform heating power. Meter deflection is proportional to the d-c feedback voltage, which,

22 Analog Meters

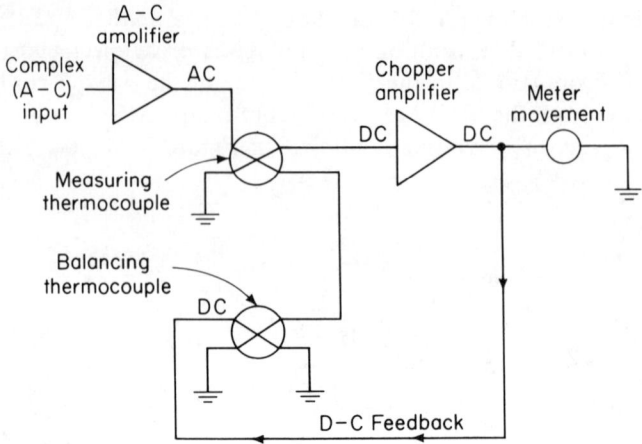

Fig. 1-25. True RMS-responding voltmeter for measurement of complex waves (Hewlett-Packard).

in turn, is proportional to the RMS of the input voltage (no matter what the wave form). Therefore, the meter indication is linear.

It is also possible to measure peak-to-peak voltages with an electronic voltmeter. The circuit is similar to that of the peak-responding circuit of Figure 1-24. However, the input capacitor is charged to the peak-to-peak value by a full-wave rectifier. Also, the scale is calibrated to read directly in peak-to-peak values.

1-5. METER SCALES AND RANGES

Figures 1-26 and 1-27 show the operating controls and scales for a typical VOM and a typical electronic volt-ohmmeter, respectively. The following notes describe each of the scales and provide information concerning their accuracy and use.

1-5.1. Ohmmeter Scales

Note that the zero indication for the VOM ohmmeter scale is at the right and, on the electronic ohmmeter, the zero indication is at the left. Although this condition is typical, it will not be found on every make and model of meter.

Also note that the high-resistance end (or infinity end) of the ohmmeter scale is cramped on both meters. Ohmmeter scales are always nonlinear.

Meter Scales and Ranges 23

Fig. 1-26. Simpson Instruments Model 262 VOM.

Therefore, ohmmeters provide their most accurate indications at mid-scale or near the low-resistance end.

In general, the ohmmeter scale is considered to be as accurate as the d-c voltmeter scale. However, in the case of a battery-operated VOM, the condition of the battery will affect accuracy. As a battery ages and its voltage output drops, the resistance indications will be *lower* than the actual value. For example, if a battery voltage drops to 90 per cent of its *minimum* value, a 100-Ω resistor will produce a 90-Ω indication (approximately). This is true even though the ohmmeter is "zeroed" before making the measurement.

Greatest accuracy will be obtained if the ohmmeter is zeroed on each range *just prior* to making the measurement.

In some meters, the ohmmeter scale is rated in *degrees of arc* rather than percentage of full scale (as are the voltmeter and ammeter scales). For example, the d-c voltage accuracy of a meter could be ±3 per cent of full scale. On the 100-V scale, this indicates an accuracy of ±3 V. As shown in Figure 1-28, a ±3-V (or a total of 6 V) indication corresponds to a certain number of degrees of arc. In turn, this arc defines the accuracy of the ohmmeter scale. Because the ohmmeter scale is nonlinear, it is possible that the error will not be constant over the entire scale. However, the error should not exceed the *rated accuracy* at any point on the scale.

24 Analog Meters

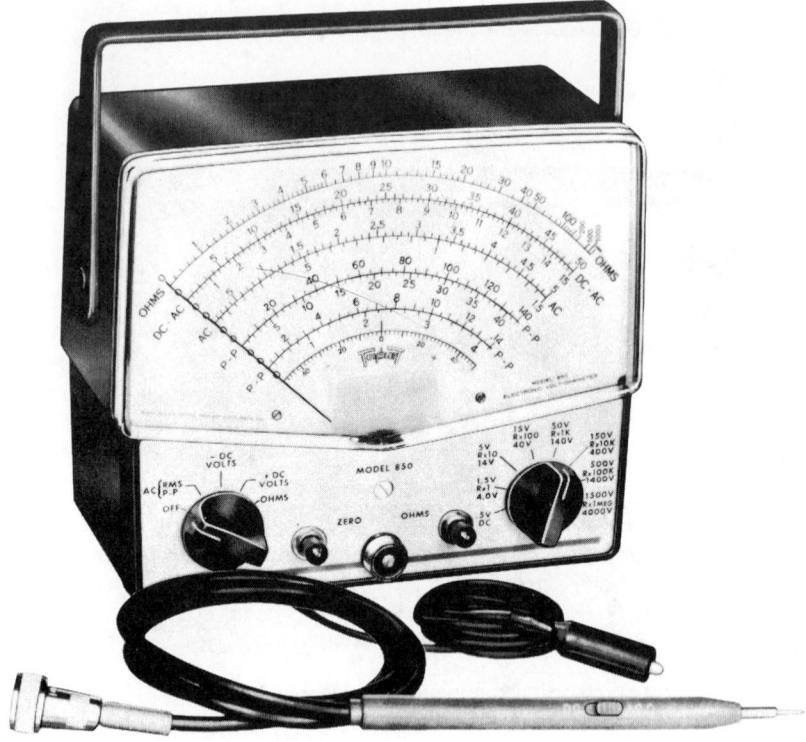

Fig. 1-27. Triplett Model 850 Electronic Volt-ohmmeter.

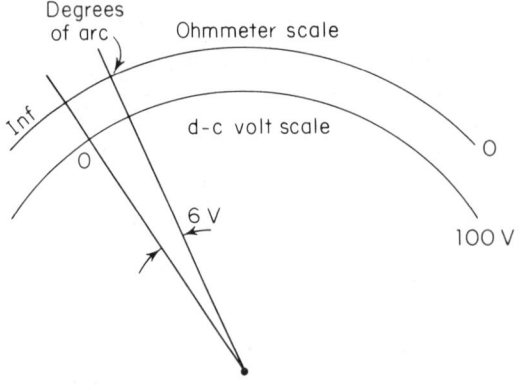

Fig. 1-28. Accuracy of ohmmeter scale (in degrees of arc) is related to accuracy of d-c voltage scale (in percentage of full scale).

1-5.2. Direct-Current Scales

The zero indication for the d-c scales is almost always on the left. Usually, the d-c scales are linear, with no cramping or bunching at either end.

Note that there are three basic d-c scales (with maximums of 8, 40, and 160) on the VOM (Figure 1-26). Each of these scales serves multiple purposes, depending on the position of the range selector. (The range selected is indicated by a dot-pointer at the scale center.) Therefore, the operator must make note of *both* the scale reading and the range-selector position, before the correct indication is obtained.

For example, the d-c readings of 60, 15, and 3 are all aligned. These readings are located on the 160, 40, and 8 scales, respectively. If the range selector were set to 1.6 V dc, then the 60 reading would apply, and the indicated values would be 0.6 V dc. If the reading were the same, but the range selector were set to 160 mA, then the 60 reading would still apply, but the indcated value would be 60 mA. Again, with the same reading but with the range selector set to 400 V dc, the 15 reading would apply, and the indicated value would be 150 V dc.

The use of meter scales is often confusing to the inexperienced operator. Therefore, the meter scales should be *studied thoroughly* before attempting to use any meter for the first time.

Accuracy of the d-c scales is dependent on the tolerance of the multiplier resistors, the accuracy of the movement, and the accuracy of the scales. In precision meters, the scales are matched to individual movements so that the combined movement and scale has a given accuracy. Usually, the multiplier (for voltage) and shunt resistors (for current) have an accuracy of ± 1 per cent or better. When this is coupled to a movement with a ± 2 per cent or better accuracy, the total accuracy is ± 3 per cent.

The accuracy of both d-c and a-c scales is usually specified as a *percentage of full scale* and not the actual reading. This is another fact often overlooked by the inexperienced operator. For example, assume that a voltage is measured on the 40-V d-c scale (Figure 1-26), that a reading of 10 is obtained, and that the rated accuracy is ± 3 per cent of full scale. The full-scale value is 40 V; therefore, the absolute accuracy is ± 1.2 V (40 $\times$ 3 per cent). The 10-V reading could thus indicate an actual value of 8.8–11.2 V. Inexperienced operators often assume (incorrectly) that the 3 per cent tolerance applies directly to the reading (or 0.3 V in the example given). To make sure, always consult the instruction manual or meter specification data about how accuracy is specified.

1-5.2.1. Differences in Direct-Current Scales

Several differences are noted between the d-c scales and ranges of the VOM (Figure 1-26) and those of the electronic volt-ohmmeter (Figure 1-27).

First, the electronic meter is provided with a zero adjustment, in addition to the ohms adjustment. Usually, this zero control must be set so that the movement's needle is aligned with zero each time the range selector is moved to another range position. There are several reasons for this. In vacuum-tube meters, there is a contact potential at the tube elements (particularly the grid) that changes with each position of the range selector. In transistor meters, there is a certain amount of leakage current that is subject to change. Also, there is a possibility of drift due to temperature change during warm-up or over a long period of operation. All of these conditions are compensated for by the zero control (which should be set to provide meter zero each time the range selector is moved, and thereafter periodically throughout operation).

Next, note that the meter function switch shows a minus d-c volts position, as well as a plus d-c volts position. In a simple VOM, positive and negative voltages are usually adjusted for by reversing the leads. Usually, the black lead is connected to negative (−) and the red lead is connected to positive (+) when a positive d-c voltage is to be read. The leads are then reversed for a negative d-c voltage. Thus the voltage indication is also a *polarity* indication. (This should be verified, however, by consulting the instruction manual. If a manual is not available, check for correct lead connections with a battery or other d-c source where polarity is known.)

In the electronic meter, the alligator clip of the "common" lead is connected to ground and the circuit voltage is measured with the probe tip. The test leads are never reversed. Thus it is necessary to have a negative and positive position for measurement of d-c voltages. On some electronic meters, the probe rather than the meter itself is provided with a switch for polarity reversing.

Next, note that the electronic meter does not have any current-measuring function. This is common for most electronic meters in general use. However, there are certain electronic meters that will measure current.

Finally, note that the electronic meter has a special *zero-center* scale (bottom scale). This scale permits both positive and negative voltage indications to be displayed on either side of a zero center without reversing the function switch or the test leads. In use, the function switch is set to plus d-c volts, and the zero knob is adjusted until the meter needle is aligned with the zero center (with no voltage applied). A positive voltage will deflect the needle to the right (+), and a negative voltage will move the needle to the left (−). On most electronic meters, the scale divisions are arbitrary and have no relation to actual voltage. However, they do provide a relative measure of voltage on either side of zero. The zero-center scale permits the meter to be used as a null meter in a bridge circuit, as an FM-receiver detector alignment indication tool, and so on.

1-5.3. Alternating-Current Scales

The zero indication for the a-c scales is almost always on the left. Usually, the a-c scales are somewhat nonlinear with some crimping or bunching on the low end. This is because the rectifier circuits required for a-c measurements are nonlinear. Both half-wave and full-wave rectifier circuits are nonlinear. Nonlinearity will be more pronounced when the multiplier resistance is small (low-voltage ranges) than when the multiplier resistance is large (high-voltage ranges). This condition is known as *swamping* effect.

The a-c scales of a meter are never more accurate and usually less accurate than the d-c scales. This is because the inaccuracy of the a-c rectifier circuit must be added to the inaccuracy of the d-c circuit (multiplier, movement, and scales). Accuracies for a typical meter are ± 3 per cent of full scale for dc and ± 5 per cent of full scale for ac. An electronic meter will sometimes have the same accuracy for both a-c and d-c scales (typically ± 3 per cent).

The effects of *frequency* must be considered in determining the accuracy of a-c scales. A typical VOM will provide accurate a-c voltage indications from 15 Hz up to 10 kHz, possibly up to 15 or 20 kHz, but rarely beyond that frequency. This means that VOM readings in the high audio range may be inaccurate.

It should be noted that an a-c meter may provide readings well beyond its maximum rated frequency, but these readings will not necessarily be accurate. A typical electronic meter will provide accurate a-c voltage indications from 15 Hz up to about 3 MHz. Of course, the frequency range of a meter can be extended by use of an r-f probe. However, the probe and meter must be calibrated together, as described in Chapter 9.

As shown in Figure 1-27, many electronic meters permit a-c voltages to be read as an RMS value or a peak-to-peak value, whichever is convenient. Note that the meter circuit responds to the average value in either case but that the scales provide for RMS and peak-to-peak indications.

Note that the RMS scales are accurate *only for a pure sine wave*. If there is any distortion or if the voltage contains any component other than a pure sine wave, the readings will be in error. On the other hand, peak-to-peak readings will be accurate on any type of wave form, including sine waves.

Alternating-current voltage measurements are usually made with a *blocking capacitor* in series with one of the test leads. On a typical VOM, the test lead is connected to a terminal marked "output" or some similar function. This blocks any dc present in the circuit being measured from passing to the meter circuit. Such dc may or may not damage the meter, depending on conditions.

Alternating-current voltage can also be measured on most meters without the blocking capacitor. On a typical VOM, this is accomplished by connecting the test leads to the same terminals as for d-c voltage measurements (usually marked plus and minus or common and plus).

Some older meters have half-wave rectifier circuits or use a half-wave probe circuit. On such meters, the a-c voltage readings may be affected by a condition known as *turnover*. This occurs when there are *even harmonics* present in the voltage being measured. Turnover will show up as different a-c voltage readings when the test leads are reversed. Turnover should not occur when the harmonics are odd, when there are no harmonics, or when a full-wave rectifier is used.

1-5.4. Decibel Scales

Most VOMs are provided with decibel (dB) scales. Actually, the a-c voltage circuit is used in the normal manner, except that the readout is made on the dB scales. Inexperienced operators are often confused by the dB scales. The following notes should clarify their use.

The dB scales represent *power ratios* and not voltage ratios. In most cases, 0 dB is considered as the power of 1 mW (0.001 W) across a 600-Ω pure-resistive load. This also represents 0.775 V RMS across a 600-Ω pure-resistive load. The term *decibel meter*, or dBM, is sometimes used to indicate this system (1 mW across 600 Ω).

The dB scale is related directly to *only one* of the a-c scales, usually the lowest scale. The VOM range selector must be set to that a-c scale, if readings are to be taken directly from the dB scale. If another a-c scale is selected by the range selector, a certain decibel value must be added to the indicated value. For example, in the VOM of Figure 1-26, the dB scale is related directly to the 3-V a-c scale. If the range selector is set to 3-V a-c, the dB scale may be read out directly. Note that the 0 dB is aligned with the 0.775-V point on the 3-V a-c scale. If the range selector were set to 8, 40, or 160 V ac, it would be necessary to add 8.5, 22.5, or 34.5 dB to the indicated dB-scale reading. These values are printed on the meter face (lower right-hand corner) and are applicable to that meter only. Always consult the meter face (or instruction manual) for data regarding the dB scales.

Note that dB-scale readings will not be accurate if

1. The voltages are other than pure sine waves
2. The load impedances are other than pure-resistive
3. The load is other than 600 Ω

If the load is other than 600 Ω, it is possible to apply a correction factor. The decibel is based on this mathematical function:

$$dB = 10 \log \frac{\text{power output}}{\text{power input}}$$

Since the power will change by the corresponding ratio when resistance is changed (power will increase if resistance decreases and voltage remains the same), it is possible to convert the function to $10 \log R_2/R_1$, where R_2 is 600 Ω and R_1 is the resistance value of the load.

For example, assume that the load resistance is 500 instead of 600 Ω, and a 0-dB indication is obtained (0.775 V RMS):

$$\frac{600}{500} = 1.2$$

$$\log 1.2 = 0.0792$$

$$10 \times 0.0792 = 0.792$$

Therefore, 0.792 (or 0.8 for practical purposes) must be added to the 0-dB value to give a true reading of 0.8 × dB.

Table 1-1 lists correction factors to be applied for some common load

TABLE 1-1. CORRECTION FACTORS FOR DECIBEL READINGS ACROSS LOADS OTHER THAN 600 Ω

Resistive Load At 1000 Hz	dBM
500	+0.8
300	+3.0
250	+3.8
150	+6.0
50	+10.8
15	+16.0
8	+18.8
3.2	+22.7

impedance values. This table shows the amount of decibel correction to be added to the indicated decibel value when the load impedance is other than 600 Ω. For example, if the load impedance is 300 Ω, +3 dB must be added to the indicated value. This can be verified by using the previous equation:

$$\frac{600}{300} = 2$$

$$\log 2 = 0.3010$$

$$10 \times 0.3010 = 3.01$$

Therefore, 3.01 (or 3.0 for practical purposes) must be added to the indicated decibel value to give a true reading.

Figure 1-29 shows the relationship between a-c voltages (RMS), decibels,

30   Analog Meters

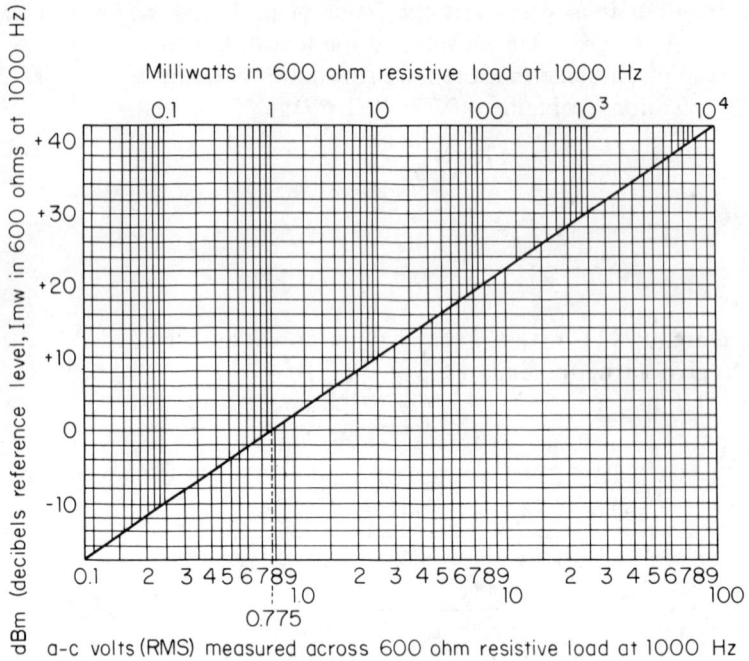

Fig. 1-29. Graph for conversion of RMS voltage into dBM values (RCA).

and power (in milliwatts across a 600-Ω pure-resistive load). This illustration can be used when a particular meter does not have a dB scale but does show RMS voltages. Of course, the correction factor of Table 1-1 must be applied to the value if the load is not 600 Ω.

1-5.4.1. Using Decibel Scales

The following note should clarify the use of decibel scales when measuring the input-output relationship of a particular circuit.

If input and output load impedances are 600 Ω (or whatever value is used on the meter scale), no problem should be found. Simply make a decibel reading at the input and at the output (under identical conditions), subtract the smaller decibel reading from the larger, and note the decibel gain (or loss).

For example, assume that the input shows 3 dB, with 13 dB at the output. This represents a 10-dB gain. If the output had been 3 dB, with 13-dB input, there would have been a 10-dB power loss.

The decibel gain (or loss) can be converted to a power ratio (or voltage or current ratio) by means of the Decibel Conversion Chart in the Appendix. Using the example of a 10-dB gain (or loss), this represents a power ratio of 10 and a voltage or curent ratio of 3.1623.

If the input and output load impedances are not 600 Ω, but are equal to each other, the relative decibel gain or loss is correct, even though the absolute decibel reading is incorrect.

For example, assume that the input and output load impedances are 50 Ω and that the input shows 3 dB with 13 dB at the output. Table 1-1 shows that 10.8 dB must be added to the input and output readings to obtain the correct decibel absolute value. However, there is still a 10-dB difference between the two readings. Therefore, the circuit shows a 10-dB gain, and the power (or voltage/current) ratios of the Decibel Conversion Chart still hold.

If the input and output load impedances are not equal, the relative decibel gain or loss indicated by the meter scales will be incorrect.

For example, assume that the input impedance is 300 Ω, the output impedance is 8 Ω, the input shows +7 dB, and the output shows +3 dB on the scales of a meter using a 600 Ω reference.

There is an apparent loss of 4 dB (7-dB input − 3-dB output). However, by referring to Table 1-1, it will be seen that the 300-Ω input (7 dB) requires a correction of +3dB (giving a corrected input of +10 dB), and the 8-Ω output (3 dB) requires a correction of +18.8 dB (giving a corrected output of 21.8 dB). Thus, there is an actual gain of +11.8 dB.

1-6. METER-PROTECTION CIRCUITS

Most modern meters are provided with some form of protection against overloading, accidentally connecting the wrong functions to a particular circuit (such as connecting the ohmmeter to a voltage), and similar occurrences. There are several types of meter-protection circuits. However, they are usually one of the following three types.

1-6.1. Fuse

A fuse can be inserted in the common line of the meter circuit to protect the movement, shunt, or multiplier resistors. A fuse is usually effective only against large surges of current. A delicate meter movement can be damaged even by small current surges, if they are beyond the maximum capability of the movement. Note that fuses have resistance values that affect the accuracy of the ohmmeter segment of a VOM. If the fuses are to be replaced, an *exact replacement* must be used.

1-6.2. Varistor Diode

A varistor diode can be placed across the meter movement, as shown in Figure 1-30. A varistor diode is usually made of silicon and has a high forward resistance until a certain forward voltage is reached. At this voltage (usually a fraction of 1 V), the forward resistance drops almost to zero. Therefore, if the voltage across the meter movement increases to the dangerous level, the forward resistance of the diode drops to near zero, and all of the current is passed through the diode. This type of diode can be added to most VOMs and is sold as a modification component or part of a modification kit.

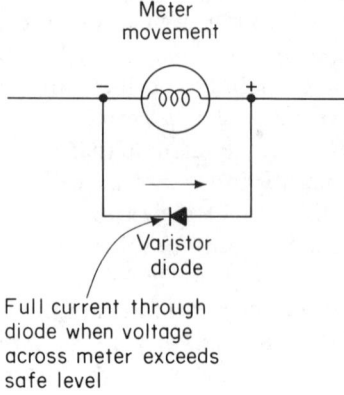

Fig. 1-30. Meter protection circuit using a varistor diode.

1-6.3. Relay

Some meters are provided with a relay that will open the movement circuit, should there be an overload. A typical relay protection circuit

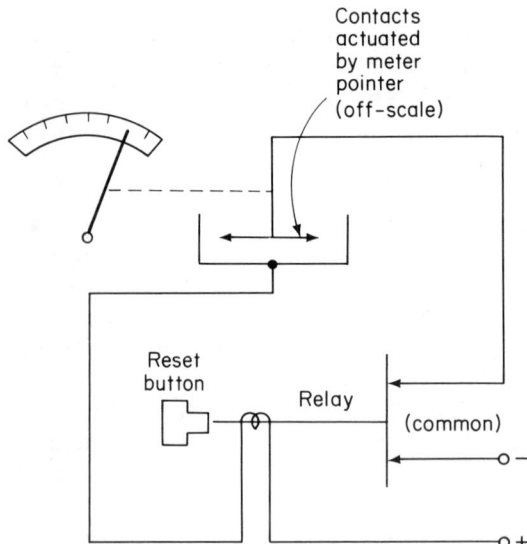

Fig. 1-31. Meter protection circuit using an overload relay.

is shown in Figure 1-31. Note that the relay is actuated when contacts are closed by the pointers being driven off-scale in either direction. When the relay is actuated, the "common" line or the power line is opened. Once the relay has been actuated, it will remain tripped until it is reset by a mechanical reset button.

1-7. METER ACCURACY

There are many factors that can affect the accuracy of meter readings. Some of these have to do with the operator; others are dependent on the meter. The following are some typical problems in meter accuracy.

1-7.1. Parallax Problems

In meters, parallax is an error in observation that occurs when the operator's eye is not directly over the pointer, as shown in Figure 1-32. This will cause the reading to appear at the right or left of the actual indication. Some manufacturers minimize this problem by placing a mirror behind the pointer on the scale. Such a meter is shown in Figure 1-33.

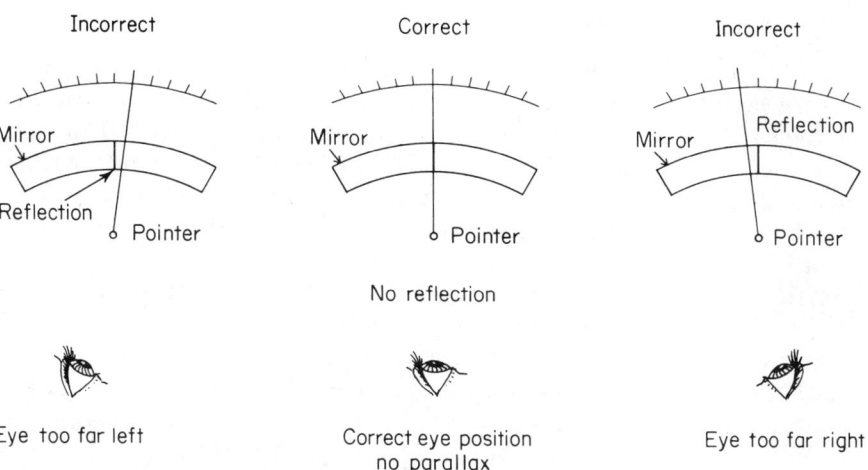

Fig. 1-32. Using anti-parallax mirrored scales (B & K Manufacturing).

To use a mirrored scale most effectively, close one eye, then observe that the pointer and its reflection appear to coincide, as shown in Figure 1-32.

34   Analog Meters

Fig. 1-33. Simpson Instruments Model 270 (with the Mirrored Scale).

1-7.2. Meter Movements

No meter can be any more accurate than its basic movement. As the permanent magnet of a movement ages, the magnetic field weakens, and the indications will be in error (usually, they will read low). This will be true even though the precision shunt and multiplier resistors may not have changed. Likewise, as a meter is subjected to shock, vibration, and overloads, the mechanical balance is disturbed and the point moves from its zero position.

Meter movements are provided with maintenenace adjustments of various types. Usually, both electrical and mechanical adjustments are provided.

Figure 1-34 shows a typical meter movement electrical-compensation circuit. The purpose of this circuit is to provide a constant reading for a given voltage and current, despite any weakening of the movement magnet, loss of tension in the movement springs, and so on. Resistor R_1 is connected in shunt across the meter movement, and resistor R_2 is in series with the movement. (Note that both of these variable resistors are *internal* meter adjustments and are not operating controls. These resistor settings should not be touched except by an experienced meter technician.)

If the movement magnet weakens, shunt resistor R_1 is adjusted (increased)

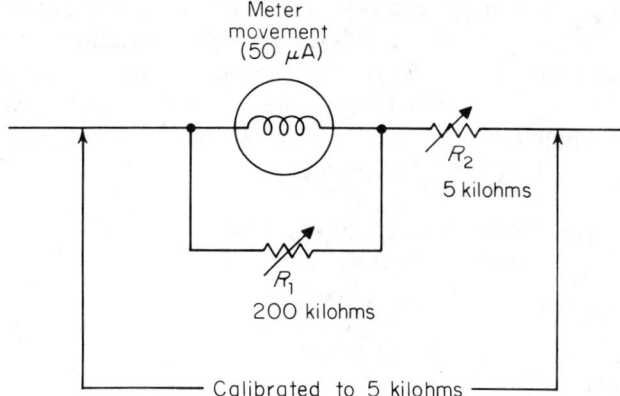

Fig. 1-34. Typical meter-movement electrical compensation circuit (Simpson Instruments).

so that more current will flow through the movement coil for a given voltage. This will compensate for a low-reading movement. After adjustment of the shunt R_1, the series resistor R_2 is adjusted (decreased in this case) so that the *total resistance* of the meter movement circuit remains at the correct value, thus providing the correct reading for a given current.

Figure 1-35 shows two typical meter movements. Note that these movements are provided with a "mechanical-zero" adjuster. This adjustment is used to set the movement to its *mechanical zero*, at which no current is flowing in the movement. The mechanical-zero adjustment is actuated by a screwdriver-type adjustment accessible from the meter front panel (shown directly

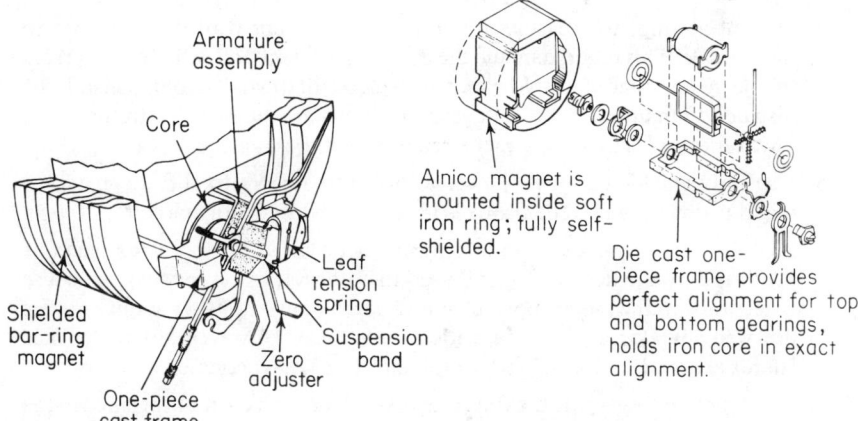

Fig. 1-35. Suspension-type and bar-ring-type meter movements (Triplett).

below the meter face in Figures 1-26, 1-27, and 1-33). This mechanical-zero adjustment is not to be confused with the zero-ohm adjustment or the "zero" adjustment of an electronic voltmeter. The mechanical adjustment permits the pointer to be set at zero *when no power is applied to the meter* and test leads are not connected.

1-8. BASIC METER-OPERATING PROCEDURES

The following notes describe the basic operating procedures for a VOM or electronic meter (resistance, voltage, current, and decibel measurements).

1-8.1. Meter-Operating Precautions

In addition to the general safety precautions described in the Introduction, the following specific precautions should be observed when operating any type of meter.

1. Even if you have had considerable experience with meters, always study the instruction manual of any meter with which you are not familiar.
2. *Never* measure a voltage with the meter set to measure current or resistance. To do so will damage the meter movement. Likewise, *never* measure a current with the meter set to measure resistance.
3. Always start voltage and current measurements on the highest voltage or current scale. Then switch to a lower range as necessary to obtain a good center-scale reading.
4. Do not attempt to measure a-c voltages or current with the meter set to measure dc. This could damage the meter movement and will produce errors in the meter readings. No damage will result (usually, but consult the instruction manual) if d-c voltages or currents are measured with the meter set to measure ac. However, the readings will be in error.
5. Use only shielded probes. Never allow your fingers to slip down to the metal probe tip when the probe is in contact with a "hot" circuit.
6. Avoid operating a meter in strong magnetic fields. These fields (such as produced by the degaussing coil used in color television service) can cause inaccuracy in the meter movement and can damage it. Most quality meters are well shielded against magnetic interference. However, the meter face is still exposed and is subject to the effects of magnetic fields.
7. Most meters have some maximum input voltage and current specified in the instruction manual (or on the meter scales). Do not exceed this maximum. Also, do not exceed the maximum line voltage or use a different power frequency on those meters that operate from line power.

8. Avoid vibration and mechanical shock. Like most electronic equipment, a meter is a delicate instrument.
9. Do not attempt repair of a meter unless you are a qualified instrument technician. If you must adjust any internal controls, follow the instruction manual.
10. Study the circuit under test before making any test connections. Try to match the capabilities of the meter to the circuit. For example, if the circuit has a range of measurements to be made (ac, dc, rf, modulated signals, pulses, or complex waves), it may be necessary to use more than one instrument. Most meters will measure dc and low-frequency ac. If an unmodulated r-f carrier is to be measured, use an r-f probe. If the carrier to be measured is modulated with low-frequency signals, a demodulator probe must be used. If pulses, square waves, or complex waves (combinations of ac, dc, and pulses) are to be measured, a peak-to-peak reading meter will provide the only meaningful indications.
11. Remember that all voltage measurements are made with the meter in *parallel* or across the circuit, and that all current measurements are made with the meter in *series* with the circuit.

1-9. BASIC OHMMETER (RESISTANCE) MEASUREMENTS

The first step in making a resistance measurement is to zero the meter on the resistance range to be read. The meter can be zeroed on other ranges and on some meters will remain constant for all ranges. On other meters the ohmmeter zero will change for each range.

The meter is usually zeroed by touching the two test prods together and adjusting the zero-ohms or ohms control until the pointer is at ohmmeter zero. This is usually at the right end of the scale for a VOM and at the left end for an electronic meter. (See Figure 1-36A.)

Once the ohmmeter is zeroed, connect the test prods across the resistance to be measured. (See Figure 1-36B.) Read the resistance from the ohmmeter scale. Make certain to apply any multiplication indicated by the range-selector switch. For example, if an indication of 3 is obtained with the range selector at $R \times 10$, resistance is 30 Ω. It should be possible to set the range selector to $R \times 1$ and obtain a direct reading of 30 Ω. However, it may or may not be necessary to zero the ohmmeter when changing ranges.

Two major problems must be considered in making any ohmmeter measurements.

First, there must be no power applied to the circuit being measured. Any power in the circuit could damage the meter and cause an incorrect reading. Remember, capacitors often retain their charge after power is turned off. With power off, short across the circuits to be measured with a screwdriver to discharge any capacitance. Then make the resistance measurement.

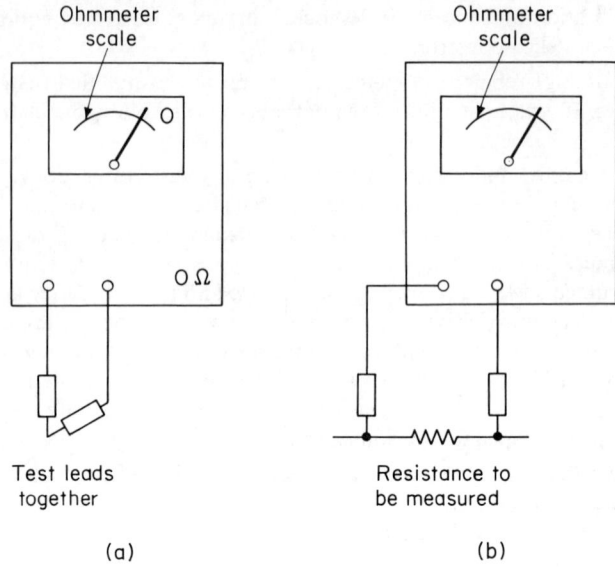

Fig. 1-36. Basic resistance measurement procedure.

Next, make certain that the circuit or component to be measured is not in parallel with (shunted by) another circuit or component that will pass direct current. For example, assume that the value of resistor R_1 in Figure 1-37

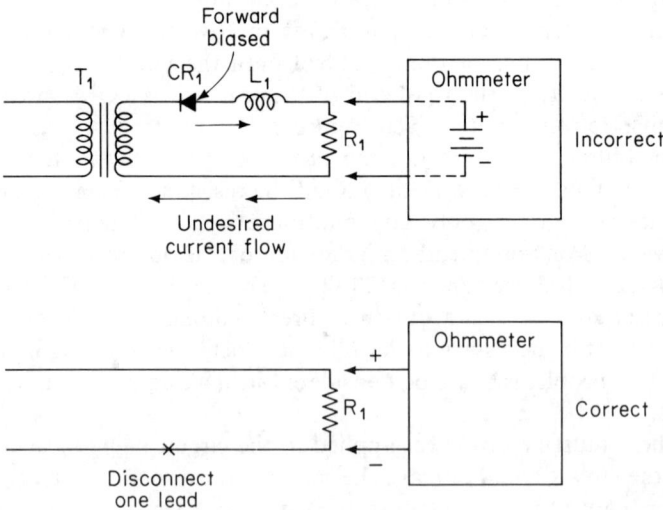

Fig. 1-37. Avoiding errors in resistance measurements due to parallel resistance.

is to be measured. If the battery in the ohmmeter were connected such that the diode CR_1 were forward-biased, current could flow through the transformer T_1 winding, diode CR_1, and choke coil L_1. All of these components have some d-c resistance, the total of which would be in parallel with resistor R_1.

The simplest method to eliminate the parallel resistance is to disconnect one lead of the resistance.

1-10. BASIC VOLTMETER (VOLTAGE) MEASUREMENTS

The first step in making a voltage measurement is to set the range. Always use a range that is higher than the anticipated voltage. If the approximate voltage is not known, use the highest voltage range initially, and then select a lower range so that a good mid-scale reading can be obtained.

Next, set the function selector to ac or dc, as required. In the case of dc, it may also be necessary to select either plus or minus by means of the function switch. On simple meters, polarity is changed by switching the test leads.

On an electronic voltmeter, the next step is to zero the meter. This should be done after the range and function have been selected. Touch the test leads together and adjust the zero control for a zero indication on the voltage scale to be used.

Remember that all voltage measurements (ac, dc, plus, minus, and decibels) are made with the meter in parallel across the circuit and voltage source, as shown in Figure 1-38. This means that some of the current normally passing through the circuit under test will be passed through the meter. In a VOM where the total meter resistance (or impedance) is low, considerable current

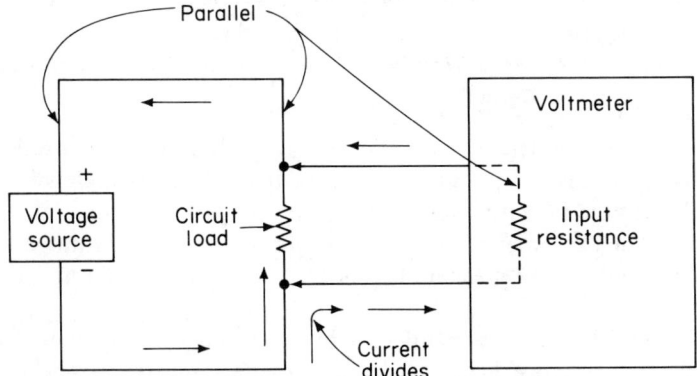

Fig. 1-38. Voltage measurements are made with meter in parallel across circuit and voltage source.

may pass through the meter. This may or may not affect the circuit operation. For example, an oscillator that develops a small voltage over a high circuit impedance can be prevented from oscillating if a VOM is used to measure the voltage. (The low-impedance VOM draws excessive current, dropping the voltage to a point where oscillator feedback cannot occur.)

This problem of parallel-current drain does not occur in an electronic voltmeter, except when a voltage is measured across a high-impedance circuit. A typical electronic meter will have an input impedance of 10–15 MΩ. If the circuit impedance is near this value, the current will divide itself between the circuit and the meter, possibly resulting in an erroneous reading.

1-10.1. Measuring Alternating Current in the Presence of Direct Current

If it is desired to measure ac only but dc is also present in the circuit, the "output" or a-c only function can be selected, thereby switching a coupling capacitor into the meter input. On a VOM, this is done by connecting the free test lead to the output terminal. In an electronic voltmeter, ac is often selected by means of a switch on the probe. In some meters, ac is always measured with a coupling capacitor at the input. In any event, the dc is blocked, and the ac is passed.

Use of the output function can present another problem. The coupling capacitor and the meter resistance form a high-pass filter and may attenuate low-frequency a-c voltages. However, most meters will provide accurate a-c indications above 15 or 20 Hz. It is also possible that the coupling capacitor and the meter movement coil will form a resonant circuit and increase the a-c signals at some particular frequency (usually about 30–60 kHz).

Always consider the frequency problem when making any a-c voltage measurements.

1-10.2. Measuring Direct Current in the Presence of Alternating Current

If it is desired to measure dc only but ac is also present in the circuit, there are several possible solutions. If the ac is of high frequency, it is possible that the meter movement will not respond and there will be no a-c indications when the meter is set to measure dc. If the a-c voltage is low in relation to the dc being measured, it is also possible that the meter will not be affected.

One solution, if the meter is affected by the presence of ac, is to connect a capacitor across the test leads. This will provide a bypass for the ac but will not affect the dc. However, the capacitor may affect operation of the circuit. Also, remember that the capacitor will be charged to the full value of the dc.

In some cases, it is possible to use a high-voltage or attenuator probe to measure dc in the presence of ac. The series resistance of the probe, combined with the natural capacitance between the probe's inner and outer conductor or shield, forms a low-pass filter. This filter action will have no effect on dc but will reject ac.

The fact that electronic voltmeters usually use some form of probe makes these instruments better suited to measure dc (in the presence of ac) than VOMs.

1-11. BASIC AMMETER (CURRENT) MEASUREMENTS

The first step in making a current measurement is to set the range. Always use a range that is higher than the anticipated current. If the approximate current is not known, use the highest-current range initially, then select a lower range so that a good mid-scale reading can be obtained.

(Note that most electronic voltmeters do not have a provision for measuring current, primarily because of their high input impedance. Since current must pass through the meter input circuit, there is a voltage drop across the meter. In an electronic meter, the voltage drop could be very high. In some electronic meters, current is measured by connecting directly to the meter and shunts, thus bypassing the high-impedance input.)

In many meters, selecting a curent range involves more than positioning a switch. A typical VOM requires that the test leads be connected to different terminals. For example, the high-current range could require that the test leads be connected to the -10-A and $+10$-A terminals. On the low-current range, the common and 50-μA terminals could be used. On all other current ranges, the common and plus terminals could be used. No matter the terminal arrangement, the range selector must be set to the appropriate range in all cases.

Once the current range has been selected, set the function selector to ac or dc, as required. (Many VOMs will not measure a-c current, so either plus or minus dc must be selected.)

Note that when the *lowest-current scale* is selected, such as 50 μA, the meter is actually functioning as a *voltmeter*. The meter movement is placed (without a shunt) in series with the circuit. Therefore, any sudden surges of current can damage the meter movement. This is especially a problem when there is both ac and dc in the circuit being measured. If the ac is of a higher frequency, it will probably have little effect on the meter movement. Lower-frequency ac can combine with the dc and possibly cause reading errors or meter movement burnout.

Remember that all current measurements (dc, plus, and minus) are made with the meter in *series* with the circuit and power source, as shown in Figure 1-39. This means that all of the current normally passing through the circuit under test will be passed through the meter. This may or may not affect circuit operation.

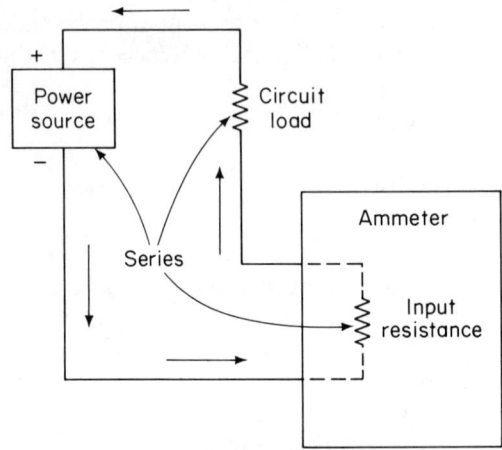

Fig. 1-39. Current measurements are made with meter in series with circuit and power source.

1-12. BASIC DECIBEL MEASUREMENTS

The procedures for measurement in decibels is similar to that for a-c voltage measurement, except that

1. The output function is always used for decibel measurements
2. The dB scales are used instead of the a-c RMS or peak-to-peak scales

When making decibel mesurements, use the basic voltage measurement procedures of Section 1-9, and observe the precautions concerning decibel scales described in Section 1-4.4.

1-13. OHMMETER TEST AND CALIBRATION

The accuracy of an ohmmeter can be checked by measuring the values of precision resistors. If the indicated resistance values are within tolerance, the ohmmeter can be considered operating properly and ready for use. The following points should be considered when making ohmmeter accuracy checks.

The resistors should have a ± 1 per cent or better rated accuracy tolerance. In any event, the resistor accuracy must be greater than the rated ohmmeter accuracy. A typical ohmmeter (VOM or electronic) will have a ± 2 or ± 3 per cent accuracy.

Select resistor values that will provide mid-scale indications on each ohmmeter range. Make certain to zero the ohmmeter when changing ranges.

To find *distribution error*, select precision test resistors that will give 25 and 75 per cent scale indications in addition to the 50 per cent (mid-scale) indication. Accuracy will not be the same on all parts of the scale due to nonuniform meter movements. However, accuracy should be within the rated tolerance on all parts of the scale and on all ranges.

Often, ohmmeter accuracy will be rated in degrees of arc rather than a percentage of full scale. It is sometimes difficult to relate pointer travel in degrees of arc to a percentage. For practical work, remember that the ohmmeter accuracy is approximately equal to the accuracy of the d-c scale. For example, assume that the d-c scale is rated as accurate to within ± 2 small divisions (the scale has 100 small divisions with a ± 2 per cent accuracy). Then the ohmmeter scale will also be accurate within the same degree of pointer travel, or arc, as it takes for the point to move ± 2 small divisions on the d-c scale.

1-14. VOLTMETER TEST AND CALIBRATION

Both the a-c and d-c scales of a voltmeter must be checked for voltage accuracy. In adition, the a-c scales must be checked for accuracy over the entire rated frequency range.

The obvious method to check the accuracy of a voltmeter is to measure a known voltage or series of voltages and check that the indicated voltage values are within tolerance.

1-14.1. Comparison Against a Standard Voltmeter

The most convenient method of voltmeter test is to compare the voltmeter to be tested against a standard voltmeter of known accuracy. It is common practice in laboratory work to have one standard voltmeter against which all meters are compared. The standard voltmeter is never used for routine work but only for a test and is sent out for calibration against a *primary standard* at regular intervals. The primary standard, in turn, is checked against the voltage standards maintained by the National Bureau of Standards (in the United States). These standards are similar to those for time and frequency, discussed in Chapter 8.

The voltmeter to be tested is connected in parallel with the standard voltmeter and a variable-voltage source, as shown in Figure 1-40. The source is then varied over the entire range of the voltmeters under test, and their voltage indications are compared with those of the standard voltmeter. In the case of a-c meters, the source is set to a given voltage; then the frequency is varied over the entire range of the meters under test.

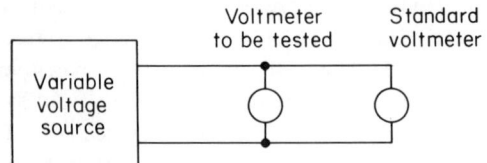

Fig. 1-40. Voltmeter calibration by comparison against a standard voltmeter.

In such a test, accuracy is dependent entirely on the accuracy of the standard meter and not on the source voltage or frequency. This is very convenient, since it is more practical to maintain a meter of known accuracy than a source of known accuracy, especially a source that can be varied over the entire voltage and frequency range of a typical VOM or electronic meter.

1-14.2. Comparison Against a Standard Cell

The next most popular method of checking a meter's voltage accuracy is to measure a voltage source of known accuracy. Usually, this voltage source is a *standard cell*. Inexperienced operators often check the voltmeter accuracy against a common dry cell or series of dry cells. This is satisfactory for rough shop work but is not accurate enough for precision work. A single dry cell rarely produces 1.5 V, as is often assumed. The accuracy of a dry cell voltage is usually less then the accuracy of a typical VOM.

A *mercury cell*, or series of mercury cells, provides much better accuracy than a common dry cell. The voltage output of a typical mercury cell is 1.35 V. The mercury cell will maintain this voltage for a long time over a wide temperature range and with an accuracy greater than that of a typical meter.

The most accurate source is a standard cell, such as the *Weston cell*. There are two types of Weston cells: the *normal* or *standard* cell and the *student* or *shop* cell. The normal cell is the most accurate, providing a voltage of 1.0183 V at 20°C. The student or shop cell provides a voltage from 1.0185 to 1.0190 V at 20°C. Therefore, the student cell can be off by as much as 0.5 mV. Usually, this is greater accuracy than is required for all but precision laboratory meters.

Also, a student cell is less affected by changes in temperature. Once the accuracy of a student cell is established, it can be considered as remaining at that voltage over the normal range of room-temperature operation.

1-14.3. Voltage Comparison Using the Balance Method

Inexperienced operators often connect a meter to be tested directly to a standard cell. Although this will provide an accurate indication, it will also place a damaging curent drain on the standard cell. Laboratories generally use some form of calibration circuit with their standard cells. Most of these calibration circuits use the *balance method* and require a galvanometer (refer to Section 1-2.7).

A calibration circuit using the balance method is shown in Figure 1-41. Operation of the circuit is as follows. Switch S_1 is left in the open position. Potentiometers R_1 and R_2 are adjusted so that the voltage at the tap of R_1 is approximately equal to that of the standard cell. Switch S_1 is then set to the standard cell position, and the potentiometer R_1 is adjusted until the galvanometer reads zero. At this point, the voltage at the center terminals of S_1 is equal to the voltage of the standard cell. Initially, switch S_2 is open to place protective resistor R_3 in series with the galvanometer. Most galvanometers are quite sensitive. If there is a large voltage difference between the standard cell and the output of R_1, this difference could damage the galvanometer. Once approximate balance is reached, switch S_2 is closed, and the voltage is adjusted for exact zero on the galvanometer. Circuits of this type make it possible to obtain a balance while drawing less than 0.1 mA from the standard cell.

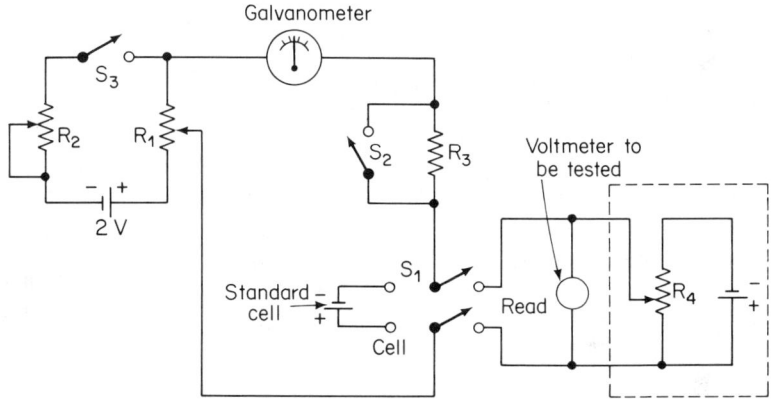

Fig. 1-41. Voltmeter calibration by comparison against a standard cell.

Once balance is obtained, switch S_1 is set to the "read" position. This removes the standard cell from the circuit and places the voltage across the meter under test. If a student or shop Weston cell were used in the circuit, the meter under test should read between 1.0185 and 1.0190 V.

Note that a typical VOM will not read out five-place values. These calibration circuits and procedures are for precision instruments.

A further variation of the calibration circuit is shown in Figure 1-41, in dotted lines. This variation is used to protect the sensitive galvanometer. After the galvanometer circuit has been balanced against the standard cell and switch S_1 has been set to "read," resistor R_4 is adjusted until the galvanometer again shows a balance, indicating that the voltage at the tap of R_4 is equal to that of R_1 (and the standard cell). Then switch S_1 is set to "open," leaving only the voltage from R_4 across the meter under test.

1-14.4. Voltage Comparison Using a Voltage Divider

It is obvious that the calibration circuit of Figure 1-41 would provide a good indication on the 2.5-V scale of a meter, but it would be of little value on the 250-V range or any range substantially higher than 2.5 V. Likewise, the circuit could not be used at all on a lower range, such as 250 mV. (Many of the newer meters have very low voltage ranges, which are necessary to measure the small voltage differences found in transistor circuits.)

These problems can be overcome by means of a *voltage-divider* circuit. The basic voltage divider can be any form of precision variable resistor from which the *exact* resistance (fraction of $1 \Omega \times 10^{-4}$, or better) can be read out or from which the *exact* ratio of full resistance to tap resistance can be shown by an external indicator. These voltage dividers have various names, such as volt box, decade-ratio potentiometer, ratiometer or decade box.

Voltage dividers can be used to provide precision voltage that can be calibrated against a standard cell, even though the voltages are much higher or lower than the cell voltage.

For example, assume that a supposed 100-V source is placed across a precision voltage-variable divider of 100 Ω. Each 1-Ω tap or position on the voltage divider should then show a corresponding voltage (the 3-Ω position would show 3 V, the 33-Ω position would show 33 V, and so on). The divider could then be set to the 1.0185- or 1.0190-V position, and the resulting voltage could be balanced against a standard cell, as shown in Figure 1-42. This will establish the accuracy of the voltage source.

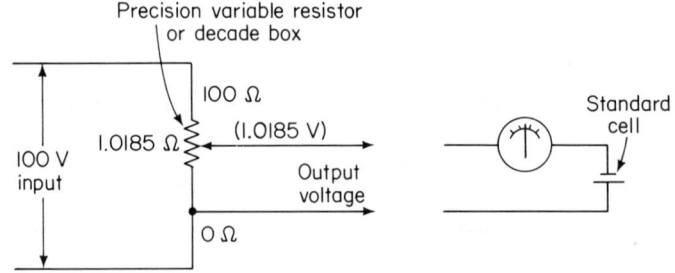

Fig. 1-42. Precision variable resistor, or decade box, used as a voltage divider.

For example, assume that the circuit shows a balance against the 1.0190-V standard cell when the resistance shows 1.000 Ω, or approximately 2 per cent off. This means that the voltage source across the divider is 2 per cent off (2 per cent high) and that all of the voltage readings will be 2 per cent off. Note that the accuracy of the voltage-divider system depends on the accuracy of the voltage divider itself and not on the accuracy of the voltage source.

1-15. FREQUENCY TEST AND CALIBRATION

There are a number of problems in checking frequency response of shop meters, even though the procedure is quite simple. The meter is connected to a signal generator capable of providing signals over the entire frequency range of the meter and/or probe, as shown in Figure 1-43. The voltage output of the generator is set to provide a good mid-scale indication on the meter. Then, without changing the voltage output, the generator frequency is varied over the full range of the meter.

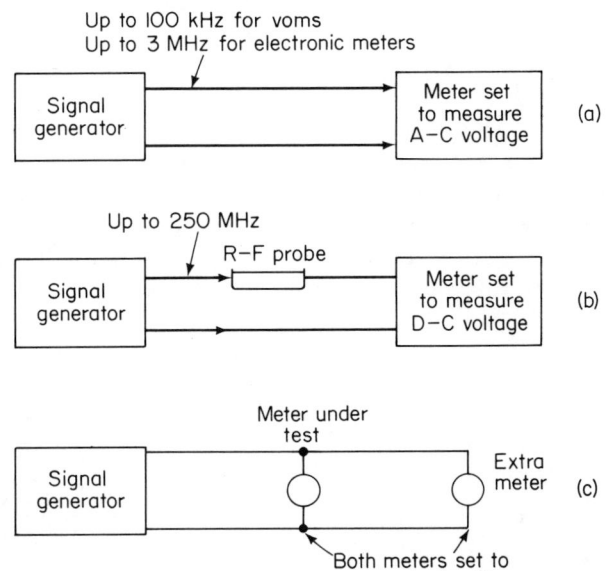

Fig. 1-43. Measuring frequency response of meters.

A typical VOM has a range from about 15 Hz to 100 kHz, whereas an electronic meter's range is from 15 Hz to 3 MHz. If an r-f probe (as shown in Figure 1-43B) is used, the frequency range is extended to approximately 250 MHz. In any case, the voltage indication should remain constant (within a given tolerance) over the entire frequency range. Each of the scales can be checked separately with different levels of voltage from the generator.

1-15.1. Precautions in Making Frequency Response Tests

Shop generators rarely have a flat output over a wide frequency range. This may lead the inexperienced operator to false conclusions regarding the meter. The problem can be minimized (but not completely eliminated) by connecting two meters simultaneously (in parallel, as shown in Figure 1-43C).

If the generator output appears to change at some particular frequency on both meters, even if by different amounts on each meter, the generator output is not flat. For example, if the voltage output appears to drop at a particular frequency, the likelihood is greater that the generator output is dropping than that both meters are giving an identical frequency response. (It is possible for two meters to have identical frequency responses, but it is not likely.)

Even if the generator's output is flat over a given frequency range, it is likely that the meter components (movement inductance, multiplier resistance, and stray capacitance) will form filter circuits and resonant circuits at certain frequencies. These circuits can cause a false indication on the meter. A typical example occurs when the movement inductance and the stray capacitance combine to form a resonant circuit. This causes an apparent rise in voltage at the resonant frequency.

This particular problem can be minimized by measuring the same voltage on two ranges. Measure the voltage at the high end of one range, then change to the low end of the adjacent range (or vice versa) and check for the same voltage indication. The effects of stray capacitance and multiplier resistance are changed when meter ranges are changed. If the voltage indication changes, it is likely to be caused by a problem in the meter circuits.

Note that there may be some vibration at low frequencies. This is typical for shop meters and is caused by inertia in the meter movement.

Also note that the input circuits of most electronic meters are provided with frequency-compensation capacitors. A typical circuit is shown in Figure 1-44.

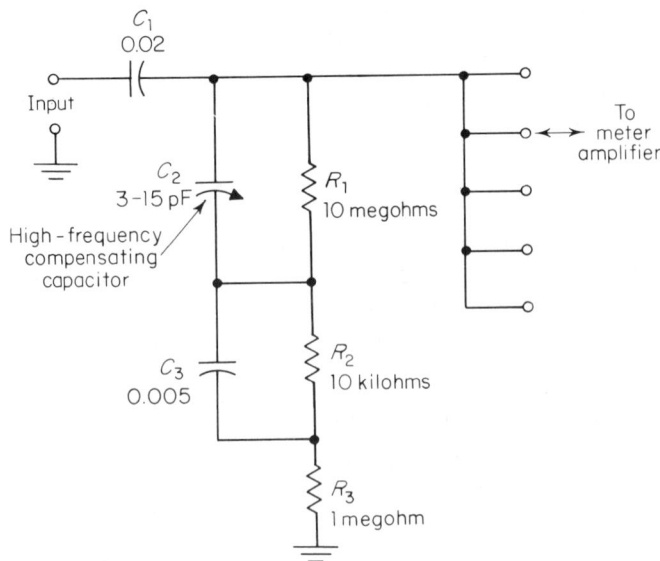

Fig. 1-44. Typical high-frequency compensation circuit for electronic meter.

The capacitor (usually in the order of 3–15 pF) is in parallel with the input resistance (typically, 10 MΩ). The capacitor is adjustable to provide a flat frequency response at the high end of the meter frequency range. As in the case of other internal meter adjustments, it is not recommended that the frequency-compensation capacitors be calibrated except by a qualified instrument technician.

1-15.2. Frequency Response in Meters with Demodulator Probes

Checking the frequency range of a *demodulator probe* requires a special test circuit, as shown in Figure 1-45. A demodulator probe produces both a d-c output (by rectifying the rf) and an a-c output (or pulsating dc-

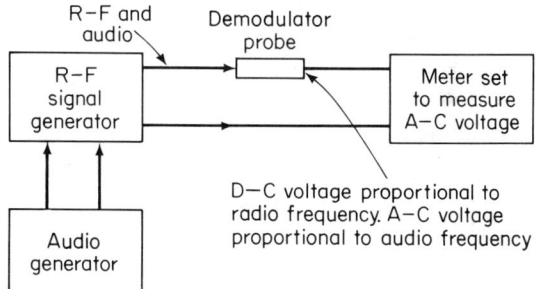

Fig. 1-45. Checking audio-frequency range of demodulator probe.

output), which is the modulation signal. The d-c output can be checked in the same way as any r-f probe (Figure 1-43). The a-c or demodulated output must be checked on the a-c scales of the meter. For a thorough test, the modulation frequency should be varied over the full range of the probe. This usually requires that the r-f signal generator be modulated by an external audio generator.

With the equipment connected as shown in Figure 1-45, set the meter to measure a-c voltage. Set the signal generator to maximum voltage output at any frequency well within the r-f range of the probe. If necessary, temporarily set the meter to measure d-c voltage and note the indication produced by the r-f signal. Vary the frequency of the external audio generator over the demodulation range of the probe. Typically, a probe will demodulate signals up to about 20 kHz. However, the a-c voltage indication will usually start dropping off at about 17 kHz.

1-16. AMMETER TEST AND CALIBRATION

The scales of an ammeter must be checked for accuracy of the current indication. A typical VOM will have only d-c current ranges, and a typical electronic meter will have no current ranges (unless the high-impedance input is bypassed and current is measured directly at the meter). Therefore, the primary concern is with d-c current accuracy. However, if a meter is provided with a-c current scales, they must also be checked for accuracy over the entire rated frequency range.

The most convenient method of checking current accuracy is to compare the ammeter to be tested against a standard ammeter of known accuracy. The ammeter to be tested is connected in series with the standard ammeter and a variable current source, as shown in Figure 1-46. The source is then

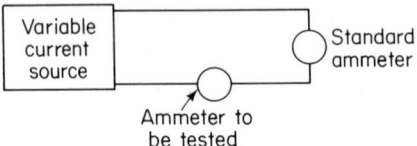

Fig. 1-46. Ammeter calibration by comparison against a standard ammeter.

varied over the entire range of the ammeter, and the current indications are compared. The accuracy of this test is dependent entirely on the accuracy of the standard meter and not on the current source.

The next most popular method for checking current accuracy is to use a precision voltmeter and precision resistance. The ammeter to be checked and the precision resistance are connected in series with a variable power source, as shown in Figure 1-47. The precision voltmeter is connected across the

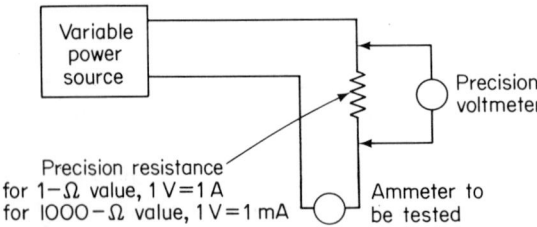

Fig. 1-47. Ammeter calibration circuit using a precision resistance and precision voltmeter.

precision resistance. Current through the circuit is computed by Ohm's law ($I = E/R$).

The value of the precision resistor is chosen so that the voltage indicated on the precision voltmeter can be related directly to current. For example, if a 1-Ω resistance is used, the voltage across the resistance can be read directly in amperes (3 V = 3 A, 7 V = 7 A, and so on). If a 1000-Ω precision resistor is used, the voltage can be read directly as milliamperes (3 V = 3 mA, and so on).

Note that the accuracy of this test method is dependent on the accuracy of both the voltmeter and series resistance. The tolerance of both components must be added. For example, assume that both the voltmeter and resistance have a 1 per cent tolerance. Then accuracy of the circuit could be no greater than 2 per cent. In any event, the combined accuracy must be greater than that of the meter to be tested.

2 DIGITAL AND DIFFERENTIAL METERS

2-1. INTRODUCTION TO DIGITAL CIRCUITS

To understand operation of digital meters or any digital test instrument, it is necessary to have a full understanding of digital logic circuits. These include gates, amplifiers, switching elements, delay elements, binary counting systems, truth tables, registers, encoders, decoders, digital-to-analog converters, analog-to-digital converters, adders, scalers, and so on. Information on the operation of such circuits is contained in the Appendix. These data will serve as a refresher for those already familiar with digital equipment and as a basic guide for the student who has no knowledge of digital logic.

In practice, most digital equipment (meters, counters, computers, and so on) is made up of logic "building blocks" (gates, registers, and so on) that are interconnected to perform various functions (mathematical operations, conversion, readout, and so on). If the student understands operation of the individual building blocks he can then understand operation of the complete instrument. Thus operating principles of complex digital equipment (meters, counters, and so on) can be presented in block form, or in logic-diagram form, with blocks and logic symbols representing the building blocks.

This method of presentation is followed by most manufacturers of digital test equipment in their instruction manuals. Therefore, if the student understands the general digital information in the Appendix plus the specific information in this chapter, he should have no difficulty in understanding operation of any digital meter.

It should be further noted that knowledge of *electronic counters* is necessary to fully understand digital meters. This is because a digital meter performs two basic functions: (1) conversion of voltage (or other quantity being measured) to time or frequency, and (2) conversion of the time or frequency data to a digital readout. In effect, a digital meter is a conversion circuit (voltage-to-time, and so on) plus an electronic counter for readout. Operation of electronic counters is discussed in Chapter 6.

2-2. BASIC DIGITAL METER

Digital instruments are available to measure a-c and d-c voltages, d-c currents, resistance, and ratio. Other physical variables can also be measured by use of suitable transducers, as described in Chapter 9. Many digital instruments have outputs that can be used to make permanent records of measurements with printers, card and tape punches, and magnetic tape equipment. With data in digital form, it may be processed with no loss of accuracy.

The most popular digital meter is the digital voltmeter, or DVM, although there are a number of digital instruments for resistance and current measurement. Such instruments display measurements as *discrete numerals*, rather than as a pointer deflection on a continuous scale commonly used in analog devices described in Chapter 1. Direct numerical readout reduces human error, eliminates parallax error, and increases reading speed. Automatic polarity and range-changing features on some digital meters reduce operator

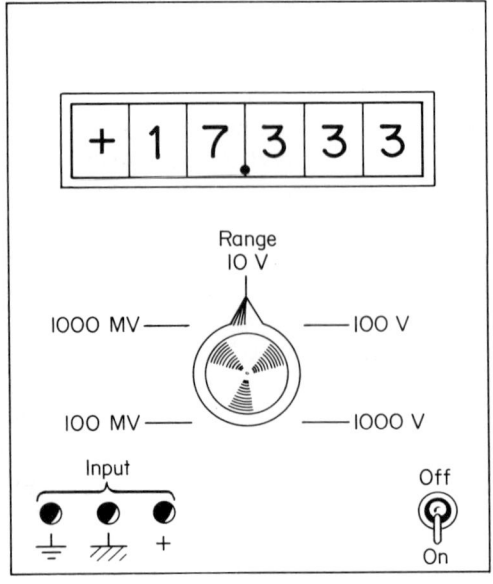

Fig. 2-1. Typical digital voltmeter panel controls and indicators.

training, measurement error, and possible instrument damage through overload.

Figure 2-1 shows the operating controls and readout of a typical DVM. Note the simplicity of controls. Once the power is turned on, the operator has only to select the desired range and connect the meter to the circuit. The readout will be automatic. On some digital meters, the range is changed automatically, further simplifying operation.

2-3. RAMP-TYPE DIGITAL METER

The operating principle of the ramp-type digital voltmeter is to measure the time required for a linear voltage ramp to change from a value equal to the voltage being measured to zero (or vice versa). The *time period* is measured with an electronic time-interval counter (Chapter 6) and is displayed on a decade readout. The ramp-type meter is essentially a voltage-to-time converter plus a counter and readout.

Conversion of a voltage to a time interval is illustrated by the timing diagram in Figure 2-2. The block diagram of a typical ramp-type DVM is shown in Figure 2-3.

At the start of a measurement cycle (there are usually two or three measurement cycles per second), a ramp voltage is generated. The ramp voltage is compared continuously with the voltage being measured. At the instant the two voltages become equal, a coincidence circuit generates a pulse that opens a gate. The ramp continues until a second comparator circuit senses that the ramp has reached 0 V. The output pulse of this comparator closes the gate.

The time duration of the gate opening is proportional to the input voltage. The gate allows pulses to pass to the counter circuit, and the number of pulses counted during the gating interval is a measure of the voltage.

As shown in Figure 2-3, the voltage ramp is generated by a ramp or sawtooth generator that, in turn, is triggered by the sample rate oscillator. This oscillator serves as a time base and produces pulses at the rate of 2 or 3 Hz. These pulses are also sent to the counter circuits to clear any readings (return the readings to zero) at the same rate. This makes it possible to monitor a steadily changing voltage.

The ramp voltage is compared with both the input or unknown voltage (in the input comparator) and zero voltage (in the zero or ground comparator). The output of the two comparators is connected to an AND gate (for stop pulses) and an OR gate (for start pulses). The AND and OR gates control operation of the counter circuit oscillator. Coincidence of the ramp voltage with either the voltage being measured or with 0 V) starts the oscillator. (The AND gate is disabled, and the OR gate is enabled.) With

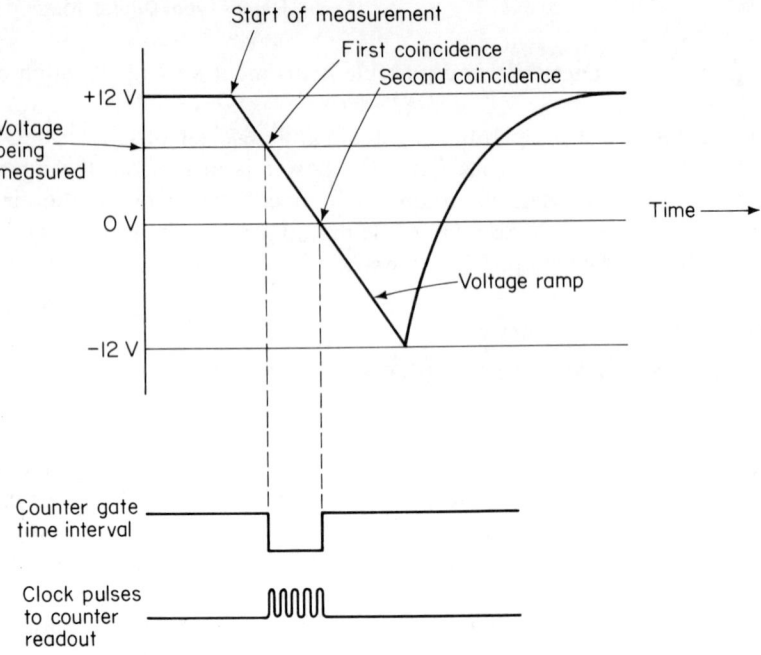

Fig. 2-2. Timing diagram showing voltage-to-time conversion (Hewlett-Packard).

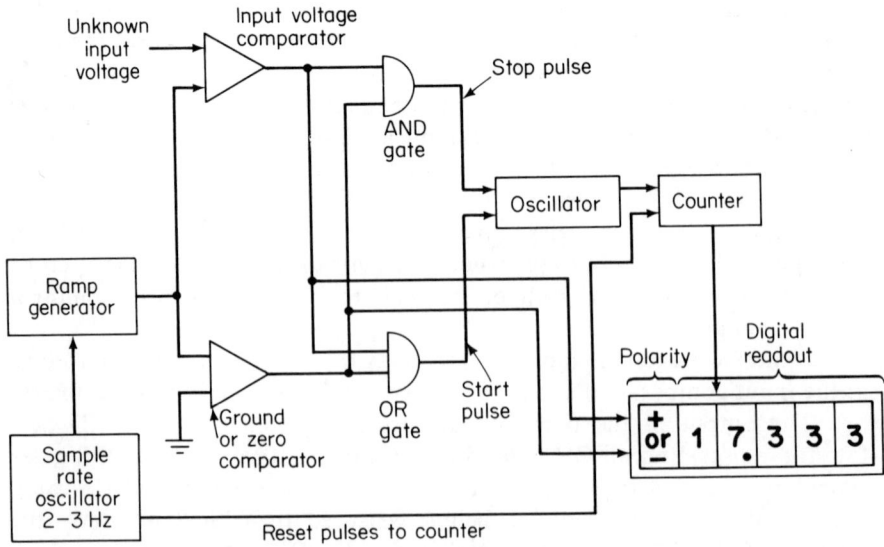

Fig. 2-3. Basic ramp-type digital voltmeter circuit (Hewlett-Packard)

the oscillator on, the electronic counter starts the count. When the second coincidence occurs, the AND gate is enabled and the oscillator stops, as does the count.

The elapsed time, as indicated by the count, is proportional to the time the ramp takes to travel between the unknown voltage and 0 V (or vice versa, in the case of a negative input voltage to be measured). Therefore, the count is equal to the input voltage. The order in which pulses come from the two comparators indicates the polarity of the unknown voltage. This triggers a readout that indicates plus or minus, as required.

2-3.1. Staircase Ramp Digital Meter

The *staircase ramp* type of digital meter is an improvement over the basic ramp type. Figure 2-4 is a block diagram of a typical staircase ramp instrument.

This meter makes voltage measurements by comparing the input voltage to an internally generated staircase ramp voltage, rather than a linear ramp. When the input and the staircase ramp voltages are equal, the comparator generates a signal to stop the ramp and the count. The instrument then displays the number of counts that were necessary to make the staircase ramp equal to the input voltage. At the end of the sample (the sample rate is fixed at two samples per second by the 2-Hz sample oscillator) a reset pulse resets the staircase ramp to zero, and the measurement starts over.

2-4. INTEGRATING-TYPE DIGITAL METER

One of the problems with a ramp digital meter is that measurements at the end of the time interval could occur simultaneously with a noise burst (where dc must be measured in the presence of other signals). These noise signals near the second coincidence of the ramp voltage could lengthen (or shorten) the time interval, thus making the count incorrect. This problem can be overcome by means of an *integrating* digital meter that makes its measurement on the basic of *voltage-to-frequency* conversion, rather than voltage-to-time, as used in the ramp meter.

A typical voltage-to-frequency integrating converter is shown in Figure 2-5. The circuitry functions as a feedback-control system that governs the rate of pulse generation, making the average voltage of the rectangular pulse train equal to the d-c input voltage.

As shown, a positive voltage at the input results in a negative-going ramp at the output of the integrator. (The use of amplifiers as integrators and voltage-to-frequency converters is discussed further in Chapter 7.) The ramp continues until it reaches a voltage level that triggers the level detector,

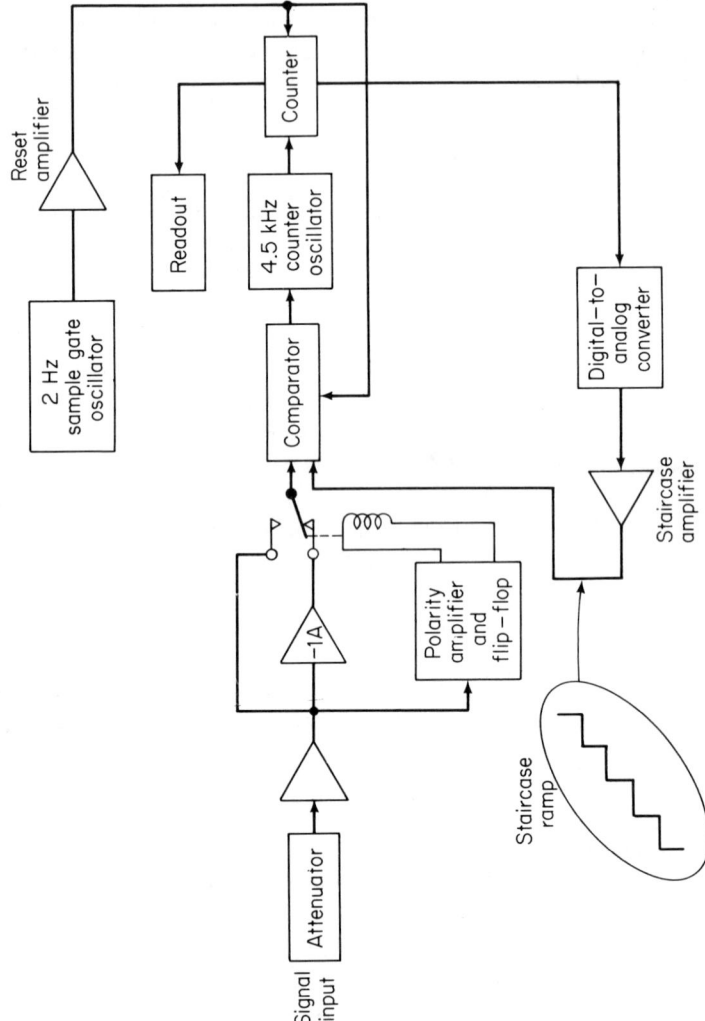

Fig. 2-4. Basic staircase-ramp digital voltmeter circuit (Hewlett-Packard).

56

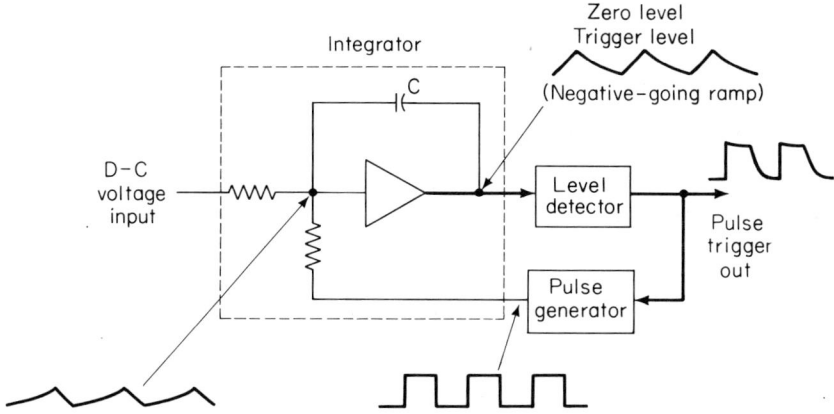

Fig. 2-5. Typical voltage-to-frequency conversion circuit (Hewlett-Packard).

which, in turn, triggers a pulse generator. The pulse generator produces a rectangular pulse with closely controlled width and amplitude just sufficient to draw enough charge from capacitor C to bring the input of the integrator back to the starting level. The cycle then repeats.

The ramp slope is proportional to the input voltage. (For example, a higher voltage at the input would result in a steeper slope, resulting in a shorter time duration for the ramp.) As a result, the pulse repetition rate would be higher. Since the pulse repetition rate is proportional to the input voltage, the pulses can be counted during a known time interval to find a digital measure of the input voltage.

Although a voltage ramp is generated in this type of DVM, the amplitude is only a fraction of a volt, and the accuracy of the analog-to-digital conversion is determined not only by the characteristics of the ramp but also by the area of the feedback pulses.

The primary advantage of this type of analog-to-digital conversion is that the input is "integrated" over the sampling interval, and the reading represents a *true average* of the input voltage. The pulse repetition frequency "tracks" a slowly varying input voltage so closely that changes in the input voltage are accurately reflected as changes in pulse repetition rate. The total pulse count during a sampling interval therefore represents the average frequency and thus the average voltage. This is important when noisy signals are encountered since the noise can be averaged out during the measurement.

Another advantage of the integrating circuit is that the measurement circuit can be completely isolated (by shielding and transformer coupling) from the counter and readout circuits. This technique is shown in Figure 2-6. (The guard shield technique is discussed further in Section 2-9.)

58 Digital and Differential Meters

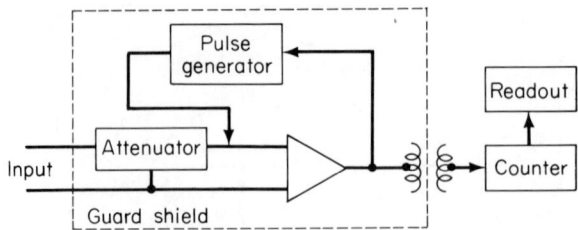

Fig. 2-6. Simplified block diagram of guard-circuit technique.

2-5. INTEGRATING/POTENTIOMETRIC-TYPE DIGITAL METER

An integrating/potentiometric meter combines the continual measurement of true average input voltage with accuracy from precision resistance ratios and stable reference voltages (refer to Section 1-14.4).

A block diagram of a typical integrating/potentiometric meter is shown

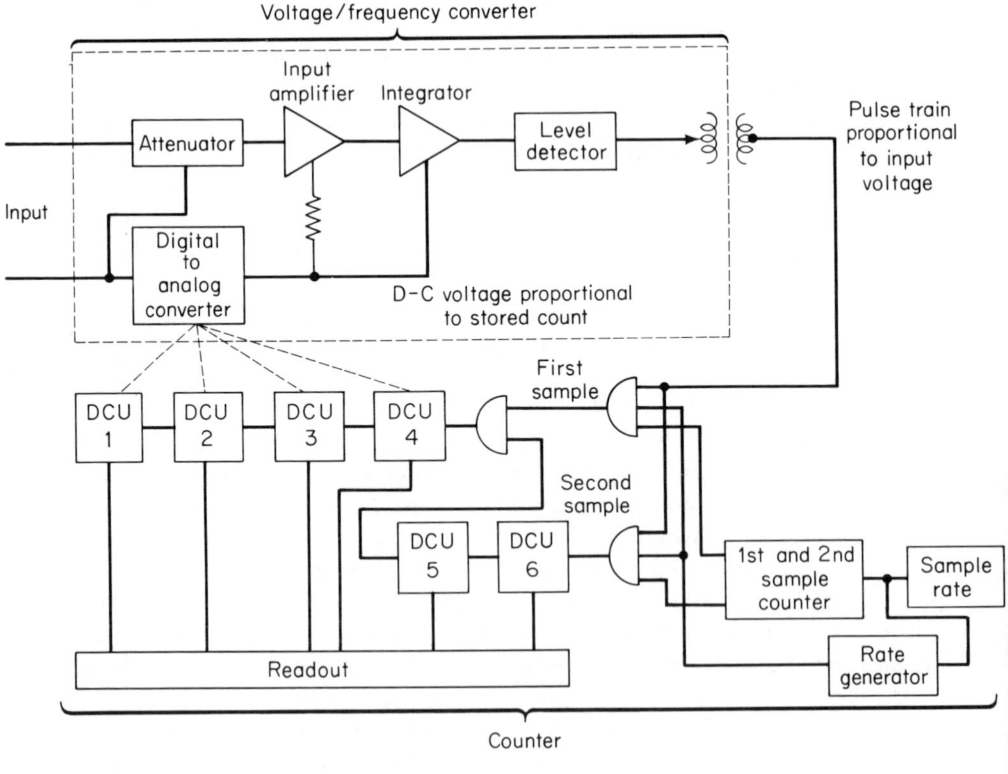

Fig. 2-7. Basic integrating/potentiometric type digital voltmeter (Hewlett-Packard).

in Figure 2-7. Note that the instrument is divided into three sections: a voltage-to-frequency (V/F) converter, a counter, and a digital-to-analog (D/A) converter.

As in the case of the conventional integrating meter, the voltage-to-frequency converter generates a pulse train with a rate exactly proportional to the input voltage. This pulse train is gated for a precise time interval and is fed to the first four places in a six-digit counter. The stored (undisplayed) count is transferred to the D/A converter, which produces a highly accurate d-c voltage proportional to the stored count. This voltage is subtracted from the unknown voltage at the input to the V/F converter.

The pulse train from the V/F converter is again gated, this time to the last two places in the six-digit counter. At the end of the second gate period, the total count is transferred to the six display tubes. The counter display is indicative of the integral of the input voltage.

Accuracy of this instrument is dependent on the accuracy of the D/A converter, as well as the V/F converter. As discussed in the Appendix, operation of a D/A converter depends on precision reference voltages and selection of precision resistors by electronic switches.

2-6. DUAL-SLOPE-INTEGRATION-TYPE DIGITAL METER

The dual-slope-type meter uses an entirely different principle to measure d-c voltage (or other values, such as resistance or a-c voltage). A block diagram of a dual-slope meter is shown in Figure 2-8. This instrument measures d-c voltage by the use of an integrator that produces a time interval proportional to the average value of the applied d-c voltage. The time interval determines the gate time of the counter and therefore the number of pulses totalized. Thus the number of pulses is proportional to the average of the d-c voltage measured.

During a precisely controlled time period ($\frac{1}{10}$ or $\frac{1}{60}$ sec, as selected by a front-panel control) the integrator charges up to a value proportional to the average value of the d-c input voltage. This charging voltage is the "up-slope" of the integrator output. After a time period, a precise reference voltage of opposite polarity is switched to discharge the integrator. This discharge is the "down-slope" of the integrator. The zero crossing of the output voltage is detected by a zero detector circuit. The counter is enabled to totalize pulses from the crystal oscillator during the discharge time or down-slope of the integrator. Since the discharge time is proportional to the stored voltage, the number of pulses totalized is proportional to the input voltage.

After completion of the integration cycle, the input amplifier is disconnected and automatically zeroed before the next measurement is taken.

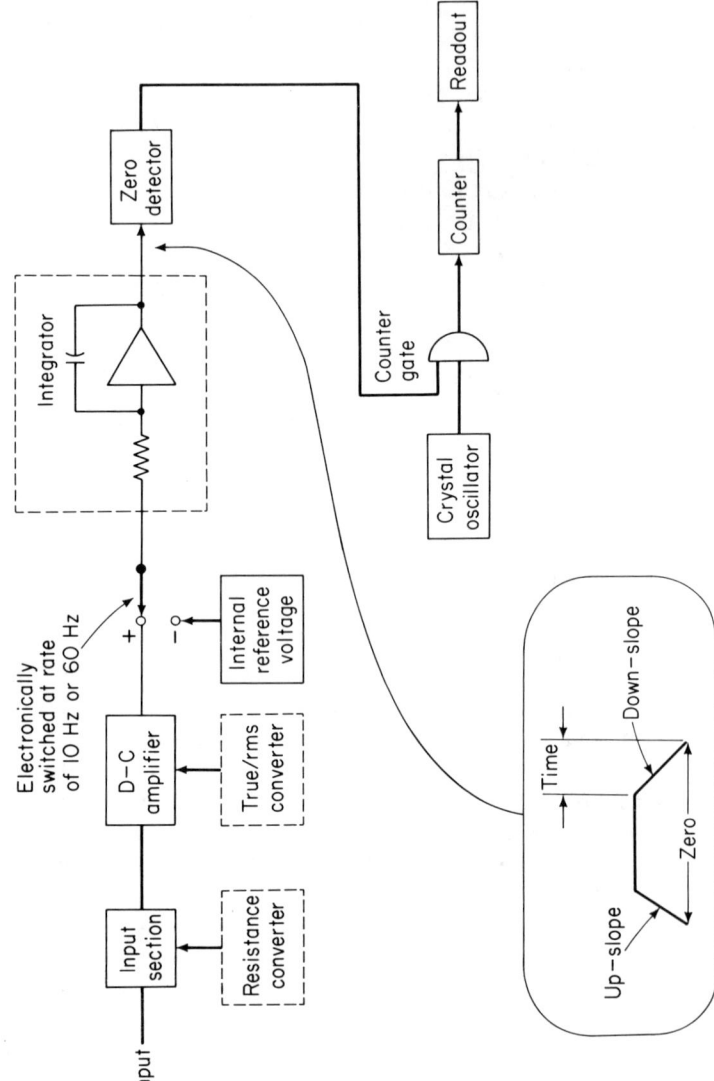

Fig. 2-8. Basic dual-slope integration type digital voltmeter (Hewlett-Packard).

This autozeroing effectively compensates for d-c drift and eliminates the need for a chopper amplifier and front-panel zero controls.

2-6.1. Alternating-Current Voltage Measurements with Dual-Slope Meter

As shown in dotted lines on the block of Figure 2-8, the dual-slope meter can be used to measure ac, as well as dc. However, this requires that the ac be converted to dc before application to the integrator. Either average or RMS values of the ac can be measured.

The average value converter circuit is shown in Figure 2-9. Note that this circuit is similar to the average-responding voltmeters described in Chapter 1, except for the negative feedback circuit. This negative feedback limits the output to 1 V dc, no matter what the a-c input value. Typically, the circuit produces a d-c output voltage between 0 and 1 V proportional to the average value of the applied a-c voltage.

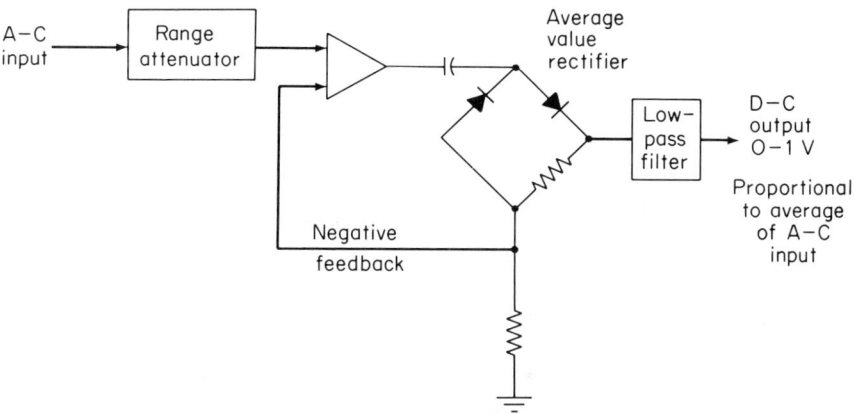

Fig. 2-9. Basic average-value a-c/d-c converter.

The true RMS converter is shown in Figure 2-10. The input circuitry consists of an operational amplifier whose gain is accurately controlled to achieve attenuation of the input signal. (Gain-controlled operational amplifiers are discussed in Chapter 7.) An a-c output from the input amplifier is sent to the modulator, and a second output is used as a trigger for the sync-generator. The sync-generator produces a 5-Hz square wave and is used to synchronize the modulator and demodulator to the input signal.

A 1-kHz oscillator drives the d-c-to-square-wave converter that converts the d-c output of the a-c converter to a reference square wave. The amplitude of the square wave is proportional to the d-c output. The output of the mod-

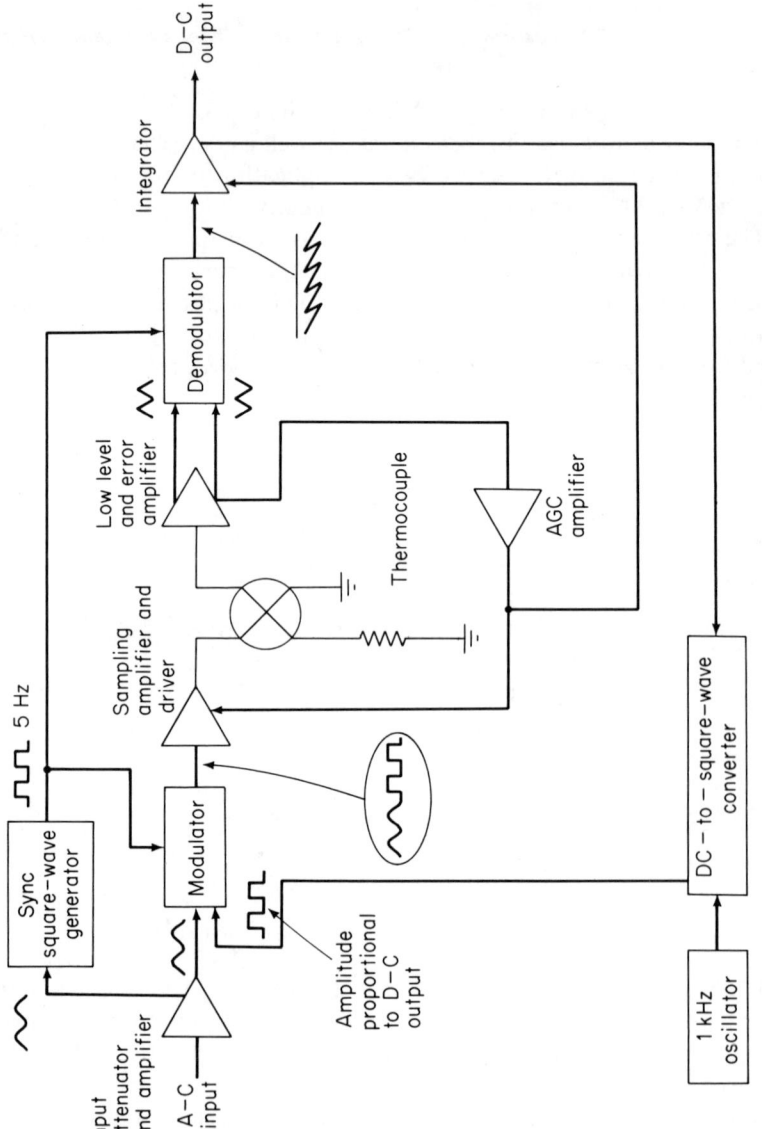

Fig. 2-10. Basic true-RMS a-c/d-c converter (Hewlett-Packard).

ulator, at a nominal 5 Hz, consists of a composite signal made up of one-half input signal and one-half reference square wave (as shown in Figure 2-10).

The automatic gain control (AGC) amplifier controls the gain of the sampling amplifier and the integrator. This keeps the RMS value of the signal applied to the thermocouple constant and holds the gain of the system constant, regardless of the level of the input signal. The output of the thermocouple varies between the RMS value of the input and the reference signal. This error signal is amplified, and two 180° out-of-phase signals are sent to the demodulator. The demodulator acts as a full-wave rectifier. The output pulses are amplified and integrated to develop the positive d-c voltage output. This d-c voltage is continuously corrected at nominally 5 times/sec (set by the 5-Hz sync-square-wave generator) to ensure a d-c voltage output proportional to the RMS value of the input signal.

2-6.2. Resistance Measurements with Dual-Slope Meter

As shown in dotted lines on the block of Figure 2-8, the dual-slope meter can also be used to measure resistance (in ohms). However, this requires that the resistance be converted to dc before application to the integrator.

The resistance (or ohms) converter circuit is shown in Figure 2-11. The resistance measurements are made by feeding a constant current through the unknown resistor and measuring the resultant voltage across the resistor. The current source supplies three constant currents of 1 mA, 10 μA, and 1 μA, and an open-loop voltage of 17 V maximum. The current is held

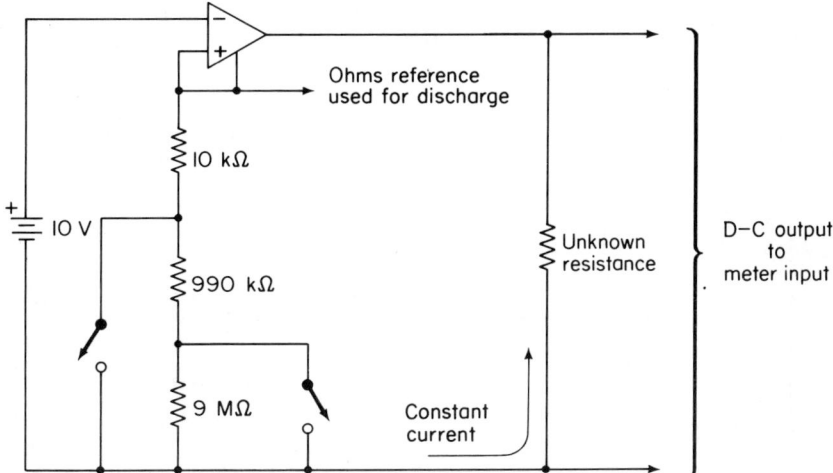

Fig. 2-11. Basic resistance converter circuit (Hewlett-Packard).

constant by means of an operational amplifier and feedback resistors. (The use of operational amplifiers as current sources is discussed in Chapter 7.) To increase measurement accuracy, the internal reference voltage (Figure 2-8) is disabled, and the ohms reference voltage (Figure 2-11) is used to discharge the integrator.

2-7. BASIC DIFFERENTIAL METER

The basic concept of differential voltage measurements is to apply an unknown voltage against one that is accurately known and to measure the difference between the two on an indicating device. If the known voltage is adjusted to the exact potential of the unknown voltage, one can determine the unknown quantity being measured as accurately as the known voltage (or reference standard).

Measurements made by the differential voltmeter technique (sometimes called a potentiometric or manual voltmeter) are recognized as one of the most accurate means of relating an unknown voltage to a known reference. In practice, these measurements are made by adjusting a *precision resistive divider* to divide down an accurately known reference voltage. The divider is adjusted to the point where the divider output equals the unknown voltage, as shown by the null voltmeter (Figure 2-12).

The unknown voltage is determined to an accuracy limited only by the accuracies of the reference voltage and the resistive divider. Meter accuracy is of little consequence, since the meter serves only to indicate any residual differential between the known and unknown voltages.

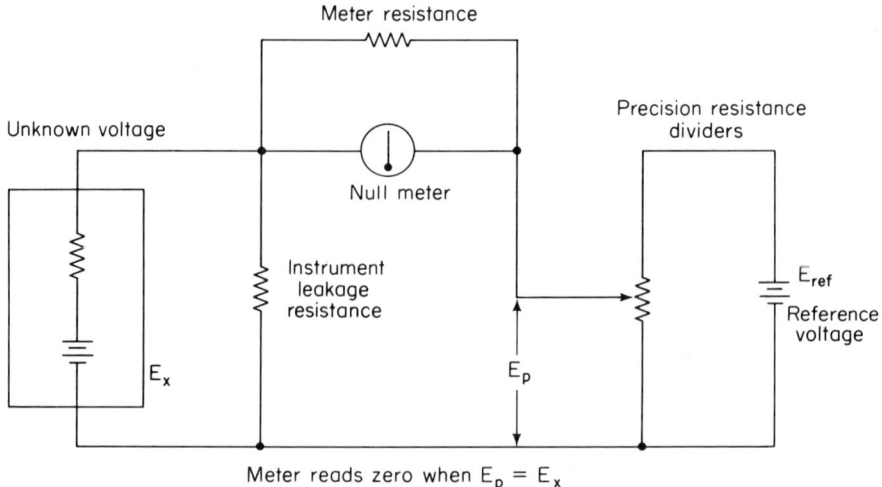

Fig. 2-12. Basic differential voltage measurement (Hewlett-Packard).

A high-voltage standard is required to measure high voltage. This need may be overcome by inserting a voltage divider between the source and the null meter, as shown in Figure 2-13. This, however, results in relatively low input resistance for voltages higher than the reference standard. This low input resistance is undesirable because accurate measurement may not be obtained if substantial current is drawn from the source being mearesud. Most differential voltmeters used today offer input resistance approaching infinity only at a null condition, and then only if an input voltage divider is not used.

To overcome these limitations, Hewlett-Packard has developed a differential voltmeter with an *input isolation stage*. A simplified diagram of the circuit is shown in Figure 2-14. Isolation is accomplished by means of an operational amplifier (Chapter 7) between the measurement source and the measurement circuits. The amplifier ensures that the high input impedance is maintained regardless of whether the instrument is adjusted for a null reading.

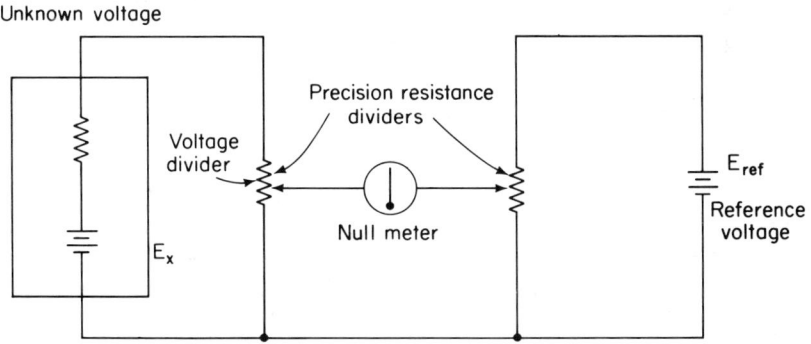

Fig. 2-13. Potentiometric method of measuring unknown voltages (Hewlett-Packard).

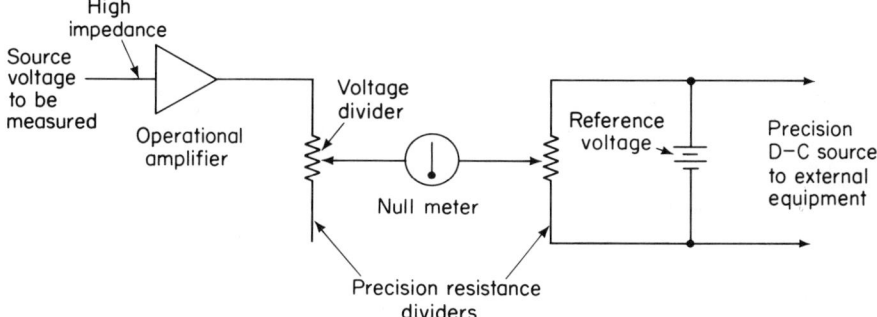

Fig. 2-14. Potentiometric voltmeter with high-impedance input amplifier (Hewlett-Packard).

A further advantage provided by the amplifier is that the resistive voltage divider that enables voltages as high as 1000 V to be compared to a precision 1-V reference may be placed at the output of the amplifier, rather than being in series with the measured voltage source. This isolation permits the instrument to have high impedance on all ranges.

As shown in Figure 2-14, the instrument also serves as a precision d-c source. The need for precision d-c sources with differential meter measurements are discussed in Section 2-10.

2-8. DIFFERENTIAL METER CIRCUITS

Differential meter circuits are used for measurement of both a-c and d-c voltages, as well as voltage ratios. When making d-c voltage measurements, there are cases where the absolute value of the voltage is of little interest. Instead, the point of interest is its value in relationship to some other voltage level or the ratio of it to some other level. This ratio, expressed as $N = V_a/V_b =$ Ratio, appears often in engineering work.

2-8.1. Alternating-Current Voltage Measurements with Differential Voltmeter

Highest accuracy in a-c voltage measurements is accomplished by using an a-c differential voltmeter. Figure 2-15 is a block diagram of a typical differential voltmeter in the a-c voltage mode of operation. This circuit uses a precision rectifier to convert the unknown ac directly to dc (equivalent to the average value of the ac), and the resulting dc is read to five-place resolution by the potentiometric voltmeter technqiue.

The measurement is straightforward in that the ac remains connected to the converter at all times and can be monitored continuously.

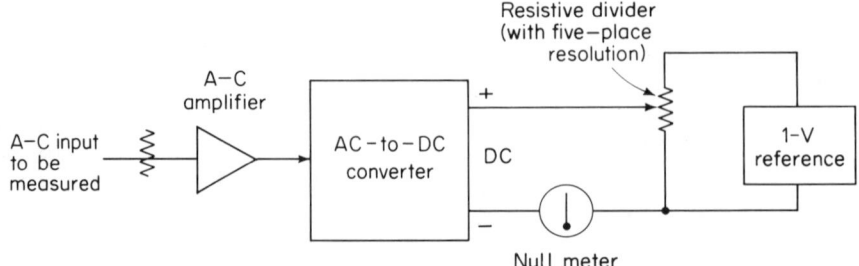

Fig. 2-15. Basic differential voltmeter in a-c voltage mode of operation (Howlett-Packard).

2-8.2. Direct-Current Voltage Measurements with Differential Voltmeter

Figure 2-16 is a block diagram of a typical differential voltmeter in the d-c voltage mode of operation. To accurately measure a d-c voltage using the differential principle, the unknown voltage is connected across the series combination of the electronic voltmeter and the 500-V reference supply. The reference voltage is then adjusted with five voltage-readout dials (mechanically coupled to a Kelvin-Varley voltage divider) until the reference voltage matches the unknown voltage, as indicated by a null on the electronic voltmeter. The unknown voltage is then read from the five reference voltage-readout dials.

The 500-V reference is composed of a well-regulated 500-V supply, a range divider, and a Kelvin-Varley 5-decade attenuator. The output of the 500-V power supply is applied directly to the Kelvin-Varley attenuator for the 500-V d-c range. In the 50-, 5-, and 0.5-V ranges, the range divider reduces the voltage to 50, 5, and 0.5 V before it is applied to the Kelvin-Varley attenuator.

The five Kelvin-Varley decade resistor strings and associated voltage-readout dials A through E (coupled to corresponding switches) provide a means of making the four precision voltages (500, 50, 5, and 0.5) adjustable. Note that each string, with the exception of the first, parallels *two resistors* of the preceding string. For example, between the two switch contacts of dial A there is a total resistance of 40 kΩ (80 K from the dial A string in parallel with 80 K from the string controlled by dial B).

Since there is a total of 200 K in the string of dial A, a total voltage of 100 V will appear across the dial A switch contacts (no matter what the position of dial A) when the range switch is in the 500-V position: $\frac{40}{200} = \frac{1}{5}$; 500 $\times$ $\frac{1}{5} = 100$. However, the voltage at the dial A switch contact will be 100, 200, 300, 400, or 500 *in reference to ground*, depending on the position of dial A. This reference value is used to balance against the unknown voltage being measured.

Dial B selects the reference voltage between 10 and 100 V in 10-V increments. Dial C selects the reference voltage between 1 and 10 V in 1-V increments. Dial D selects the reference voltage between 0.1 and 1 V in 0.1 increments. Dial E selects between 0.01 and 0.1 V in 0.01-V increments.

All of these voltages are reduced by a factor of 10 for *each lower voltage range* (50, 5, and 0.5). For example, if the range switch is set to 0.5 V, dial E will select a reference voltage between 0.00001 and 0.0001 V in 0.00001-V increments.

The Kelvin-Varley arrangement of decade resistors is also used in other test equipment, such as precision bridges, discussed in Chapter 3. These resistors must be matched for both temperature coefficient and tolerance, thus providing a typical overall attenuator accuracy of 0.002 per cent.

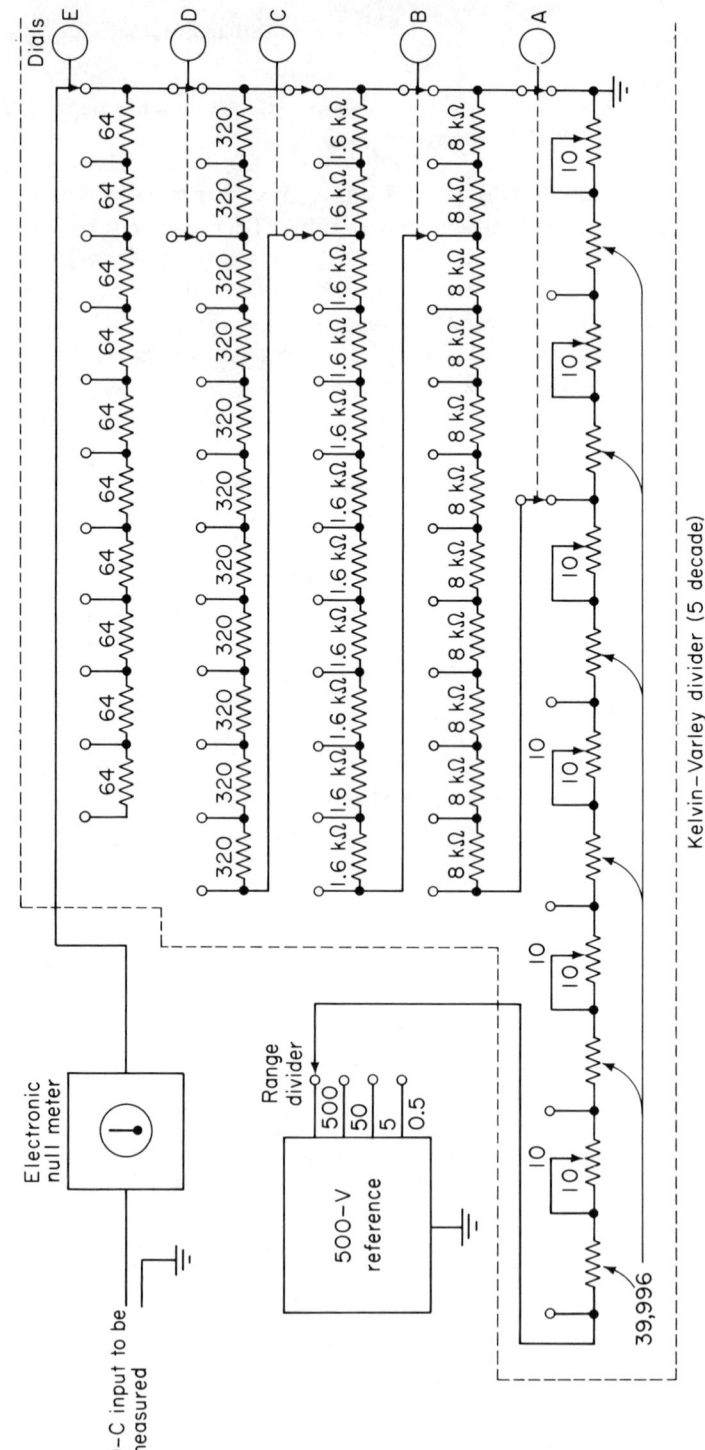

Fig. 2-16. Basic differential voltmeter in d-c voltage mode of operation, using Kelvin-Varley divider.

2-9. COMMON-MODE SIGNAL PROBLEMS

One of the problems in any instrument, particularly those that operate on the differential principle (such as the differential voltmeter), is an effect known as *common-mode insertion*. This is essentially a form of noise or undesired signals appearing at the input of an instrument, due to *circulating ground currents* between the instrument and signal source.

One of the major causes of common-mode signals is induced ground currents, usually at the a-c power-line frequency (typically 60 Hz). These signals can generate a potential of several volts between the signal source ground and the instrument's chassis ground. Unless bypassed, these currents will cause a voltage to appear at the input. This voltage could be larger than the signal itself. In any event, it will be added to the signal, resulting in an improper reading or output.

A typical common-mode problem is shown in Figure 2-17. Here, a bridge-type transducer is connected to the input of a differential meter (or any meter with a differential input). The output across R_H and R_L of the bridge is applied to the measuring circuit input and results in desired reading.

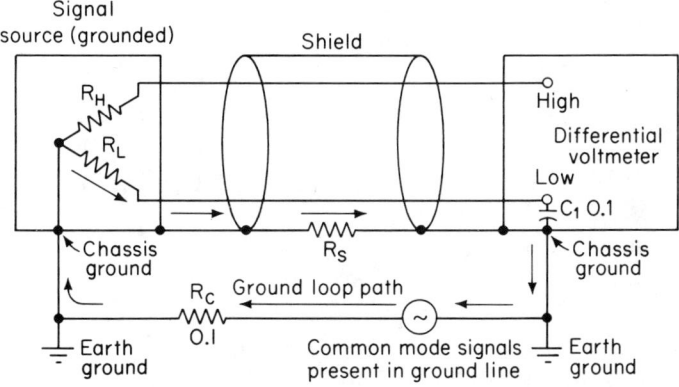

Fig. 2-17. Typical common-mode signal problem (Hewlett-Packard).

Note that there is another circuit (or a-c path) through R_L, C_1, and R_C. Capacitor C_1 is the input capacitance between the LO terminal of the measuring circuit and the chassis ground (typically 0.1 μF). Resistor R_C is the resistance between the earth grounds of the transducer and measuring instrument (typically 0.1 Ω or less). Although the values of C_1 and R_C are low, they do exist and provide an a-c path for any signals or currents present in the ground line (known as the ground loop). These currents or signals cause an undesired voltage to be developed across R_L, which is mixed with the desired transducer output and applied to the measuring circuit. This spurious signal current is a source of error.

The most effective means of reducing common-mode ground currents is some form of *guard circuit*, such as that shown in Figure 2-18. Here, the measuring circuit is placed within a guard shield that breaks up the input capacitance (C_1) into C_2 and C_3. Typically, the values of C_2 and C_3 are 0.002 μF, with a C_1 capacitance of about 2–4 pF. This provides two paths for any ground currents. One path is through R_C, R_S (cable shield resistance), and C_3. The other path is through R_C, R_L, and C_1 (or the combination of C_2 and C_3). Since the path through C_3 and R_S is very low resistance, most of the ground currents take this path and are bypassed away from the measuring circuit input. Any currents through the other path of R_L and C_1 are limited by the low value of C_1, resulting in a high common-mode rejection.

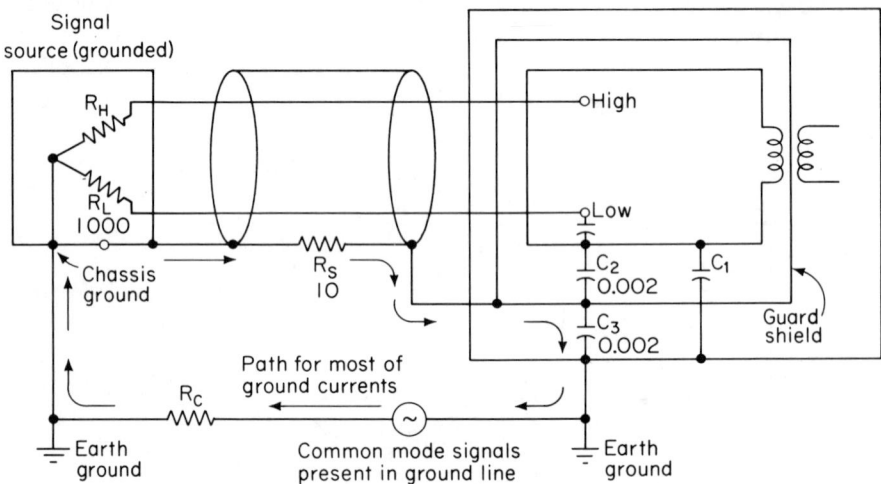

Fig. 2-18. Elimination of common-mode signal problem with guard shield (Hewlett-Packard).

The capacity of an instrument to reject a common-mode signal and thus reduce the undesired signal currents is called *common-mode rejection*. Usually, common-mode rejection is specified in decibels at some frequency or range of frequencies. For example, a typical common-mode rejection figure is 130 dB min at 60 Hz. This means that the undesired signal effect of a given common-mode signal is reduced by 130 dB or more. For example, the effect of a 100-V, 60-Hz common-mode signal is reduced to that produced by a 33-μV maximum equivalent signal.

2-10. MEASUREMENT OF RESISTANCE WITH A DIFFERENTIAL VOLTMETER

A differential voltmeter is well suited for the measurement of precision resistances. There are two basic procedures, one of which is best suited for measurement of resistances above 1 MΩ, while the other is better suited for precision measurement of lower resistance values. The following sections describe how a differential voltmeter can be used to make such resistance measurements.

2-10.1. Measurement of Resistance Below 1 MΩ

It is possible to make accurate resistance measurements of lower-value (1 MΩ or less) resistors by comparing the *voltage ratio* between a standard resistor and the unknown resistor to be measured.

If a standard resistor R_s is connected in series with the unknown resistor R_x, as shown in Figure 2-19, and if the current through both resistors is the same, then the ratio of the voltage drops, as determined by the differential voltmeter, is the same as the ratio of their resistances.

The accuracy of resistance measurements obtained by using this method

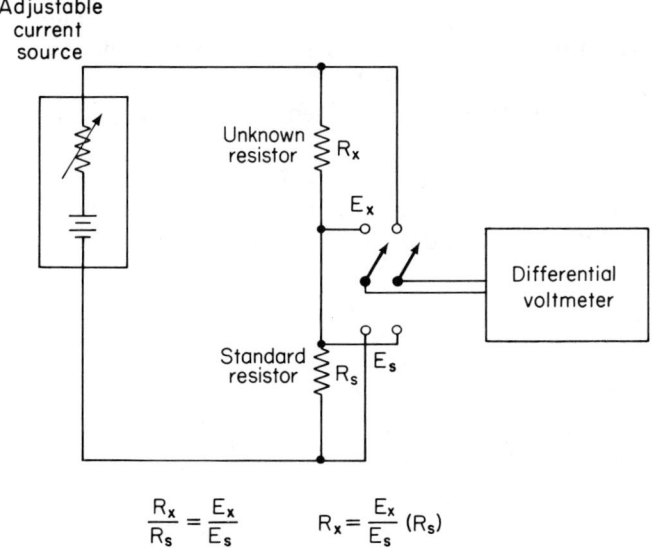

Fig. 2-19. Measurement of resistances below 1 megohm with differential voltmeter.

depends on the accuracy of the voltmeter and standard resistor. Standard resistors can be obtained that typically provide calibration accuracies of ±0.005 per cent or better, and that will remain within 50 ppm of their certified value over a temperature range of 20° to 30° C. The *resolution* of the voltmeter must also be considered. (Resolution is defined as the smallest quantity discernable in using a given measurement device. The resolution of any measurement should always be greater than the accuracy expected.)

If the ratio of the resistors is such that the voltage drop across R_S and R_x is measured on the *same voltage range*, then the basic error of the voltmeter will cancel out. Therefore, measurements should be made on the same voltage range, where possible.

The following example is intended to give an idea of the results obtainable by measuring resistors with precision voltmeters.

Assume that a 10-Ω standard resistor is used, the voltage drop across the standard is 1.0000 V, and the voltage drop across the unknown is 1.0035 V. Using the equation of Figure 2-19,

$$R_x = \frac{E_x}{E_S}(R_S) = \frac{1.0035}{1.0000}(10.000) = 10.0035 \; \Omega$$

2-10.2. Measurement of Resistance Above 1 MΩ

Differential voltmeters provide a feature that allows them to be used for the direct measurement of high resistances. The use of a differential voltmeter for this purpose provides a convenient and rapid method for measuring resistances from 1 to 250,000 MΩ. The applications for this type of measurement include the measurement of leakage resistance of capacitors, transformers, and insulators.

Figure 2-20 shows the basic circuit. Note that this circuit consists of a known power supply voltage that can be selected by the voltage range switch (such as the 500-, 50-, 5-, and 0.5-V settings of a typical differential voltmeter)

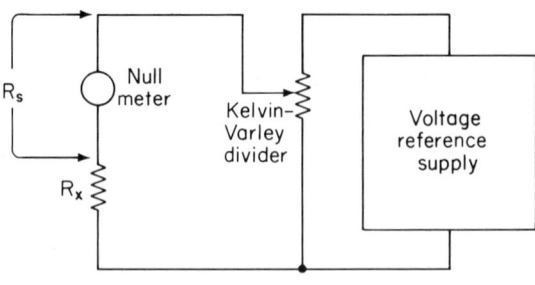

Fig. 2-20. Measurement of resistances above 1 megohm with differential voltmeter.

and a Kelvin-Varley divider (also part of the differential voltmeter). The setting of the Kelvin-Varley divider determines the exact voltage that is applied across the null detector connected in series with the unknown resistance R_X. The input impedance of the null detector (typically in the order of 1 or 10 MΩ, depending on the range setting and the model of the instrument) is used as the standard resistor R_S.

The measurement of resistance with a differential voltmeter is a direct application of Ohm's law (the measurement of the current through the unknown resistance and the voltage drop across it). With an unknown resistor connected to the input terminals of a differential voltmeter, the current through the unknown resistance is equal to the voltage indicated by the null detector, divided by the null detector input resistance. Likewise, the voltage drop across the unknown resistance is equal to the voltage measured on the Kelvin-Varley divider, minus the voltage indicated by the null detector.

Since $R = E/I$, the voltage drop across the unknown resistance divided by the current through the unknown resistance is equal to the unknown resistance value.

For practical measurements, it is recommended that resistance measurements up to 50,000 MΩ be made with the null detector meter needle at end scale (full scale). In a true differential voltmeter application, where known and unknown voltages are compared, the reading is made with the null detector at null (or zero). Since the measurement of resistance does not involve an unknown external voltage, the reading must be made with the meter indicating some voltage, preferably full-scale.

Assume that the null detector has an input resistance of 10 MΩ, that the Kelvin-Varley dials indicate 1.1 V, and that the null detector indicates an end-scale reading of 1.0 V. Using the following equation,

$$R_X = \frac{R_N}{E_M}(E - E_M)$$

where R_X is the unknown resistance in megohms
E is the voltage indicated on the Kelvin-Varley dials
E_M is the voltage indicated on the null detector
R_N is the resistance of the null detector in megohms
The value of R_X can be computed by

$$R_X = \frac{10}{1}(1.1 - 1.0) = 10 \times 0.1 = 1 \text{ M}\Omega$$

3 BRIDGE-TYPE TEST EQUIPMENT

Many quantities in electronics, such as impedance, admittance, capacity, inductance, conductance, and so on, are measured by means of bridge circuits and bridge-type test instruments. Some manufacturers, such as Hewlett-Packard and General Radio, produce *universal bridges* that will measure more than one quantity. These instruments operate on the *balance or null principle*, or on the principle of *comparison against a standard*. This chapter is devoted to test equipment that operates on null or comparison techniques.

3-1. WHEATSTONE BRIDGES

Null methods have long been used as the most precise and convenient way to measure all types of impedance—resistive and reactive, inductive and capacitive, from low frequencies to ultra-high frequencies. Most null or balance-type instruments are evolved from the basic Wheatstone bridge, still the fundamental circuit for measuring d-c resistance.

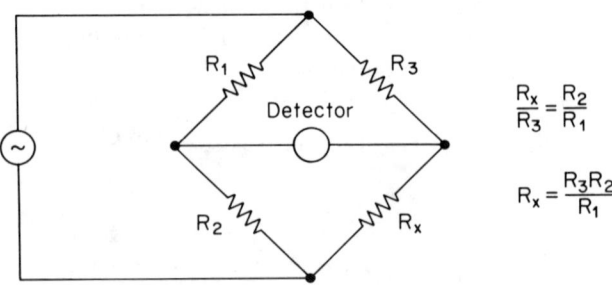

Fig. 3-1. The general Wheatstone bridge circuit.

The Wheatstone bridge measures an unknown resistance R_x in terms of calibrated standards of resistance, as shown in Figure 3-1.

The relationship of resistance elements is

$$R_x = \frac{R_3 R_2}{R_1}$$

which is satisfied when the voltage across the detector terminal is zero.

A more complete Wheatstone bridge circuit for the measurement of d-c resistance is shown in Figure 3-2. Resistors R_A and R_B are fixed and of known value. R_S is a variable resistor with the necessary calibration arrangement to read the resistance value for any setting (usually a calibrated dial coupled to the variable resistance shaft). The unknown resistance value R_x is connected across terminals B and C. A battery or other power source is connected across points A and C.

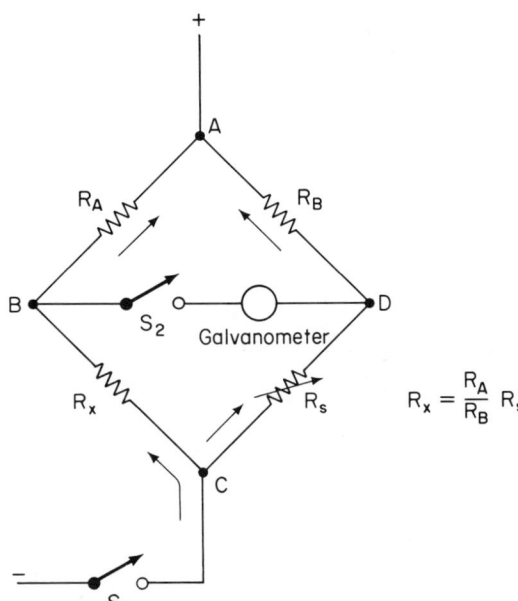

Fig. 3-2. Basic Wheatstone resistance bridge.

When switch S_1 is closed, current flows in the direction of the arrows, and there is a voltage drop across all four resistors. The drop across R_A is equal to the drop across R_B (provided that R_A and R_B are of equal resistance value). Variable resistance R_S is adjusted so that the galvanometer reads zero (center scale) when switch S_2 is closed. At this adjustment, R_S is equal to R_X in resistance. By reading the resistance of R_S (from the calibrated dial), the resistance of R_X is known.

76 Bridge-Type Test Equipment

When the variable resistance R_S is equal to R_X, the difference of potential between points B and D will be zero, and no current will flow through the galvanometer. If R_S is not equal to R_X, the B and D are not at the same voltage, and current will flow through the galvanometer, moving the pointer away from zero (center scale).

The equation shown in Figure 3-2 is used when the values of R_A and R_B are not equal. In some commercial bridges, the value of R_A is 10 times that of R_B. Thus the actual value of R_X is 10 times the indicated value of R_S. This permits a large value of R_X to be measured with a low-value R_S. In other bridges, the value of R_B is 10 times that of R_A. Thus the actual value of R_X is one-tenth the indicated value of R_S. This permits a small value of R_X to be measured with a high-value R_S. Commercial Wheatstone bridges can be used to make measurements from 1 Ω to 1 mΩ with accuracy of ±0.25 per cent.

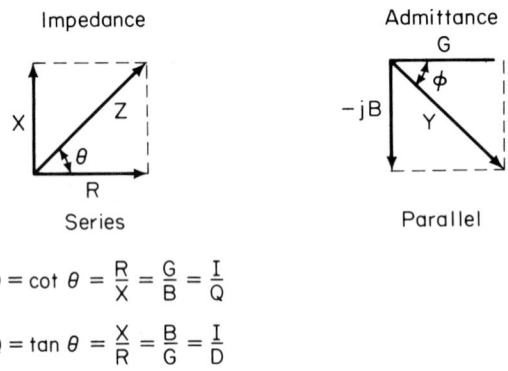

Fig. 3-3. Definitions of D and Q.

3-2. ALTERNATING-CURRENT BRIDGES

The Wheatstone bridge circuit is easily adapted to a-c measurements. With complex impedances, two balance conditions must be satisfied, one for the resistive component and one for the reactive component. Likewise, both the conductive and susceptance components must be satisfied for complex admittances.

An important characteristic of an inductor or a capacitor, and often a resistor, is the ratio of resistance to reactance, or of conductance to susceptance. The ratio is called dissipation factor D and its reciprocal is storage factor Q. These terms are defined in Figure 3-3. Note that dissipation factor is directly proportional to energy dissipated, and storage factor to energy stored, per cycle.

In practical bridge circuits, dissipation factor D, which varies directly with power loss, is commonly used for capacitors. Storage factor Q is more often used for inductors. Q can also be used for resistors, in which case the value is usually quite small. However, most bridge circuits permit measurement of D or Q for capacitors, inductors, and resistors.

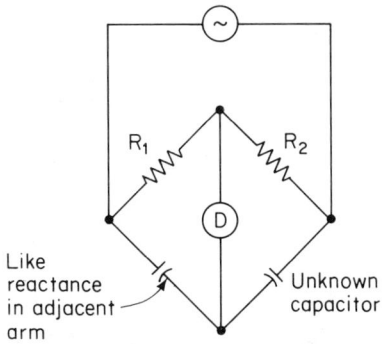

As in the case of d-c bridges, balance for the resistive component in an a-c bridge is obtained by a precision variable resistance. Balance for the reactive component can be obtained from a similar reactance in an adjacent arm of the bridge or an unlike reactance in the opposite arm, as shown in Figure 3-4. In most practical circuits, the reactance is supplied by means of a fixed, precision capacitor in series or parallel with a variable resistance.

In some a-c bridges, the unknown is connected in series or in parallel with the main adjustable component, and balances are made before and after the unknown is connected. The magnitude of the unknown then equals the change made in the adjustable component, since the total impedance of the unknown arm remains constant. The main advantage of this substitution

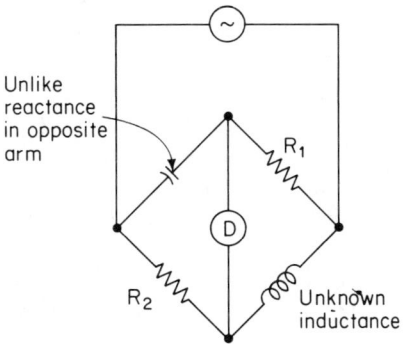

Fig. 3-4. Bridge circuits showing balance for reactive component.

technique is that its accuracy depends only on the calibration of the adjustable arm and not on the other bridge arms (as long as they are constant). The substitution principle can also be used to advantage with any bridge if the balances are made with an external, calibrated, adjustable component.

Transformer-ratio arms are used in some a-c bridge circuits. The basic circuit is shown in Figure 3-5. In practical circuits, the transformer windings are tapped on the standard side in decimal steps (usually from 0.1 to 1), and on the unknown side in decade steps (usually from 1 to 0.001). Several fixed-capacitance standards are used, one for each decade step. In a typical transformer-type capacitance bridge, the standard values range from 0.0001 to 1000 pF. Such a combination of internal standards and transformer voltage ratios makes possible a measurement range of $10^9:1$.

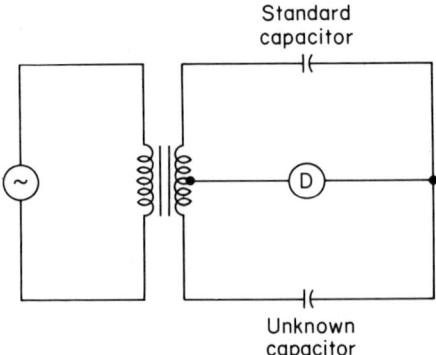

Fig. 3-5. Basic circuit for transformer-type capacitance bridge.

3-3. UNIVERSAL BRIDGES

Analysis of capacitors, inductors, and resistors for low-frequency applications is commonly made with a universal bridge. Universal bridges have considerable versatility, being able to measure not only resistance, capacitance, conductance, and inductance over wide ranges, but also Q and D.

A generalized universal bridge circuit is shown in Figure 3-6. The bridge is driven by an a-c source across the corners OQ. When the voltage across arm OP equals the voltage across arm OS, the output voltage, expressed across the detector connected to P and S, is zero. With the bridge balanced, or nulled, the product of the impedance across OS and that across PQ is equal to the product of the impedance across SQ and that across OP. At balance, the value of any of the four impedances can be calculated if the other three are known.

In addition to the basic bridge circuit, a typical universal bridge will contain an audio oscillator as an a-c driving source and a null detector circuit.

The audio oscillator is similar to the precision, distortion-free oscillators described in Chapter 8. The null detector includes an amplifier circuit (to measure extremely low voltages at or near null), a fixed or tunable filter or frequency-selective circuit (to remove any signals but the audio driving source), and rectifiers (to convert the a-c signal to a d-c voltage for display on the meter).

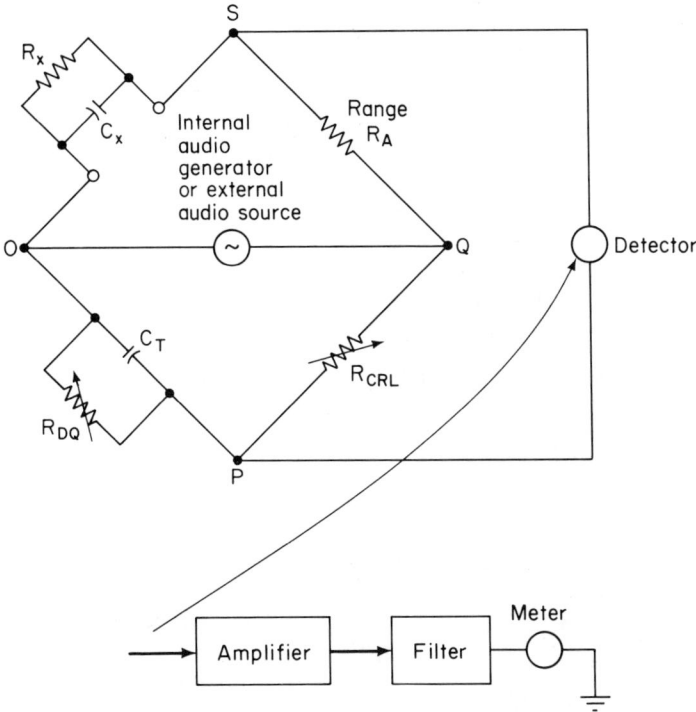

Fig. 3-6. Generalized universal bridge circuit (Hewlett-Packard).

In some universal bridges, the a-c driving source is tunable over a range from about 50 Hz to 10 kHz, while in other circuits the driving source is fixed (usually at 1 kHz). A typical universal bridge will also contain a d-c supply (batteries) for measurement of resistance. Many universal bridges are provided with connectors that permit external driving sources and external standards to be used with the internal bridge circuit.

Figures 3-7–3-12 show typical universal bridge circuits for the measurement of capacity, inductance, resistance, and conductance, as well as D and Q. Note that these circuits give the user the option of measuring the unknown in terms of either its series or parallel equivalents. The choice is a matter of convenience for the problem at hand. For example, if Q is 10 or more (or if

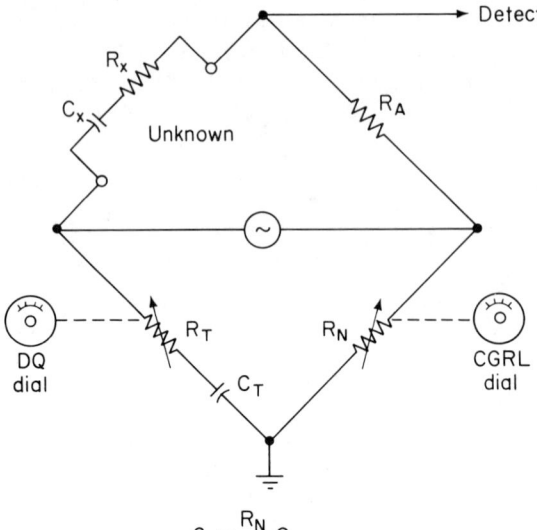

Fig. 3-7. Universal bridge circuit for measurement of series capacitance (General Radio).

$$C_x = \frac{R_N}{R_A} C_T$$
$$D_x = \omega R_x C_x = \omega R_T C_T$$

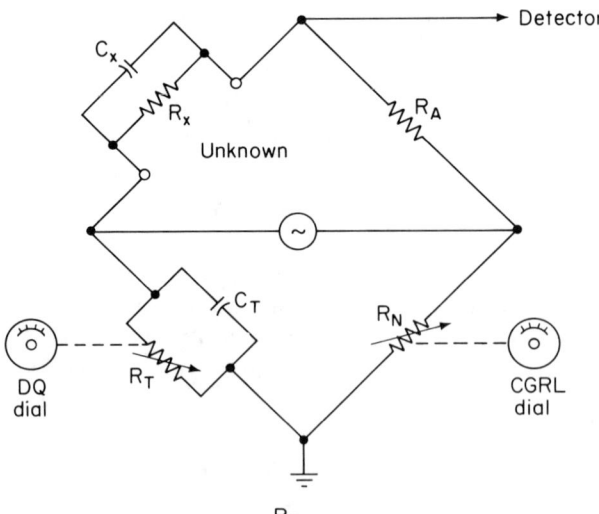

Fig. 3-8. Universal bridge circuit for measurement of parallel capacitance (General Radio).

$$C_x = \frac{R_N}{R_A} C_T$$
$$D_x = \frac{1}{\omega R_x C_x} = \frac{1}{\omega R_T C_T}$$

Universal Bridges 81

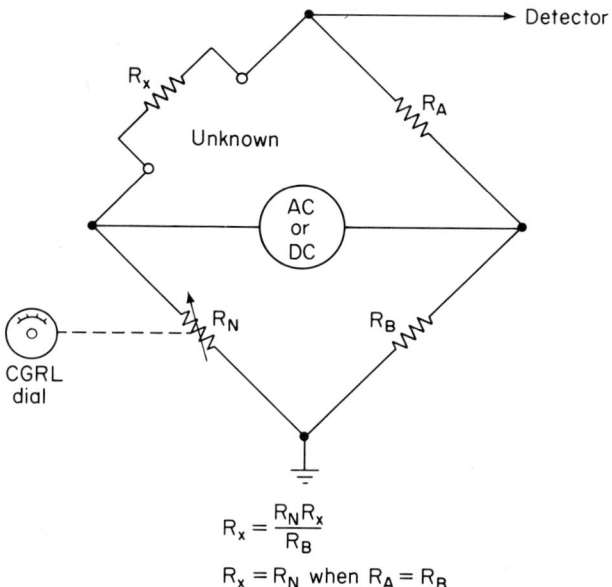

Fig. 3-9. Universal bridge circuit for measurement of resistance (General Radio).

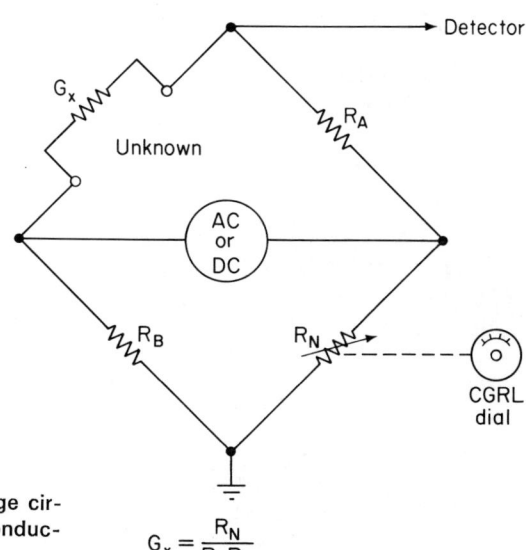

Fig. 3-10. Universal bridge circuit for measurement of conductance (General Radio).

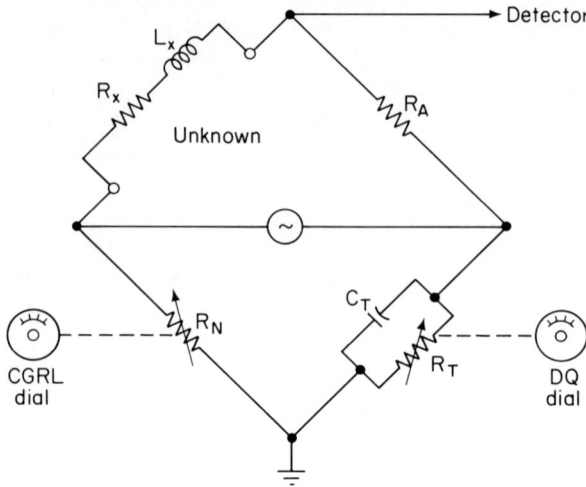

$$L_x = R_N R_A C_T$$
$$Q = \frac{\omega L_x}{R_x} = \omega R_T C_T$$

Fig. 3-11. Universal bridge circuit for measurement of series inductance (General Radio).

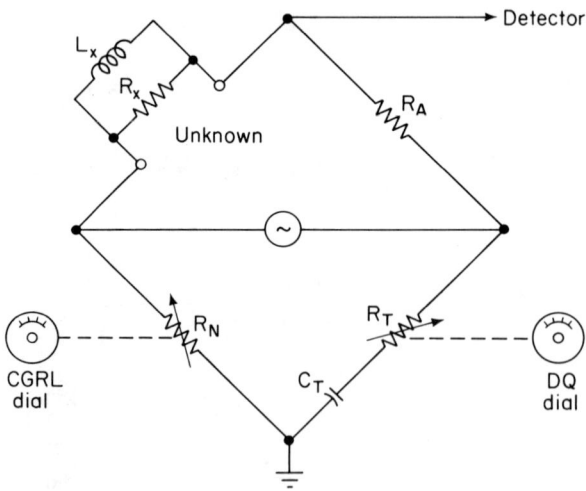

$$L_x = R_N R_A C_T$$
$$Q = \frac{R_x}{\omega L_x} = \frac{1}{\omega R_T C_T}$$

Fig. 3-12. Universal bridge circuit for measurement of parallel inductance (General Radio).

D is 0.1 or less), the difference between series and parallel reactance is no more than 1 per cent. However, for very low Q (or high D) the difference is substantial (with a Q of 1, parallel reactance is twice that of series reactance).

Operation is essentially the same for each circuit. In use, the DQ and $CGRL$ dials are adjusted for a null (the DQ dial is not used for resistance and conductance measurements). Then, is the value read from the corresponding dial indication. In most cases, the dials are not direct reading. Instead, the dial reading must be multiplied by some factor (a multiplier control setting in the case of the $CGRL$ dial, and frequency in the case of the DQ dial).

Note that, in these universal bridge circuits, both capacitance and inductance are measured in terms of impedance they present at a given frequency, rather than by comparison against a standard, even though the readout is in terms of capacity and inductance value (μF, μH, and so on). This is the major difference between the universal bridge circuit and the standard capacitance or inductance bridge described in Section 3-4.

Most universal bridges are provided with a connector that permits application of an external bias voltage to the component under test. For example, it may be advantageous to measure a capacitor with the full working voltage applied. Many circuits are used to apply the external bias. A typical circuit is shown in Figure 3-13.

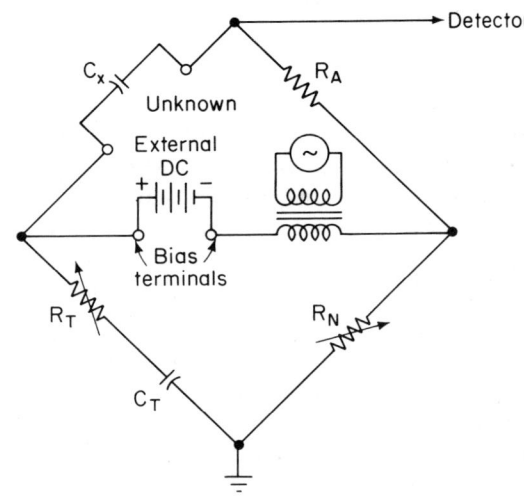

Fig. 3-13. Universal bridge circuit for application of external d-c bias during capacitance measurement (General Radio).

3-4. STANDARD CAPACITANCE AND INDUCTANCE BRIDGES

Capacitance can be measured on a standard capacitance bridge, rather than on a universal bridge. Standard-type bridges provide greater accuracy (in general) than univeral bridges but are limited in the variety of measurements that can be made. Standard inductance bridges are also available but are not in such common use as capacitance bridges.

Standard bridges operate on the principle of comparing an unknown against a standard. The basic standard bridge circuit for capacitance measurement is shown in Figure 3-14.

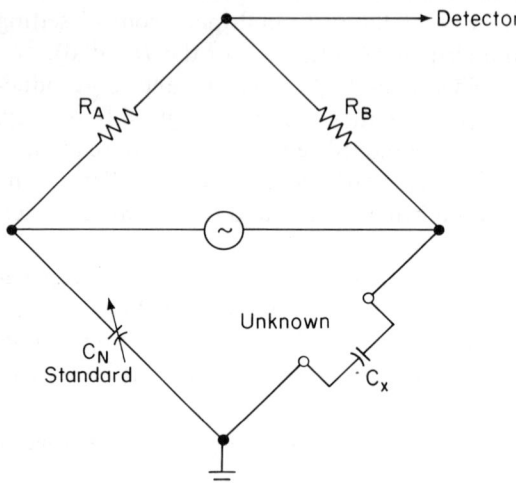

Fig. 3-14. Basic standard capacitance bridge circuit.

The capacitance of the unknown C_X is balanced by a calibrated, variable, standard capacitor C_N, or by a fixed standard capacitor and a variable ratio arm, such as R_A. Such bridges with resistive ratio arms and with calibrated variable capacitors or resistors can be used over a wide range of both capacitance and frequency and with a direct-reading accuracy that seldom exceeds 0.1 per cent.

For higher accuracy, resolution, and stability in capacitance measurements at audio frequencies, a bridge with transformer-ratio arms has many advantages, and increasing use of transformer-ratio-arm bridges is being made in the measurement of many types and sizes of capacitors. The basic transformer-type bridge is shown in Figure 3-5 and is discussed in Section 3-2. Figure 3-15 shows three possible ways of balancing a simple transformer-ratio capacitance bridge. For simplicity, the generator and primary are not shown, but it is assumed that the two secondaries have 100 turns each and are excited so that there is 1 V/turn. The capacitor in the unknown arm is assumed to be 72 pF.

In Figure 3-15A, the two ratio arms are equal and the bridge is balanced in the conventional way, with a variable standard capacitor that is adjusted to 72 pF.

The detector current can also be adjusted by a variation in the voltage applied to a fixed, standard capacitor. In Figure 3-15B, the standard capacitor is fixed at 100 pF, and this is balanced against the 72-pF unknown connected to the 100-V end of the transformer by connection of the standard to 72 V of the opposite phase, obtained from suitable taps on the transformer windings.

The inductive divider shown has a winding of 100 turns with taps every 10 turns and, on the same core, another winding of 10 turns tapped every turn. If, as shown, the second winding is connected to the 70-V tap on the first winding and the capacitor to the 2-V tap on the second winding, the required 72 V are applied to the capacitor. Six or more decades for high precision can be obtained in a similar fashion with more turns on one core and the use of

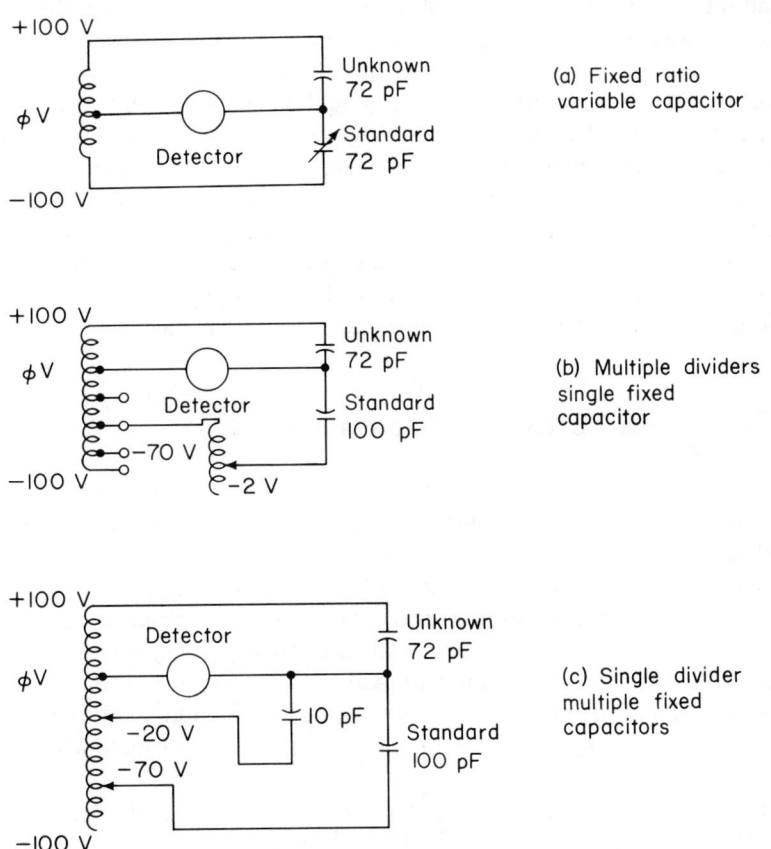

Fig. 3-15. Methods of balancing capacitance in a transformer-ratio bridge (General Radio).

additional cores driven from the first. Such inductive dividers have very accurate and stable ratios, but the errors increase with the number of decades because of loading effects.

Another method of balance by voltage variation is shown in Figure 3-15C, where a single decade divider is used in combination with multiple fixed capacitors. The 100-turn secondary is tapped every 10 turns to provide 10-V increments. If a 100-pF capacitor is connected to the 70-V tap and a 10-pF capacitor to the 20-V tap, the resulting detector current balances that of the 72-pF unknown connected to 100 V.

Such a bridge can be given six-figure resolution by the use of six fixed capacitors in decade steps from 100 to 0.001 pF, each of which can be connected to any one of the taps on the transformer.

In any of these bridges, the bridge ratio can also be varied by use of taps on the unknown side of the transformer to vary the voltage applied to the unknown capacitor. For example, if the unknown capacitor were connected to a 10-turn or 10-V tap on the upper half of the transformer, then a capacitance of 720 pF (instead of 72 pF) would be balanced by the standard capacitor shown. The range of the bridge can thus be extended to measure capacitors that are much larger than the standards in the bridge.

These advantages of transformer-ratio arms and dividers make possible a bridge of wide range and high accuracy, because not only are the ratios stable and accurate but, when only a few fixed capacitors are required as standards, the standards can be constructed to have high stability and accuracy. Such a bridge can also operate over a wide range of frequencies.

3-5. AUTOMATIC BRIDGES

The universal and standard bridges discussed thus far require that controls be adjusted for balance as each capacitor (or other device being tested) is connected to the bridge. In effect, they are manual bridges. In recent years, a number of automatic bridges have been developed. These bridges provide an automatic readout without adjustment of balance controls. In some cases, the automatic bridges will also provide a binary-coded-digital (BCD) readout to external equipment. Automatic bridges are similar in operation to the digital meters discussed in Chapter 2. To understand their operation, it is necessary to understand logic and digital circuit methods, such as are outlined in the Appendix.

A typical automatic bridge circuit is shown in Figure 3-16. The circuit is a transformer–ratio–arm bridge for the automatic measurement of capacity. The circuit is in balance when the currents through the standard capacitor and the unknown capacitor are equal so that the current in the phase detector is zero. The range is chosen automatically by relays that select decade taps on the ratio transformer. The phase detector determines whether the current passing through the unknown arm of the bridge is higher or lower than that through the standard arm and produces an error signal that indicates whether more or less voltage is required on the standard capacitor to reach a balance.

This information is used by a counter decade that controls, through electronic switching circuits, the voltage on the standard capacitor. The counter then counts in a direction to minimize the error signal until a balance is reached. At balance, the value of the unknown is displayed on an in-line digital readout that indicates capacitance. This information is also presented in binary-coded-decimal form for use with printers and other data-handling equipment.

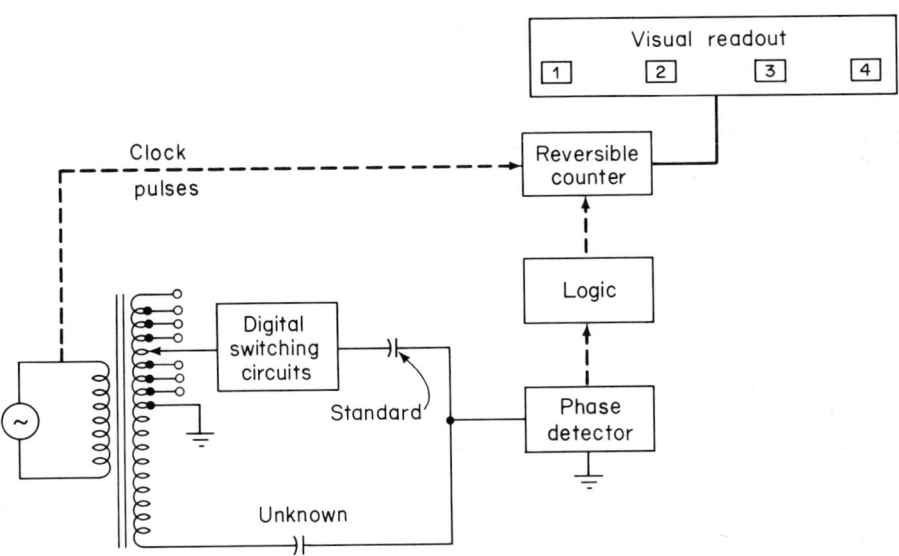

Fig. 3-16. Typical automatic bridge circuit (General Radio).

3-6. Q METERS

Although it is possible to measure the Q of a coil with a universal bridge or a standard bridge, it is more common to use a Q meter. The basic circuit of a typical Q meter is shown in Figure 3-17.

The Q of a coil or complete low-current (L-C) circuit is the ratio of resistance to reactance. In a series circuit, the Q is determined as the ratio of reactance divided by resistance. Circuit Q can also be determined by the ratio of reactive voltage to resistive voltage, shown as E_1 and E_2, respectively, in Figure 3-17.

In a practical Q-meter circuit, the resistive voltage E_1 is not actually measured but is held constant. The reactive voltage E_2 is then measured but is read out in terms of Q. For example, if E_1 is 1 V, a 15-V E_2 reading indicates a Q of 15, a 20-V E_2 reading indicates a Q of 20, and so on.

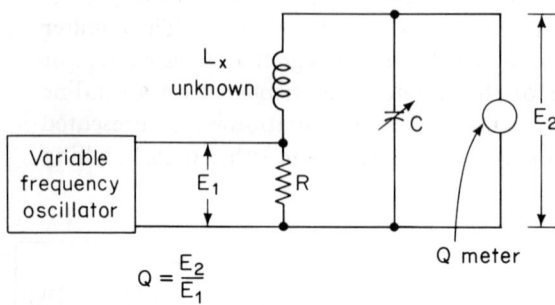

$$Q = \frac{E_2}{E_1}$$

Fig. 3-17. Basic Q meter circuit (Hewlett-Packard).

In most Q-meter circuits, the resistive voltage is supplied by a *variable frequency oscillator* (Chapter 4) that can be tuned over a wide range of frequencies (50 kHz to 50 MHz, for example). Likewise, the internal capacitor is variable and can be adjusted by a panel control to provide resonance across the frequency range.

In operation, the coil to be tested is connected into the circuit, the variable frequency oscillator is adjusted to the frequency of interest, and the variable capacitor is adjusted until the circuit is at resonance. The Q is then read directly on the meter.

In many Q meters, provisions are made for inserting low impedances in series with the coil or high impedances in parallel with the capacitor. Thus parameters of unknown circuits or components can be measured in terms of their effect on the circuit Q and resonant frequency. For example, assume that a coil is to be used as part of an L-C circuit (say an r-f tank circuit) and that the L-C circuit will be used with various load impedances. The coil Q can be measured without load, then with load, to find what effect the change will have on the coil Q.

3-7. RX METERS

An RX meter is used to measure the separate resistive and reactive component of a parallel impedance network. This is especially useful in the design of r-f tank circuits but can be used with any parallel impedance. The basic circuit of a typical RX meter is shown in Figure 3-18.

As shown, there are two variable frequency oscillators that track each other at frequencies 100 kHz apart. The output of 0.5–250 MHz oscillator F_1

is fed into a bridge. When the impedance network to be measured is connected across one arm of the bridge, the equivalent parallel resistance and reactance (capacitive or inductive) unbalance the bridge, and the resulting voltage is fed to the mixer. The output of the 0.6–250.1-MHz oscillator F_2, tracking 100 kHz above F_1, also is fed to the mixer. This results in a 100-kHz difference frequency *proportional in level* to the bridge unbalance. The difference frequency signal is amplified by a filter-amplifier combination and is applied to a null meter. When the bridge resistive and reactive controls are nulled, their respective dials accurately indicate the parallel impedance components of the network under test. For example, if balance is achieved with 50 Ω of resistance and 300 Ω of reactance, then the network under test has these same values.

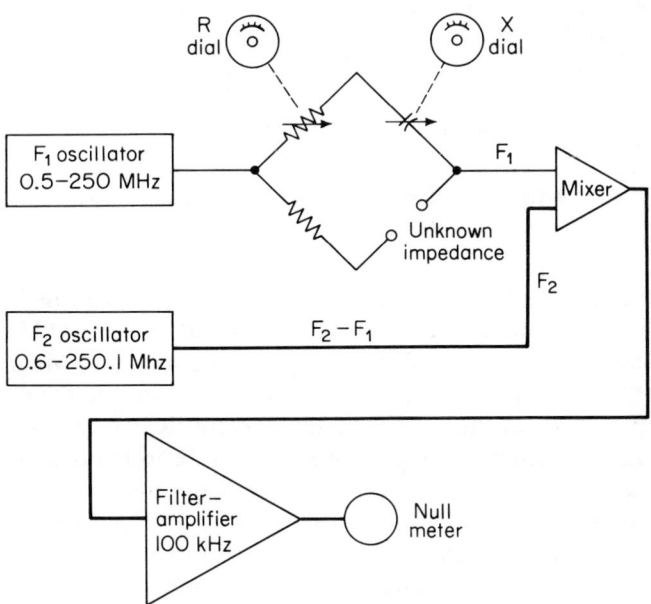

Fig. 3-18. Basic R_X meter circuit (Hewlett-Packard).

3-8. ADMITTANCE METERS

In some cases, it is more convenient to measure characteristics in terms of admittance instead of impedance. This is particularly true when characteristics must be measured at high frequencies, such as the measurement of microwave or ultra-high-frequency antennas and transmission lines. Just as an impedance can be broken down into its real (resistive) and imaginary (reactive) components, admittance can be broken into conductance and susceptance by means of an admittance meter.

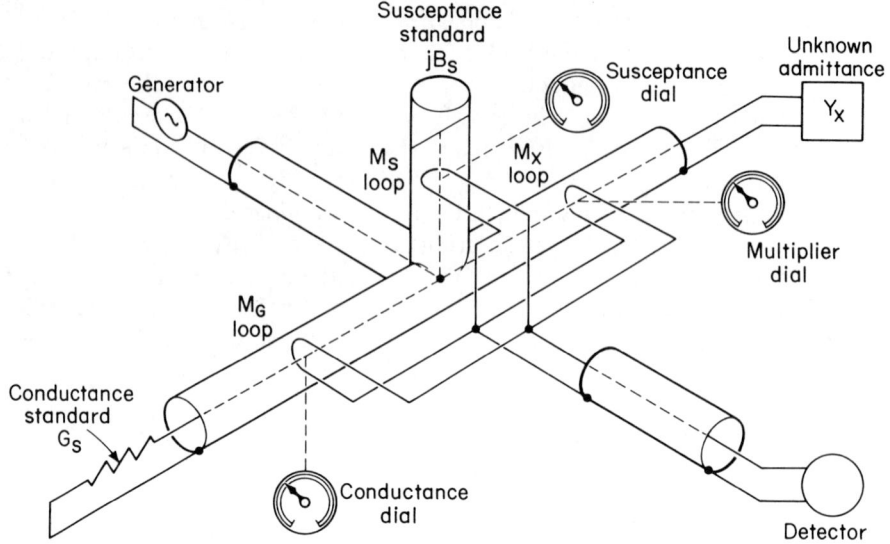

Fig. 3-19. Basic admittance meter circuit (General Radio).

The basic circuit of a typical admittance meter is shown in Figure 3-19. Such a meter will normally be used at frequencies between 40 MHz and 1.5 GHz but can also be used at frequencies down to about 10 MHz. As a null instrument, the admittance meter can be used to measure the conductance and susceptance of an unknown circuit directly. It can also be used as a comparator to indicate equality of one admittance to another on degree of departure of one from the other. An admittance meter can also be used to measure voltage standing wave ratio (VSWR) and reflection coefficient. (VSWR and reflection coefficient are discussed in Chapter 11.)

Although the admittance meter makes use of a null indication for measurement, it is not a true bridge. In the admittance meter, the currents flowing in three coaxial lines fed from a common source at a common junction point are sampled by three adjustable loops that couple to the magnetic field in each line, as shown in Figure 3-19.

The coupling of each of the loops can be varied by rotation of the loop. One of the coaxial lines is terminated in a *conductance standard* G_s, which is a pure resistance equal to the characteristic impedance of the line. Another line is terminated in a *susceptance standard* jB_s, which is a short-circuited length of coaxial line. The third line is terminated by the unknown circuit or network Y_x. The outputs of the three loops are combined by connecting all three loops in parallel. When the loops are properly oriented, the combined output is zero, as indicated by a null on the detector. The admittance meter therefore balances in the same manner as a bridge.

At balance, the vector sum of the voltages induced in the three loops is proportional to the mutual inductance M and to the current flowing in the corresponding line. Since all three lines are fed from a common source, the input voltage is the same for each line, and the current flowing in each line is proportional to the input admittance.

At balance, the sum of the three voltages is zero. Therefore, the amount of coupling required to produce balance is related directly to the amount of susceptance and conductance in the unknown circuit. The amount of coupling is indicated by corresponding dials attached to the three coupling loops (M_G, M_B, and M_X).

In operation, a signal source, a detector, and the unknown circuit are connected as shown in Figure 3-19. The three coupling loops are adjusted for a null indication on the detector, and the susceptance, conductance, and multiplying factor are read from the three dials. Various signal sources and detectors can be used with the admittance meter. Usually, the signal source is an unmodulated r-f signal generator that produces a signal at the frequency of interest. The detector is a diode and meter combination, possibly with some amplification.

3-9. VECTOR IMPEDANCE METERS

A vector impedance meter provides a direct readout of both impedance and phase angle for an unknown circuit or component. These measurements are made by simply connecting the unknown to the instrument and setting the frequency as desired. The readout gives the driving-point impedance of the unknown over the frequency range of interest on two meters, one for impedance and one for phase angle.

Driving-point impedance is defined as the ratio of applied voltage to current entering one point of a network or component. The impedance may be represented vectorially, either as a point in the impedance plane having Cartesian coordinates $R \pm jX$ or polar coordinates $(Z) \angle \theta$, where θ is the angle between the voltage and current vectors.

Operation of the vector impedance meter is based directly on the fundamental definition of driving-point impedance, and the impedance vector is read out in the polar coordinates $(Z) \angle \theta$.

Figure 3-20 is a simplified block diagram of a typical vector impedance meter. In operation, a variable frequency oscillator applies an unmodulated signal to an amplifier with leveled output; that is, the amplifier output remains constant despite changes in frequency or load. Current from the ampli-

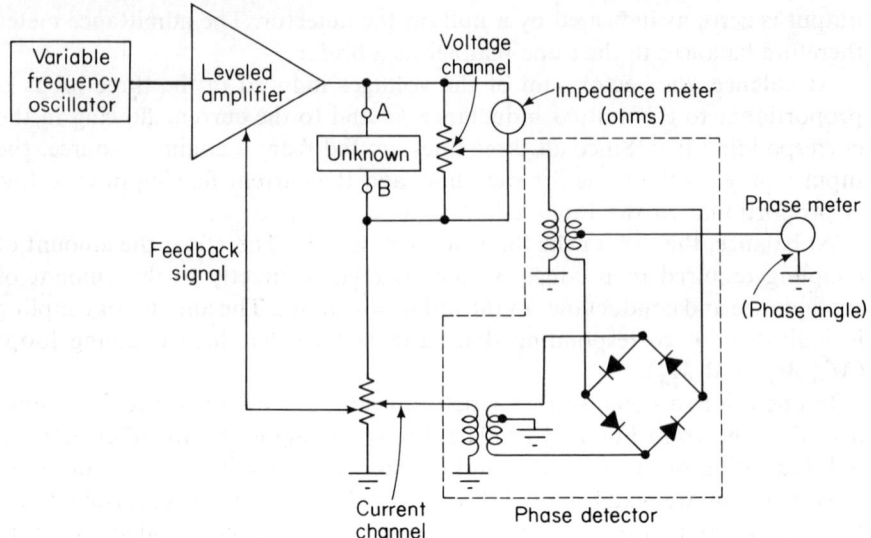

Fig. 3-20. Basic vector impedance meter circuit (Hewlett-Packard).

fier passes through the unknown component, mounted across terminals A and B, that is to be measured. Current passes from the B terminal through a load to ground. The current through the unknown is used to generate a feedback signal that, in turn, levels the output of the amplifier. The amplifier-feedback arrangement then holds the current through the unknown at a constant level.

Since Z (driving-point impedance) $= E/I$ and I is now a constant, Z is directly proportional to the voltage across the unknown. Therefore, a high-impedance broadband voltmeter is placed across terminals A and B, and the output readout is calibrated directly in impedance.

To measure phase angle, a-c outputs from both the voltage channel and current channel are used to trigger a zero-crossing phase detector. The output of the phase detector is displayed on a meter calibrated directly in phase angle.

3-10. L-C METERS

An L-C meter is actually a reactance meter, but it reads out in terms of capacitance or inductance. L-C meters are used primarily to measure the value of very small capacitors and coils. A typical L-C meter will provide a full-scale meter reading with capacitances of 0–3 pF and inductances of 0–3 μH. Such a meter will also read capacitances up to 300 pF and inductances up to 300 μH.

In operation, the capacitor or coil to be measured is connected into the resonant circuit of an oscillator whose frequency is compared to that of a reference oscillator. A counter circuit measures the difference in frequency and produces a meter reading in µH or pF.

Figure 3-21 is the block diagram of a typical L-C meter. Figures 3-22 and 3-23 show the oscillator circuit arrangement for inductance and capacitance measurements, respectively.

Note that the variable oscillator circuit operates in a different manner for capacitance and inductance measurements. With capacitance measurements, the unknown capacitor is connected in parallel with the tank circuit capacitor. This increases the capacitance and lowers the oscillator frequency from its nominal 140 kHz. With inductance measurements, the unknown coil is connected in series with the tank circuit coil, also reducing oscillator frequency from 140 kHz.

When capacitance is measured but the capacitor is not yet connected, the mixer receives two identical 140-kHz signals. The oscillators are identical circuits with equal temperature compensation, so their output frequency changes with temperature are essentially the same after warm-up. Connecting a capacitor to the variable oscillator reduces the frequency to some point between about 125 and 140 kHz. A larger capacitance value will produce a larger drop from 140 kHz. When this occurs, the mixer produces an output frequency equal to the *difference* between the two oscillator frequencies. If the variable oscillator frequency has been reduced a large amount by a large capacitance value, the difference frequency will also be large.

The filter prevents any 140-kHz signals from passing from the mixer to the counter circuits. Difference frequency signals (in the range from 0 to 15 kHz)

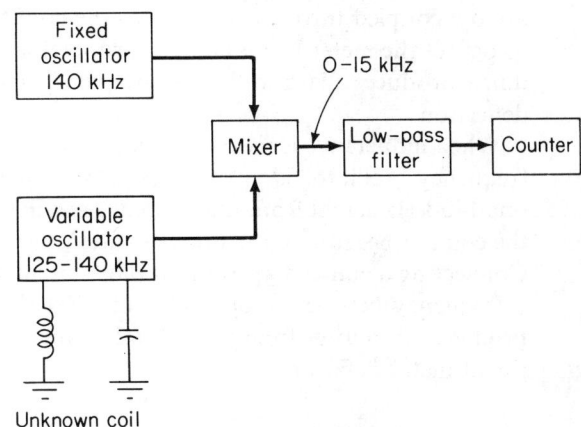

Fig. 3-21. Basic L-C meter circuit (Tektronix).

94　Bridge-Type Test Equipment

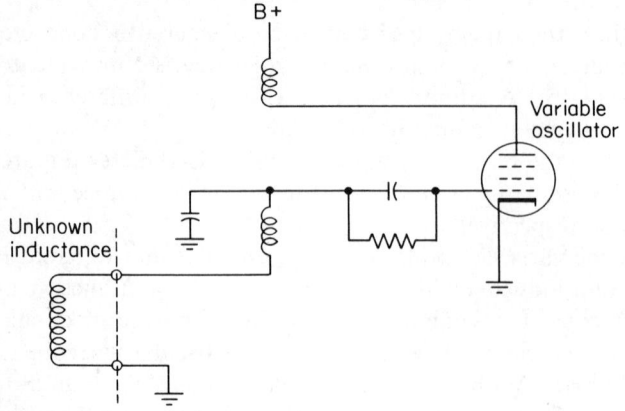

Fig. 3-22. Variable oscillator circuit for inductance measurement. Oscillator d-c grid return is completed through the unknown inductance (Tektronix).

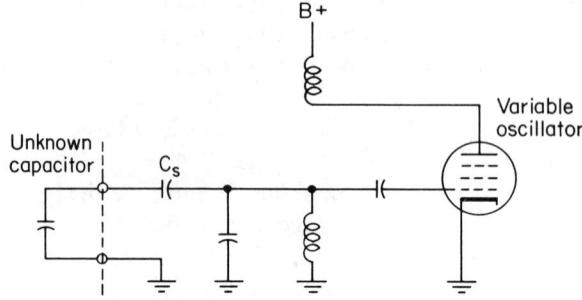

Fig. 3-23. Variable oscillator circuit for capacitance measurement. Series capacitor C_S permits unknown to be d-c biased, if required.

are d-c coupled through the filter to the counter. The counter output is used to deflect the meter by an amount proportional to frequency. A large capacitance produces a large difference frequency and a correspondingly large meter deflection.

When inductance is measured but the coil is not yet connected, the variable frequency oscillator is not yet operative. Therefore, the mixer receives only one 140-kHz signal from the reference oscillator. This signal does not pass to the counter because of the low-pass filter, so the counter and meter read zero. Connecting a coil energizes the variable oscillator, which produces a signal at a frequency between about 125 and 140 kHz. When this occurs, the mixer produces an output frequency equal to the difference, resulting in a proportional meter deflection.

4 SIGNAL GENERATORS

Next to a meter, a signal generator is the most useful tool in electronic service and laboratory work. Without some form of signal generator, technicians would be entirely dependent on signals broadcast from transmitting stations or on signals available from equipment containing oscillators. Under these circumstances, the technician would have no control over the frequency, amplitude, or modulation of such signals.

With a signal generator of the appropriate type, the technician can duplicate transmitted signals or produce special signals required for alignment or test of a particular equipment. Likewise, the frequency, amplitude, and modulation characteristics of these signals can be controlled.

4-1. SIGNAL GENERATOR BASICS

An oscillator (audio, rf, pulse, and so on) is the simplest form of signal generator. At the most elementary level of electronic servicing, a single-stage audio or r-f oscillator can serve the purpose of providing a signal source. An example of this is the simple probe-type oscillator used to trace signals through an audio amplifier or broadcast receiver. Such oscillators produce a broad range of signal frequencies that can be injected into the various stages of the amplifier or receiver.

Except for basic service work, such an oscillator has many obvious drawbacks. For example, to check the selectivity or tuning range of a receiver, the signal source must be variable over a range of frequencies. To check the sensitivity of a receiver or other instrument, the signal source must be variable in amplitude. To check the detector or audio portions of receivers, the oscillator should be capable of internal and/or external modulation. As a result, even the least expensive shop-type (or kit-type) generators will have many refinements to the basic oscillator circuit.

4-2. RADIO-FREQUENCY SIGNAL GENERATORS

Radio-frequency signal generators are found in both shop and laboratory work. There are no basic differences between the shop and laboratory generators; that is, both instruments will produce r-f signals capable of being varied in frequency and amplitude and capable of internal or external modulation. However, the laboratory instruments incorporate several refinements not found in shop equipment, as well as a number of quality features. (This accounts for the wide difference in price.) The following is a summary of the differences between shop and laboratory r-f generators.

Output Meter. In most shop generators, the r-f output amplitude is either unknown or approximated by means of dial markings. The laboratory generator will incorporate an output meter. This meter is usually calibrated in microvolts so that the actual r-f output can be read directly.

Percentage of Modulation Meter. Most shop generators have a fixed percentage of modulation (usually about 30 per cent). Laboratory generators provide for a variable percentage of modulation and a meter to indicate this percentage. Some generators have two meters (one for output amplitude and one for modulation percentage), while other generators use the same meter for both functions.

Output Uniformity. Shop generators vary in output amplitude from band to band, and (usually) cover their range by means of harmonics or beat notes. Laboratory generators have a more uniform output over their entire operating range and cover the range with pure fundamental signals.

Wideband Modulation. Generally, the oscillator of a shop generator will be modulated directly. This can result in undesired frequency modulation, just as it does with a transmitter with a directly modulated oscillator. The oscillator of a laboratory generator is never modulated directly (unless it is designed to produce an FM output). Instead, the oscillator is fed to a wideband amplifier, where the modulation is introduced. Thus the oscillator is isolated from the modulating signal.

Frequency Drift. Because a signal generator must provide continuous tuning across a given range, a variable frequency oscillator (VFO) of some type must be used. As a result, the output is subject to frequency drift, instability, modulation (by noise, mechanical shock, power supply ripple), and other problems associated with VFO units. Frequency instability does not present too great a problem in shop work. Also, it is possible to calibrate a signal generator at or near the most used frequency points. However, laboratory signal generators must provide a *known degree* of frequency accuracy over their entire operating range.

In laboratory generators, the output is less subject to frequency drift due to the incorporation of temperature-compensating capacitors. Likewise, the effects of line voltage variation are offset by regulated power supplies. The better generators also have more elaborate shielding, especially for the output attenuator circuits where rf is most likely to leak. The leakage of rf from signal generators is somewhat of a problem in many laboratory applications (such as receiver sensitivity tests). Shop generators often leak like sieves.

Bandspread. Shop generators usually have a minimum number of bands for a given frequency range; thus they cover a large part of the frequency range in each band. This makes the tuning dial or frequency control adjustments more critical, as well as difficult to see. Laboratory generators usually have a much greater bandspread; that is, they cover a smaller part of the frequency range in each band.

4-2.1. Typical Radio-frequency Generator Circuits

Figure 4-1 is the basic circuit of a shop r-f generator and Figure 4-2 is the corresponding circuit for a laboratory r-f generator.

In the shop generator (Figure 4-1), the r-f oscillator band is selected by changing the tuned tank circuit coil (one coil for each band). The actual frequency (within the band) is controlled by the setting of the tuned tank capacitor (usually geared or directly coupled to a frequency dial). The amplitude of the r-f output is controlled by coarse and fine attenuators. The audio

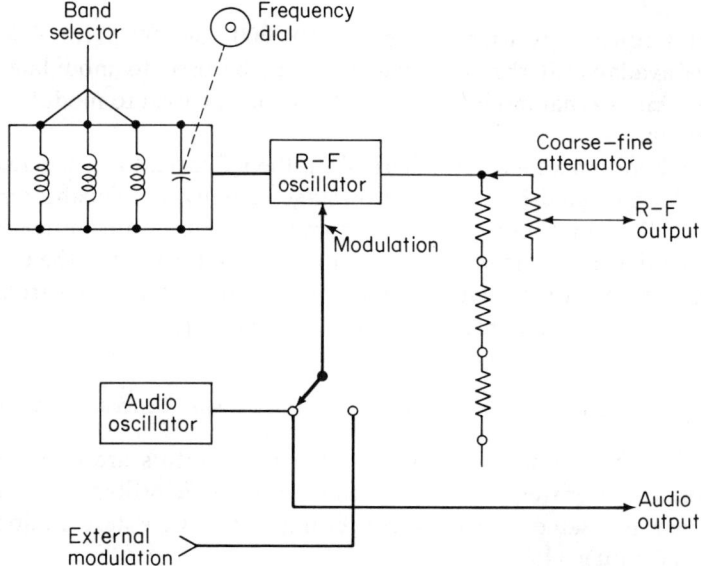

Fig. 4-1. Typical shop-type RF generator circuit.

98 Signal Generators

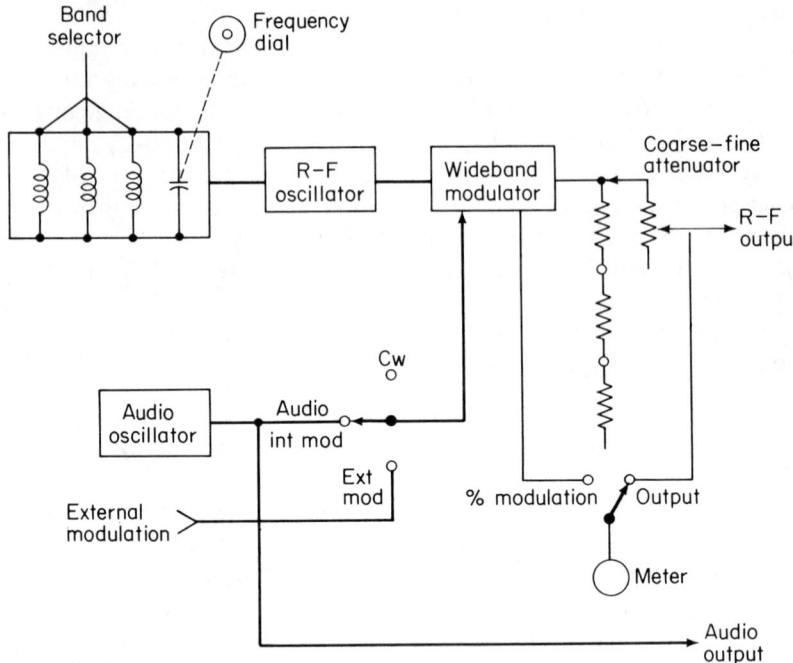

Fig. 4-2. Typical laboratory-type RF generator circuit.

oscillator frequency is usually fixed at 60, 400, or 1000 Hz. This audio output is available at the front panel or can be used to modulate the r-f oscillator. An external modulating signal can also be used to modulate the r-f oscillator.

In the laboratory generator (Figure 4-2), the r-f oscillator is isolated from the modulating signal by a modulator stage, usually a wideband amplifier. The r-f oscillator may also be isolated from the output by an emitter follower or some similar buffer stage. The meter circuit monitors either the r-f output or the percentage of modulation, depending on the position of a front-panel control. Some generators use two meters for this purpose.

4-2.1.1. Microwave Radio-frequency Signal Generator Circuits

The circuits for microwave r-f generators are quite different from those of lower frequencies, even though the basic problems and characteristics are the same. All data concerning microwave test equipment is covered in Chapter 11.

4-2.2. Typical Radio-frequency Generator Outputs

The outputs of a typical r-f generator are as follows:

1. An r-f signal, variable in frequency from about 100 kHz to 250 MHz (up to 1000 MHz for some laboratory generators). Usually, this frequency range is covered in four to eight bands, The r-f signal output is variable from 0 to 100,000 μV (up to 1 or 2 V for laboratory generators).
2. An audio signal, usually 60, 400, or 1000 Hz. With most signal generators this audio signal can be used to modulate the r-f output or is available as an output, or both.

4-2.3. Typical Radio-frequency Generator Controls and Indicators

Figure 4-3 shows the front-panel controls of a typical r-f generator. The following are descriptions of the control functions.

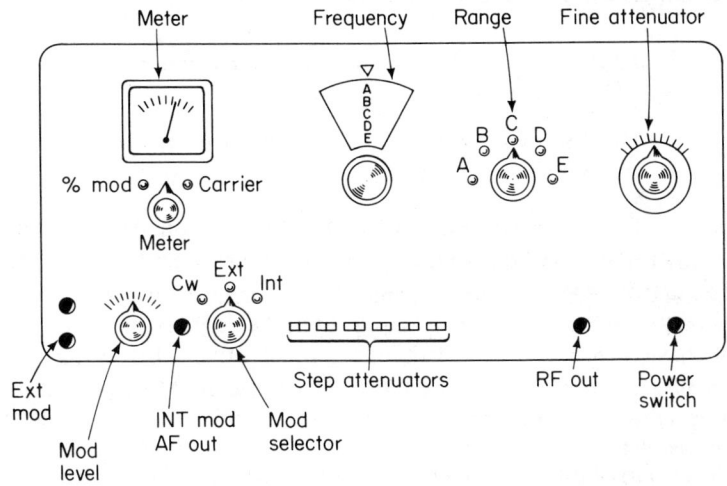

Fig. 4-3. Typical RF generator controls and indicators.

1. POWER switch. Applies and removes power to the instrument.
2. FREQUENCY control. Continuous dial scale assembly, used in conjunction with the range switch to select output frequency.
3. STEP ATTENUATORS. Six slide switches that have a total capability of 96-dB attenuation on the r-f output.
4. FINE ATTENUATOR. Continuous dial, used in conjunction with the STEP ATTENUATORS to adjust amplitude of the output signal.

5. RF OUT. The modulated or unmodulated r-f output from the generator is available at the connector. The output connector accommodates a coaxial cable (usually supplied with the instrument).
6. EXT MOD. External modulation signal to be superimposed on the r-f output is fed into these terminals. The MOD SELECTOR switch must be set to EXT MOD for these terminals to be placed in operation.
7. INT MOD/AF OUT. The internal modulation signal (usually 60 or 400 Hz) is available at this connector when the MOD SELECTOR switch is set to INT.
8. MOD SELECTOR. In the INT MOD position, the generator output is modulated by the internal 60- or 400-Hz signal. Simultaneously, this modulating signal is available at the INT MOD/AF OUT connector. In the EXT MOD position, the generator output is modulated by external signals applied to the EXT MOD terminals. In the CW (continuous wave) position, all modulation is removed.
9. METER. Indicates carrier output level (in microvolts) or percentage of modulation, depending on position of METER switch.
10. MOD LEVEL. Sets level or percentage of modulation.

4-2.4. Typical Radio-frequency Generator Operating Procedures

The following steps describe the basic operating procedures for a typical r-f signal generator.

1. Observe the General Safety Precautions described in the Introduction.
2. Set the POWER switch to OFF. Connect the power cord to an a-c outlet.
3. Set the POWER switch to ON, and allow 15 minutes (or as recommended by the manufacturer) for warm-up. All but the very best signal generators will have some tendency to drift in frequency as they warm up.
4. Connect the test cable to the RF OUT connector. Most generators are supplied with a cable for the r-f output. These cables are specially designed for use with the instrument and should always be used in preference to other types of cables. Such cables are shielded throughout their lengths to prevent excessive radiation of the output signal and to minimize hum pickup. The cables are usually terminated in a 50- or 75-Ω impedance and are usually unbalanced. Some signal generators designed specifically for television service have a balanced 300-Ω output.
5. Set the RANGE switch to the appropriate frequency range or band.
6. Set the FREQUENCY dial to the exact frequency. Make sure that you read the frequency on the correct band, as selected by the RANGE control.
7. Set the MOD SELECTOR switch to INT MOD (for internal modulation), CW (for no modulation), or EXT MOD (for external modulation), as desired.

8. Set the STEP ATTENUATORS (coarse attenuation) and FINE ATTENUATOR control, as necessary. Set the METER switch to CARRIER, then adjust the attenuators for the desired r-f output amplitude (usually specified in microvolts). If no value is specified for a particular test, use the following. Set the attenuator controls to give the *smallest output* necessary to obtain a trace of the desired heights on an oscilloscope (or a mid-scale indication on a meter). If too strong a signal is injected into the circuit under test, it is possible that overloading may cause distortion. The vertical gain of the oscilloscope should be set at or near the maximum gain point so that the oscilloscope furnishes a good share of the signal amplification. If a meter is used to monitor the circuit under test, the lowest scale should be used that will provide a good mid-scale reading.
9. Set the METER switch to % MOD. Adjust the MOD LEVEL control for the desired percentage of modulation.
10. Clip the ground lead of the r-f output cable to the ground of the circuit under test. Connect the probe or clip of the r-f output cable to the point of signal injection.
11. In some signal generators, the internal circuits are connected directly to the output terminals. A connection between the output terminals or cable and an external circuit carrying d-c voltage may result in damage to the signal generator. If it becomes necessary to connect a generator (without an internal blocking capacitor) to a power circuit, an external blocking capacitor of suitable value should be used in series with the output terminal.

4-2.5. Radio-frequency Signal Generator Frequency Calibration

An r-f signal generator can be checked against a local frequency standard or against a government standard broadcast, as described in Chapter 8.

4-2.6. Radio-frequency Signal Generator Frequency Stability

Quality r-f signal generators will maintain the frequency output despite changes in line voltage. This condition can be checked if the shop or laboratory is equipped with a Variac.

1. Connect the equipment as shown in Figure 4-4.
2. Tune the receiver to a station signal.
3. Tune the signal generator for a zero beat (Chapter 8) against the station carrier.
4. Adjust the input voltage to the generator with the Variac. If the zero beat remains constant with variations in input voltage, the signal generator output is stable. Generators with well-designed power supplies will maintain a constant output with input voltages from about 105 to 115 V.

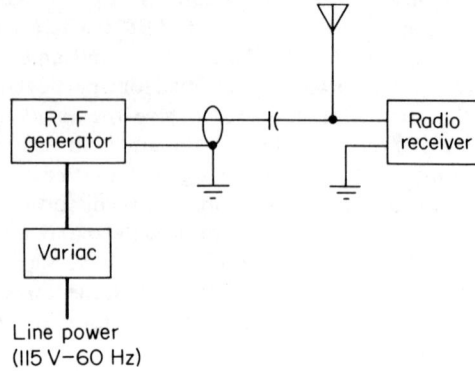

Fig. 4-4. Checking RF signal generator frequency stability with line voltage variations.

4-2.7. Radio-frequency Signal Generator Output Uniformity

Quality r-f signal generators will maintain the output signal level (amplitude) over a wide frequency range. If the generator is equipped with an output meter, the output can be monitored over the entire frequency range without external test equipment. If the generator is not provided with an output meter, connect a voltmeter (preferably a VTVM or electronic voltmeter) with an r-f probe (Chapter 9) to the generator output as shown in Figure 4-5, and use the following procedure.

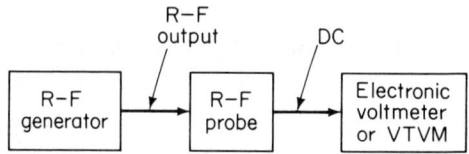

Fig. 4-5. Checking RF signal generator output uniformity.

1. Make certain to terminate the generator output in its own characteristic impedance (usually 50 or 72 Ω). If the generator is not terminated in its own impedance, the output may appear nonuniform as the frequency changes. This is due to reflected standing waves, which occur when an antenna is incorrectly matched to a transmission line (Chapter 11). The generator output cable will be purely resistive if the termination iscorrect.

2. In some cases where the meter is not sufficiently sensitive to measure the r-f output, it may be necessary to use an amplifier between the probe and meter. Typically, an r-f generator will have less than 1 V output maximum (usually 0.1 V for most shop generators). Therefore, the meter should have a scale that will provide a good, readable indication with about 0.05 V (50,000 μV).

3. Do not use any form of modulation for this test. With an unmodulated r-f output, the output of the probe will be a d-c voltage representing the r-f signal strength.
4. Set the meter to an appropriate *d-c voltage* range.
5. Tune the signal generator throughout its entire range of frequencies. Leave the generator output (or attenuator) at one setting. If the meter variation is *small* throughout the entire tuning range, the generator can be considered as having *excellent uniformity* of output. If there is no variation throughout the frequency range, check the meter (the needle is probably stuck). It should be noted that many generators will show some change in output from band to band, even though the output is fairly constant from one end of the band to the other.

4-2.8. Radio-frequency Signal Generator Modulation Percentage Uniformity

Those generators where the percentage of modulation can be adjusted are usually provided with a panel meter to show the percentage. Often, there is a SET line on the meter to indicate a given percentage (typically 30 per cent). In use, the meter selector is set to read precentage, and the MODULATION LEVEL control is adjusted until the meter needle is at the SET position. Each time the frequency is changed, the percentage of modulation must be adjusted (or at least checked).

To make a true percentage of modulation test, it is necessary to display the modulation envelope on an oscilloscope, as described in Chapter 10. This is recommended for all laboratory generators. For shop generators (where the exact percentage is not critical), a test of *modulation uniformity* is usually sufficient.

Modulation uniformity can be checked in essentially the same way as r-f output uniformity (Section 4-2.7), except that the meter is set to measure a-c voltage. With a modulated r-f output from the generator, the output of the probe will be a d-c voltage, with an a-c voltage (representing the modulation) superimposed. By setting the meter to measure a-c voltage, only the modulation will be indicated. Modulation uniformity should be checked throughout the entire frequency range of the generator.

If there is a variation noted with a modulated output, recheck the generator with an unmodulated output. If the unmodulated output has no variation, a problem exists in the modulation circuits.

This same test can be used if the signal generator is modulated from an external source. However, the amplitude of the external-modulating voltage must remain constant if the percentage of modulation is to remain at a fixed value.

4-2.9. Radio-frequency Signal Generator Attenuator Operation

Several methods are aviable to check the attenuator for proper operation. First, the attenuators can be checked simply for attenuation in steps. This is usually sufficient for most shop work, and can be accomplished using the same test connections as for r-f output uniformity (Section 4-2.7 and Figure 4-5). If the generator has its own output meter, the accuracy can be checked against that of the external meter. If the generator does not have a built-in output meter, the external meter can be used to calibrate (or check the calibration of) the attenuators for a given output voltage.

1. Connect the equipment as shown in Figure 4-5.
2. Tune the signal generator to an often used frequency.
3. Set the attenuators for maximum output (no attenuation). Set the meter range so that a nearly full-scale reading is obtained. (It is best to use an unmodulated output for test of the attenuators.)
4. Taking each step in turn, reduce the attenuator to progressively lower settings. Note the meter reading at each step. Change the meter scales as necessary. If the generator attenuators are marked in decibels, but the meter readout is in volts, use the Decibel Conversion Chart in the Appendix.
5. If the generator is not provided with an output meter, note and record the meter readings for each attenuator setting.
6. If the attenuators are operating properly, the meter reading will be reduced by a corresponding amount as the attenuator setting is reduced. If the meter reading remains the same (or increases) as the attenuator is moved to a position of more attenuation, the attenuator is defective. This is often a result of r-f leakage around the attenuator and is most common at higher frequencies. It should be noted that most attenuators are reactive (or have some reactive component in addition to the pure resistance), especially at higher frequencies. Therefore, the attenuator calibration does not remain constant over a wide range of frequencies. (This is another advantage of having a generator with an output meter.)

4-2.10. Checking for Line Power Hum

It is possible that a minor defect in a signal generator will cause power-line hum (usually 60-Hz) to appear at the output. This hum could be caused by leakage, defective or poorly designed power supply filtering, and so on. The hum can appear in two forms: as an independent signal or as a modulation source for the r-f output. Either way, the hum will be transferred to the circuit under test and can result in erroneous readings or plain annoyance. The presence of hum can be checked as follows.

Independent Signal. Connect the equipment as shown in Figure 4-6. With the generator connected to the audio amplifier, an r-f output modulated by a

60-Hz signal should not be audible. However, an independent 60-Hz signal will pass and be amplified.

Hum Modulation. Connect the equipment as shown in Figure 4-7. With the generator connected to the receiver, and the generator and receiver both tuned to the same frequency, hum modulation will pass through the r-f, i-f, and detector stages and will be heard on the receiver loudspeaker. However, unmodulated (independent signal) hum will not pass and should be inaudible.

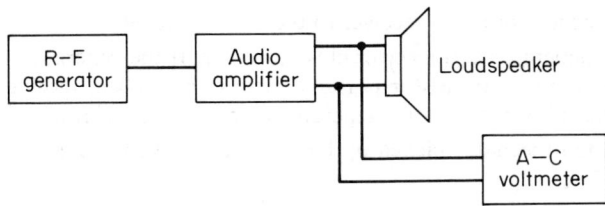

Fig. 4-6. Checking RF signal generator for line power hum (as an independent signal).

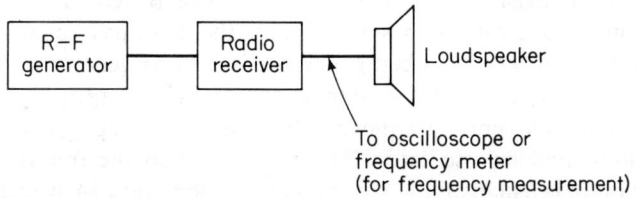

Fig. 4-7. Checking RF signal generator for line power hum (as hum modulation).

If there is no receiver capable of measuring the signal generator output frequency (such as with high-frequency laboratory generators), a meter with r-f probe can be used. The signal generator should be unmodulated, and the meter should be set to measure a-c voltage. Thus any indication on the meter is the result of hum modulation.

If any form of hum is present, its frequency can be checked by using an oscilloscope (Chapter 5) or, possibly, a frequency bridge (Chapter 3). This will confirm that the modulation or other undesired signal is at the line frequency.

4-2.11. Radio-frequency Signal Generator Leakage Tests

Any signal generator is subject to possible r-f leakage. The better generators have improved shielding and filtering to minimize this leakage. One particular weak point is the power cord. To provide maximum protection, the line cord should be filtered with series chokes and bypass capacitors. Another weak point is where control shafts pass through the panel. The presence of r-f leakage can be checked as follows:

1. Remove the output cable from the generator and cap the connector so that the output terminal is completely surrounded by grounded metal. Most laboratory generators are provided with such a cap.
2. Set the generator for modulated r-f output.
3. Tune the generator and a receiver to the same frequency.
4. Place the generator and receiver close together. If the generator covers the broadcast range, a transistor portable makes an excellent receiver, since it can be moved over the entire generator case in search of leakage.
5. If even a faint signal is picked up by the receiver, there is some r-f leakage.

4-3. SWEEP AND SWEPT-FREQUENCY GENERATORS

The main purpose of a sweep generator is the sweep-frequency alignment of television and FM receivers. Sweep generators are used in conjunction with oscilloscopes to display the bandpass characteristics of the receiver under test. A sweep generator is an FM generator. When set to a given frequency, this is the center frequency. The output varies back and forth through this center frequency. Basically, a sweep generator is a frequency-modulated r-f oscillator. The rate at which the frequency modulation takes place is usually 60 Hz for most television and FM sweep generators.

Other sweep rates could be used, but since power lines usually have a 60-Hz frequency, this frequency is both convenient and economical for the sweep rate. The sweep width, or the amount of variation from the center frequency, is determined by a control, as is the center frequency.

A swept-frequency generator (as they are sometimes called) is a laboratory version of the sweep generator. Both generators function to sweep the signal across a given band of frequencies at a given rate. There are a great number of uses for a sweep-frequency generator in laboratory work. Some of the basic uses are discussed in Chapter 11.

The main difference between the shop and laboratory models is that the laboratory model has some provision for maintaining the output level constant over the entire sweep-frequency range. Although this is desirable in any sweep generator, it is critical in laboratory work. Therefore, most laboratory sweep generators have an automatic leveling control (ALC) circuit, as discussed in Chapter 11.

As in the case of r-f signal generators, sweep generators are also available in various prices. In some instances this price differential is due to quality features (well-regulated power supplies, better shielding, and so on) previously discussed. In other cases, the better sweep generators incorporate special features.

For example, a *marker generator* is included with some sweep generators. As discussed in later sections of this chapter, marker signals are necessary to pinpoint frequencies when making sweep-frequency alignments. Although sweep generators are accurate in both center frequency and sweep width, it is almost impossible to pick out a particular frequency along the spectrum of frequencies being swept. Therefore, fixed-frequency "marker" signals are injected into the circuit along with the sweep-frequency generator output. On those sweep generators that do not incorporate a built-in marker generator, markers can be added by means of an absorption-type marker adder. These markers either can be built in or are available as accessories and will provide markers on the output.

A *fixed or variable bias* source is incorporated on some sweep generators. During alignment of most receivers, it is necessary to disable the AVC-AGC (automatic volume control or automatic gain control) circuits. This can be accomplished by applying a bias signal to the receiver's AVC-AGC line.

A *blanking circuit* is another feature found on some sweep generators. When the sweep generator output is swept across its spectrum, the frequencies go from low to high, then return from high to low. With the blanking circuit actuated, the return or retrace is blanked off. This makes it possible to view a zero reference line on the oscilloscope during the retrace period.

4-3.1. Typical Sweep Generator Circuit

Figure 4-8 is the block diagram of a typical sweep generator. Operation of the unit is as follows.

Transistor Q_1 serves as the swept oscillator that supplies an r-f output signal to broadband amplifier Q_2. The r-f oscillator bands are selected by changing inductances. Diode CR_1 is a voltage-variable capacitor (VVC) that receives a sawtooth sweep voltage from sweep generator Q_3. The sawtooth output from Q_3 is at a frequency of 60 Hz and an amplitude dependent on the setting of SWEEP WIDTH control. The sawtooth sweep from Q_3 causes the value of CR_1 to vary, thus sweeping the r-f output across a range of frequencies at a rate of 60 Hz. If the sawtooth output is increased in amplitude by the SWEEP WIDTH control, the range of frequencies covered in each sweep is increased by a corresponding amount.

Part of the r-f output is sampled by diode CR_2, which provides a bias for amplifier Q_2. If the r-f output increases, the bias increases accordingly and reduces the amplifier gain. A drop in r-f output causes the amplifier gain to

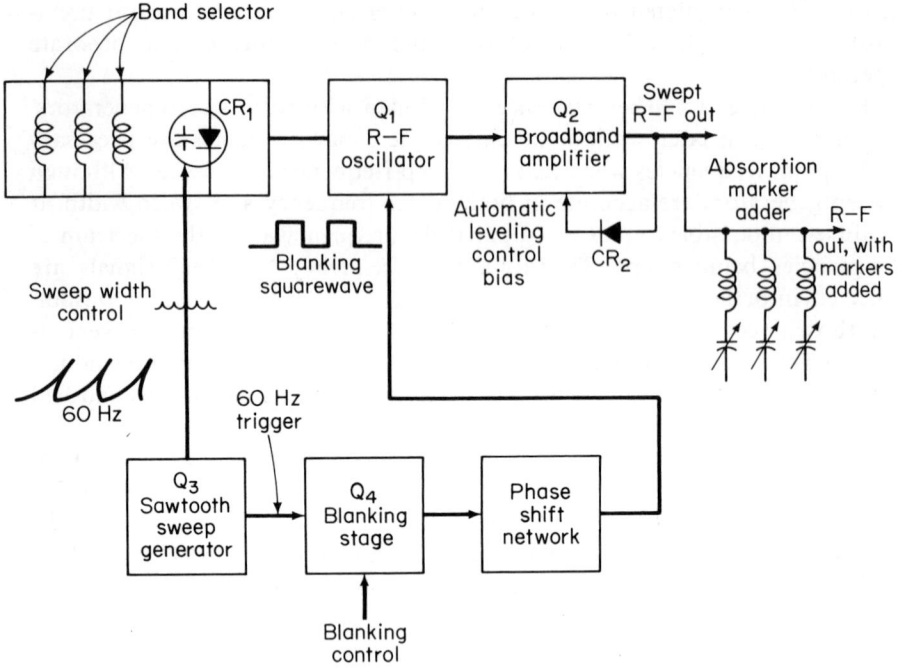

Fig. 4-8. Typical sweep generator circuit.

increase. Any variation in r-f output is offset, and the output is leveled automatically, thus providing automatic-leveling control (ALC).

Transistor Q_4 is a blanking stage that provides return-trace blanking of the sweep signal when the BLANKING control is set to ON. A blanking pulse, generated and clipped to resemble a square wave, is applied to sweep oscillator Q_1. Transistor Q_4 receives its trigger pulse from sweep generator Q_3 and is synchronized with the sweep frequency (60 Hz). An adjustable blanking phase-shift network permits proper phasing of the blanking pulse with the sweep output. Usually, about 90° phase shift is required.

Some sweep generators incorporate absorption-type marker adder circuits. As shown in Figure 4-8, these are tuned series tank circuits, placed across a sample of the r-f output. The series tank circuits absorb signals of the frequency to which they are tuned. This reduces the r-f output amplitude at these selected frequencies. If an oscilloscope is used to monitor the circuit under test, the drops in output amplitude appear as "pips," or "birdies," or (properly) "markers" of the corresponding frequency. This makes it easy to identify frequencies along the sweep spectrum.

4-3.2. Typical Sweep Generator Outputs

A typical TV/FM service-type sweep generator will provide r-f signals variable in frequency across both the TV and FM bands (54–216 MHz). Sweep width is approximately 10–12 MHz.

Laboratory sweep generators will provide sweep signals from a few hertz on up to 40 GHz.

In some sweep generators, the r-f output is continuously variable in several bands, while in other units the r-f output is selected for specific frequencies (such as the television channels or selected frequencies in the FM range of 88–108 MHz).

4-3.3. Typical Sweep Generator Controls and Indicators

Figure 4-9 shows the front-panel controls of a typical sweep generator. The following are descriptions of the control functions.

1. POWER switch. Applies and removes power to the instrument.
2. BAND selector. Selects desired operating band or channel. In a TV/FM sweep generator, the bands or channels would include the television channels (2–13), the FM band (88–108 MHz), and possibly the UHF television channels.
3. R-F OUT connector. The r-f sweep output from the generator is available at this connector. For television and FM work, the r-f output impedance is 300 Ω balanced. Other sweep generators have a 50- or 72-Ω output.
4. SWEEP WIDTH. Adjusts the sweep width of the output voltage at RF OUT connector. Some sweep width control dials are marked with the amount of sweep width for a given setting. Usually, however, the numbered positions of the SWEEP WIDTH dial are for reference or logging purposes.
5. RF OUT ATTENUATOR. Adjusts the amplitude of the output signal.
6. SWEEP OUT connector. The sawtooth sweep output is available at this connector. Such an output is used to synchronize the oscilloscope horizontal sweep. Not all sweep generators provide a sawtooth output.
7. RF SAMPLE OUT. A fixed-value r-f sample voltage (not affected by the RF OUT ATTENUATOR) from the swept oscillator is available at this connector for use with internal or external marker adder units. Usually, this type of connector provision is found only on TV/FM sweep generators.
8. BLANKING. When set at ON, this control provides a zero reference line on the oscilloscope for use in checking voltage amplitude of sweep response curves.
9. BIAS terminals. Most TV/FM sweep generators provide a bias source to disable the receiver's AVC-AGC line. The bias voltages are adjustable by means of the ADJ (adjustment) control over a range of 0–12 V (typical).

110 Signal Generators

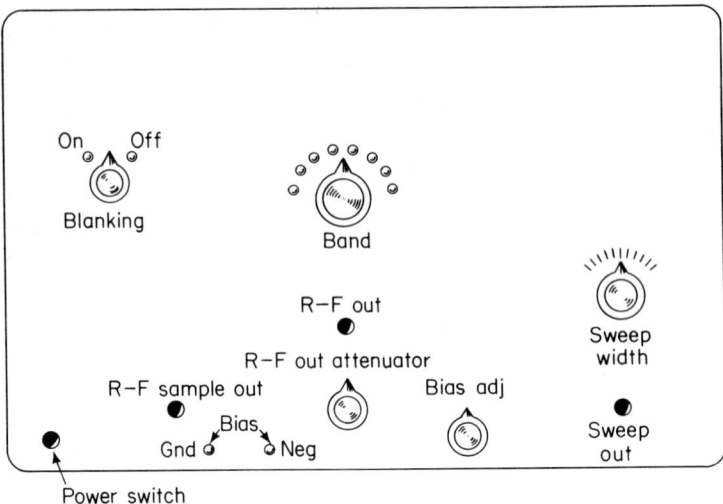

Fig. 4-9. Typical sweep generator controls and indicators.

4-3.4. Typical Sweep Generator Operating Procedures

The following steps describe the basic operating procedures for a typical sweep generator.

1. Observe the General Safety Precautions described in the Introduction.
2. Set the POWER switch to OFF. Connect the power cord to an a-c outlet.
3. Set the POWER switch to ON, and allow 15 min (or as recommended by the manufacturer) for warm-up.
4. Connect the r-f output cable to the R-F OUT connector. Cap the R-F SAMPLE OUT connector (if any) when not in use. On sweep generators with more than one output connector, do not leave unused cables connected to the generator. Some energy may be radiated form the end of the cable not in use and interfere with the test procedure.
5. Select the desired band or channel with the BAND selector.
6. Adjust the sweep to the desired width with the SWEEP WIDTH control.
7. Set the BLANKING control ON or OFF as required. As discussed in later sections of this chapter, it is important that the generator and oscilloscope be adjusted so that blanking is correct and the sweep of both instruments is in phase. Blanking may be applied by means of the BLANKING switch. The phase of the sweep is adjusted with the phasing control of the oscilloscope (although some sweep generators have phasing controls). If the phase adjustment is not properly set, the sweep curve on the oscilloscope may be prematurely cut off, or the curve may appear as a double or mirror image. These effects are shown in Figure 4-10.

Fig. 4-10. Sweep response curves showing effects of improper phasing control adjustment.

Double or mirror image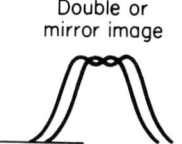

Sharp cutoff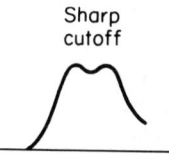

8. If bias voltages are required, connect the BIAS terminals to the circuit under test. Usually, this is the AVC-AGC line. The GND (ground) terminal is connected to the chassis, while the NEG (negative) terminal is connected to the AVC-AGC line. Connect a voltmeter across the BIAS terminals, and adjust the bias to the correct voltage, using the ADJ control.
9. Set the RF OUT ATTENUATOR as required. In general, the attenuator should be set to give only the amount of output signal necessary to obtain a sweep curve of the desired height on the oscilloscope. If too much signal is injected into the test circuit, it is possible that overloading may cause distortion of the curve and result in an erroneous picture of alignment. The oscilloscope vertical gain control should be set at or near the maximum gain point so the oscilloscope furnishes a good share of the signal amplification.

4-3.5. Basic Sweep Generator-Oscilloscope Procedures

The following steps describe the basic procedure for using a sweep generator with an oscilloscope.

1. Place the oscilloscope in operation. Refer to Chapter 5 and the oscilloscope instruction manual.
2. Place the sweep generator in operation. Refer to Section 4-2.4 and the sweep generator instruction manual.
3. Connect the equipment as shown in Figure 4-11 or 4-12, as required.
4. If the equipment is connected as shown in Figure 4-11, the oscilloscope horizontal sweep is triggered by the sawtooth output from the sweep generator. The oscilloscope's internal recurrent sweep is switched off, and the oscilloscope sweep selector and sync selector are set to EXTERNAL. Under these conditions, the oscilloscope horizontal sweep should be obtained from the sweep generator, and the length of horizontal sweep should represent total sweep spectrum. (See Figure 4-13.) For example, if the sweep is from 10 to 20 kHz, the left-hand end of the horizontal trace represents 10 kHz and the right-hand end represents 20 kHz. Any point along the horizontal trace will represent a corresponding frequency. For example, the midpoint on the trace would represent 15 kHz. If a rough approximation of frequency is desired, the horizontal gain control can be adjusted until the trace occupies an exact number of scale divisions on the oscilloscope screen

(such as 10 cm for the 10–20-kHz sweep signal). Each centimeter division would then represent 1 kHz.

5. If the equipment is connected as shown in Figure 4-12, the oscilloscope horizontal sweep is triggered by the oscilloscope internal circuits (both the sweep selector and sync selector are set to INTERNAL). Two conditions must be met to use the test connections of Figure 4-12. First, the sweep generator must be swept at the same frequency as the oscilloscope horizontal sweep (usually the line power frequency of 60 Hz). Second, the oscilloscope, or the sweep generator, must have a phasing control so that the two sweeps can be synchronized. (The phasing problem is discussed in Section 4-3.4.) This alternate method is used where the sweep generator does not have a sweep output separate from the r-f signal output or when it is not desired

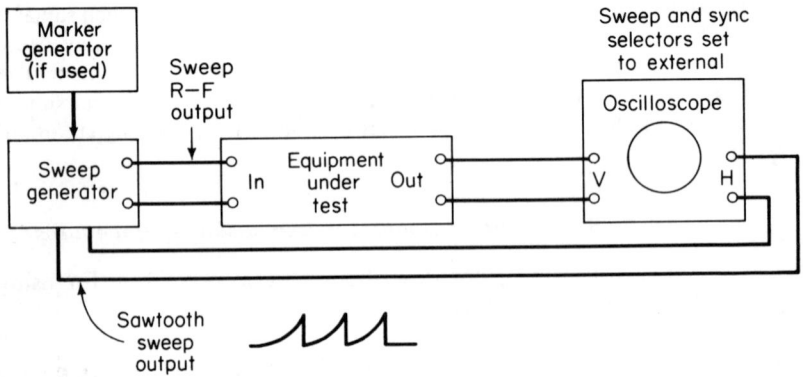

Fig. 4-11. Basic sweep generator/oscilloscope test circuit (using horizontal sweep from generator).

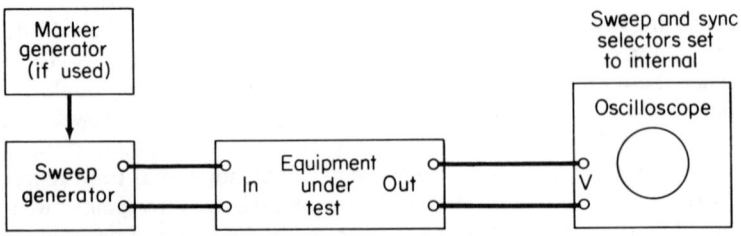

Fig. 4-12. Basic sweep generator/oscilloscope test circuit (using oscilloscope internal horizontal sweep).

to use the sweep output. Blanking of the trace (if any blanking is used) is controlled by the oscilloscope circuits.

6. If accurate frequency measurement is desired, a marker generator must be used (Figure 4-13). The marker generator output frequency is adjusted until the marker pip is aligned at the desired point on the trace. The frequency is then read from the marker generator frequency. If the sweep generator has built-in markers, pips will appear at the corresponding frequencies.

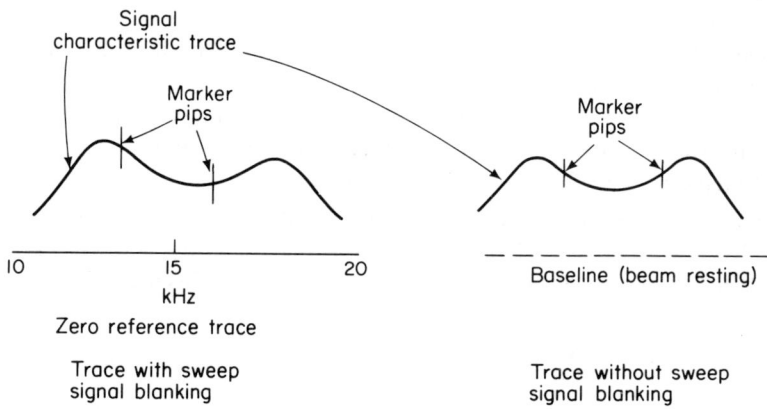

Fig. 4-13. Basic sweep generator/oscilloscope displays.

7. The response curve (oscilloscope trace) depends on the device under test. If the device has a large passband (as do most receiver circuits) and the sweep generator is set so that its sweep is wider than the passband, the trace will start low at the left, rise toward the middle, and then droop off at the right, as shown in Figure 4-13. The sweep generator-oscilloscope test method will tell at a glance the overall passband characteristics of the device (sharp response, flat response, irregular response at certain frequencies, and so on). The exact frequency limits of the passband can be measured with the marker generator pip.

4-3.6. Checking Sweep Generator Output Uniformity

Many sweep generators do not have a uniform output; that is, the output voltage is not constant over the swept band. In some tests, this can lead to false conclusions. In other tests, it is only necessary to know the amount of nonuniformity and then make allowances. For example, a sweep generator can be checked for flatness before connection to an a-c circuit, and

114   Signal Generators

any variation in output noted. If the output remains the same after it is connected to the circuit, even though it may have variation, the circuit under test is not at fault.

Sweep generator output can be checked as follows:

1. Connect the equipment as shown in Figure 4-14. The r-f or demodulator probe shown in Figure 4-14 can be omitted if the sweep generator output signal frequency is within the passband of the oscilloscope vertical amplifier. This is not usually the case with shop-type oscilloscopes, so some form of demodulator is necessary.
2. Place the oscilloscope in operation. Refer to Chapter 5 and the oscilloscope instruction manual.
3. Place the sweep generator in operation. Refer to Section 4-2.4 and the sweep generator instruction manual. Set sweep width to maximum.
4. Switch off the oscilloscope internal recurrent sweep.
5. Set the oscilloscope sweep selector and sync selector to external so that the horizontal sweep is obtained from the generator sawtooth sweep output.
6. Switch the sweep generator blanking control on or off as desired. With the

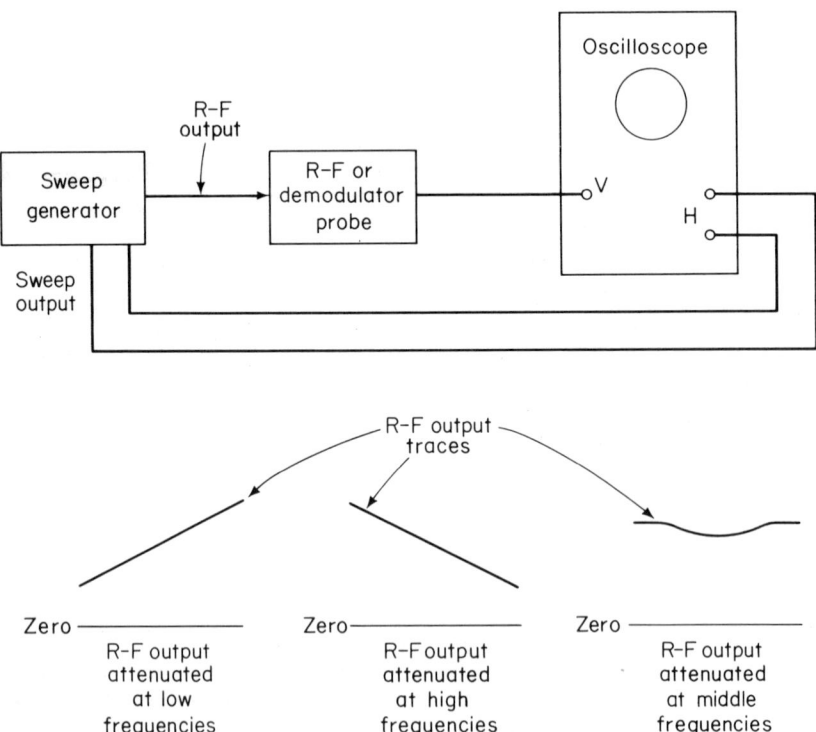

Fig. 4-14. Checking sweep generator output uniformity.

blanking function not in effect, only one trace will appear. This should be deflected vertically from the normal trace resting position. With the blanking function on, there will be two traces. The upper trace is the generator output characteristic; the lower trace is the no-signal trace.

7. Check the sweep generator output trace for *flatness*. If the right-hand side of the trace drops or slopes, this indicates that the sweep output is reduced at the high-frequency end of the sweep. A slope to the left indicates a reduced output at the low-frequency end of the sweep. If the trace drops off suddenly or dips in the middle, this indicates an uneven output. A perfectly flat trace or (more realistically) a trace that has only a slight curvature at the ends indicates an even output across the entire swept band.
8. Leave the sweep width at maximum, but adjust the center frequency of the sweep generator over its entire range. Check that the output is flat, or at least that any variations are consistent, across the range of the sweep generator.

4-3.7. Checking Sweep Generator Sweep Width

Although not usually critical, it is often convenient to know the actual sweep width for various positions of the sweep width control. Sweep width can best be checked with a marker generator, either built in or external, as follows:

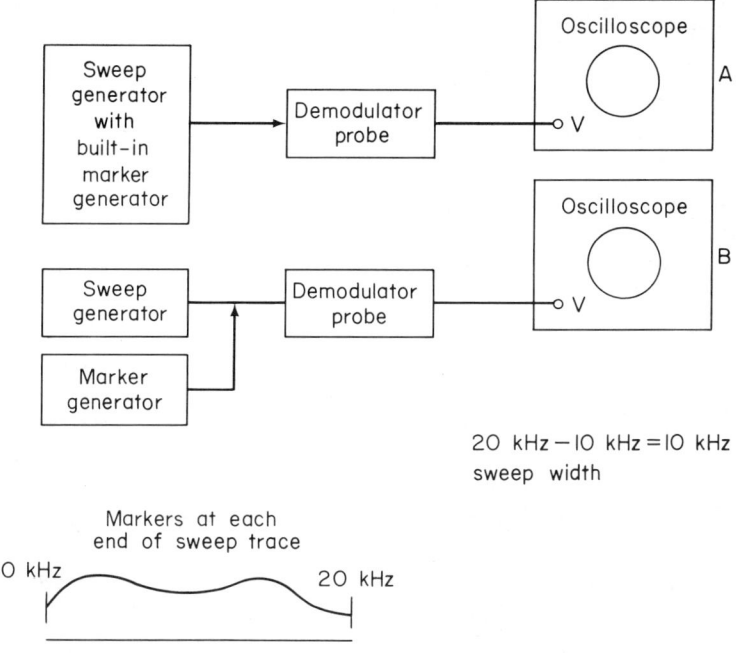

Fig. 4-15. Checking sweep generator sweep width.

1. Connect the equipment as shown in Figure 4-15. Figure 4-15A is for a sweep generator with built-in marker generator, while Figure 4-15B is for separate sweep and marker generators.
2. With both generators operating, the marker will appear on the swept trace when the marker generator is tuned to a frequency within the swept band.
3. Set the sweep width control to maximum or to some other reference point on the sweep width dial.
4. Adjust the marker generator until the marker is at one end of the sweep trace. Note the marker generator frequency. Run the marker generator to the opposite end of the trace. Note the marker generator frequency. The *difference* between the two marker generator readings is the *sweep width*.

4-4. MARKER GENERATOR AND TIME BASE GENERATOR

A marker generator is used in conjunction with a sweep generator and oscilloscope for alignment of TV and FM receivers. Basically, a marker generator is an r-f signal generator with highly accurate dial markings that can be calibrated precisely against internal or external signals. Usually, the internal signals are crystal controlled. In sweep-frequency alignment, the sweep generator is tuned to sweep the band of frequencies passed by the wideband circuits (tuner, if, video, and so on) in the television receiver, and a trace representing the response characteristics of the circuits is displayed on the oscilloscope. The marker generator is used to provide calibrated markers along the response curve for checking the frequency settings of traps, adjustment of capacitors and coils, and for measuring overall bandwidth of the receiver.

When the marker signal from the marker generator is coupled into the test circuit, a vertical "pip" or marker will appear on the curve. When the marker generator is tuned to a frequency within the passband accepted by the receiver, the marker will indicate the position of that frequency on the sweep trace. The receiver circuits can then be adjusted to obtain the desired wave shape, using the different frequency markers as checkpoints.

A *time base generator*, often called a *time mark generator*, is quite similar to the marker generator. Whereas the marker generator has a variable r-f signal with three or four crystal-controlled marker signals, the time base (or time mark) generator has several crystal-controlled, fixed-frequency signals. Usually, the signals are multiples and submultiples of a single crystal-controlled oscillator. A typical time mark generator will have r-f outputs that are multiples of the basic oscillator, and pulse outputs that have a repetition rate that is a submultiple of the basic oscillator frequency. For example, assuming that the basic oscillator frequency is 1 MHz, a typical series of r-f signals would be 5, 10, and 50 MHz, while typical pulse output would be at

1, 10, and 100 Hz and 1, 10, and 100 kHz. Both the r-f signals and the pulses can be used as markers.

4-4.1. Typical Marker Generator Circuit

Figure 4-16 is the block diagram of a typical marker generator. Operation of the unit is as follows:

The main marker signal source is obtained from variable frequency oscillator Q_1. This oscillator is tunable over a band of frequencies from about 1 to 260 MHz. This band is divided into eight overlapping r-f ranges. Output from Q_1 is applied to the input of modulation stage Q_2. Any internal or external modulation is mixed with the r-f signal in Q_2. Output from the Q_2 modulator stage is fed to the attenuator network.

Internal crystal-controlled calibrating markers are generated by crystal oscillator stage Q_4, which generates 10-MHz harmonic signals; 1-MHz calibrating markers are generated by a 1-MHz oscillator Q_5. The output of Q_5 is mixed with Q_4 in mixer Q_6.

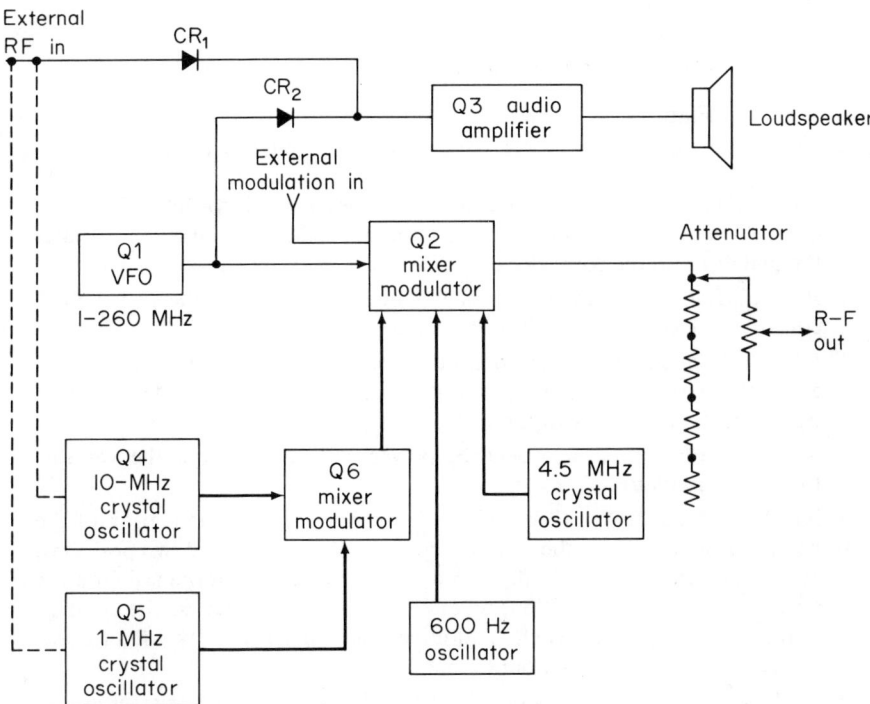

Fig. 4-16. Typical marker generator circuit.

The output of Q_6 is applied to modulation stage Q_2. A 4.5-MHz signal generated by Q_8 and a 600-Hz signal produced by Q_7 are also applied to Q_2. Likewise, any signal fed into the MOD IN terminals is fed to Q_2. Transistor Q_2 therefore functions as a mixer and modulator.

Crystal diodes CR_1 and CR_2 operate as a beat-frequency detector. In this circuit, the 1-MHz, 10-MHz, or any external signals are beat with the VFO signal from Q_1 to produce an audio-beat signal from the loudspeaker. Transistor Q_3 is an audio amplifier.

4-4.2. Typical Marker Generator Outputs

The outputs of a typical marker generator are as follows:
A service-type marker generator will provide r-f signals, variable in frequency from about 1 to 260 MHz. Usually, this frequency range is covered in several bands. The r-f output is variable or can be attenuated in steps. Maximum r-f signal is usually 100 mV.

Crystal-controlled calibrator signals at various frequencies (such as 1, 3.58, 4.5, and 10.7 MHz) that will permit the variable r-f signal to be calibrated accurately against the generator tuning dial.

4-4.3. Typical Marker Generator Controls and Indicators

Figure 4-17 shows the front-panel controls of a typical marker generator. The following are descriptions of the control functions.

1. RF RANGE. Selects one of eight r-f ranges from 1 to 260 MHz. This control simultaneously switches the internal oscillator circuits, and rotates the dial drum so the corresponding frequency scale is in view.
2. RF TUNING. Fine-tuning control used in conjunction with the RF RANGE control to select the desired output frequency.
3. OUTPUT. The modulated or unmodulated r-f output from the generator is available at this connector. The output connector accommodates the r-f output scale supplied with the instrument.
4. RF ON-OFF. When set to OFF, power is removed from the variable frequency oscillator.
5. RF ATTENUATION. Five slide switches provide step attenuation of the r-f output signal. When the slide switches are placed in their down positions, attenuation in steps of 15, 10, or 5 dB is provided. When the switches are set to their up or out position, no attenuation occurs. Because the attenuation provided by each switch is additive, the switches may be used to provide attenuation in any amount from 0 to 60 dB.
6. RF IN. External r-f signals are fed in here when the marker generator is used as a heterodyne frequency (zero-beat) meter. (Refer to Chapter 8 for a discussion of zero-beat and heterodyne frequency methods.) This terminal

Marker Generator and Time Base Generator 119

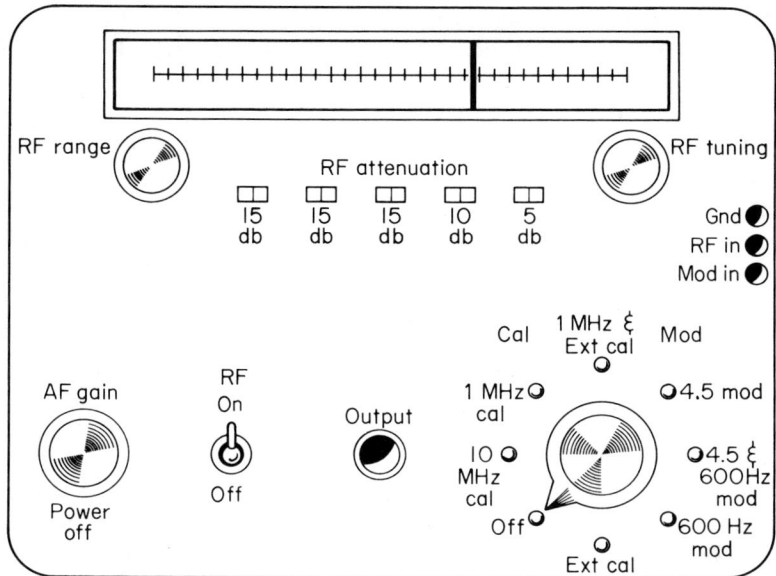

Fig. 4-17. Typical marker generator controls and indicators.

connects to the r-f detector. When an external signal is beat with the VFO or 1- or 10-MHz oscillator signal, the zero-beat condition can be heard on the loudspeaker.

7. AF GAIN-POWER OFF. Turns power off when set to POWER OFF. Increases volume level from loudspeaker when turned clockwise.

8. MOD IN. External modulation signal to be superimposed on the r-f output from the marker generator is fed into this terminal. The MOD IN terminal connects directly to the internal modulator-mixer stage. When the RF ON-OFF control is set to OFF and the CAL/MOD (calibrate/modulation) control is set to 600 Hz, the internal 600-Hz signal is available separately from this terminal.

9. CAL/MOD. Selects type of modulation applied internally or externally to the r-f output signal or made available at the output terminal. When this control is set to OFF, no modulation is applied. When set to one of the seven remaining positions, the corresponding modulation or calibrating signals are available.

4-4.4. Typical Marker Generator Operating Procedures

To prepare the marker generator for use, proceed as follows. The following steps describe the basic procedure for calibrating and using a marker generator.

1. Observe the General Safety Precautions described in the Introduction.
2. Set the AF GAIN-POWER OFF switch to POWER OFF. Connect the power cord to an a-c outlet.
3. Turn the AF GAIN-POWER OFF control clockwise from the POWER OFF position and allow 15 min (or as recommended by the manufacturer) for warm-up.
4. Connect the output cable to the OUTPUT terminal.
5. Set the RF RANGE control to the desired frequency range, as indicated on the dial scales.
6. Set the CAL/MOD control to the 10-MHz CAL position.
7. Set the RF ON-OFF control to ON.
8. Adjust the RF TUNING control to position the dial pointer at the desired 10-MHz calibration point on the tuning dial. A strong beat note should be heard from the loudspeaker. Carefully adjust the RF TUNING to obtain zero beat.
9. Observe the position of the dial pointer. If the pointer does not coincide exactly with the 10-MHz calibration mark on the tuning dial, slide the pointer to the left or right as required to line up the pointer with the dial-scale calibration mark. Leave the RF TUNING control set to the zero-beat setting while making this adjustment.
10. After setting the dial pointer at the 10-MHz check point that is closest to the desired frequency, set the CAL/MOD control to the 1-MHz CAL position and check for proper calibration.

    It should be noted that not all marker generators incorporate two or three calibration markers (some have more). Also, the method for zero-beating the calibration signal varies with each type of generator. However, no matter what calibration method is used, *always calibrate with the beat point that is nearest the frequency to be used.*
11. Once calibration is complete, the instrument is ready for use. It is usually recommended that a marker generator (or any precision generator) be calibrated immediately before each use.

4-4.5. Direct Injection Versus Postinjection

There are two basic methods for injection of marker signals into a sweep generator-oscilloscope display.

With *direct injection*, as shown in Figure 4-18, the sweep generator and marker generator signals are mixed before they are applied to the circuit under test. This method is sometimes known as *preinjection*.

With *postinjection*, as shown in Figure 4-19, the sweep generator output is applied to the circuit under test. A portion of the sweep generator output is also mixed with the marker generator output in a mixer-detector circuit known as a *postinjection marker adder*. The mixed and detected output from both generators is then mixed with the detected ouptut from the circuit under

Marker Generator and Time Base Generator   121

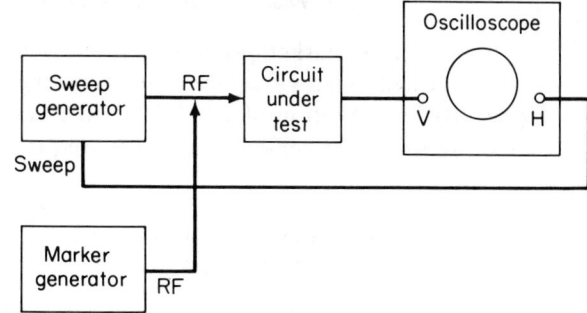

Fig. 4-18. Direct injection of marker signals.

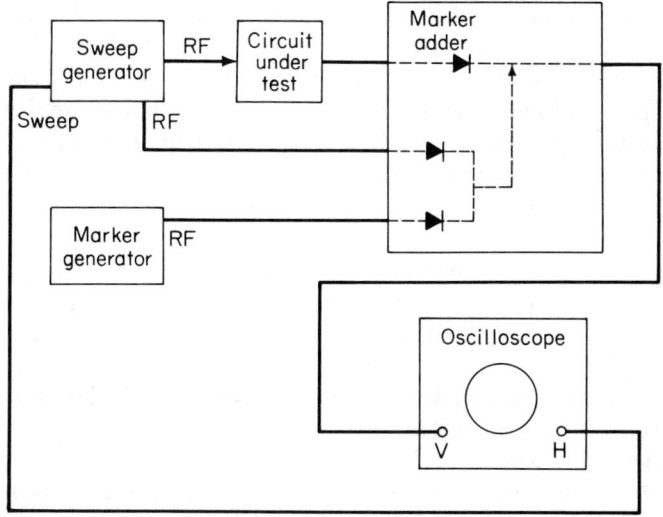

Fig. 4-19. Postinjection (or bypass injection) of marker signals.

test. Thus the oscilloscope vertical input represents the detected values of all three signals (sweep, marker, and circuit output).

Some marker generators (or combined sweep and marker generators) have built-in postinjection marker adder circuits. Postinjection marker adders are also available as separate units. Postinjection marker adders are not to be confused with the absorption-type marker adders discussed in Section 4-3.1, that provide fixed markers at various frequencies.

The postinjection (sometimes known as bypass injection) method for adding markers is usually preferred when servicing television and FM receivers. This is because postinjection minimizes the chance of overloading the circuits under test and permits use of a narrowband oscilloscope.

4-4.6. Marker Generator Tests and Calibration

A marker generator (or time mark generator) can be tested and calibrated as an r-f signal generator, since the primary output is an r-f signal. Refer to Sections 4-2.5–4-2.11. Also, most marker generators contain some provision for self-calibration of frequency, such as a built-in crystal.

4-5. AUDIO GENERATORS AND FUNCTION GENERATORS

Audio generators (also known as audio oscillators by some oldtimers in the electronics field) are particularly useful in testing all types of audio amplifiers, as well as the audio sections of other equipment (such as television and FM/AM receivers). Audio generators can also be used as modulation sources for r-f signal generators. As in the case of r-f signal generators, audio generators in their simplest form are essentially audio oscillators. For practical purposes, the audio output is tunable in frequency over the entire audio range (and beyond), and is variable in amplitude.

Early audio generators produced only sine waves. However, in recent times most commercial audio generators also produce square waves at audio frequencies.

Some laboratory audio generators are referred to as *function generators*, since they produce various functions: sine waves, square waves, triangular, and/or sawtooth waves. All of these functions are produced at audio frequencies.

The major differences in audio generators are quality differences rather than special features. Examples follow.

Laboratory audio generators have a more uniform output over their entire operating range. Shop audio generators may vary in amplitude from band to band.

Laboratory audio generators are less subject to frequency drift and line variation. The effects of hum or other line noises are minimized by extensive filtering. Accuracy and dial resolution are generally better for laboratory-type generators. This makes the tuning dial or frequency control adjustments less critical.

4-5.1. Typical Audio Generator Circuits

Figure 4-20 is the block diagram of a typical shop-type audio generator. Operation of the unit is as follows.

The basic oscillator circuit is comprised of transistors Q_1 and Q_2. The oscillator frequency is determined by the RC time constants in a feedback circuit. The frequency range is selected by changing resistance values (the R

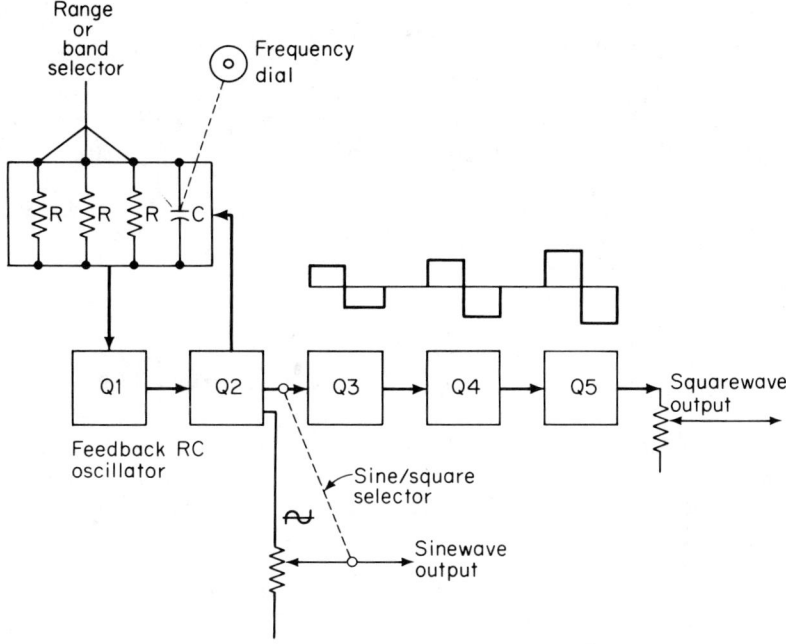

Fig. 4-20. Typical shop-type audio generator circuit.

of the RC circuit), while tuning within a given range is controlled by a variable capacitance (the C of the RC circuit).

With the SINE-SQUARE selector in the SINE position, the output is taken from the emitter of Q_2 and fed through the output divider network to the output jack.

With the SINE-SQUARE selector in the SQUARE position, the oscillator output is taken from the collector of Q_2 and fed to Q_3, where the square-wave-shaping action begins. The sine wave is reshaped to a square wave by the amplification and limiting action of Q_3, Q_4, and Q_5. The square-wave signal is taken from the emitter of Q_5 and fed through the output divider network to the output jack.

Figure 4-21 is the basic circuit of a laboratory audio generator. There are a number of audio generator circuits that qualify as laboratory equipment and provide the high-frequency stability and low-distortion characteristics. However, the Wien bridge RC oscillator has become the standard.

As shown in Figure 4-21, the basic circuit is a two-stage amplifier with both negative and positive feedback. (Refer to Chapter 7 for a discussion of amplifiers and feedback characteristics.) Positive feedback for sustaining oscillations is applied through the frequency-selective network $R_1 C_1$-$R_2 C_2$ of the Wien bridge. The amplitude response is maximum at the same frequency at

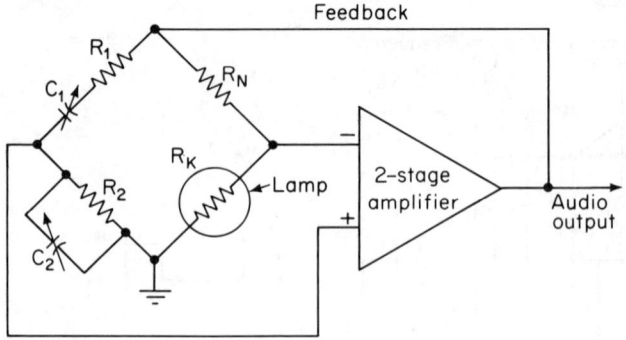

Fig. 4-21. Typical laboratory-type audio generator circuit.

which the phase shift through the network is zero. Oscillations are therefore sustained at this frequency.

The resonant frequency F_0 is expressed by the equation

$$F_0 = \frac{1}{6.28\,RC}$$

when $R_1 = R_2$ and $C_1 = C_2$.

Unlike L-C circuits used in r-f generators, where the frequency varies inversely with the square root of C, the frequency of the RC Wien bridge oscillator varies inversely with C. Thus frequency variations greater than 10:1 are possible with a single sweep of an air-dielectric tuning capacitor. Range switching usually is accomplished by switching the resistors.

The negative feedback loop involves the other pair of bridge arms R_N and R_K. In a Wien bridge oscillator, R_K is often a temperature-sensitive resistor with a positive temperature coefficient. Generally, it is an incandescent lamp operated at a temperature level lower than its illumination level. This lamp, shown as B in Figure 4-21, is sensitive to the amplitude of the driving signals and therefore adjusts the voltage division ratio of the branch accordingly. Thus as the amplitude of oscillations increases, the resistance of R_K increases, reducing the gain of the amplifier and restoring the amplitude to normal.

A different type of amplitude stabilization is used in solid-state oscillators. Because the current drawn by a lamp would be incompatible for use with transistors and battery power sources, these instruments use an output or peak detector circuit that provides a bias voltage proportional to the oscillator output voltage. Such a circuit is shown in Figure 4-22. As the amplitude of the amplifier changes, the detector circuit sends an error signal to the automatic gain control (AGC), which contains a field-effect transistor (FET). The purpose of the AGC is to control the oscillator gain continuously to maintain unity loop gain. The resistance of the AGC circuit can be varied

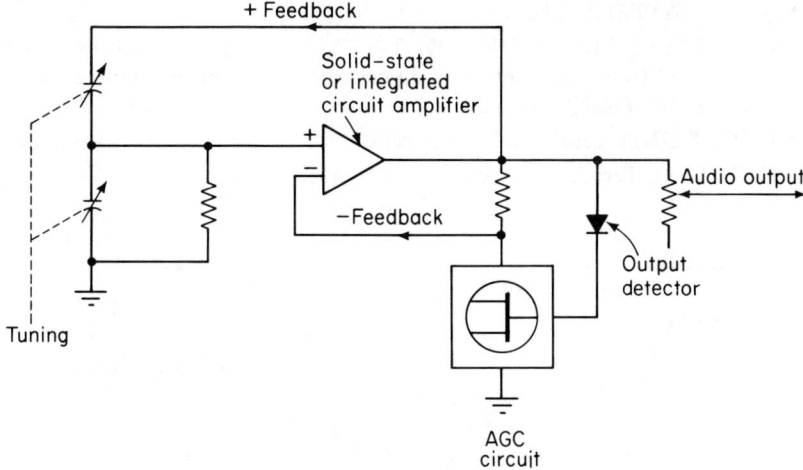

Fig. 4-22. Typical solid-state audio generator circuit.

slightly to change the divider ratio of the negative feedback network. An error in output voltage is detected by the detector and sent to the AGC field-effect transistor. This changes the resistance ratio in the negative feedback loop, thus bringing the output back to a constant level.

4-5.2. Typical Audio Generator Outputs

The outputs of a typical audio generator are as follows:

1. A sine wave, variable in frequency from about 20 Hz to 200 kHz, or possibly as high as 1 MHz. Usually, this frequency range is covered in several bands. The sine-wave output is variable from 0 to about 8 or 10 V.
2. A square wave, variable in frequency and amplitude over the same range as the sine wave.

4-5.3. Typical Audio Generator Controls and Indicators

Figure 4-23 shows the front-panel controls of a typical audio generator. The following are descriptions of the control functions.

1. FREQ RANGE. Selects the multiple to be applied to TUNING CONTROL dial frequency setting.
2. SINE-SQUARE ATTENUATOR. Selects either sine or square output and the output voltage range.
3. TUNING CONTROL. Continuous, single-scale dial, used in conjunction with the FREQ RANGE switch to select output frequency.

4. OUTPUT. Adjusts amplitude of the output signal within range of SINE-SQUARE ATTENUATOR.
5. OFF-ON/LINE FREQUENCY OUTPUT. Applies power to instrument when rotated from OFF position and controls line-frequency voltage supplied to LINE-FREQ terminals.
6. OUTPUT-GND. Terminals for connection to the selected frequency output.
7. LINE FREQ. Terminals provide a line-frequency voltage adjustable up to 6V.

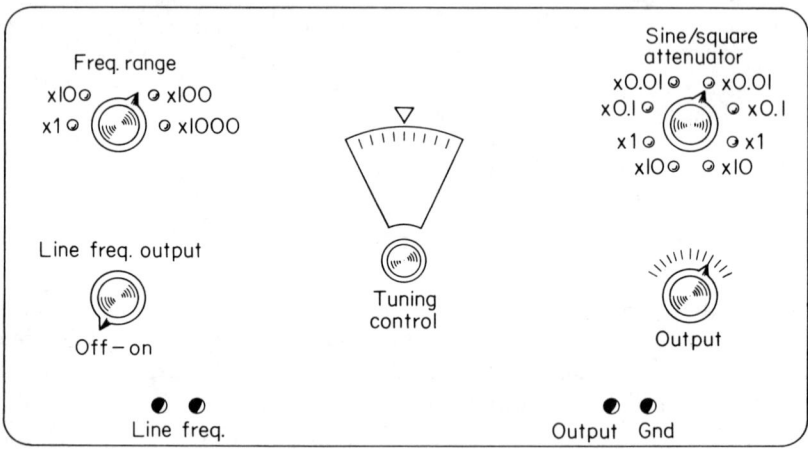

Fig. 4-23. Typical audio generator controls and indicators.

4-5.4. Typical Audio Generator Operating Procedures

The following steps describe the basic operating procedures for a typical audio generator.

1. Observe the General Safety Precautions described in the Introduction.
2. Turn the OFF-ON/LINE FREQ OUTPUT switch to the OFF position. Connect the power cord to an a-c outlet.
3. Turn the OFF-ON/LINE FREQ OUTPUT switch to the ON position and allow 15 min (or as recommended by the manufacturer) for warm-up. The OFF-ON/LINE FREQ OUTPUT control can be used for further regulation of the voltage at the LINE FREQ terminals.
4. Connect a set of test leads to the OUTPUT and ground terminals.
5. Adjust the tuning dial to the basic setting for the desired frequency.
6. Set the FREQ RANGE switch to the appropriate multiple. For example, for a tuning range of 20–200 Hz, set the FREQ RANGE switch to ×1. For a tuning range of 200–2000 kHz, set the FREQ-RANGE switch to ×10, and so on.

7. Set the SINE-SQUARE function switch to obtain the desired wave form and amplitude. The four left-hand settings are for sine wave, and the four right-hand settings are for square wave. The OUTPUT control can be used for further regulation of the output voltage.
8. Clip the cable attached to the GND terminal of the audio generator to the ground of the circuit under test. Connect the probe or clip from the OUTPUT terminal to the point of signal injection.

4-5.5. Audio Generator Frequency Calibration

An audio generator can be checked against a local frequency standard or against a government standard broadcast, as described in Chapter 8.

4-5.6. Audio Generator Frequency Stability

Quality audio generators will maintain the frequency output despite changes in line voltage. This condition can be checked if the shop or laboratory is equipped with a Variac.

The test is essentially the same as for frequency calibration (Section 4-5.5), except that a-c power is applied through the Variac to the audio generator. To make the test, adjust the input voltage to the generator with the Variac. If the zero beat (or other frequency comparison indication) remains constant with variations in input voltage, the audio generator output is stable. Generators with well-designed power supplies will maintain a constant output with input voltages from about 105 to 115 V.

4-5.7. Audio Generator Output Uniformity

Quality audio generators will maintain the output signal level (amplitude) over a wide frequency range. This condition can be checked as follows:

1. Connect the equipment as shown in Figure 4-24.
2. Tune the audio generator throughout its entire range of frequencies. Leave the generator output (or attenuator) at one setting. Adjust the meter controls for a good mid-scale reading.
3. If the meter variation is small throughout the entire tuning range, the generator can ge considered as having excellent uniformity of output. How-

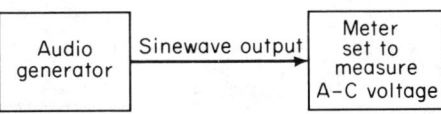

Fig. 4-24. Checking audio generator output uniformity.

ever, it should be noted that many generators will show some change in output from band to band, even though the output is fairly constant from one end of the band to the other.

4-5.8. Audio Generator Attenuator Action

The output control of most audio generators is in the form of an attenuator. Several methods are available to check the attenuator for proper operation. First, the attenuator can be checked simply for attenuation in steps. Second, the output can be checked for actual value (calibrated in volts).

1. Connect the equipment as shown in Figure 4-24.
2. Tune the audio generator to various points throughout its frequency range.
3. Set the attenuator for maximum output (no attenuation). Set the meter range so that a nearly full-scale reading is obtained.
4. Taking each step in turn, reduce the attenuator to progressively lower settings. Note the meter readings at each step. Change the meter scales as necessary.
5. The meter reading should be reduced by a corresponding amount as the attenuator setting is reduced. For example, if the full-output meter reading is 10 V, the meter reading at the ×0.1 setting should be 1 V.
6. If the meter reading remains the same, increases, or changes to an incorrect value, the attenuator is defective.
7. If it is desired to calibrate the output, note the meter reading at each step of the attenuator.

4-5.9. Checking Audio Generator Output Hum

A minor defect or poor design can cause a 60-Hz hum to appear at the output of an audio generator. The presence of this hum can be checked as follows:

1. Connect the equipment as shown in Figure 4-25.
2. Operate the generator and oscilloscope controls (Chapter 5) to display either a sine-wave or square-wave display.
3. Set the generator output to near 60 Hz, but not at 60 Hz.

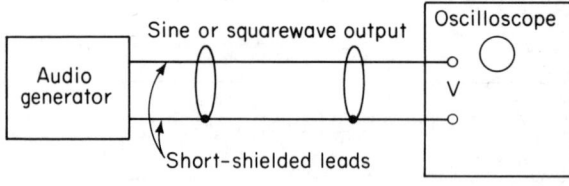

Fig. 4-25. Checking audio generator output hum.

4. If the display is unsteady and appears to wriggle about, this indicates 60-Hz hum. If there is any doubt, change the generator frequency to something other than 60 Hz. If the wriggling stops, this is an indication of hum.

Make certain that the hum is not being picked up from some outside source. Use very short, shielded leads between the generator and oscilloscope. Also, ground the chassis of both instruments to a common point.

4-5.10. Checking Audio Generator Output Distortion

An audio generator produces harmonics as does almost any oscillator. These harmonics can combine with the fundamental in such a way that they distort the output.

A full discussion of harmonic distortion, as well as procedures for measuring such distortion, will be found in Chapter 10. The audio generator should be submitted to these procedures.

If the distortion appears to be high, beyond the manufacturer's specification, try reducing the output in amplitude. Distortion is often the result of driving an audio generator to extremes. If a reduction in amplitude causes a corresponding reduction in harmonic distortion, this fact can also be noted for future use.

4-5.11. Checking Audio Generator Load Sensitivity

The outputs of audio generators can be sensitive to the shunt capacitance of circuit loads; that is, the output will drop off when the load is applied. This is especially true at higher frequencies and lower generator outputs.

If a square-wave audio generator is load sensitive, it may produce distorted square waves when connected to a shunt capacitance. It is essential that these conditions be noted when using an audio generator to test audio amplifiers.

1. To check the sine-wave output for load sensitivity, connect the equipment as shown in Figure 4-26.
2. Set the audio generator for maximum output. Adjust the meter for a good mid-scale indication. Set the audio generator to various frequencies throughout its range, starting with a frequency of about 15 kHz.

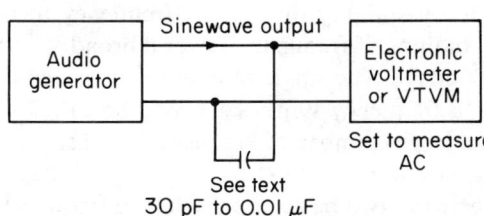

Fig. 4-26. Checking sinewave output of an audio generator for load sensitivity.

3. Shunt progressively larger capacitors across the output terminals of the audio generator. The actual capacitor values are not important.
4. If the meter indication shows a drop in output with a very small value of shunt capacitance, the output is load sensitive.
5. If the meter shows little or no drop in output, try reducing the generator output to a point near its minimum. Reset the meter controls as necessary for a mid-scale reading. Again, add the shunt capacitance in steps. Also, repeat the test with higher generator frequencies. For practical purposes, it is only necessary for a shop-type audio generator to have immunity to load capacitance changes up to about 15 kHz, since most audio amplifiers are used below this range.
6. To check the square-wave output for load sensitivity, connect the equipment as shown in Figure 4-27.

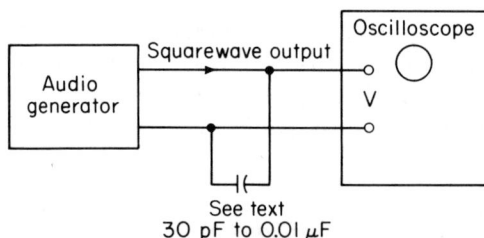

Fig. 4-27. Checking square-wave output of an audio generator for load sensitivity.

7. Set the audio generator for maximum square-wave output. Adjust the oscilloscope for a good display of the square wave. Set the audio generator to various frequencies throughout its range, starting with a frequency of about 15 kHz.
8. Shunt progressively larger capacitors across the generator output terminals until the corners of the square waves are noticeably rounded.
9. Repeat the test with the generator output near its minimum and at higher output frequencies. These are the two conditions most likely to produce load sensitivity.

4-6. FREQUENCY SYNTHESIZERS

A frequency synthesizer is an instrument that translates the frequency stability of a single frequency to any selected one of thousands, even billions of frequencies over a broad spectrum that may extend from d-c to 500 MHz. The single frequency generated is of frequency standard quality. Thus a frequency synthesizer may be called a *frequency standard* that furnishes a large number of frequencies. (Refer to Chapter 8 for a discussion of frequency standards.)

There are two basic approaches to frequency synthesis: direct and indirect.

The *direct-frequency synthesis* (which is generally preferred) performs a series of arithmetic operations (division, multiplication, and mixing) on the signal from the frequency standard to produce the desired output frequency.

The indirect method uses tunable oscillators, phase-locked to harmonics of a standard frequency to derive a desired output frequency.

We will discuss only the direct method of frequency synthesis, since it has all but replaced the indirect method.

4-6.1. Typical Frequency Synthesizer Circuits

A typical frequency synthesizer consists of a driver section, a high-frequency section, and an ultra-high-frequency section. The driver furnishes basic signals required for the high-frequency and ultra-high-frequency synthesizers. A basic driver unit can furnish signals for up to four synthesizers.

Figure 4-28 shows the technique for generating 22 signals from a single standard signal in the synthesizer driver. The standard signal can be either a 1- or 5-MHz signal from an external source or the 1-MHz signal from the internal crystal-controlled oscillator. Either way, accuracy of the frequency synthesizer will be no greater than that of the basic standard signal. If a 5-MHz external standard is selected, the signal is divided by a 5:1 divider circuit to produce a 1-MHz signal at the selector switch. (Refer to Chapters 2 and 6 and the Appendix for a discussion of frequency divider and multiplier circuits.)

The 1-MHz signal is filtered by a crystal filter circuit. The filter circuit consists essentially of a quartz crystal cut to resonate at 1 MHz. Signals at frequencies on either side of 1 MHz will not cause the crystal to oscillate. Thus the crystal filter output is always at 1 MHz. This output is applied to an output jack for external use (such as for synchronizing other equipment or monitoring against a frequency standard). The output is also applied to a *spectrum generator*. The spectrum generator is a step recovery diode that has the property of producing a broad range of signal frequencies whenever a single signal frequency is applied. The spectrum generator is sometimes known as a *comb* generator. This is due to the appearance of its output signals on a spectrum analyzer. (Refer to Chapter 10 for a discussion of spectrum analyzers.) The output of the spectrum generator is applied to a series of filters.

In the circuit of Figure 4-28, the output of the spectrum generator is fed to filters that select 20-, 24-, and 30-39-MHz signals. These filters are very effective, with adjacent 1-MHz signals 105 dB down. The 30-39-MHz components are also applied to decade frequency dividers yielding a low-frequency output of 3.0-3.9 MHz in 0.1-MHz steps. Thus 22 signals are available to the high- and ultra-high-frequency sections.

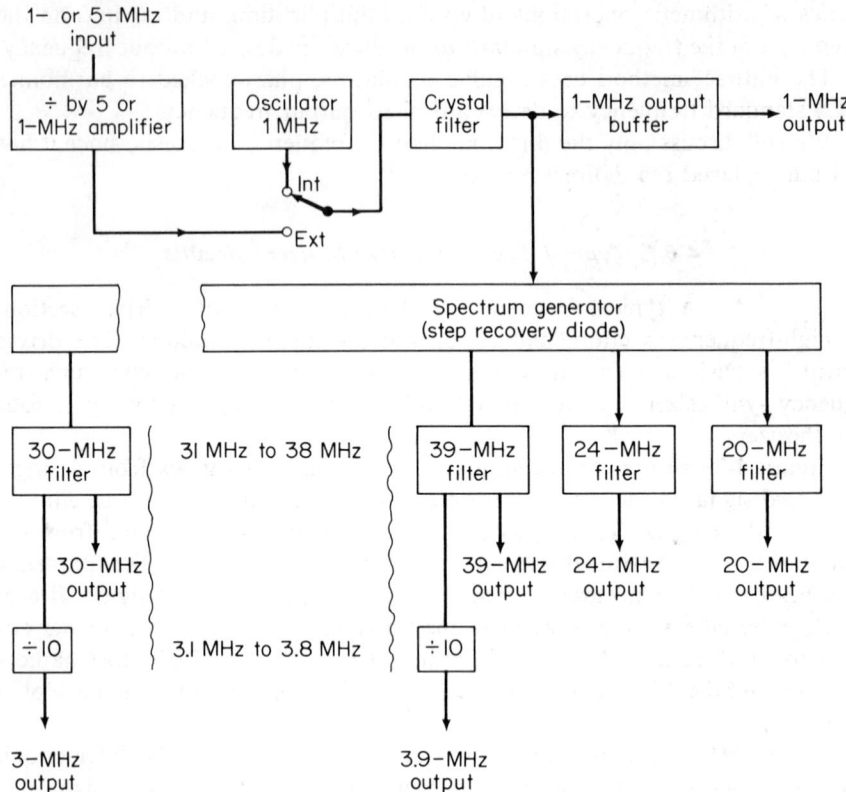

Fig. 4-28. Basic frequency synthesizer driver circuit for generating 22 signals from a single standard (Hewlett-Packard).

Figure 4-29 shows the relationship of the driver signals to the other sections (HF and UHF). Note that the HF section uses the 24-MHz signal, as well as the 3.0–3.9-MHz signals. Figure 4-30 is a block diagram of the H-F section. Note that the 3.0–3.9-MHz signals are fed into a *diode switch matrix* controlled by d-c voltages. Control is provided either by front-panel pushbuttons or remotely through the rear-panel connectors.

The basic diode matrix switch circuit is shown in Figure 4-31. There is one pushbutton for each diode switch circuit. When the corresponding pushbutton is pressed, −12.6 V are applied to the circuit. This biases the diodes so that the signal will pass.

Note that there are eight separate outputs from the diode switch matrix in the HF section (Figure 4-31). Each of these outputs may be individually selected between 3.0 and 3.9 MHz. Each output is connected to a decade module. The first seven decades (decades 2–3) are composed of two balanced

Frequency Synthesizers 133

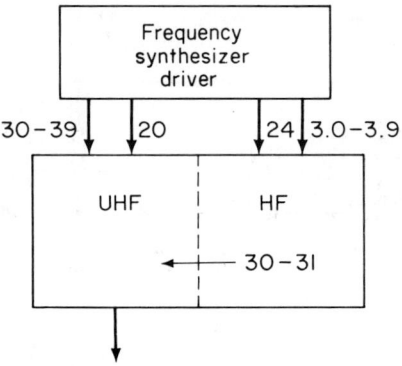

Fig. 4-29. Relationship of driver to HF and UHF synthesizer sections (Hewlett-Packard).

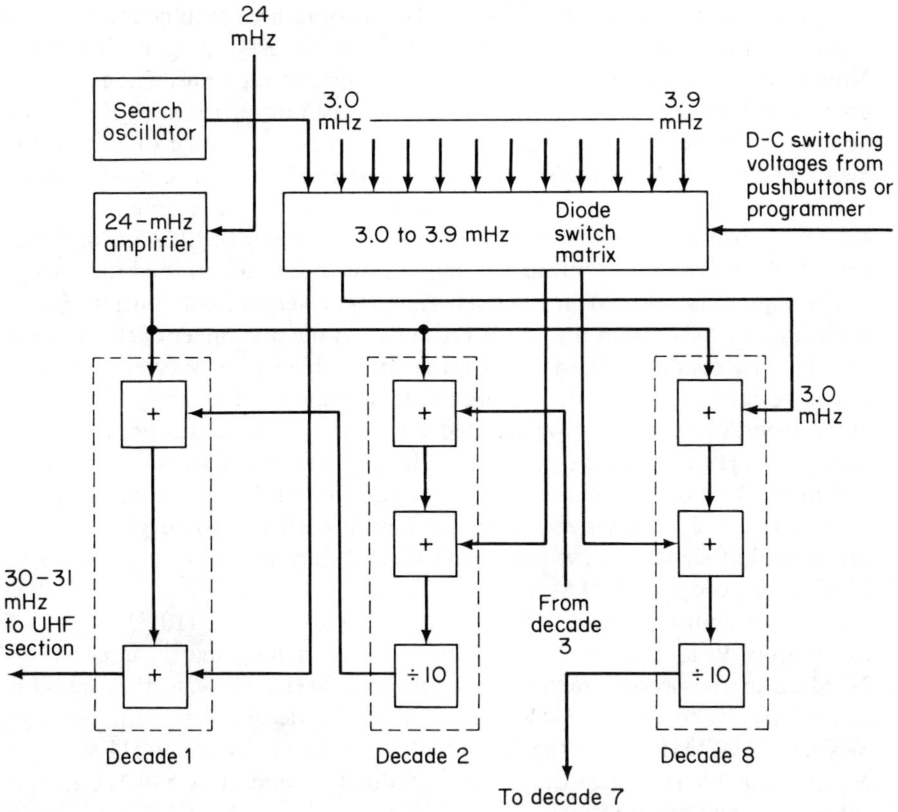

0.1 Hz to 500 mHz in 0.1-Hz increments

Fig. 4-30. Block diagram of synthesizer HF section (Hewlett-Packard).

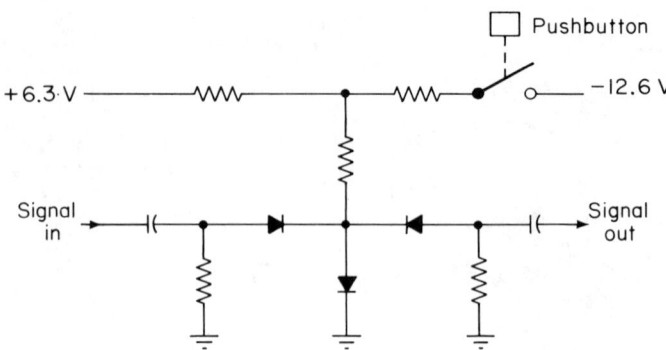

Fig. 4-31. Basic diode switch matrix circuit (Hewlett-Packard).

mixers and one decade divider. The remaining decade (decade 1) is similar to the other decades except that it lacks the divider.

Figure 4-32 shows how the HF and UHF sections are combined to produce a given output frequency, as selected by the corresponding pushbuttons. Note that the pushbuttons are arranged in vertical columns. Each column goes from 0 to 9 (bottom to top) with one pushbutton for each digit. The number of vertical columns depends on the frequency range of the instrument. The pushbuttons are interlocked mechanically so that only one pushbutton can be pressed in each vertical column at any given time. In Figure 4-32, the pushbuttons are shown for a selected frequency of 12,345,678.90 Hz. Only the required pushbuttons are shown; all others are omitted for clarity.

The digit 0 is selected in the least significant vertical column (10^{-2}); 24 MHz and 3.0 MHz from the driver are added in the first mixer of the selected decade. The resultant 27-MHz signal is then added to a selected 3.0-MHz signal (resulting from the depressing of the pushbutton 0 in the 10^{-2} column) in the second mixer. (Had we selected the digit 2 in the least significant column, 3.2 MHz would have been added to 27 MHz in the second mixer.)

Returning to the original example, the addition of 27 and 3.0 MHz results in 30 MHz. This synthesized signal is now divided by ten to give 3 MHz. The resultant signal is now fed to the second decade, where it is added to 24 MHz to obtain 27.00 MHz.

If the pushbutton column controlling the second decade (10^{-1}) is now set to the digit 9, as shown, 3.9 MHz from the switching matrix is added to 27 MHz in the second mixer to obtain 30.9 MHz. In turn, this signal is divided by 10 to obtain 3.090 MHz. The third decade first adds the synthesized 3.090 MHz from the output of the second decade to 24 MHz to obtain 27.090 MHz; then 3.8 MHz are added to obtain 30.890 MHz. This value is divided by 10 to yield 3.0890 MHz.

This process is repeated, successively adding digits, starting with the least

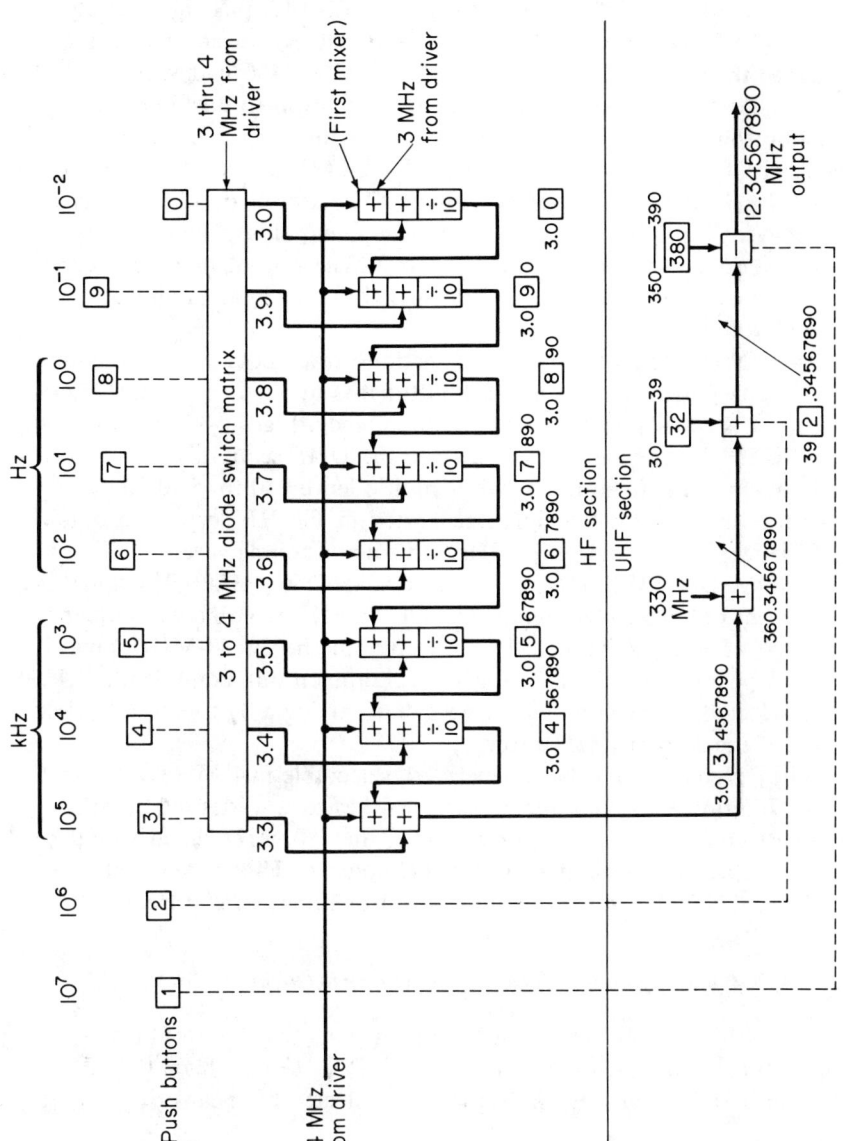

Fig. 4-32. Block diagram of how HF and UFH sections are combined to synthesize a 12.34567890 MHz output (Hewlett-Packard).

significant, and then dividing by 10. After the least significant columns (10^{-2}, 10^{-1}, 10^0, 10^1, 10^2, 10^3, and 10^4) have been processed, the result is a synthesized signal of 3.04567890 MHz. In the 10^5 decade, this signal is again added to 24 MHz and the selected 3.3-MHz signal. However, since there is no divider in this decade, it provides an output of 30.34567890 MHz.

The eight least significant columns have, in addition to pushbuttons 0–9, an S pushbutton. If this position is selected in any of these columns, a *search oscillator*, with frequency variable between 3 and 4 MHz, is substituted for the 3.0–3.9-MHz signals in the selected column. The search oscillator may be varied between 3 and 4 MHz to provide *continuous tuning*, and it may be frequency-modulated from an external source. The output from the search oscillator is controlled by either the front-panel dial or by the application of an external voltage.

The 30–31-MHz signal from the HF section (at a frequency representing the eight digits punched into the pushbuttons) is applied to the UHF section, where it is translated to the desired frequency. Figure 4-33 shows how the 12.34567890-MHz signal is generated in the UHF section.

The 33-MHz signal from the driver is multiplied by 10 to yield 330 MHz. This signal is mixed with the signal generated in the HF section to obtain a signal between 360 and 361 MHz. In our example, we would add 30.34567890 MHz to 330 MHz, yielding 360.34567890 MHz. This signal is mixed with one of ten selected frequencies (30–39 MHz) to obtain a frequency in the range of 390–400 MHz. The frequency in the 30–39-MHz range is chosen by a pushbutton (digit) in the 10^6 column. In our example, 32 MHz would be selected, corresponding to the 2 digit, giving a 392.34567890-MHz signal out of the second UHF mixer.

Thus far, a frequency has been generated, selectable in 1-MHz steps, with 0.01-Hz resolution. All that remains to synthesize our desired frequency is the most significant digit. In our example, the 380-MHz signal would be selected by depressing the 2 digit in the 10^7 column. This, when subtracted from 392,345,678.90 Hz, gives the final output of 12,345,678.90 Hz.

4-6.2. Typical Frequency Synthesizer Outputs

Typical outputs from a frequency synthesizer would be spectrally pure (no sidebands, modulation, or noise) r-f signals from 0.01 Hz to 500 MHz in 0.01-Hz steps. Each frequency is selected by means of pushbuttons.

In addition to the fixed-frequency signals, most synthesizers provide a search oscillator for continuously variable frequency selection. This permits a variable frequency output over a selected portion of the frequency spectrum. For example, if the search oscillator function had been selected for the two least significant digits (by pressing the S buttons in the 10^{-2} and 10^{-1} columns),

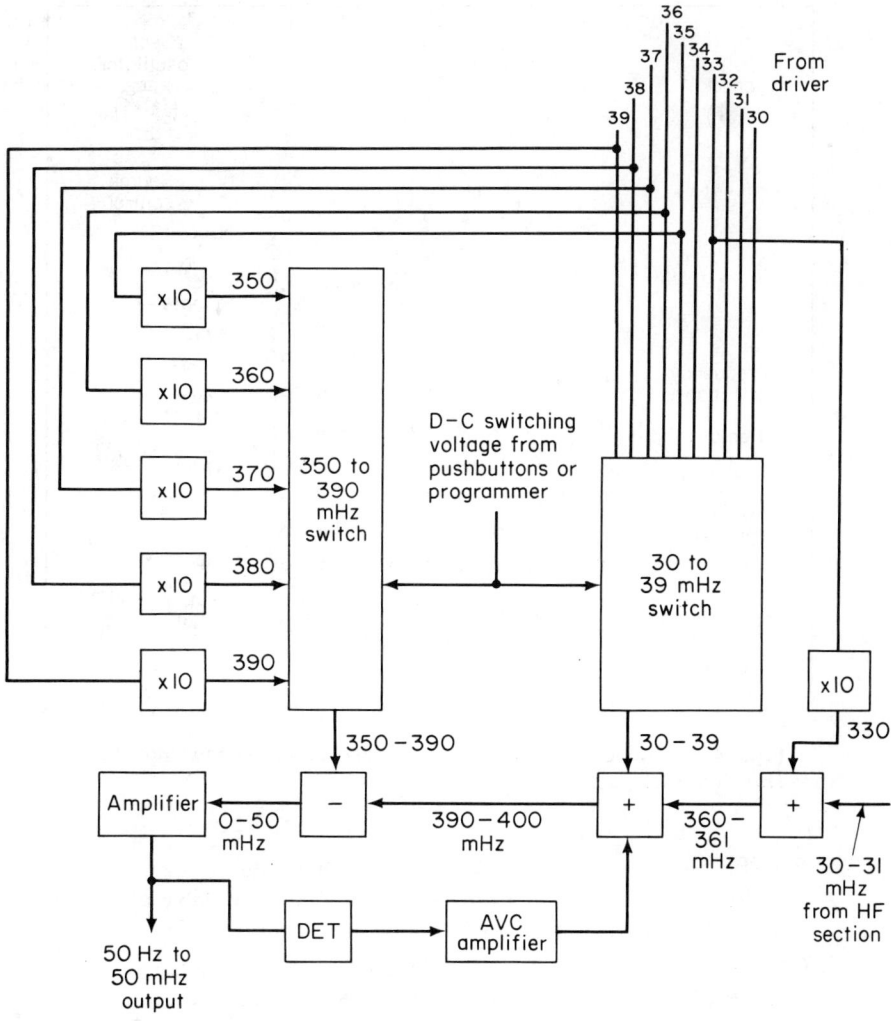

Fig. 4-33. Block diagram of synthesizer UHF section (Hewlett-Packard).

the r-f output would be continuously variable from 12,345,678.00 Hz to 12,345,678.90 Hz.

4-6.3. Typical Frequency Synthesizer Controls and Indicators

Figure 4-34 shows the front-panel controls of a typical frequency synthesizer. The following are descriptions of the control functions.

138   Signal Generators

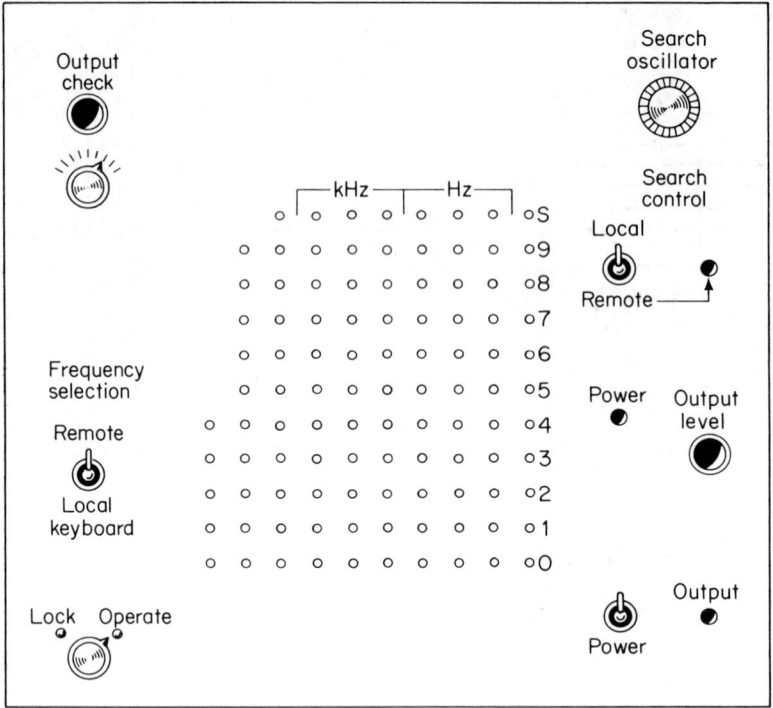

Fig. 4-34. Typical frequency synthesizer controls and indicators (Hewlett-Packard).

1. POWER. Applies or removes power to the unit.
2. OUTPUT. The r-f output from the synthesizer is available at this connector. The output connector accommodates a coaxial cable.
3. OUTPUT LEVEL. Sets level of r-f output.
4. PUSHBUTTONS. Permit selection of r-f output frequency.
5. LOCK/OPERATE switch. In LOCK, all pushbuttons are locked and cannot be pressed. This prevents accidental change in frequency. In OPERATE, any frequency can be selected, but no more than one pushbutton in each vertical column can be selected (at one time). If two pushbuttons in the same vertical column are pressed simultaneously, neither button will hold.
6. FREQUENCY SELECTION. In LOCAL KEYBOARD, frequency selection is determined by the front-panel pushbuttons. In REMOTE, frequency is determined by voltages from an external source, such as a programmer or remote keyboard.
7. OUTPUT CHECK. This meter permits internal circuits to be checked, as selected by the corresponding switch.

8. SEARCH OSCILLATOR. Provides for variable frequency output. When SEARCH CONTROL switch is in LOCAL, frequency is variable (depending on selection of S pushbuttons and position of SEARCH OSCILLATOR dial). In REMOTE, frequency is variable (depending on value of external voltage applied to REMOTE connector).

4-6.4. Frequency Synthesizer Operating Procedures

Operation of a frequency synthesizer is usually described quite thoroughly in the operating manual. Therefore, no detailed operating procedures will be provided here.

4-6.5. Frequency Synthesizer Test and Calibration

A frequency synthesizer can be considered an r-f signal generator and can be tested as described in Section 4-2. However, a frequency synthesizer is usually considered a frequency standard and should be tested accordingly. Refer to Chapter 8 for data on calibrating frequency standards.

4-7. PULSE GENERATORS AND DIGITAL DELAY GENERATORS

A pulse generator is quite similar to the square-wave generator described in Section 4-5. However, there are two basic differences. First, a laboratory pulse generator will have a repetition rate (or output frequency) well beyond the audio range. Many laboratory pulse generators have an output up to 50 MHz. The second fundamental difference between pulse and square-wave generators concerns the *duty cycle*. Square-wave generators have equal on and off periods (or equal positive and negative periods), even though the repetition rate and amplitude may be varied.

On the other hand, the duration of a pulse generator on period is independent of pulse repetition rate. The duty cycle of a pulse generator can be made quite low so that these instruments are generally able to supply more power during the on period than square-wave generators. Short pulses (even of higher amplitude) reduce power dissipation in a component or system under test. For example, measurements of transistor gain are made with high-amplitude pulses short enough to prevent junction heating (and the consequent effect of heat on transistor gain).

In a typical laboratory pulse generator, the pulse duration (or width), pulse amplitude, pulse polarity, and repetition rate are all controllable.

A *digital delay generator* is a form of pulse generator. Generally, a digital delay generator is for highly specialized applications where two pulses are required. These pulses must be of a given amplitude, duration, polarity, and repetition rate, and there must be a precise time interval between the pulses. In some cases the time interval is fixed, while in other cases the interval can be varied, even to the point of superimposing the two pulses.

Because of the great variety of laboratory pulse generators available, and because many of these are special purpose instruments, no typical circuits or typical operating controls will be given here. Likewise, no test procedures will be described. Generally, the instruction manuals for laboratory pulse generators are quite thorough. In brief, a pulse generator is tested in a similar manner to that of a square-wave generator, as described in Section 4-5. The pulse output of the generator is displayed on an oscilloscope. This permits the pulse to be checked for shape, amplitude, duration, and repetition rate. The basic procedures for making these measurements are described in Chapter 5.

In the remaining paragraphs of this section, the general techniques required for use of pulse generators with related test equipment are described. For example, in almost all applications, a pulse generator is used in conjunction with an oscilloscope so that the pulse wave form may be observed (either at the input of the circuit under test, at the output, or both). Therefore, it is important to understand the interconnections between pulse generators and oscilloscopes. Such data are often omitted from the operating-instruction manuals for pulse generators.

4-7.1. Pulse Definitions

The following terms are commonly used in describing pulse characteristics. The terms are illustrated in Figure 4-35. The input pulse represents an ideal input wave form for comparison purposes. The other wave-forms represent typical output wave-forms. To show the relationships, the terms are defined as follows:

Rise time t_r is the time interval during which the amplitude of the pulse voltage changes from 10 to 90 per cent of the rising portion of the pulse. (Note that the rising portion may be negative, as well as positive.)

Fall time t_f is the time interval during which the amplitude of the pulse voltage changes from 90 to 10 per cent of the falling portion of the waveform.

Time delay t_d is the time interval between the beginning of an input pulse (time zero) and the time when the rising portion of the output pulse attains an arbitrary amplitude of 10 per cent from the base line.

Storage time t_s is the time interval between the end of an input pulse (trailing edge) and the time when the falling portion of the output pulse drops to an arbitary amplitude of 90 per cent from the base line.

Pulse width (or pulse duration) t_w is the time duration of the pulse measured between two 50 per cent amplitude levels of the rising and falling portions of the wave-form.

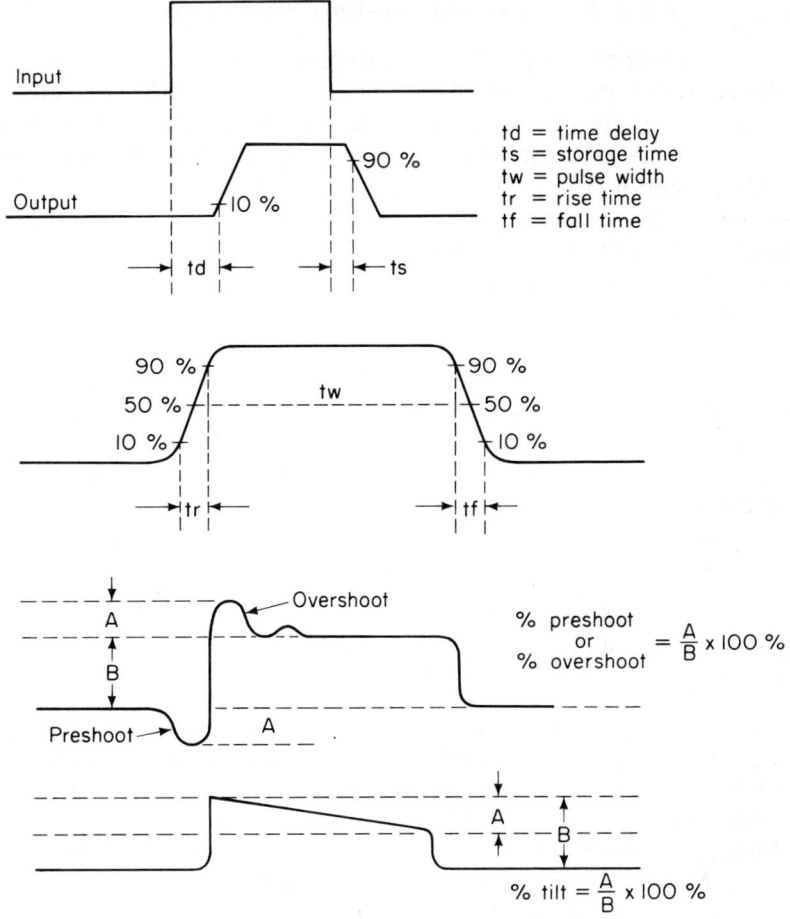

Fig. 4-35. Pulse definitions.

Tilt (also known as sag or droop) is a measure of the tilt of the full amplitude (flat-top) portion of a pulse. The tilt measurement is usually expressed as a percentage of the amplitude of the rising portion of the pulse.

Overshoot is a measure of the overshoot generally occuring above the 100 per cent amplitude level. This measurement is also expressed as a percentage of the pulse rise.

Preshoot is a measure of the preshoot generally occuring below the base line (or zero line). This measurement is usually expressed as a percentage of the 100 per cent amplitude level.

These definitions are for guide purposes only. When pulses are very irregular (such as excessive tilt, droop, sag, preshoot, overshoot, ringing, oscillation, and so on), the definitions may become ambiguous.

4-7.2. Rise-Time and Fall-Time Measurements

Rise-time and fall-time measurements are of special importance in pulse generator testing. The relationship between the pulse generator rise time and the rise time of the device in test must be taken into account. The accuracy of rise-time measurements can be no greater than the rise time of the pulse generator. For example, if a pulse generator with a 20-nsec rise time is used when measuring the rise time of a 15-nsec oscilloscope, the measurements would be hopelessly inaccurate. Likewise, if the same pulse generator and oscilloscope were used to measure the rise time of another system, the fastest rise time for accurate measurement would be something greater than 20 nsec.

As a rule of thumb, if the equipment or signal being measured has a rise time 10 *times slower* than the pulse generator and the oscilloscope, the error will be 1 per cent or less. This amount is small and can be considered negligible. If the equipment under test has a rise time 3 *times slower* than the test equipment, the error is about 5 or 6 per cent.

4-7.3. Rules for Using Pulse Generators

In addition to the General Safety Precautions described in the Introduction, the following rules should be followed when using pulse (and square-wave) generators.

1. Use proper cables, terminals, attenuators, and impedance-matching networks. It is important that cables used for test setup match the impedance of the pulse generator and that they be terminated in their own characteristic impedance. This will prevent reflections and standing waves. (Refer to Chapters 10 and 11 for a discussion of reflections and standing waves.)
2. Keep unshielded wires of uncertain impedance short so that reflection and/or cross-coupling effects are not introduced. Keep all ground return paths short and direct.
3. Shield measuring equipment leads to prevent undesired coupling to other parts of the circuit. Shielding is especially required where pulse radiation is a problem and where high-impedance dividers or circuits are involved.
4. If for some reason it is not possible or practical to terminate the output of a pulse generator in its characteristic impedance and an impedance-matching network cannot be used, one possible solution is to use a long coaxial cable between the pulse generator and the load. This will delay the load's reflection until after the time of interest. The reflection will appear at a time equal to twice the output lead delay plus the pulse length.

4-7.4. Impedance-Matching Pulse Generator Output

If it becomes necessary to match a pulse generator to a unit or circuit with a different impedance characteristic, observe the following procedure. A simple resistive impedance-matching network that provides minimum attenuation is shown in Figure 4-36, together with the applicable equations. Note that the matching network can be located at the pulse generator end of the interconnecting cable or at the circuit end. However, there is usually less attenuation and signal radiation if the matching network is placed at the circuit end. This requires that the interconnecting cable be of the same impedance as the pulse generator, not the circuit under test. If the matching network must be placed at the pulse generator end, then the interconnecting cable must be of the same impedance as the circuit under test.

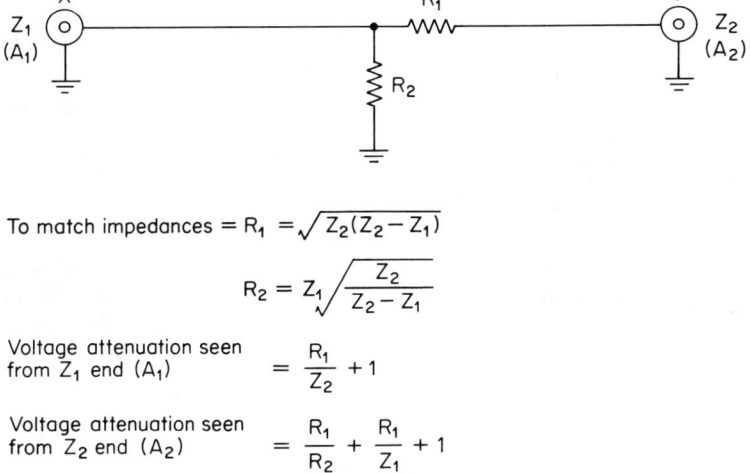

To match impedances $= R_1 = \sqrt{Z_2(Z_2 - Z_1)}$

$$R_2 = Z_1 \sqrt{\frac{Z_2}{Z_2 - Z_1}}$$

Voltage attenuation seen from Z_1 end (A_1) $= \dfrac{R_1}{Z_2} + 1$

Voltage attenuation seen from Z_2 end (A_2) $= \dfrac{R_1}{R_2} + \dfrac{R_1}{Z_1} + 1$

Fig. 4-36. Resistive impedance-matching network.

For example, to match a pulse generator with a 50-Ω output impedance to a 125-Ω system, connect the pulse generator to a 50-Ω interconnecting cable, connect the cable to connector X of the matching network, then connect the Y connector to the circuit under test.

As shown by the equations of Figure 4-36, the values of R_1 and R_2 would be

$$R_1 = 125\sqrt{(125 - 50)} = 96.8 \ \Omega$$

$$R_2 = 50\sqrt{\frac{125}{125 - 50}} = 64.6 \ \Omega$$

Although the network of Figure 4-36 provides a minimum attenuation for a purely resistive impedance-matching device, the attenuation seen from one end does not equal that seen from the other end. A signal applied from the lower-impedance source (Z_1) encounters a voltage attenuation A_1 that may be determined by the equations of Figure 4-36.

$$A_1 = \frac{96.8}{125} + 1 = 1.77 \text{ attenuation}$$

A signal applied from the higher-impedance source Z_2 will produce an even greater voltage attenuation A_2 that may be determined by the equations of Figure 4-36.

$$A_2 = \frac{96.8}{64.6} + \frac{96.8}{50} + 1 = 4.44 \text{ attenuation}$$

4-8. FM STEREO GENERATOR

The monaural portion of an FM stereo receiver can be aligned and tested by using a meter or an oscilloscope. (The detailed procedures are described in the author's *Handbook of Meters: Theory and Application* and the *Handbook of Oscilloscopes: Theory and Application*, published by Prentice-Hall, Inc.)

The stereo portion of the FM stereo receiver requires a stereo generator. (Some FM generators incorporate both monaural and stereo test circuits.) A stereo generator produces signals that simulate the signals transmitted by an FM station during the broadcast of a stereo program. Therefore, it is necessary to understand the FM stereo system to understand fully the basic operating principles of an FM stereo generator.

4-8.1. Basic FM Stereo Transmission

Monaural FM signals are produced by using the audio signal to frequency-modulate an r-f carrier. Stereo FM, however, must contain two separate audio signals, comprising the left and right channels. The two-channel audio information has to be transmitted in a matter that enables it to be received as a stereo signal by a stereo FM receiver or as a monaural signal by a monaural FM receiver. A technique called *multiplexing* is used to obtain this stereo FM signal. Multiplexing permits additional signals to be transmitted on an r-f carrier through the use of a *modulated subcarrier*.

Figure 4-37 shows the composition signal that frequency-modulates the r-f carrier to produce a stereo FM signal.

The $L + R$ portion of the stereo multiplex signal is formed by combining the left and right channels in phase. This $L + R$ signal corresponds to a regular monaural audio signal with a frequency range of 50 Hz–15 kHz. It is this portion of the stereo FM signal that is received and detected as a monaural FM transmission by a monaural FM receiver.

The left and right audio signals are also combined with the right signal shifted 180° out of phase. This, in effect, subtracts the right signal from the left signal, producing a difference signal called $L - R$. This $L - R$ information is used to *amplitude-modulate* a 38-kHz subcarrier in such a mannaer that two sidebands are formed, with the 38-kHz subcarrier suppressed or removed. These sidebands containing the $L - R$ information extend from 23 to 53 kHz, as shown in Figure 4-37.

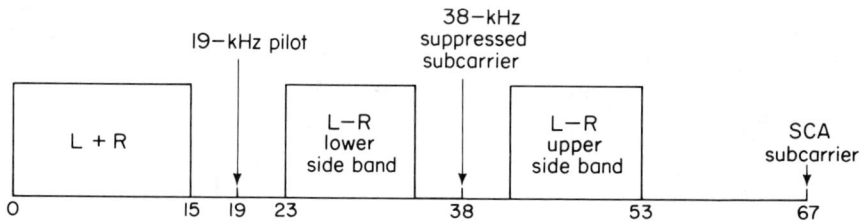

Fig. 4-37. Frequency distribution of composite stereo FM signal.

An unmodulated 19-kHz pilot subcarrier is also included in the stereo signal. This pilot signal is used in the stereo FM receiver to synchronize the stereo detector circuit. Note that the 38-kHz subcarrier is the second harmonic of this 19-kHz pilot signal.

In both monaural and stereo FM transmission, some stations also include "storecasting" or SCA (subsidiary carrier assignment) information. The SCA modulation extends from 60 to 74 kHz, with a center subcarrier frequency of 67 kHz. This SCA signal must be removed or "trapped" in a stereo FM receiver since it would interfere with the operation of the stereo detector circuit.

Thus, although a monaural FM transmission consists simply of an r-f carrier frequency-modulated by the audio program material, a stereo FM transmission consists of an r-f carrier frequency-modulated with a complex wave-form that includes an $L + R$ or monaural signal, two $L - R$ sidebands around a 38-kHz suppressed carrier, a 19-kHz pilot signal, and possibly a 67-kHz SCA signal.

4-8.2. Basic FM Stereo Receiver

An FM stereo receiver is a conventional FM receiver with an additional audio amplifier (two channels for stereo) and a multiplex circuit (used to demodulate and break the composite signal back down to left and right channels). The basic multiplex and amplifier circuit is shown in Figure 4-38.

The composite FM signal is detected in the regular way and fed into the multiplex section. Here the 19 kHz signal is picked off and amplified. The 19-kHz signal is then doubled to 38 kHz (in some receivers) or is used to control a 38-kHz oscillator (in other receivers). The signal from the 38-kHz oscillator

146   Signal Generators

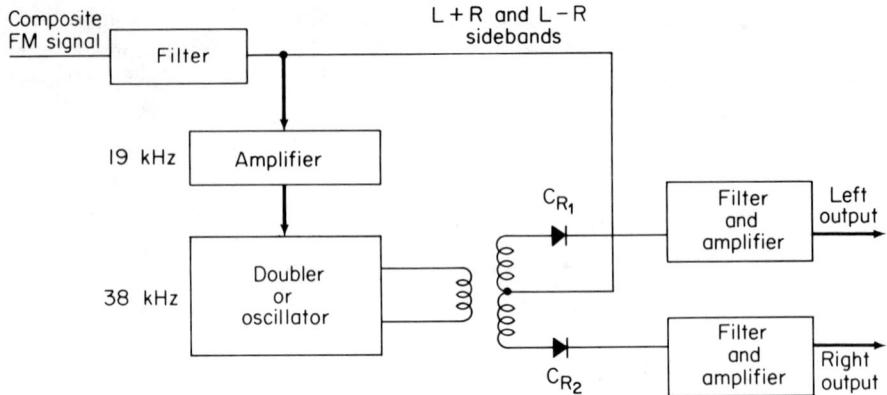

Fig. 4-38. Basic multiplex and amplifier circuit for FM stereo receiver.

or doubler is fed to a diode demodulator. The $L - R$ and $L + R$ signals are fed to the center tap of the transformer secondary at the output of the demodulator. These signals are mixed with the 38-kHz signal and combined to form the left and right channels. In effect, the signal at the multiplex diodes CR_1 and CR_2 is the 38-kHz carrier, amplitude-modulated by the left and right information (with left being 180° out of phase in relation to right). This amplitude-modulation data is detected by diodes CR_1 and CR_2 in the normal manner and amplified by separate left and right amplifiers, then applied to the left and right loudspeakers.

4-8.3. Typical FM Stereo Generator Circuit

Figure 4-39 is the block diagram of a typical FM stereo generator. Operation of the unit is as follows.

The FM stereo generator is essentially a miniature FM stereo transmitter, providing both left and right signals plus a 19-kHz pilot signal.

Transistor Q_1 is a crystal-controlled oscillator used to control multiplexing of the generator. The 76-kHz output of Q_1 is applied to shaping stage Q_2, where the signal is shaped into pulses and then applied to a 38-kHz multivibrator Q_3 and Q_4.

The 38-kHz multivibrator divides the frequency of the 76-kHz signal by one half to 38 kHz, and produces symmetrical square waves 180° out of phase with each other. These square waves are used to control the left and right signal switches Q_9 and Q_{10}.

Transistor Q_6 is a phase-shift oscillator, producing a 1-kHz signal that is

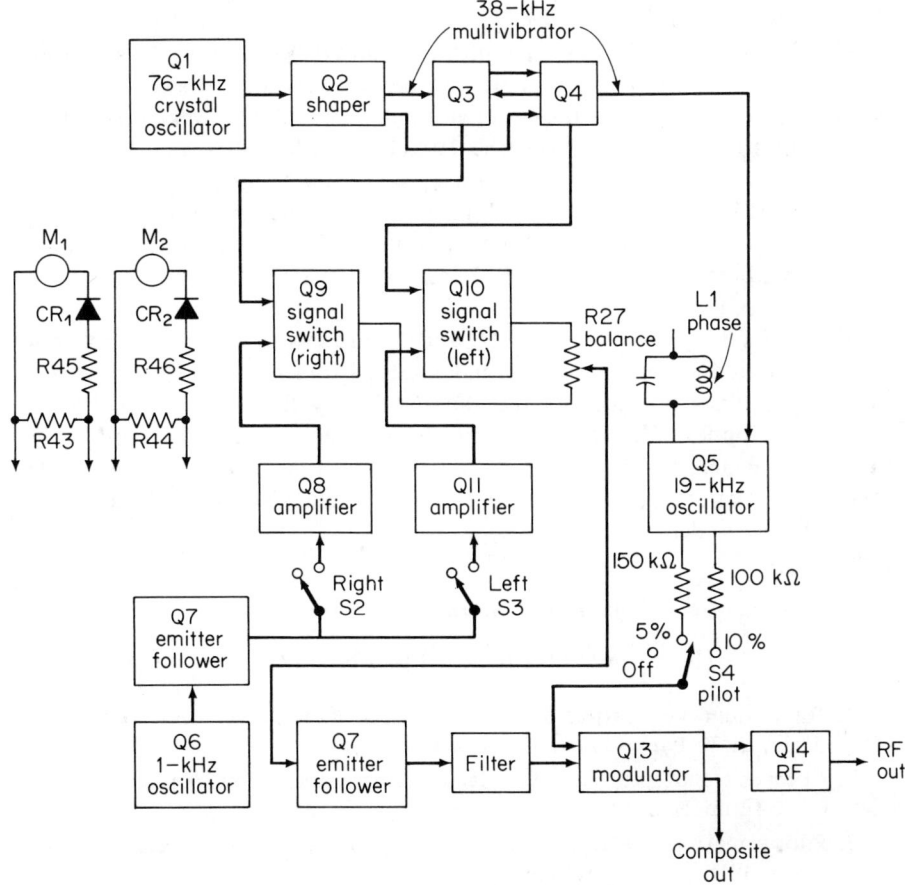

Fig. 4-39. Typical FM stereo generator circuit (Sencore).

applied to emitter follower Q_7. The output of Q_7 is applied through the LEFT and RIGHT switches S_2 and S_3 to emitter followers Q_8 and Q_{11}.

When S_2 and S_3 are both on OFF, supply voltage is removed from Q_6 and Q_7 to prevent interference from the 1-kHz signal. Transistor Q_8 couples the RIGHT signal in the right signal switch Q_9, and Q_{11} couples the LEFT signal into the left signal switch Q_{10}. Transistors Q_9 and Q_{10} turn the left and right signals off and on at the 38-kHz rate, thus producing the multiplex signal. This signal is taken from the center arm of the balance control R_{27} and is coupled to balanced emitter follower Q_{12}. The output of Q_{12} is coupled to a filter (to remove any 114-kHz signals) and is applied to modulator Q_{13}.

The 19-kHz pilot oscillator Q_5 is controlled by a pulse from one side of the 38-kHz multivibrator (Q_3 and Q_4). The phase relationship between the

19- and 38-kHz signals is adjusted by L_1. The output of Q_5 is coupled through PILOT switch S_4 to modulator Q_{13}.

The PILOT switch S_4 has three positions: in OFF position, Q_5 is disabled to remove the 19-kHz pilot signal; in 5 per cent position, the 19-kHz pilot signal is passed through a 150-KΩ resistance to simulate a weak stereo pilot signal; in 10 per cent position, the 19-kHz pilot signal is passed through 100-kΩ resistance to simulate a stronger stereo pilot signal.

Both the 19-kHz pilot signal and the multiplex signal are applied to modulator Q_{13}, which serves to frequency-modulate r-f oscillator Q_{14}. The output of Q_{14} is adjustable across the FM broadcast range (88–108 MHz) or a portion of the range.

The LEFT and RIGHT meters M_1 and M_2 are separate from the remaining circuits. Both are identical a-c voltmeters. Resistors R_{43} and R_{44} are used as loads in the absence of loudspeakers so that there will always be a load on the receiver amplifier. The receiver output signals are rectified by diodes CR_1 and CR_2, which supply the necessary direct current to operate meters M_1 and M_2. Resistors R_{45} and R_{46} are selected so that 3 V RMS will deflect the meter to full scale.

4-8.4. Typical FM Stereo Generator Outputs

The outputs of a typical FM stereo generator are as follows:

1. Radio-frequency output tuned to 100 MHz (adjustable from 90 to 105 MHz) at 3000 μV (300-Ω impedance).
2. Composite multiplex signal (available at front-panel jacks) 2 V peak-to-peak (1000-Ω impedance).
3. Pilot signal of 19 kHz (also available at panel jacks) 0.2 V peak-to-peak (when 10 per cent is selected).
4. Both the multiplex signal and 19-kHz signals modulate the r-f output signal.

4-8.5. Typical FM Stereo Generator Controls and Indicators

Figure 4-40 shows the front-panel controls and indicators of a typical FM stereo generator. The following are descriptions of the control functions:

1. PILOT switch. In OFF position, 19-kHz pilot signal is removed, but rf and multiplex remain on for monaural reception. In 5 per cent position, 19-kHz signal is present but at a reduced level to simulate a weak FM station. In 10 per cent position, the 19-kHz signal is at a level specified by the FCC for standard FM stereo broadcasting.
2. OFF-ON switch. Controls power to the generator.

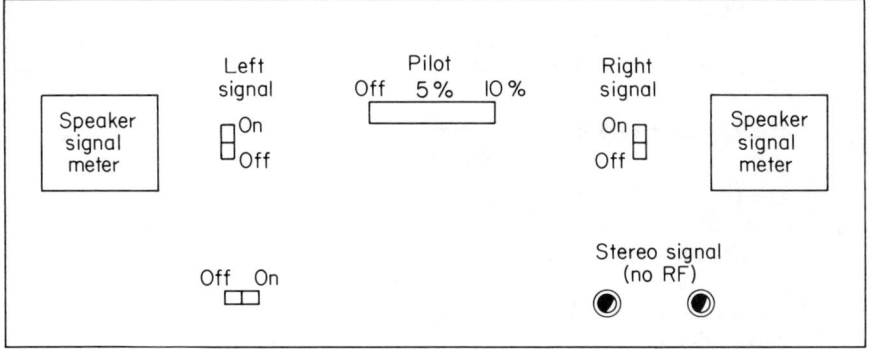

Fig. 4-40. Typical FM stereo generator controls and indicators (Sencore).

3. STEREO SIGNAL output jacks (no rf). Provides a composite multiplex signal plus pilot signal, but without rf, for injection directly into the FM detector to isolate stereo troubles.
4. LEFT and RIGHT SPEAKER SIGNAL meters. These meters monitor the left and right outputs of the audio amplifiers in the stereo receiver. These meters take the place of the loudspeakers or can be used with the loudspeakers. The meters are calibrated in LO, GOOD, or HI separation and also in decibels.
5. LEFT SIGNAL and RIGHT SIGNAL switches. These switches control 1-kHz audio modulation of the left and right channels, respectively.

4-8.6. FM Stereo Generator Operation

The operating procedures for FM stereo generators are not as standardized as those of generators previously described. This is because the circuitry used to achieve the various outputs differ considerably from model to model. For that reason, it is recommended that the readers make direct reference to the instruction manual for their particular make or model. However, the following is a summary of typical FM stereo generator operating procedures.

1. The r-f output of the generator is connected to the antenna input of the receiver. The receiver is tuned to the operating frequency of the generator (between 88 and 108 MHz).
2. The channel meters are connected at the loudspeaker output at the receiver. The loudspeakers can be disconnected if desired.
3. The r-f output is modulated by the multiplex audio tone but not by the 19-kHz pilot signal. This checks monaural reception.
4. The r-f output is then modulated by both the multiplex and 19-kHz pilot

signals. This checks stereo reception. The STEREO indicator lamp on the receiver should light, and a stereo signal should be heard in both loudspeakers and/or monitored on both meters. The meter indications should be the same for both channels.

5. The 19-kHz pilot signal is reduced in strength. This simulates a weak stereo signal. Stereo reception should remain the same in a good receiver.
6. The 19-kHz pilot signal is restored to full strength. The LEFT audio modulation is removed. Under these conditions, the RIGHT channel should show normal reception, while the LEFT channel should (theoretically) drop to zero. Any signal in the LEFT channel is due to poor stereo separation in the receiver. The RIGHT channel can be checked in the same manner (Right channel modulation off, LEFT modulation on).
7. To isolate receiver troubles to either the regular monaural circuits (rf, if, limiter, detector, audio, and so on) or the multiplex circuits, most stereo generators provide a composite 19-kHz pilot signal and multiplex signal separate from the r-f output. This composite output can be fed directly to the receiver multiplex circuit (since the composite is essentially the same signal that the multiplex circuit sees in normal operations, detected with no rf). If the multiplex circuit works properly with direct signal injection, the receiver trouble is most likely in the monaural section.

4-8.7. FM Stereo Generator Test and Calibration

The procedures for test and calibration of FM stereo generators differ considerably with each model and manufacturer. Therefore, no attempt has been made to cover specific procedures. In general, the r-f and composite signals are checked for frequency, uniformity of output, amplitude, and so on, as are corresponding r-f and audio generators. The FM modulation can be checked by using a spectrum analyzer (Chapter 10). However, this is usually not required. In general, the most critical test and calibration procedure involved is checking the phase relationship between the 19-kHz pilot signal and the multiplex signal. This relationship determines synchronization between the left and right channels.

4-9. COLOR TELEVISION SIGNAL GENERATORS

Color television signal generators produce color signals that simulate those transmitted by color television stations. These signals are applied to color television receivers and provide a means of checking the receiver's response to the signals. Some color television generators include circuits for test and service of the black and white portion of the television receiver, as well as the color portion; that is, such generators provide sweep and marker outputs in addition to color signals. Other color generators provide only those signals necessary for service of the color circuits in the television receiver.

There are two basic types of color television generators: these are the rainbow generator and the NTSC (National Television Systems Committee) generator. The rainbow generator produces signals that appear as a rainbow of 10 vertical bars or stripes on the receiver screen. Most rainbow generators also supply a dot pattern, crosshatch pattern, horizontal lines, vertical lines, and possibly a gray scale display. The same is true of the NTSC generators, except they they supply individual colors instead of (or in addition to) the rainbow display.

To understand operation of any color generator it is also necessary to understand operation of the color television system, both the transmitted signals and the receiver circuits. The following paragraphs describe the color system. It is not intended that these paragraphs provide a complete course in color television, but a summary.

4-9.1. Make-up of Color Television Signal

The color signal is made up of two components: brightness and color information. The brightness portion contains all the information pertaining to the details of the picture and is commonly referred to as the Y or luminance signal. The color portion is called the chrominance signal and contains the information pertaining to the hue and saturation of the picture.

Figure 4-41 shows the basic color television transmitting system. The

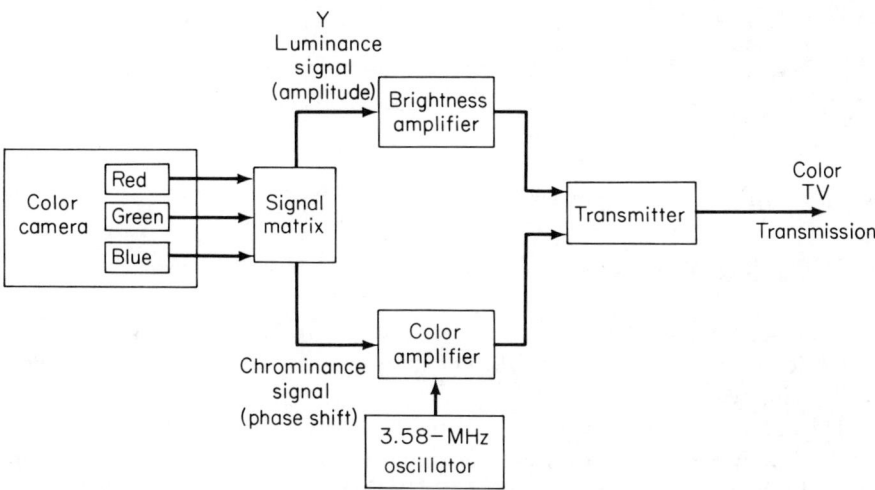

Fig. 4-41. Basic color TV transmitting system.

camera in Figure 4-41 contains three separate image pickup tubes, one tube for each of three colors, red, green, and blue. The camera tubes divide the scene being scanned into three colors (red, yellow, and blue). The relative intensity of the scene is also contained in this signal. The output signal from the camera contains the luminance and chrominance information.

The signal is separated and the individual components are amplified. The color information (chrominance) is impressed on a 3.58-MHz subcarrier (in the form of phase shift) and is then transmitted together with the Y signal (which is in the form of instantaneous amplitude shift or level). This method provides a compatible signal that can be reproduced on black and white as well as color receivers (the black and white receivers require only the luminance or Y signal).

A compatible color signal must fit within the standard 6-MHz television channel and have horizontal and vertical scanning rates of 15,750 and 60 Hz plus a video bandwidth not in excess of 4.25 MHz. The color information is interleaved with the video information and transmitted within the 6-MHz television channel on the 3.58-MHz subcarrier, as shown in Figure 4-42.

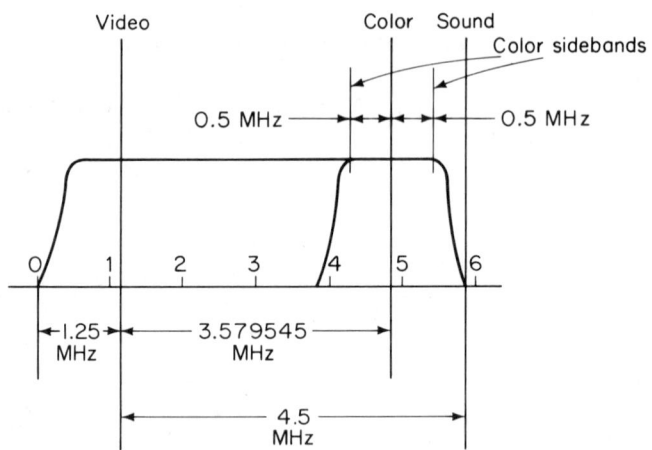

Fig. 4-42. Location of color subcarrier in 6-MHz TV channel.

The location of the color subcarrier was selected to be 3.579545 MHz above the video carrier, since this would not interfere with the reception on black and white receivers.

The color signals are synchronized by transmitting approximately 8 cycles of the 3.58 MHz subcarrier oscillator signal (of the correct phase) along with the color signal. This signal is called the *burst* sync signal and is added to the "back porch" of the horizontal blanking bar, as shown in Figure 4-43.

Color Television Signal Generators   153

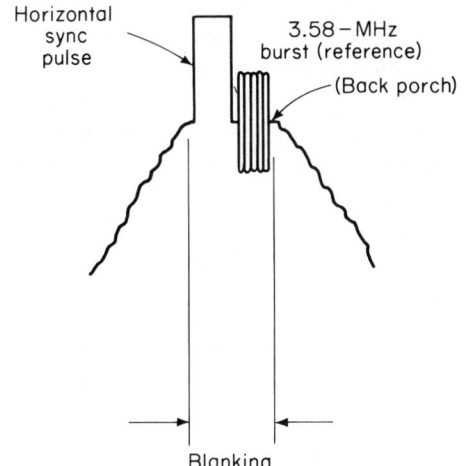

Fig. 4-43. Location of 3.58-MHz reference burst on horizontal blanking signal.

The location of the burst signal will not interfere with horizontal sync on black and white sets, since it is part of the horizontal blanking bar.

4-9.2. Black and White Versus Color Signals

A black and white presentation requires only that the amplitude or luminance (or Y) signal be transmitted. A color broadcast also requires that phase-shift information (representing colors) be transmitted. In color television, the hue of color is determined by the phase angle difference between the reference burst and the phase of the color signal, while the intensity (sometimes known as saturation) of the color is determined by the instantaneous amplitude of the signal.

Figure 4-44 shows the phase and amplitude required to produce each of six basic colors. For example, at any given instant, if the amplitude is at 0.64

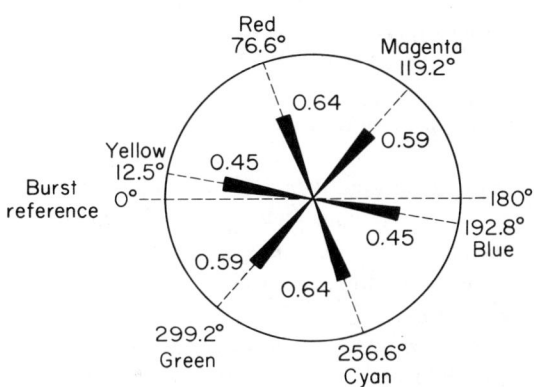

Fig. 4-44. Phase and amplitude relationships of NTSC color signals.

(with an amplitude of 0 for black and 1.0 for white) and the phase of the 3.58-MHz signal is 76.6° (with the burst signal considered to be at 0°), the receiver picture tube will produce pure red. If the amplitude is changed to 0.45 and the phase to 192.9°, a pure blue will be produced. If the amplitude is changed to 0.59 and the phase to 119.2°, the picture tube will produce magenta (which is a combination of red and blue). If the phase of the 3.58-MHz signal were swept from 0 through 360°, a complete rainbow of colors would result.

4-9.3. Color Television Receiver Operation

Figure 4-45 is the block diagram of a typical color receiver. As shown, the circuits are very similar to those of a black and white set, with two exceptions: the luminance or Y channel and the chrominance channel.

4-9.3.1. Luminance Channel Operation

The luminance or Y channel consists of the detector, first video amplifier, delay line, and second video amplifier. (Note that a separate sound detector is used prior to the video detector in most color receivers. This provides good separation of the sound and video signals.) The composite video signal is detected and amplified in the first stage of the luminance channel. A portion of this signal is fed to the sync and color stages. The rest of the signal continues on to the delay line. This delay is required so that the luminance information (amplitude) arrives at the picture tube at the same time as the color information (phase relationship). The narrow bandwidth of the color circuit causes the chrominance signal to take longer to reach the picture tube. Additional amplification is given to the luminance signal in the second video amplifier before it is applied to the cathodes of the picture tube.

4-9.3.2. Chrominance Channel Operation

The composite video signal is fed to the bandpass amplifier (sometimes called the color IF) from the first video amplifier. The chrominance signal is separated from the composite color signal, is amplified, and is fed to the grids of the demodulators. The burst amplifier removes the burst signal from the chrominance signal and applies the burst signal to the color phase detector.

A color killer stage is used to prevent high-frequency signals in the vicinity of 3.58 MHz from passing through the color circuit during the reception of black and white telecasts. In the absence of a color broadcast, the color killer biases the bandpass amplifier off, thus disabling all of the chrominance channel circuits.

The 3.58-MHz oscillator provides a locally generated reference signal for demodulation of the composite color signal. The color phase detector com-

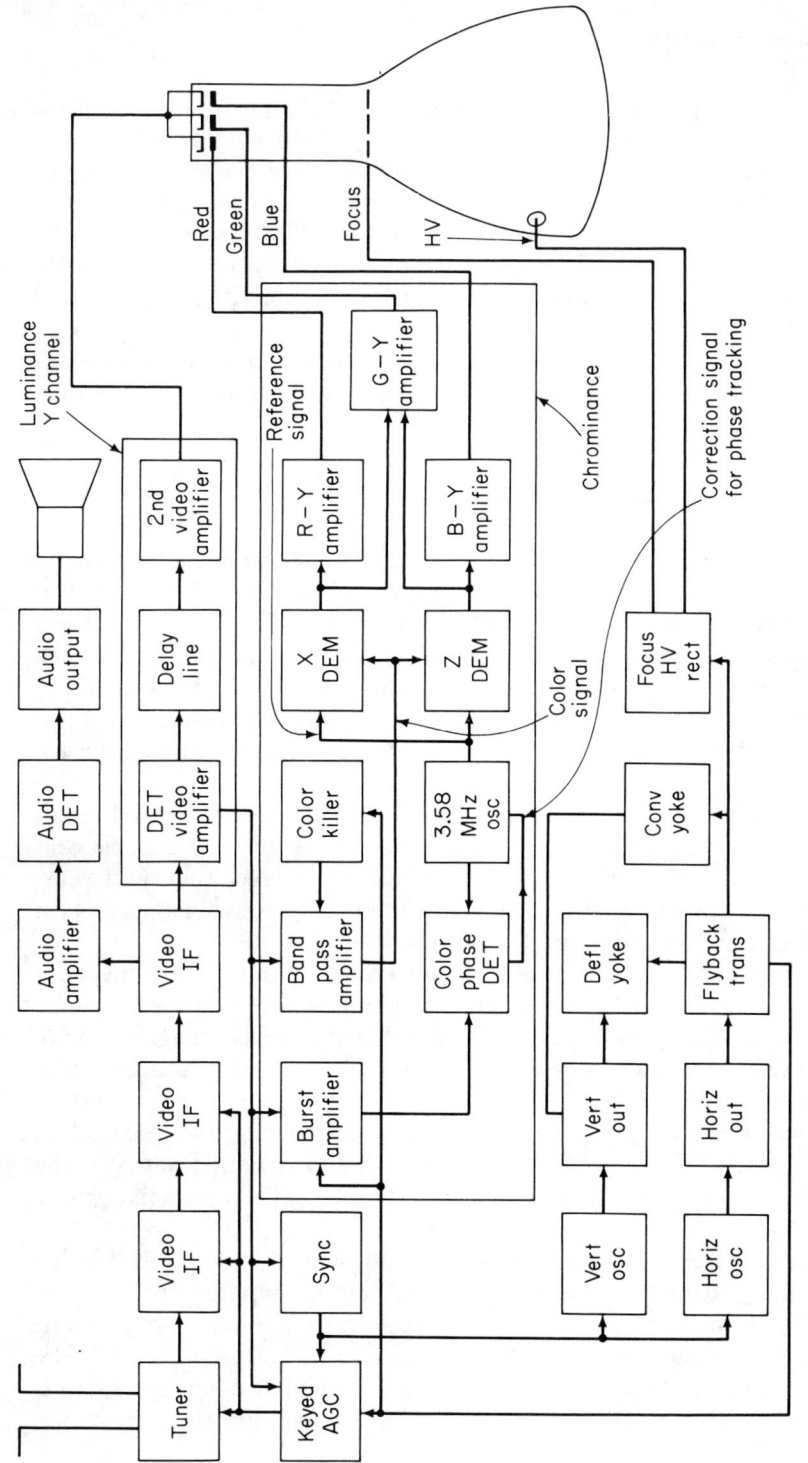

Fig. 4-45. Typical color receiver circuit.

pares the 3.58-MHz oscillator signal with the burst signal and develops a correction voltage to keep the oscillator locked in phase with the burst signal.

The X and Z demodulators detect the amplitude and phase variations of the chrominance signal to recover color information. At any given instant, the X and Z demodulators receive two signals: a 3.58-MHz signal from the reference oscillator (locked in phase to the reference burst) and a 3.58-MHz signal from the bandpass amplifier (at a phase and amplitude representing the color at that instant).

The R-Y (red-luminance), B-Y (blue-luminance), and G-Y (green-luminance) amplifiers amplify the color signals and apply them to the picture-tube grids.

4-9.4. Make-up of Colors

Color picture tubes produce most colors by mixing three colors (red, blue, and green). (Operation of the color picture tube is described in later sections.) Any color can be created by the proper blending of these three colors; that is, any color can be created if the three colors are present, each at a given level of intensity or brightness. This is due to a characteristic of the human eye. When the eye sees two different colors of the same brightness level, they appear to be of different brightness levels. For example, the eye is more sensitive to green and yellow than to red or blue.

The combination of any two primary colors will produce a third color. In color television, the process of mixing colors is *additive* and is dependent on the self-illuminating properties of the picture-tube screen (not on the surrounding light source, as found with the subtractive color-mixing process familiar to paints and printing).

If two primary colors, such as red and green, are combined, they will produce a secondary color of yellow. The combination of green and blue makes cyan, and the combination of blue and red produces magenta. Also, a secondary color added to its complementary primary will produce white. For example, yellow + blue = white, cyan + red = white, and magenta + green = white. Therefore, any color can be produced if the three colors (red, blue, and green) are mixed in the proper proportions (that is, if the amplitudes of the three signals at the picture-tube grids are at the proper levels).

For the purposes of our discussion, it is sufficient to understand that the three picture grids will receive signals of the correct amplitude proportions (identical to the proportions existing at the color camera for any given instant) provided that (1) the receiver 3.58-MHz oscillator is locked in phase to the burst signal and (2) the demodulators receive a 3.58-MHz color signal that is of a phase and amplitude corresponding to the color proportions existing at the color camera.

4-9.5. Color Picture Tubes

Figure 4-46 illustrates a color picture tube and its related components. The color picture tube is a cathode-ray tube, similar to that used for black and white receivers and for oscilloscopes. (Refer to Chapter 5 for a further discussion of cathode-ray tubes.) However, the color picture tube has three complete electron guns, a shadow mask, and a three-color phosphor dot screen within the vacuum glass envelope.

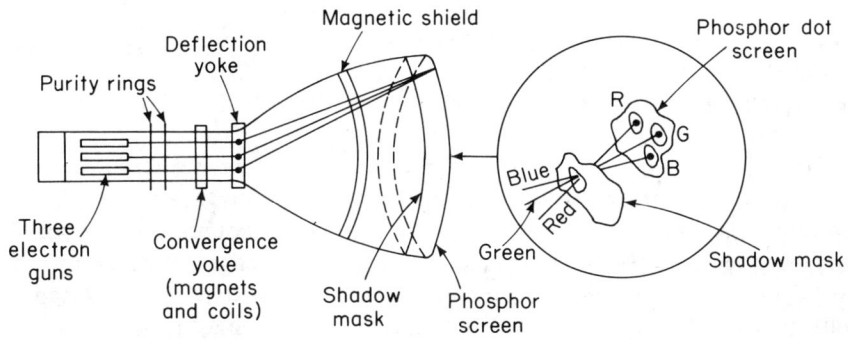

Fig. 4-46. Color picture tube.

The *screen* consists of several million phosphor dots of the primary television colors (red, green, and blue). These dots are arranged in trios and produce colored light when struck by electrons. Even though the dots are placed next to each other, they appear (to the human eye) to be superimposed so that the colors are (apparently) blended.

The *shadow mask* is a thin plate of metal that has a number of small holes, each centered over a phosphor dot trio. This mask is positioned in front of the screen and controls the landing of the electron beams on the phosphor dots.

The *three electron guns* (one for each color) are positioned approximately 120° apart, and are precisely aimed so the individual electron beams will pass through the shadow mask at different angles to strike their corresponding phosphor dots. The control grid controls the amount of light leaving the gun, and thus regulates the amount of colored light emitted from the phosphor screen.

4-9.5.1. Color Purity

For the color picture tube to reproduce color pictures correctly, the individual electron beams must strike phosphors of only one color. Each electron beam is then capable of producing a pure field of either red, blue, or green. *Purity* refers to the uniformity of the hue and brightness over the entire area of each color field and over the entire area of the combination of all three. For exmple, red is pure when the red display is a uniform red with no contamination from blue or green.

Color purity is determined by where the three electron beams strike the screen and can be adjusted by means of the purity magnet and deflection yoke (Figure 4-46).

4-9.5.2. Convergence

Convergence is the registry of the three beams at the same point in the shadow mask across the entire screen. The individual aiming of the beams must be corrected so that each passes through the same holes in the mask and strike the same phosphor dot triad.

Static convergence is controlled by convergence magnets (Figure 4-46), which correct the travel of the individual electron beams. The static magnets converge the beams in the center of the screen but cannot compensate for the curvature of the screen to converge the beam at the edges. Such compensation (known as *dynamic convergence*) is accomplished by electromagnetic coils mounted on the convergence yoke. Sawtooth voltages from the vertical and horizontal circuits are fed to the coils, which then alter the static magnetic field in unison with the scanning.

4-9.6. Keyed Rainbow Generator

The basic rainbow generator is a crystal-controlled oscillator producing a 3,563,795-Hz signal. The signal can be fed directly to the color bandpass amplifier of the receiver, or it can be used to modulate an r-f oscillator that operates on the carrier frequency of a selected television channel. The rainbow generator principle is quite simple. An oscillator that is operated at a frequency of 3,563,795 Hz (the color carrier frequency of 3,579,545 Hz minus the horizontal line frequency of 15,750 Hz) will appear as a 3.58-MHz signal that is constantly changing in phase, when compared to the 3.58-MHz reference oscillator in the television receiver, so that there is a complete change of phase of 360° for each horizontal sweep. This relationship is shown in Figure 4-47. Thus a complete range of colors (rainbow) is produced during each horizontal line. Each line displays all colors similarly, since the phase between both oscillators at the beginning of the sweep is always zero. If the phase changes 360° during one sweep, then it will also be zero at the beginning of the next sweep.

The basic rainbow generator (found in some early color generators) has been replaced by the *keyed rainbow generator*, which produces a rainbow of colors of uniform spacing and width (with blanks between each color).

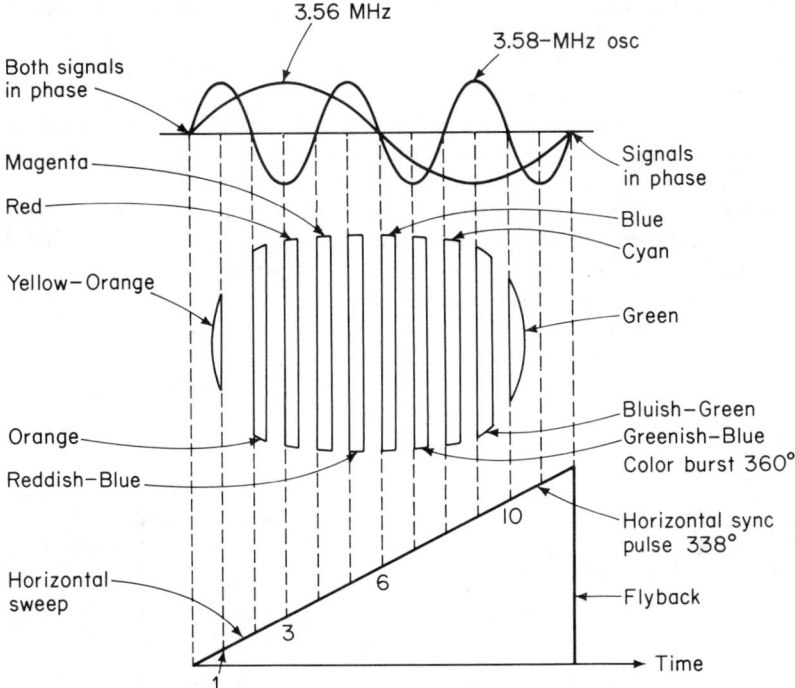

Fig. 4-47. Output signals and display of keyed rainbow generator (10 color bar generator).

The color-bar pattern is produced by gating the 3.56-MHz oscillator at a frequency 12 times higher than the horizontal sweep frequency (15,750 Hz $\times 12 = 189$ kHz). This 189-kHz gating produces color bars that are 30° apart all around the color spectrum. When viewed from the picture tube of a normal-operating television set, these bars appear as shown in Figure 4-47.

Note that of the 12 gated bursts, only ten show on the picture tube as color bars. This is because one of the bursts occurs at the same time as the horizontal sync pulse and is thus eliminated. The other burst occurs immediately after the horizontal sync pulse and becomes the color sync burst, which is used to control the 3.58-MHz reference oscillator in the television set.

4-9.6.1. Keyed Rainbow Generator Outputs

The outputs from a typical keyed rainbow generator are as follows:

1. R-F Output. The r-f signal is tuned to a specific television channel. Channels 3, 4, or 5 are used most often.

2. Modulation. Any one of five patterns; color-bar, dot, crosshatch, vertical-bar, or horizontal-bar: (Figure 4-48)
   a. Color bars. The ten color bars are generated for color alignment and troubleshooting in the color circuits of the television receiver. The color output is controlled with a separate adjustment control. Often, the colors that would be displayed on a normal color set are shown on a front-panel decal.
   b. Dots (usually about 120). Dots of small size are available, primarily for static convergence adjustments. The dot display is selected, and the static convergence magnets are adjusted so that white dots appear in the center of the screen (a result of all three guns hitting the same spot on the CRT).

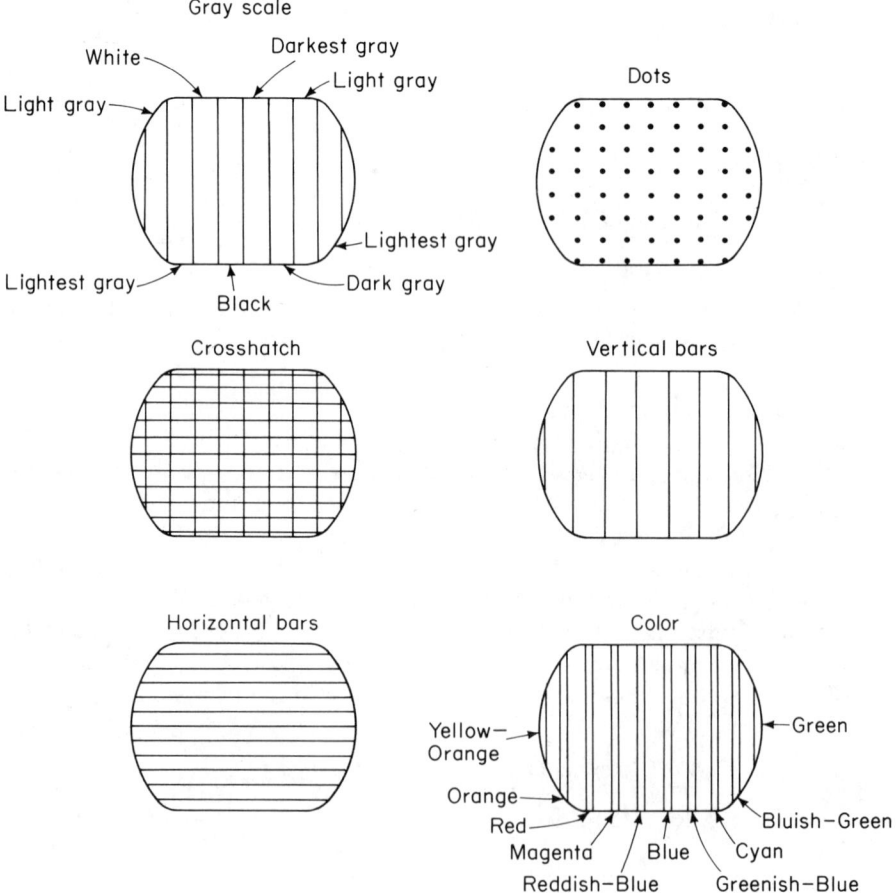

Fig. 4-48. Typical keyed rainbow generator displays.

c. Crosshatch. The crosshatch modulation display consists of nine vertical and 13 borizontal bars. The crosshatch pattern is used for dynamic convergence adjustments, overscan (height and width), and linearity adjustments. The crosshatch display is selected, and the dynamic convergence magnets are adjusted so that white lines appear at the outer edges of the screen.
d. Vertical bars. The vertical-bar modulation display consists of nine vertical bars, used primarily when adjusting dynamic vertical convergence controls and for centering.
e. Horizontal bars. The horizontal-bar modulation display consists of 13 horizontal bars, used primarily when adjusting the dynamic horizontal convergence controls and for centering.
3. Gray Scale Tracking Display. The gray scale tracking display is not found on all keyed rainbow generators, although most late-model generators now incorporate some form of gray scale. As shown in Figure 4-48, a typical gray scale display provides six levels of wide vertical gray bars plus pure black and pure white.

The compatible television system is designed so that the reception of black and white pictures will be reproduced faithfully on color receivers. In the color receiver, the three guns must be adjusted so that these telecasts will be reproduced and appear as *values of gray*. The color receiver must be able to reproduce black and white pictures within the normal usable range of the contrast and brightness controls. Adjustments using the gray scale display are often referred to as *white balance* or *screen temperature* since they pertain to the reproduction of various luminance values from black to white. Only a set correctly adjusted for gray scale tracking can, in turn, reproduce proper color when tuned to a color telecast.
4. Color-Gun Interrupters. Although not an output, color-gun interrupters are incorporated on most keyed rainbow generators. These consist of switches that ground the corresponding color gun (red, blue, or green) through fixed resistors, thus disabling that particular color.

4-9.7. NTSC-Type Generator

The NTSC-type generator differs from the rainbow types in that it will produce *single color bars*, one at a time. A typical NTSC-type generator will provide independent selection of six (or more) fully saturated colors (usually red, magenta, blue, cyan, green, and yellow) plus white, whose phase angles are permanently established in accordance with NTSC standards. These signals are selected by taps on a *linear delay line*, at the correct phase relationship. The amplitudes of the chrominance signals are also accurately set to NTSC standards, and the color reference burst is placed in its precise NTSC position, closely following the horizontal sync pulse. Therefore, the color signal produced by an NTSC-type generator is exactly the same as if the signals were being produced by a television station transmitting color.

Most NTSC-type generators also provide the same outputs as a rainbow generator (crosshatch, vertical, horizontal, dots, gray scale, and so on).

4-9.8. Color Generator Controls, Operation, Test, and Calibration

The operating procedures and controls, as well as the test and calibration procedures, are not as standardized as those of generators previously described. This is because circuitry used to achieve the various outputs differs considerably from model to model. For that reason, and because color television is a highly complex and specialized subject, it is recommended that readers make direct reference to the instruction manual for their particular make or model.

5 OSCILLOSCOPES AND RECORDERS

The cathode-ray oscilloscope (CRO) is an extremely fast X-Y plotter capable of plotting an input signal versus another signal or versus time, whichever is required. A luminous spot acts as a stylus or pen and moves over the display area in response to input voltages. The formal name *cathode-ray oscilloscope* is usually abbreviated to *oscilloscope* or, simply, *scope*. An *oscillograph* is the pictorial representation of an oscilloscope trace. Some older texts apply the word oscillograph to the complete equipment.

The X-Y recorder (or plotter) also plots an input signal versus other signals or time on a two-axis basis. However, the plotting is done on a sheet of graph or chart paper with a stylus capable of being positioned in each axis (X and Y) simultaneously. (Note that not all recorders are X-Y recorders; some plot in one axis only, as discussed in later sections of this chapter.)

In most oscilloscope or X-Y recorder applications, the Y-axis (vertical) input receives its signal from the voltage being examined, moving the luminous spot (or stylus) up or down in accordance with the instantaneous value of the voltage. The X-axis (horizontal) input is usually an internally-generated linear ramp voltage that moves the spot (or stylus) uniformly from left to right across the display screen (or chart paper). The spot (or stylus) then traces a curve that shows how the input voltage varies as a function of time.

The X-Y recorder is used to make a permanent trace of a repetitive or non-repetitive wave-form on a one-time basis. Usually, an X-Y recorder is used where the wave-form is low in frequency and nonrepetitive.

The *basic* oscilloscope is used to display fast, repetitive wave-forms. (As discussed in later sections, the *storage oscilloscope* is used to display and hold one-time wave-forms, while the *sampling oscilloscope* is used to display extremely fast wave-forms.)

If the signal under examination is repetitive at a fast enough rate, the display appears to stand still on an oscilloscope. The oscilloscope is thus a means of visualizing time-varying voltages. As such, the oscilloscope has become a universal tool in all kinds of electronic investigation and measurement.

Oscilloscopes operate on voltage signals. It is possible, however, to convert current, strain, acceleration, pressure, and other physical quantities to voltages by means of transducers (Chapter 9) and thus to present visual representations of a wide variety of dynamic phenomena on oscilloscopes.

5-1. OSCILLOSCOPE CATHODE-RAY TUBE

All circuits of an oscilloscope are built around a cathode-ray tube, or CRT. As shown in Figure 5-1, the filament heats the cathode to the degree where it emits electrons. The control grid influences the amount of current flow. Two anodes are used, each with a positive d-c voltage applied.

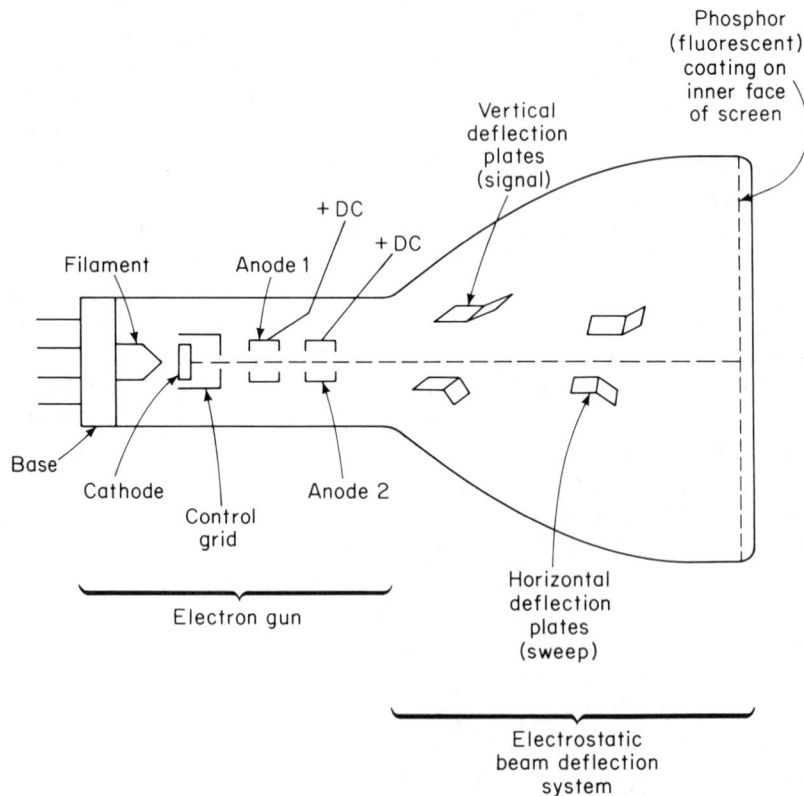

Fig. 5-1. Internal construction of cathode ray tube (CRT).

These anodes accelerate the electrons and form them into a beam. The intensity of the beam is regulated by the voltage applied to the control grid.

The grid structure, consisting of a cylinder with a tiny circular opening, controls electron flow (as does the grid in conventional vacuum tubes). The electron beam is focused into a sharp pinpoint by controlling the voltage on the first anode. The two anodes of the CRT can be compared to a glass lens system, because the anodes focus the beam to a pinpoint at the face of the tube. A high voltage is applied to the second anode so that the electron stream will attain high velocity for increased intensity and visibility when it strikes the tube face. The beam-forming section of the tube is known as an *electron gun*.

The inside of the tube face is coated with phosphor so that, when electrons strike this coating, it will *fluoresce* and emit light. After the electron stream has left the area, the fluorescing characteristics that emit light rapidly decay, and the light level is reduced. However, the emitted light persists for an appreciable interval, depending on chemical characteristics. Both long- and short-persistence CRTs are available. In general, the persistence should be long enough for the beam to leave a *complete trace* but, if the beam stops or moves, the pattern traced on the tube should disappear very rapidly.

5-2. OSCILLOSCOPE BEAM DEFLECTION SYSTEM

As shown in Figure 5-2, two sets of plates are present within the tube, beyond the second anode. These plates are for deflecting the electron beam both horizontally and vertically so that the beam will "write out" information delivered to the deflecting system. Such a system is known as *electrostatic deflection*. Magnetic deflection, once used in a few oscilloscopes, is more often used in television picture tubes and radar displays.

The electrostatic beam deflection is accomplished through the two pairs of parallel plates. For example, a voltage applied across the deflection plates will influence the beam, because the negative potential on one plate repels the electron stream, whereas the positive potential on the other plate attracts the beam. If such a voltage is a sawtooth or ramp type, the gradually rising potential of the sawtooth will gradually pull the beam toward the positive plate. Therefore, the electron beam is made to scan across the face of the tube.

Figure 5-2 shows how an oscilloscope traces out a sine wave on the tube face when such a signal is applied to the vertical input. In Figure 5-2A, the sine wave applied at the vertical input is shown. If the internal horizontal sweep generator is turned off, the rising positive potential of the first alternation of the input signal causes the electron beam to move upward, as shown in Figure 5-2B. When the negative alternation of the input signal arrives at the vertical deflection plates, the electron beam is pulled downward from the

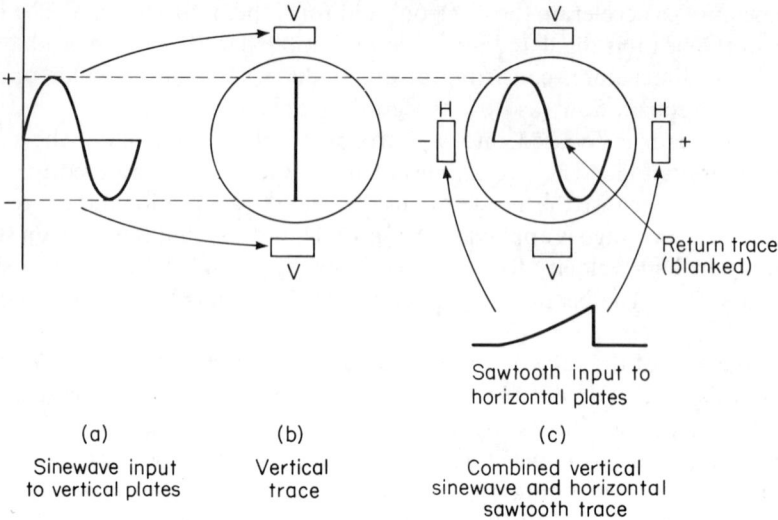

Fig. 5-2. How sinewave (vertical) and sawtooth (horizontal) signals are combined to produce an oscilloscope trace.

center. The rapid rise and fall of the signal alternation causes the electron beam to move up and down the face of the CRT very rapidly, leaving a vertical line trace.

If the horizontal oscillator within the oscilloscope is now turned on, a sawtooth of voltage, as shown in Figure 5-2C, will be applied to the horizontal deflection plates. The rising potential of the sawtooth voltage causes the right horizontal plate to become positive and the left horizontal plate to become negative. The negative left plate and the positive right plate cause the beam to move from left to right. If one sawtooth occurs from each cycle of the sine-wave signal, the beam will be pulled across the face of the tube once for each cycle. Thus, as the first alternation of the input cycle rises in amplitude, the electron beam would normally rise in a vertical plane, as shown in Figure 5-2B. The sawtooth on the horizontal plates gradually pulls the beam from left to right and traces out the input signal wave shape in visual form.

5-3. BASIC FREQUENCY MEASUREMENT WITH AN OSCILLOSCOPE

If the input signal has a frequency twice that of the sawtooth applied to the horizontal plates, 2 cycles will appear on the screen, because the beam is pulled across the screen only once for each 2 cycles of the input signal. By regulating the ratio of input signal frequency to the sawtooth sweep frequency, portions of the input signal or a number of cycles of the input signal can be made visible on the screen. By calibrating the frequency of the horizontal sweep wave-form so that its exact frequency is known, the frequency of the input signals to the oscilloscope can be calculated. For example, if 3 cycles of a sine wave appear on the screen and the sawtooth generator is set for 100 Hz, the frequency of the signal applied at the input of the oscilloscope is 300 Hz.

5-4. BASIC VOLTAGE MEASUREMENT WITH AN OSCILLOSCOPE

In addition to frequency measurements, the oscilloscope can be used for reading the peak-to-peak voltages of a-c signals, pulses, square waves, and so on, as well as d-c voltages. For reading voltages, a transparent plastic screen is attached to the face of the oscilloscope. Such a screen (also known as *grid, grid mask, mask, grating, or graticule*) is found in several forms. Usually, a screen is a transparent scale with vertical and horizontal lines. The screen is fitted against the face of the CRT, allowing time and amplitude to be read directly.

These graduated scales often have small markings that subdivide the major divisions to assist in making accurate measurements. Most laboratory oscilloscopes are calibrated in centimeters. Some shop oscilloscopes are calibrated in inches. Still other oscilloscopes use no particular standard of measurement, but are simply equal-spaced "divisions."

To voltage-calibrate an oscilloscope screen, an a-c voltmeter of known accuracy and a low-voltage a-c signal must be available. Some oscilloscopes have a terminal on the front panel that supplies such an a-c reference voltage. For example, if the reference voltage is 5 V, this voltage is applied to the vertical input of the oscilloscope. The internal horizontal sweep generator is shut off so that a vertical trace, as shown in Figure 5-2B, is visible on the screen. This vertical line represents the peak-to-peak voltage of the input signal. The vertical height control is then adjusted so that the line is five divisions high, and the control is left in this position after calibration. Knowing the RMS value of the applied a-c calibrating voltage, the peak-to-peak voltage can be calculated by multiplying the RMS value by 1.41 to obtain the peak value, then doubling the peak value to obtain the peak-to-peak value. Refer to Figure 1-13 for conversion data.

5-5. BASIC OSCILLOSCOPE CIRCUITS

Figure 5-3 is the block diagram of a typical oscilloscope. The following paragraphs describe circuit operation.

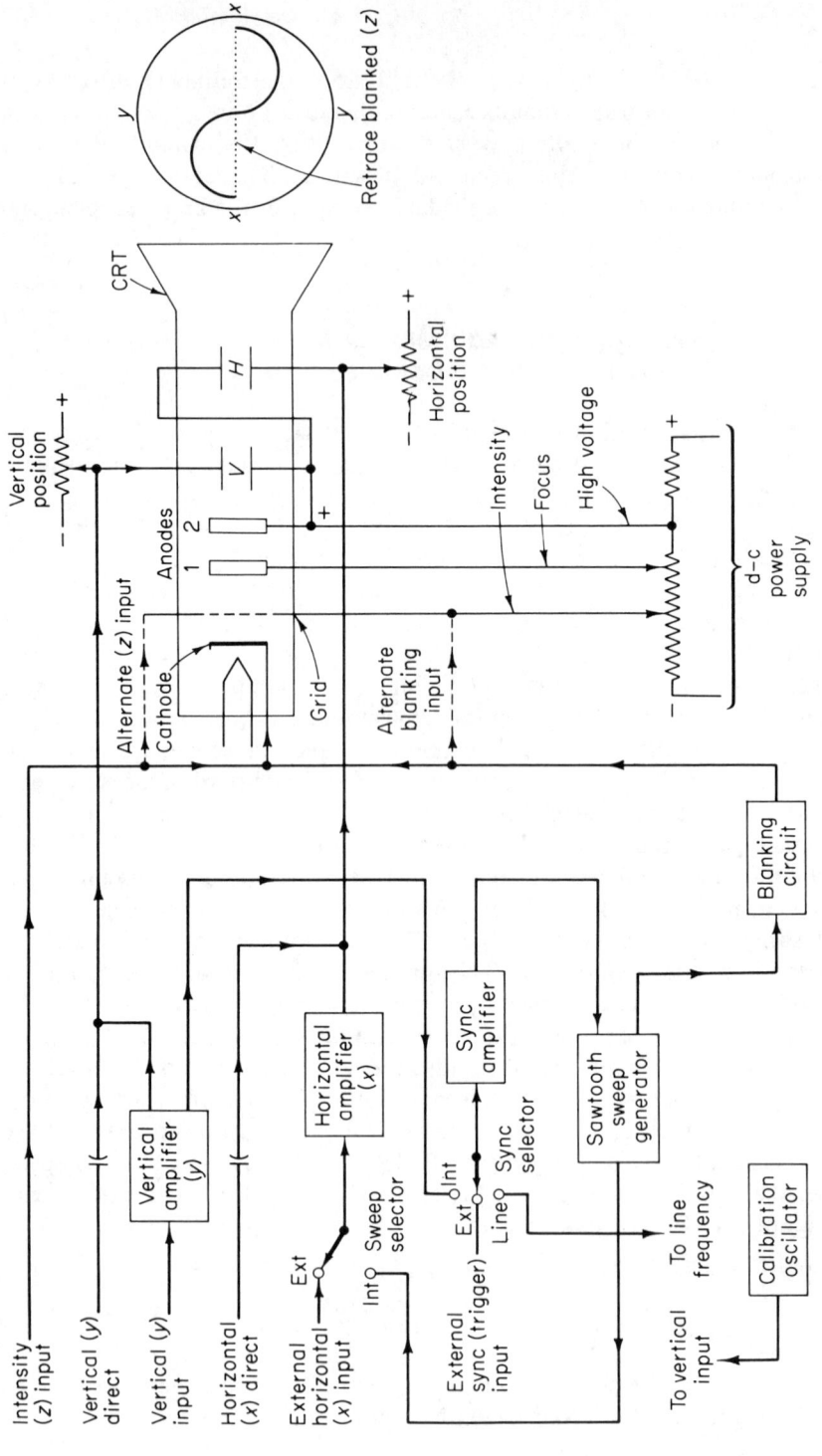

Fig. 5-3. Simplified block diagram of typical oscilloscope.

5-5.1. Vertical (Y-Axis) Channel

Signals to be examined are usually applied to the vertical, or *Y*, deflection plates through the vertical amplifier. A vertical deflection amplifier is required, since the signals are usually not strong enough to produce measurable vertical deflection on the CRT. A typical CRT requires 50 V dc to produce a deflection of 1 in. The high-gain amplifier in a laboratory oscilloscope permits a 1-in. deflection with 0.5 mV. The average shop-type oscilloscope requires at least 20 mV for a 1-in. deflection. Another difference between laboratory and shop amplifiers is the frequency response. (Refer to Chapter 7 for data on amplifiers.) The frequency response must be wide enough to pass faithfully the entire band of frequencies to be measured by the oscilloscope. For simple audio work, an upper limit of 200 kHz is sufficient. Television service requires a passband of at least 5 MHz. Some laboratory oscilloscopes provide up to 1000 MHz (1 GHz); 50 MHz are common for laboratory scopes.

When high-voltage signals are to be examined, they can be applied directly to the vertical deflection plates (on most oscilloscopes).

The vertical amplifier output is also applied to the sync amplifier through the sync selector switch (in the *internal* position). This permits the horizontal sweep circuit to be triggered by the signal being examined.

5-5.2. Horizontal (X-Axis) Channel

Usually, the horizontal deflection plates are fed a sweep voltage that provides a *time base*. The horizontal plates are fed through an amplifier, but they can be fed directly when the voltages are of sufficient amplitude. When external signals are to be applied to the horizontal channel, they can also be fed through the horizontal amplifier, via the sweep selector switch in the *external* position. When the sweep selector is in the *internal* position, the horizontal amplifier receives an input from the sawtooth sweep generator, which is triggered by the sync amplifier.

In most oscilloscopes, the sawtooth sweep voltage is generated by a multivibrator, relaxation oscillator, or pulse generator. There are four basic types of sweeps.

The *recurrent* sweep presents the display repetitively, and the eye sees a lasting pattern.

In a *single sweep*, the spot is swept once across the screen in response to a trigger signal. This is similar to the *X-Y* recorder display. The trigger can be obtained from the signal under study or from an external source.

In most cases, a *driven sweep* is used where the sweep is recurrent but triggered by the signal under test.

In special cases, some oscilloscopes provide a *nonsawtooth sweep*, such as a sine wave.

Sweep frequencies vary with the type of oscilloscope. A laboratory oscilloscope may have sweep frequencies up to several megahertz; a simple shop-type oscilloscope for audio work has an upper limit of about 100 kHz. Most television service requires a horizontal sweep frequency up to 1 MHz.

Whatever type of sweep is used, it must be synchronized with the signal being investigated. If not, the pattern will appear to drift across the screen in a random fashion. Three usual sources for synchronization can be selected by the sync selector:

Internal, where the trigger is obtained from the signal under investigation (through the vertical amplifier),

External, where an external trigger source is also used to trigger or initiate the signal being measured,

Line, where the sync trigger is obtained from the line frequency (usually 60 Hz). Line sync is often used in television service where an external sweep generator and an oscilloscope are both triggered at the line frequency, as described in Chapter 4.

The oscilloscope sweep system is also used to produce a *blanking* signal. The blanking signal is necessary to eliminate the retrace that would occur when the sweep trace snaps back from its final (right-hand) position to the starting point. This retrace could cause confusion if it were not eliminated by blanking the CRT during the retrace period with a high negative voltage on the control grid (or a high positive voltage on the CRT cathode). The blanking voltage is usually developed (or triggered) by the sweep generator.

5-5.3. Intensity (Z-Axis) Channel

Intensity modulation, sometimes known as *Z-axis modulation*, is accomplished by inserting a signal between ground and the cathode (or control grid) of the CRT. When the signal voltage is large enough it can cut off the CRT on selected parts of the trace, just as the retrace blanking signal does. Z-axis modulation is applied during the normally visible portion of the trace.

The Z axis can also be used to brighten the trace. Periodically applying positive pulse voltages to the CRT control grid (or negative pulses to the cathode) brightens the electron beam throughout its trace to give a third, or Z, dimension. These periodically brightened spots can be used as markers for time calibration of the main wave-form.

5-5.4. Positioning Controls

For many measurements, it is necessary to provide some means of positioning the trace on the CRT face. Such positioning provisions are accomplished by applying small, independent, internal d-c potentials to the deflection plates and controlling them by means of potentiometers. The positioning controls are particularly useful during voltage calibration and during enlargement of a wave-form for examination of small characteristics. The portion of interest may move off the CRT screen so that positioning controls are necessary to bring it back on again.

5-5.5. Focus and Intensity Controls

The CRT electron beam is focused in a manner similar to that of an optical lens, but the focal length is altered by changing the ratio of potentials between the first and second anodes. This ratio is changed by varying the potential on the anode with the focus control, which is a potentiometer usually located on the oscilloscope front panel. The potential on the second anode remains constant.

The intensity of the beam is varied by the intensity control potentiometer, which changes the grid potential with respect to the cathode, thus permitting more or fewer electrons to flow. Because the potentials applied to the control grid and to both anodes are taken from a common voltage-divider network, any change made in the setting of the intensity control requires a compensating change in the setting of the focus control, and vice versa.

5-5.6. Calibration Circuit

Most laboratory oscilloscopes have an internally generated and stabilized wave-form of known amplitude that serves as a calibrating reference. Usually, a square wave is used, with the calibrating signal accessible on a front-panel connector.

On shop oscilloscopes, the 60-Hz line voltage is used. Either way, the calibrating voltage can be applied to the vertical channel by running a lead between the front-panel calibration output connector and the vertical input connector. On some oscilloscopes, the calibrating voltage is applied to the vertical input through a front-panel control switch.

5-5.7. Electronic Switch and Dual-Trace Displays

In many applications, it is convenient to display two signals simultaneously. Many laboratory oscilloscopes have a *dual-trace* provision, whereby two signals can be applied directly to separate oscilloscope inputs and will be displayed simultaneously. This same type of operation can be accomplished with a shop oscilloscope when an *electronic switch*, or "chopper," is used at the input. The electronic switch acts as two gates, one for each signal, that open and close on alternate half-cycles at a predetermined frequency. One gate is open while the opposite gate is closed. The output of both gates is applied to the oscilloscope input. Therefore, both signals are displayed on the oscilloscope. In practice, the gate-switching frequency must be much higher than either signal frequency.

The basic circuit of a typical electronic switch is shown in Figure 5-4. Each signal is applied to a separate gain control and gate stage. The gate stages are alternately biased to cutoff by square-wave signals from the square-wave generator. Therefore, only one gate stage is in a condition to pass its signal at any given time. The output of both gates is applied directly to the oscilloscope input.

Figure 5-4 also shows some typical dual-trace displays. Actually, each trace is composed of a tiny bit of the corresponding signal. Since the switching rate is fast, the trace appears as a solid line. Most electronic switches are provided with some form of positioning control so that the traces can be superimposed or separated, whichever is more convenient. The positioning control is shown as R_3 in Figure 5-4. Also, so that the amplitude (or height) of both signals can be made to appear approximately the same as the display despite an actual difference in signal strength, the electronic switch is provided with separate gain controls (R_1 and R_2).

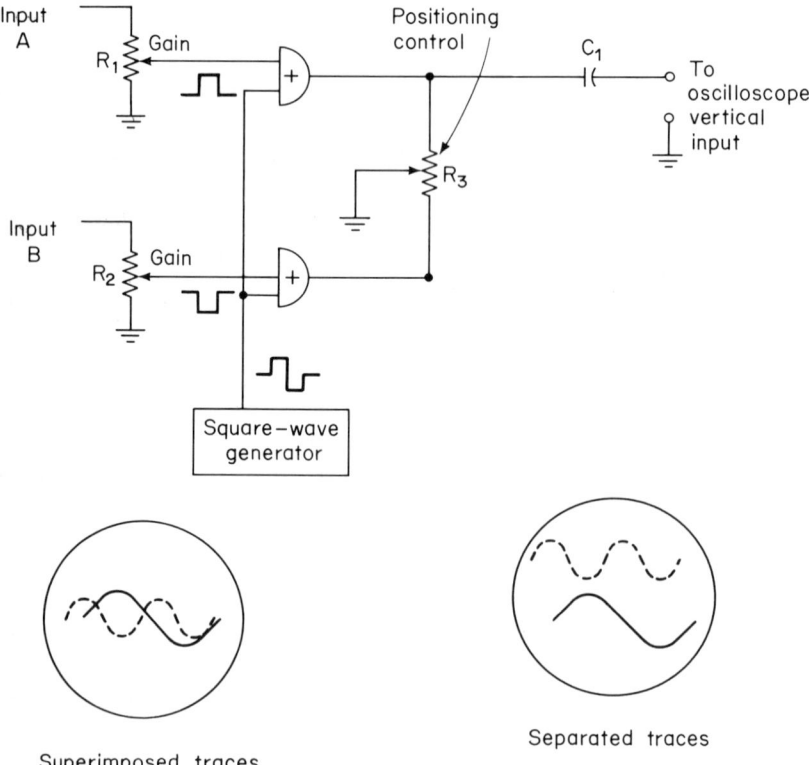

Fig. 5-4. Electronic switch circuit and typical dual-trace displays.

5-5.8. Plug-In Modules

Plug-in modules are often used to extend the usefulness of laboratory oscilloscopes. In this case, the basic oscilloscope consists of a cathode-ray tube and power circuits. The remaining circuits (such as amplifiers, sweep generators, time base generators, pulse generators, and so on) are contained in plug-in units. Operation of such circuits is identical to the circuits found in corresponding items of test equipment, as described in other chapters.

These plug-in units can be changed to perform specific tests. In laboratory work it is usual to use the basic oscilloscope and those plug-ins required for routine or essential operation (vertical and horizontal amplifiers, horizontal sweep generators). Then, as new work areas are developed, additional plug-ins are obtained to meet their needs.

5-6. OSCILLOSCOPE OPERATING CONTROLS AND INDICATORS

Figures 5-5 and 5-6 show the front-panel operating controls of two representative oscilloscopes. As will be seen, both units have some con-

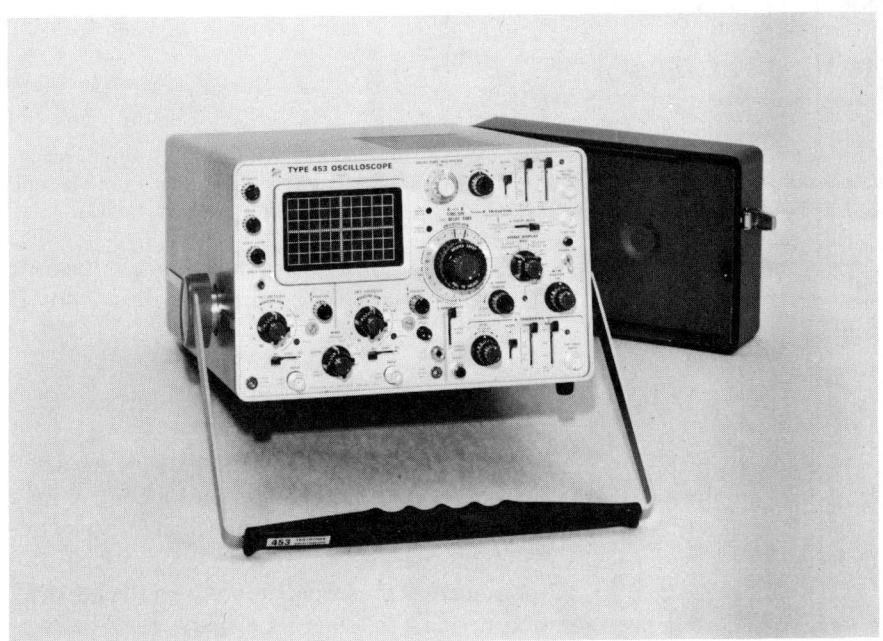

Fig. 5-5. Typical laboratory oscilloscope operating controls and indicators (Tektronix).

Fig. 5-6. Typical shop oscilloscope operating controls and indicators (Allied Radio).

trols in common (with the same or different names), but each oscilloscope has certain unique controls. The following sections describe these control functions.

5-6.1. Laboratory Oscilloscope (*Figure 5-5.*)

Intensity. Controls brightness of the display. The display can be adjusted from very bright to total darkness. The setting of the intensity control may affect the correct focus of the display. Slight readjustment of the focus control may be necessary when the intensity level is changed.

To protect the CRT phosphor of any oscilloscope, do not turn the intensity control higher than necessary to provide a satisfactory display. Also, be careful that the intensity control is not set too high when changing from a fast to a slow sweep rate.

Focus. Provides adjustment for a well-defined display. If a well-defined trace cannot be obtained with the focus control, it may be necessary to adjust the *astigmatism* control (if such a control is provided). To check for proper setting of the astigmatism adjustment, slowly turn the focus control through the optimum setting. If the astigmatism adjustment is correctly set, the vertical and horizontal portions of the trace will come into sharpest focus at the *same position* of the focus control This setting of the astigmatism adjustment should be correct for any display. It may, however, be necessary to reset the focus control slightly when the intensity control is changed.

Scale Illum. Controls screen illumination. The engraved lines of the transparent viewing screen are brightened by edge-lighting. This provides a sharp reproduction of the lines when photographs are made from the screen but does not produce an interfering glare.

Trace Finder. Returns the display to the screen, when pressed, by reducing horizontal and vertical deflection. The trace finder provides a means of locating a display that overscans the viewing area either vertically or horizontally. This type of control is found only in a precision laboratory oscilloscope.

Volts/Div. Selects vertical deflection factor. The amount of vertical deflection produced by a signal is determined by the signal amplitude, the attenuation factor of the probe (if used), the setting of the VOLTS/DIV switch, and the setting of the variable VOLTS/DIV control (if used). The calibrated deflection factors indicated by the VOLTS/DIV switches apply only when the variable control is set to the CAL (calibrate) position.

On some shop oscilloscopes, this control, called *V-gain (amplitude) control* or *Y-gain (amplitude) control*, is a variable potentiometer rather than a step-attenuator. On other oscilloscopes, the combination step-attenuator and variable control are used. In that case, the potentiometer provides continuously variable control of vertical gain in any one of the ranges provided by the step-attenuator. In some shop oscilloscopes, the step-attenuator is calibrated as a multiplier (such as $\times 0.1$, $\times 1.0$, $\times 10.0$, $\times 100.0$, $\times 1000.0$) rather than in specific values of volts-per-division.

Variable. Provides continuously variable deflection factor to at least 2.5 times setting of VOLTS/DIV switch.

Position (vertical and horizontal) controls. Control vertical and horizontal position of trace.

Gain. Screwdriver adjustment to set gain of the vertical preamplifier. This is usually an internal adjustment control on most oscilloscopes.

AC—GND—DC. Selects method of coupling input signal to input (vertical) amplifier.

In AC, the d-c component of the input signal is blocked (by a coupling capacitor inserted between the vertical input connector and amplifier).

In GND, the input circuit is grounded, but the applied signal is not connected to a d-c ground.

In DC, all components of the input signal (a-c, d-c) are passed to the input amplifier.

Step Atten Bal. Screwdriver adjustment to balance input amplifier in various positions of the VOLTS/DIV switch. This control is usually an internal adjustment on most oscilloscopes.

Input. Vertical input connector for signal.

Mode. Selects vertical mode of operation. This control (or similar control) is found only on oscilloscopes having more than one mode of operation.

Invert. When pulled out, inverts the vertical display. When observing certain signals (such as television wave-forms) the signal is often displayed upside down (signal provides negative deflection). The invert switch permits the display to be viewed right side up. On some oscilloscopes, the control (called a *trace reverser*) switch positions are labeled plus for an upright pattern and minus for an inverted display.

Ext Trig Input. Input connector for external-triggering signal. Connector also serves as external horizontal input when horizontal display switch is in external horizontal position.

Source. Selects source of triggering signal. This control (also called *sync selector*) selects the type of signal used to synchronize the horizontal sweep oscillator.

Int. For most applications, the sweep can be triggered internally. In the INT position, the trigger signal is obtained from the vertical system.

Line. The line position connects a sample of the power-line frequency to the trigger generator (and thus to the horizontal sweep oscillator). Line-triggering is useful when the input signal is time-related to the line frequency. It is also useful for providing a stable display of line-frequency component in a complex wave-form.

Ext. An external signal connected to the EXT TRIG INPUT connector can be used to trigger the sweep in the EXT position. The external signal must be time-related to the displayed signal for a stable display. An external-trigger signal can be used to provide a triggered display when the internal signal is too low in amplitude for correct triggering or contains signal components on which it is not desired to trigger.

The EXT position is also useful when signal-tracing in amplifiers, phase-shift networks, wave-shaping circuits, and so on. The signal from a single point in the circuit can be connected to the EXT-TRIG INPUT connector through a signal probe or cable. The sweep is then triggered by the same signal at all times and allows amplitude, time relationship, or wave-shape changes of signals at various points in the circuit to be examined without resetting the triggering controls.

Coupling. Determines method of coupling the triggering signal to trigger circuit. On some oscilloscopes, this function is combined with the trigger source control.

AC. The AC position blocks the d-c component of the trigger signal. Signals with low-frequency components below about 30 Hz will be attenuated. In general, a-c coupling can be used for most applications. If the trigger signal contains unwanted components, or if the sweep is to be triggered at a d-c level, one of the remaining coupling switch positions will provide a better display.

The triggering point in the AC position depends on the *average* voltage level of the trigger signal. If the trigger signal occurs in a random fashion, the average voltage level will vary, causing the triggering point to vary, also. This shift of the triggering point may be enough to make it impossible to maintain a stable display. In such cases, use d-c coupling.

LF Rej. In the LF REJ position, d-c is rejected and signals below about 30 kHz are attenuated. Therefore, the sweep signals will be triggered only by the higher-frequency components of the signal. This position is particularly useful for providing stable triggering if the trigger signal contains line-frequency components.

HF Rej. The HF REJ position passes all low-frequency signals between about 30 Hz and 40 kHz. Direct current is rejected and signals outside the given range are attenuated. When triggering from complex wave-forms, this position is useful for providing stable display of low-frequency components.

DC. Direct-current coupling can be used to provide stable triggering with low-frequency signals that would be attenuated in the AC position or with low-repetition rate signals. The LEVEL control can be adjusted to provide triggering at the desired d-c level on the wave-form. When using internal triggering, the setting of the vertical position controls affect the d-c trigger level.

Slope. The triggering SLOPE switch determines whether the trigger circuit responds on the positive-going or negative-going portion of the trigger signal. Shop-type oscilloscopes usually do not have a slope switch function.

When the SLOPE switch is in the plus or positive-going position, the display will start with the positive-going portion of the wave-form. In the minus or negative-going position, the display will start with the negative-going portion of the wave-form. This is shown in Figure 5-7.

When several cycles of a signal appear in the display, the setting of the SLOPE switch is less important. If only a certain portion of a cycle is to be displayed, correct setting of the SLOPE switch provides a display that starts on the desired slope of the input signal.

Level. The triggering LEVEL control determines the voltage level on the triggering wave-form at which the sweep is triggered. On shop oscilloscopes, this control is usually called the *sync* or *trigger amplitude* control, and its adjustment varies the amplitude of the sync voltage applied to the internal horizontal sweep generator. The sawtooth sweep locks in with the sync volt-

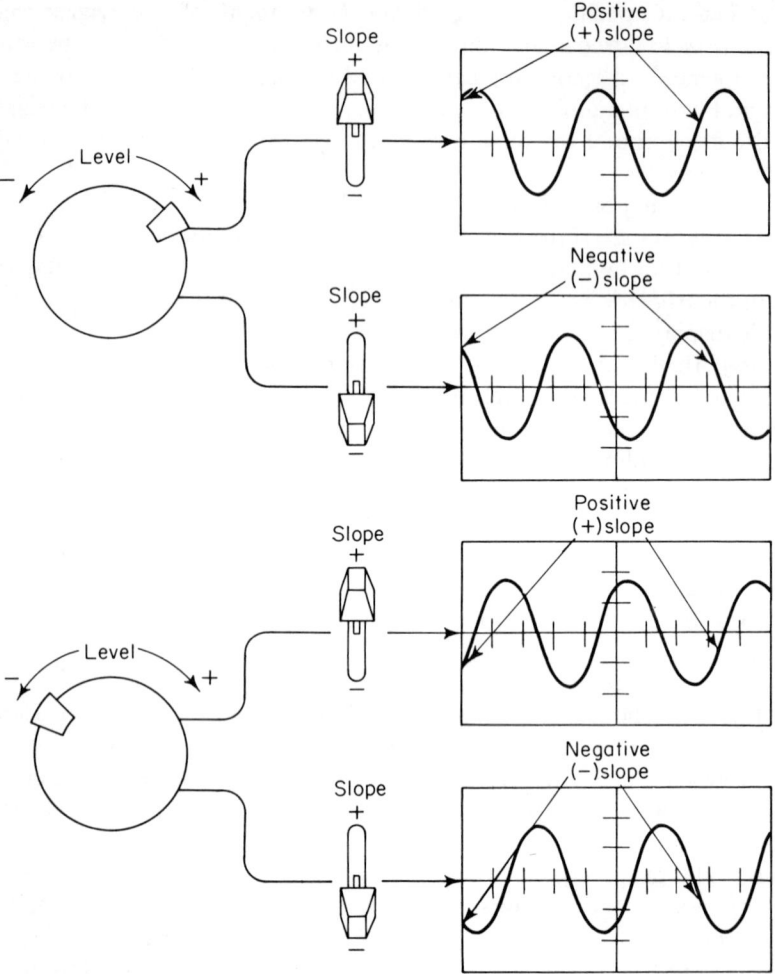

Fig. 5-7. Effects of triggering SLOPE control and LEVEL control.

age, and the display stands still when the LEVEL control is properly set.

When the LEVEL control is set in the plus region, the trigger circuit responds at a more positive point on the trigger signal. When the LEVEL control is set in the minus region, the trigger circuit responds at a more negative point on the trigger signal. Figure 5-7 illustrates this effect with different settings of the LEVEL and SLOPE switches.

Sweep Mode. Determines sweep operating mode. On shop oscilloscopes, this control is usually called the *sweep selector* and provides three basic operating modes: linear sawtooth sweep (internal), external, and line-frequency sine-wave (internal).

On oscilloscopes designed specifically for television receiver service there are usually two special sweep frequencies: 30 Hz (one-half the television vertical deflection frequency) and 7875 Hz (one-half the television horizontal deflection frequency). (Sweeps of one-half the sync frequencies are used so that 2 cycles will be displayed.)

A *single-sweep* mode is available on many laboratory oscilloscopes. The single-sweep mode is used for nonrepetitive signals (such as photographing a single wave-form).

Time/Div. The TIME/DIV switch selects calibrated horizontal sweep rates for the internal sweep generators. The VARIABLE control provides continuously variable sweep rates between the settings of the TIME/DIV switch. The calibrated sweep rates apply only when the VARIABLE control is set to the CAL position.

On some shop oscilloscopes, the TIME/DIV switch is called the *sweep range selector* or the *coarse frequency control*. In such instruments, the switch positions are calibrated in terms of *frequency* rather than time (typically, 10–100 Hz, 100–1000 Hz, 1–10 kHz, and 10–100 kHz).

In laboratory oscilloscopes, the sweep rates are expressed in *time*, since the time interval of the display is of greater importance than sweep frequency (for most laboratory measurements).

Variable. Provides continuously variable sweep rate to at least 2.5 times setting of TIME/DIV switch. On shop oscilloscopes, this control is usually called the *sweep-frequency control, fine-frequency control,* or *frequency vernier* and permits continuous variation of sweep frequency within any of the ranges provided by the sweep range selector.

Sweep Length. Adjusts the length of the horizontal sweep. In the full or maximum position, the sweep is about 11 divisions long. As the control is rotated counterclockwise, the sweep length will be reduced until it is less than 4 divisions long. On shop oscilloscopes, this control is usually called the *H-gain (amplitude) control* or *X-gain (amplitude) control*.

On some oscilloscopes, a combination step-attenuator and variable control is used. In that case, the potentiometer provides continuously variable control of vertical gain in any one of the ranges provided by the step-attenuator. Usually, the step-attenuator is calibrated as a multiplier (such as $\times 0.1$, $\times 10$, $\times 100$).

Z-Axis Input. Input connector for intensity modulation of the CRT display. A Z-axis gain control is incorporated on many oscilloscopes. Adjustment of this control (also called *intensity modulation gain* or *amplitude control*) provides continuous variation of the intensity modulation voltage. Time markers applied to the Z-axis input provide a direct time reference on the display (in the form of dots or bright spots).

1 KC Cal. Calibration signal output connector. The internal 1-kHz square-wave calibrator provides a convenient signal source for checking vertical gain and basic horizontal timing. The calibrator provides peak-to-peak square-wave voltages of 0.1–1 V. Voltage range is selected by the calibrator switch. The square-wave calibrator signal is very useful for checking and adjusting probe compensation. (Refer to Chapter 9 for data on probe compensation with square waves.) The calibrator can also be used as a convenient signal source for application to external equipment.

In shop oscilloscopes, the calibrating voltage is a line-frequency sine wave obtained from a winding of the oscilloscope power transformer and set to 1 V (or 0.1 V) peak-to-peak by means of a voltage divider. In some oscilloscopes, a calibration voltage control allows this voltage to be adjusted between 0 and 1 V peak-to-peak.

Mag. Magnifies or increases sweep rate to 10 times setting of TIME/DIV switch, thus expanding (horizontally) the center portion of the display. This permits closer observation of a part of the signal. As shown in Figure 5-8,

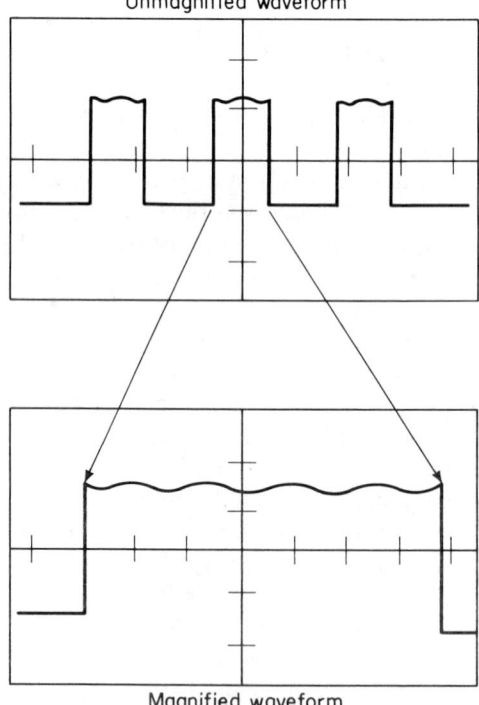

Fig. 5-8. Operation of a sweep magnifier.

the center division of the unmagnified display is the portion visible on the screen in magnified form. Equivalent length of the magnified sweep is about 100 divisions. Any ten-division portion may be viewed by adjusting the horizontal position control to bring the desired portion into the viewing area. A sweep magnifier control is usually found only on laboratory oscilloscopes.

5-6.2. Shop Oscilloscope (*Figure 5-6*)

The following paragraphs describe the function or operation of the controls and connectors of the shop oscilloscope shown in Figure 5-6. Also included are comments on how these controls relate to controls on the laboratory oscilloscope (Figure 5-5) previously described.

The functions of the astigmatism, focus, intensity, illumination, vertical position, horizontal position vertical input, trigger slope, and level controls are identical with those of the laboratory oscilloscope.

The VOLTS/CM selector and VARIABLE control are similar to the laboratory VOLTS/DIV and VARIABLE controls, except for the calibration scales.

The AC-DC switch is similar to the laboratory AC-GND-DC switch, except that there is no ground (GND) position or function.

The TRIGGER INPUT switch is similar in function to the laboratory SOURCE switch, with certain exceptions.

In the INT AC position, the d-c component of the trigger signal is blocked, and the sweep is triggered by the vertical input.

In the INT DC position, the sweep is also triggered by the vertical input, but by d-c and low-frequency signals that would be attenuated in the INT AC position.

In the EXT AC position, the d-c component of the trigger signal is blocked, and the sweep is triggered by an external signal applied to the EXT-TRIG connector.

In the EXT DC position, the sweep is also triggered by an external source, but by d-c and low-frequency signals that would be attenuated in the EXT AC position.

In the LINE position, a sample of the power-line frequency is connected to the trigger generator (and thus to the horizontal sweep oscillator).

The HORIZONTAL AMP switch is similar in basic function to the laboratory SWEEP MODE switch in that both determine the operating mode. Four positions are provided. INT ×1 and INT ×5 (internal linear sawtooth sweep), EXT SIG (external signal), and LINE (line-frequency sine wave).

The TIME/CM selector and VARIABLE control are similar to the laboratory TIME/DIV and VARIABLE controls, except for the calibration scales.

The 0.1-V peak-to-peak connector is similar to the laboratory 1 KC CAL connector, except that the calibration output voltage is 0.1-V peak-to-peak *sine wave* instead of a square wave.

The HORIZONTAL connectors provide two functions:

The center or common connector and the EXT SIG connector provide for input of an external signal to the horizontal amplifier or sweep circuit.

The common connector and the EXT CAP connector provide for connection of an external capacitor to the horizontal sweep oscillator. This provides for very slow sweeps, as determined by the capacity of the external capacitor. There is no comparable function on the laboratory oscilloscope.

5-6.3. Miscellaneous Operating Controls for Oscilloscopes

In addition to the controls described thus far, many other controls must be adjusted for proper oscilloscope performance. On some oscilloscopes, these additional functions are adjusted by front-panel controls; on other instruments, the function is set by a screwdriver adjustment.

5-6.3.1. Phasing Control

One front-panel control, found particularly on oscilloscopes used for television service, is a *phasing control*. Adjustment of this control varies the horizontal sweep when it is being driven by the line voltage at the line frequency. The phasing control is especially important when the oscilloscope is used with a sweep generator (driven at line frequency) to observe a response pattern of a tuned circuit. Sweep generators and phasing problems are discussed in Chapter 4.

The following additional controls or adjustments are found in many oscilloscopes:

Voltage Regulation. Sets the output voltage of regulated power supplies in oscilloscopes.

Calibration Voltage. Sets the calibration voltage output.

Sweep Frequency. Sets the horizontal sweep oscillator frequencies to match the sweep selector switch calibrations.

DC Balance. Balances the d-c amplifiers for trace-centering.

Linearity. Sets linear horizontal and vertical deflection on each side of the screen center.

Frequency Compensation. Sets amplifier and attenuator components (vertical and horizontal) for wideband response.

Hum Balance. Cancels power supply hum.

5-7. STORAGE OSCILLOSCOPE

The storage oscilloscope is especially useful for one-shot displays, where it is not practical to photograph the display. A storage oscilloscope will hold a display on the screen for an indefinite time. Several waveforms may be superimposed for comparison. When desired, the operator can remove the display by depressing an erase button. A storage oscilloscope can be used as a conventional oscilloscope when the storage function is disabled.

Several types of storage cathode-ray tubes are available. Storage CRTs may be classified as either *bistable* or *halftone* tubes. The stored display on a bistable tube has one level of brightness. A halftone tube has the capacity of displaying a stored signal at different levels of brightness. The brightness of a halftone tube is dependent on beam current and the time the beam remains on a particular storage element. A bistable tube will either store or not store an event. All stored events have the same brightness.

Storage CRTs may also be classified as either *direct-viewing*-type or *electrical-readout*-type tubes. An electrical-readout-type tube has an electrical input and output. A direct-viewing-type tube has an electrical input but a visual output.

Figure 5-9 is a simplified diagram of a typical storage CRT. The focus, intensity, and accelerator electrodes have been omitted for clarity.

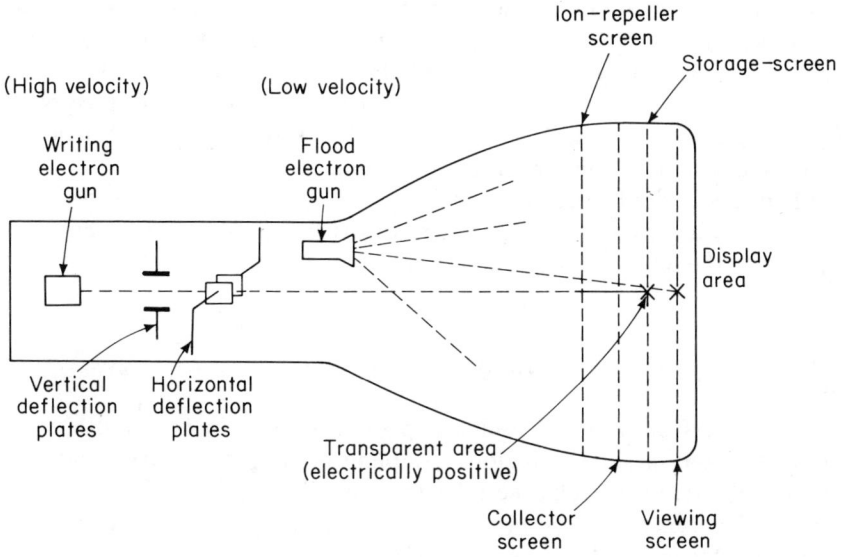

Fig. 5-9. Simplified diagram of storage cathode-ray tube.

In addition to the electron gun (known as the *writing gun* of a conventional CRT, the storage CRT also has a *flood* gun (or guns). Unlike the conventional CRT, a storage CRT has four screens behind the display area.

The *viewing screen* is a phosphor screen similar to that of a conventional CRT. Directly behind the viewing screen is a *storage screen*, usually in the form of a fine metal mesh coated with dielectric. A *collector* screen and *ion-repeller* screen are located behind the storage screen. Not all storage CRTs have the ion-repeller screen.

When signals are applied to the horizontal and vertical deflection plates, the electron beam from the writing gun is moved to trace out the corresponding wave-form, in the normal manner. When this beam strikes the storage screen, an electrically positive trace is written out, by removing electrons from the storage screen dielectric atoms where it strikes. This leaves a positive trace stored on the storage screen after the signal has been removed. The excess electrons removed from the storage screen are collected by the collector screen.

The removal of electrons from the storage screen makes the affected areas (of the dielectric) transparent to low-velocity electrons; that is, low-velocity electrons will pass through the areas written out by the high-velocity electron beam.

Any time after the trace has been written on the storage screen, the flood gun can be turned on to spray the entire storage screen with low-velocity electrons. These electrons pass through the transparent areas but will not penetrate other areas of the storage screen. Usually, the storage screen is maintained at the same voltage level as the cathode of the flood gun, thus repelling the flood gun electrons (except for the transparent areas).

Electrons passing through the storage screen reproduce the trace on the phosphor viewing screen. In effect, the storage screen acts as a "stencil" to the low-velocity electrons from the flood gun.

When it is desired to erase the display, the voltage at the collector screen is changed, causing the storage screen to discharge (all of the electrons are restored to the dielectric atoms in the storage screen).

5-8. SAMPLING OSCILLOSCOPE

Conventional oscilloscopes are limited in bandwidth to frequencies in the megahertz region. Sampling oscilloscopes have bandwidths up to about 12 GHz. This permits the sampling oscilloscope to measure signals of extremely high frequency or to monitor wave-forms of very fast rise times (0.5 nsec or faster is typical).

The sampling oscilloscope uses a stroboscopic approach to reconstruct the input wave-form from samples taken during many recurrences of the wave-form. This technique is illustrated in Figure 5-10. In reconstructing a wave-form, the sampling pulse "turns on" the sampling circuit for an extremely short interval, and the wave-form voltage at that instant is measured. The CRT spot is positioned vertically to correspond to this voltage amplitude.

The next sample is taken during a subsequent cycle at a slightly later point on the input wave-form. The CRT spot moves horizontally a short distance and is repositioned vertically to the new voltage. In this way, the oscilloscope plots the wave-form point by point (or dot by dot), as many as 1000 samples being used to reconstruct a single, recurrent wave-form.

A bright trace is obtained regardless of sampling rate, sweep speed, or wave-form duty cycle, since each CRT spot remains "on" during the full

Samples taken during nine recurrences
of the same waveform

Reconstructed waveform

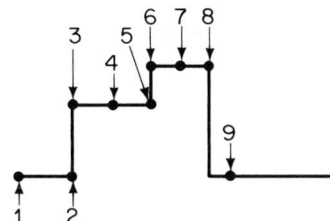

Fig. 5-10. Sampling oscilloscope/stroboscopic technique to reconstruction waveform from many samples (Hewlett-Packard).

interval between samples. For example, assume that a pulse with an extremely fast rise time were to be monitored. If the rise time were at the maximum speed of the electron beam in a conventional oscilloscope, the beam would move so quickly as to produce a weak trace on the leading and trailing edges of the wave-form. In the sampling oscilloscope, the leading and trailing edges would be sampled many times at different points from top to bottom, thus producing a bright, steady trace.

The *random sampling oscilloscope* is an improved version of the sampling oscilloscope. Operation of a random sampling oscilloscope can be divided into two separate parts: (1) timing the samples to fall somewhere within a time window (displayed portion of the wave-form or signal period), and (2) constructing the display from a series of such samples placed at random.

In a nonrandom sampling oscilloscope, the samples are taken at a specific time interval. In random sampling, there is considerable time uncertainty between the signal being sampled and the sample-taking process.

Figure 5-11 shows how the X and Y deflection signals are generated in a random sampling osciloscope. As shown, five randomly placed samples are taken of the signal. Note that these five sampls are taken on successive *repetitions* of the signal. This is known as random *equivalent-time* sampling. There is also a method of random *real-time* sampling, where many randomly placed samples are taken during a single, relatively slow, signal occurrence.

The Y component of the first sample is passed through a sampling gate

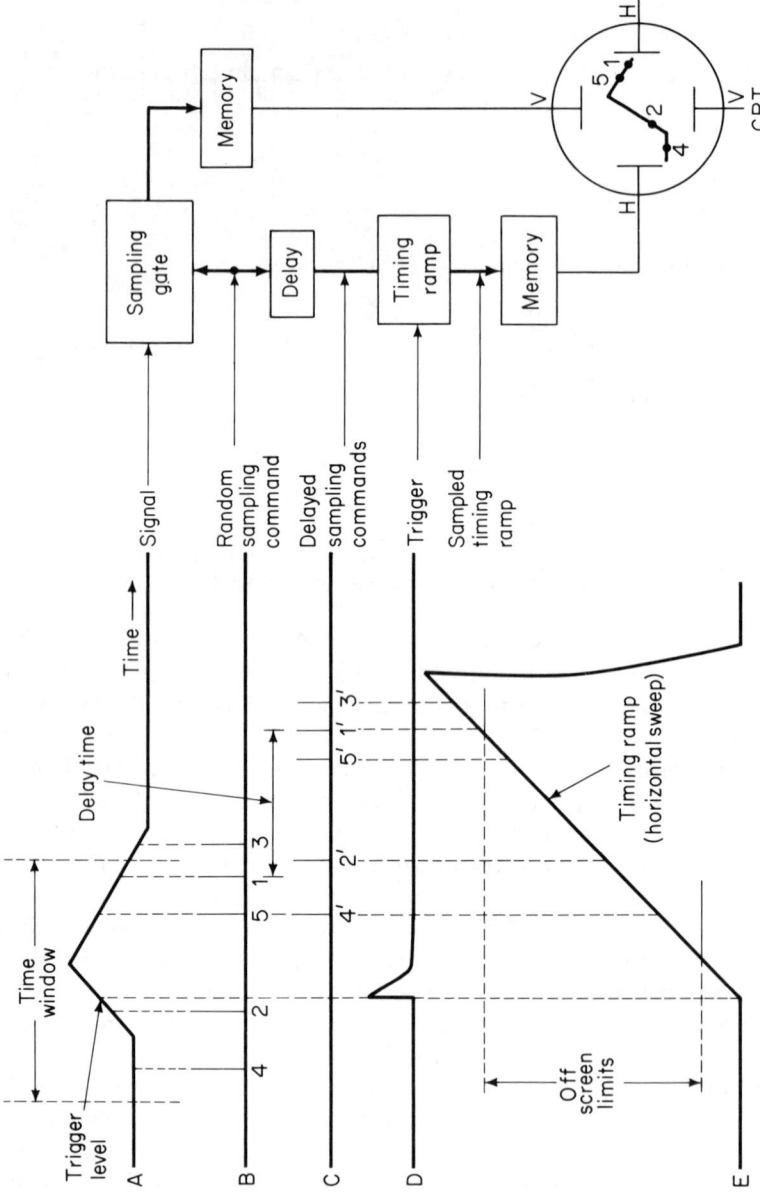

Fig. 5-11. How X and Y deflection signals are obtained in random sampling oscilloscope (Tektronix).

and held in a memory circuit for a brief period of time. The Y signal is then applied to the vertical deflection plates in the normal manner, deflecting the CRT spot vertically.

The sampling command (which opened the sampling gate) is delayed by a fixed interval, as indicated in Figure 5-11C. The delayed sampling command is used to sample a horizontal timing ramp (sawtooth sweep). The timing ramp is triggered when the input signal reaches a certain level, as shown in Figure 5-11A and D. The sample of the timing ramp is held in a memory circuit for a brief period and then applied to the horizontal deflection plates in the normal manner, deflecting the CRT spot horizontally.

The process is repeated for each of the remaining samples. Each sample supplies both vertical and horizontal information to deflect the CRT beam from dot to dot, thus constructing a display of the signal from those samples that fall within the "time window." Note that sample 3 is not displayed, even though it is sampled and available at the vertical deflection plates. This is because the horizontal sweep has already driven the beam off the screen (horizontally) before the horizontal sweep ramp is sampled.

If the delay time interval is increased, a time shift of the sampling distribution to the left (earlier in time) is necessary in order that the required information be collected for the display. Also note that although only five samples are shown, a typical random sampling oscilloscope might use 1000 samples to reconstruct such a wave-form.

Figure 5-12 is a complete operational block diagram of the random sampling oscilloscope, including those circuits that control the distribution of samples across the time window. The following is a description of the basic circuit functions.

The *trigger recognizer and holdoff* block responds to the presence of a suitable trigger (when the input signal has reached a certain level—Figure 5-11A), and immediately starts the timing ramp. The holdoff function prevents re-starting the timing ramp until it has sufficient time to complete its previous cycle.

The *timing ramp* is a linear ramp whose slope is also controlled by the TIME/DIV switch (refer to Figure 5-4). The starting and subsequent sampling of this ramp is shown in Figure 5-11E.

The sampled and stored levels of the timing ramp are available at the output of the *horizontal memory*, where they supply the horizontal (X-axis) data to the display.

The *rate-meter* block also receives an input indicating that a trigger has been recognized (the input signal has reached the trigger level) and proceeds to measure the repetition rate of such recognitions over an extremely wide range of trigger repetition rates. On the basis of this measurement and an error signal supplied by the *differential comparator*, the rate meter then supplies the *slewing ramp* with a pretrigger signal. This pretrigger is the rate

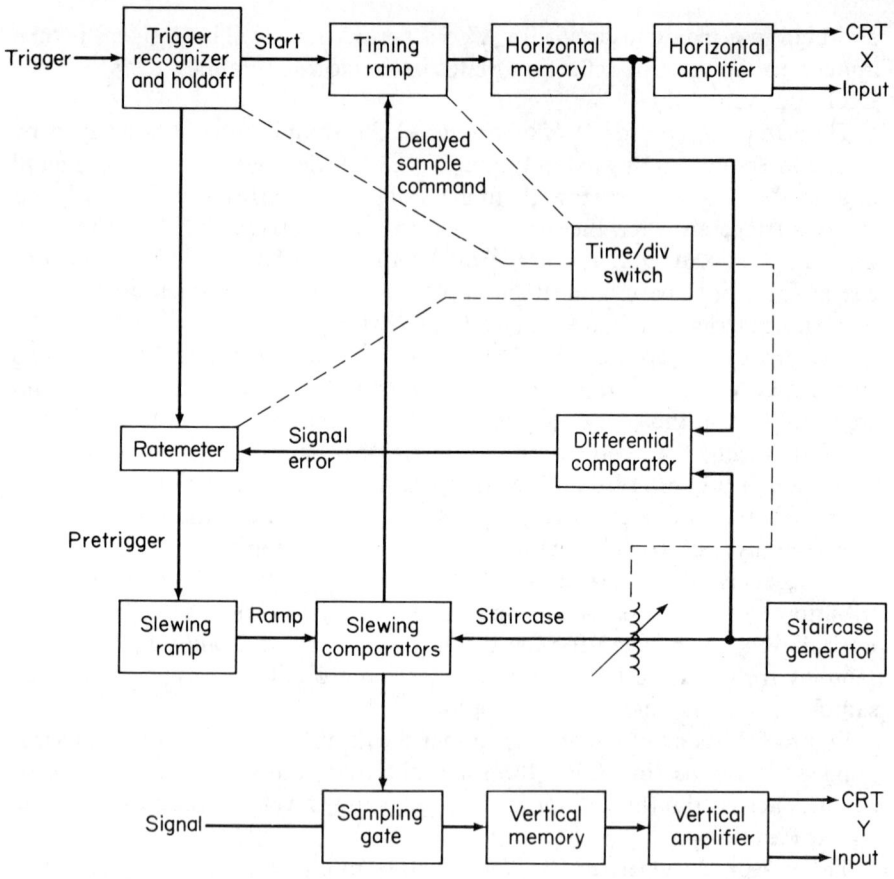

Fig. 5-12. Basic circuit of random sampling oscilloscope (Tektronix).

meter's "best guess" as to when to start the slewing ramp. The pretrigger may contain considerable time uncertainty on a sample-to-sample basis. If a number of successive samples is taken later than desired, an appropriate error signal arrives to cause the rate meter to start the slewing ramp earlier for the next few cycles.

The *slewing ramp* is a linear ramp that is started on command from the rate meter. The slope of the slewing ramp is controlled by the TIME/DIV switch.

The *slewing comparator* provides both the sampling commands and the delayed sampling commands, shown in Figure 5-11B and C. The delayed sampling command is issued when the relatively fast slewing ramp reaches the voltage level of the *staircase generator* output. Successive sweeps of the slewing ramp find the staircase at slightly higher levels. Thus the resulting comparisons and sampling comands are successively delayed, or *slewed in time*.

The delayed sampling commands are generated in a similar fashion from the slewing ramp and staircase, but a *d-c offset* added to the staircase causes these comparisons to occur later by the fixed *time interval* (Figure 5-11C).

The *differential comparator* receives both the horizontal signal and the staircase and generates an error signal when the horizontal signal does not track along with the staircase on the basis of an *average* of many samples. The staircase generates its signal at a constant rate. The horizontal signal is dependent on the average of the samples taken. Thus a "closed-loop" system is formed, causing a random sampling distribution to slew across the time window under control of the staircase generator.

The output of the slewing comparators sends a command to the sampling gate, permitting the signal to pass to the vertical memory circuit, as shown in both Figures 5-11 and 5-12.

5-9. VECTORSCOPE

A vectorscope is used in service of color television equipment. The vectorscope can be operated in conjunction with a rainbow generator (discussed in Chapter 4) to check a color television receiver's response to color signals. As discussed in Chapter 4, the *colors produced* by the receiver are dependent on the *phase relationship* of the 3.58-MHz color signal and the 3.58-MHz color reference burst. The rainbow generator produces ten color signals, each spaced 30° apart, resulting in a display of ten corresponding colors (arranged as vertical stripes or a "rainbow" across the color television screen).

A vectorscope permits the phase relationship (and amplitude) of the ten color signals to be displayed as a single pattern. The vectorscope monitors the color signals directly at the color-gun grids of the color picture tube. This makes the vectorscope adaptable to any type of color circuit. By comparing the vectorscope display against that of an ideal display (for phase relationship, amplitude, general appearance, and so on) the condition of the receiver color circuits can be analyzed.

A conventional oscilloscope can be used as a vectorscope. However, this would require special connections to the horizontal and vertical deflection plates. Most commercial vectorscopes are conventional oscilloscopes, with special deflection circuits to provide the vector display. The vectorscope provisions are usually selected by means of the oscilloscope's MODE selector, or some similar control.

5-9.1. How a Vector Display Is Obtained

A vector pattern is composed of two signals; one is applied to the horizontal plates and the other to the vertical plates. The result is a

190   Oscilloscopes and Recorders

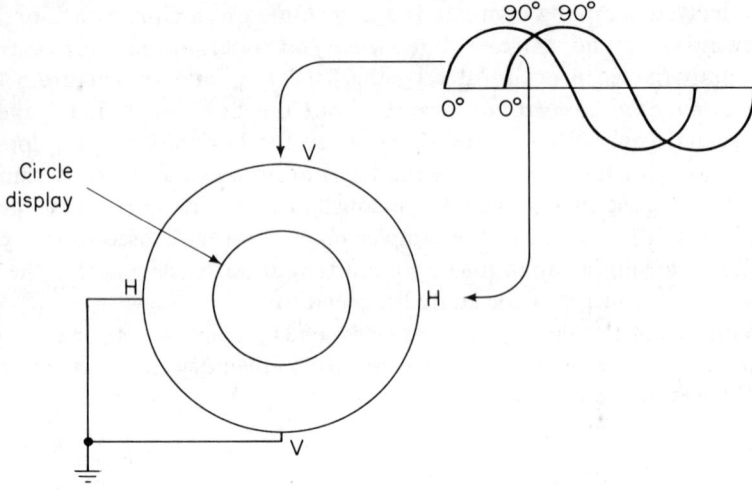

Fig. 5-13. How circular Lissajous pattern is formed (two sinewaves 90° out-of-phase).

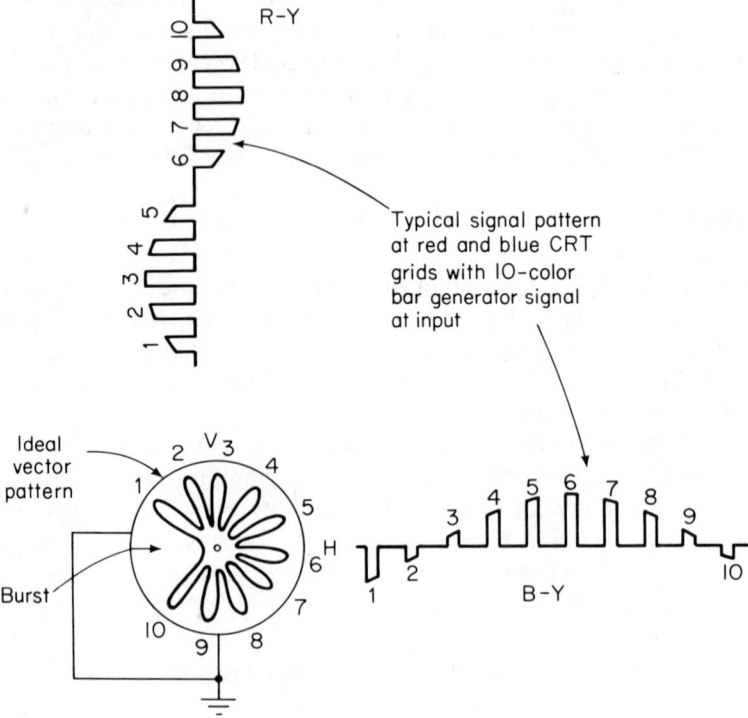

Fig. 5-14. How basic vector pattern is formed on a vectorscope (Sencore).

Lissajous pattern (as discussed in later sections of this chapter). If two sine waves are applied and one is 90° out of phase but at the same frequency, the result will be a circle, as shown in Figure 5-13.

If the patterns formed by a standard ten-color-bar generator (keyed rainbow generator) are taken from the red and blue grids of a color CRT and these signals are applied to the vertical and horizontal deflection plates of an oscilloscope, the result will be a display similar to that shown in Fig. 5-14. This pattern is referred to as a *vector pattern*. The phase between the *R-Y* (red-luminance) and *B-Y* (blue-luminance) signals can readily be seen on the screen of the oscilloscope. A complete circle represents 360°. Therefore, the screen can be marked off in degrees, and measurements can be made using the vector patterns. The length of the arm of the "petals" (patterns produced by each of the ten color signals) may vary, depending on the strength of the signal. Likewise, the entire pattern will rotate to a different position of the color receiver when tint control is improperly set. (The tint control changes colors by slightly shifting the phase relationship of the color signals to the color burst.)

5-9.2. Practical Vectorscope Connections

The horizontal and vertical inputs of a vectorscope must be connected to the red and blue grids of the color CRT. (See Figure 4-45 for a diagram showing connections between the color circuits and the CRT). On some solid-state color receivers, the color input is made at the color-gun cathodes instead of the grids. Either way, the green gun need not be connected to present a vector display.

On some vectorscopes, the leads are permanently attached and are color-coded (red lead to red color gun, blue lead to blue gun). Other vectorscopes have connectors for the color-gun leads and a switch that converts the instrument from a conventional oscilloscope to a vectorscope. A few models of vectorscopes are complete color analyzer systems, containing a color generator, as well as the vectorscope. This makes it possible to connect the output leads of the generator to the antenna input of the receiver, while the resulting color signals are monitored at the color CRT. On other instruments, the color generator provides an output at the video frequency, rather than (or in addition to) the r-f output. This makes it possible to introduce color signals directly into the color circuits, bypassing the r-f, i-f, and detector stages.

5-9.3. Interpreting Vectorscope Patterns

Figure 5-14 is an "ideal" vector pattern, and will rarely be found in practical applications. Figure 5-15 is a more realistic vector pattern. The display of Figure 5-15 is that produced by color sets using a 90° demodulator system (typical of the Zenith color system).

192 Oscilloscopes and Recorders

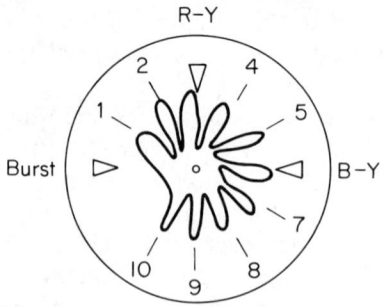

Fig. 5-15. Vectorscope display produced by color sets using a 90° demodulator system (Sencore).

In a typical vectorscope, the reference burst signal is considered as 0° (or no phase shift). This is represented by the 0, or burst mark, on the vectorscope screen. The red bar of the standard ten-color-bar pattern is the third bar and is 90° from the burst. This will be at right angles to the burst line and at the top of the vectorscope screen. The red bar is marked *R-Y*. The Nos. 1 and 2 represent the first and second bars of the pattern and are all 30° apart on the screen. The *B-Y* mark (blue) is 90° away from the *R-Y* mark and 180° from the burst.

5-9.4. Making Adjustments with Vectorscope Patterns

The vectorscope pattern is very useful for adjusting the phase angle between the red and blue demodulators for a 90°, 105°, or 116° demodulator system.

To align a 90° demodulator system (Zenith-type, Figure 5-15), adjustment is made on the color circuits until the third bar rests on the *R-Y* mark, while the sixth bar is on the *R-Y* mark.

If the set calls for a demodulator angle of 105° (General Electric-type, Figure 5-16), then the third bar should rest on the *R-Y* mark, while the sixth bar falls halfway between the *B-Y* and No. 7. Note that this pattern of Figure 5-16 is not as round as that of Figure 5-15. This is due to the increased angle of demodulation and is normal.

As the angle is increased to 116° (Motorola-type, Figure 5-17), the pattern appears more squared, as shown.

As can be seen, the overall shape of vector patterns will describe the type of demodulation used, as well as the existing state of adjustment.

The *tint range* of a color receiver can be set quite easily, using the vector pattern. By rotating the receiver tint (or hue) control from one end of its range to the other, the third bar (*R-Y*) should be able to be adjusted from the No. 2 mark, through *R-Y*, to the No. 4 mark on the oscilloscope screen. This indicates a range of 30° on each side of the proper setting. Normally, the third bar should be exactly at the *R-Y* mark when the tint control is at mid-scale.

Alignment of an off-frequency 3.58–MHz reference oscillator is a relatively simple procedure, using the vector pattern. When the correction signal from the color phase detector to the 3.58-MHz oscillator (Figure 4-45) is removed,

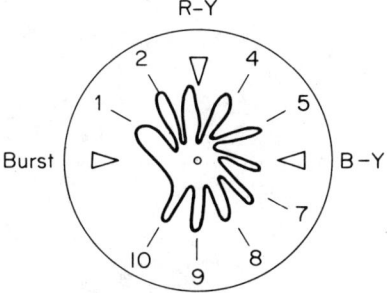

Fig. 5-16. Vectorscope display produced by color sets using a 105° demodulator system (Sencore).

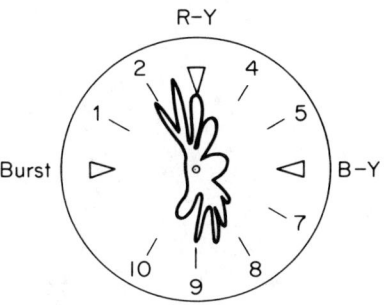

Fig. 5-17. Vectorscope display produced by color sets using a 116° demodulator system (Sencore).

the oscillator will be free-running. (Usually, the correction signal can be disabled at a test point in the color circuit.) With the reference oscillator in a free-running condition, the pattern will appear to rotate or will appear as a blurred circle, depending on how far the oscillator is off frequency. (A free-running reference oscillator can also be verified by a "barber-pole" effect on the CRT screen.) The 3.58-MHz oscillator is adjusted until the vectorscope pattern stands still, or as close as possible to a motionless condition.

It is also possible to touch up bandpass amplifier alignment using a vectorscope. However, complete alignment of any stage in a television receiver should be accomplished by using a sweep generator and conventional oscilloscope, preferably with a marker generator. Refer to Chapter 4. A poorly aligned bandpass amplifier will produce a display similar to that of Figure 5-18.

5-9.5. Troubleshooting with Vectorscope Patterns

The vectorscope is an efficient troubleshooting tool for color television circuits. For example:

A loss of the *R-Y* signal will cause no vertical deflection and the pattern will appear as shown in Figure 5-19. The *B-Y* signal will deflect the beam along the horizontal axis, producing a bright line. This indicates that the trouble lies in the *R-Y* demodulator, matrix, or difference amplifier, depending on the circuit used in the receiver. If the *R-Y* signal is weak, some deflection will be noted and an extremely distorted pattern will result, again pointing to the *R-Y* circuits.

A loss of *B-Y* signal will result in no horizontal deflection and appear as shown in Figure 5-20. The *R-Y* signal will cause the beam to deflect vertically. This indicates that the problem is in the *B-Y* difference amplifier, matrix,

194 Oscilloscopes and Recorders

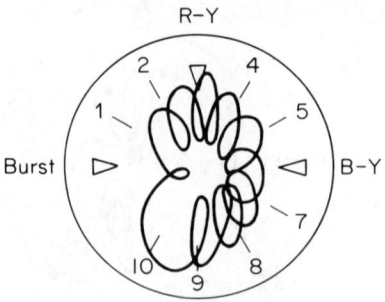

Fig. 5-18. Vectorscope display produced by color set with poorly aligned bandpass amplifier (Sencore).

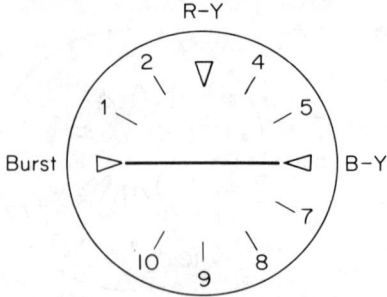

Fig. 5-19. Vectorscope display produced by color set with loss of R-Y signal; no vertical deflection (Sencore).

or demodulator circuits. Again, if the *B-Y* is weak, some deflection will be noted, and a distorted display will appear.

If there is a complete loss of color, the pattern will appear some what like that of Figure 5-21.

Keep in mind that not all color circuits produce identical vectorscope patterns. For example, as demodulation angle is increased (such as with General Electric and Motorola sets) the pattern will appear more "square" than with 90° demodulation systems (such as Zenith), even though all of the receivers are operating properly.

The patterns discussed here are for reference only and must be considered as typical. Always consult the vectorscope instruction manual and all service data on the receiver being tested.

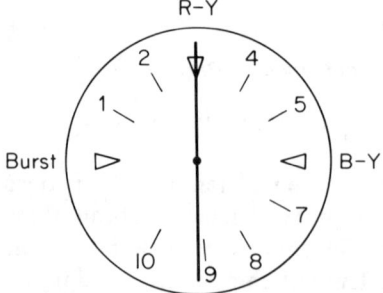

Fig. 5-20. Vectorscope display produced by color set with loss of B-Y signal; no horizontal deflection (Sencore).

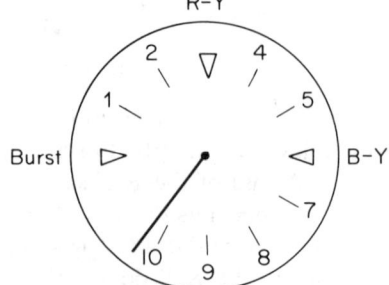

Fig. 5-21. Vectorscope display produced by color set where there is a complete loss of color (Sencore).

5-10. CURVE TRACERS

The most practical means of measuring transistor characteristics in the laboratory is to display the transistor characteristics as *curve traces* on an oscilloscope. Since the oscilloscope screen can be calibrated in voltage and current, the transistor characteristics can be read off or from the screen directly. If a number of curves are made with an oscilloscope, they can be compared with the curves drawn on transistor data sheets. (Some data sheet curves are reproductions of those obtained by tracing curves on an oscilloscope.)

There are a number of oscilloscopes (or oscilloscope adapters) manufactured specifically to display transistor and diode characteristic curves. The Tektronix transistor curve tracer is a typical unit. Some curve tracers are for only one type of transistor, while other instruments will display the characteristics for many types [field-effect transistors or FETs, unijunction transistors, thyristors, and PNPN devices], as well as diodes (signal, rectifying, Zener, tunnel, step, and so on).

A complete discussion of all curve-tracing circuits is beyond the scope of this publication. However, following are descriptions of the most common curve-tracing techniques.

5-10.1. Transistor Curve Tracers

Figure 5-22 is a functional block diagram of a typical transistor curve tracer. Such a tracer introduces changes in base current in the form of equal-value steps, that is, steps of selectable known amounts. These steps occur at the same rate as the collector supply voltage is swept between 0 V and some peak value and back to zero, producing a separate curve corresponding to each different value of base current.

When curves depict collector current versus collector voltage (for different values of base current), the change in collector current induced by one step of base current will be *proportional* to the vertical distance between adjacent curves. This change can be read directly from the scale.

Which vertical line is chosen for the scale will depend on what collector voltage was specified, because each vertical line corresponds to a particular collector voltage.

5-10.2. Transistor Gain

The basic technique for measurement of small-signal short-circuit forward-current transfer ratio (beta or gain) of a transistor, using a

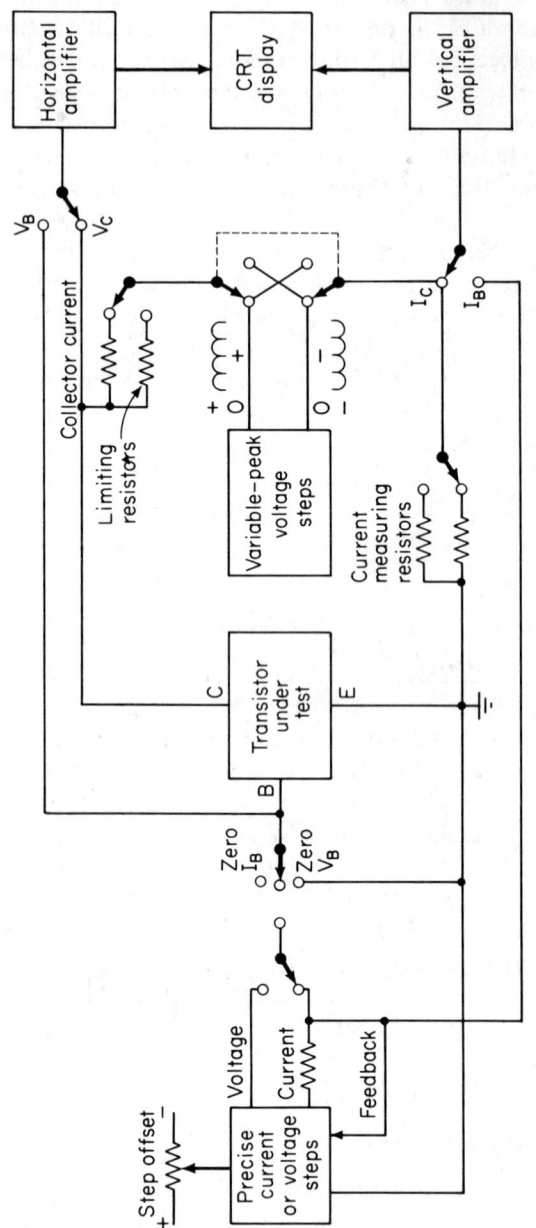

Fig. 5-22. Typical transistor curve tracer (Tektronix).

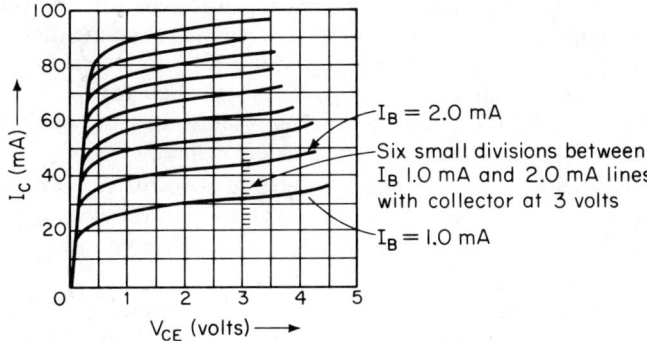

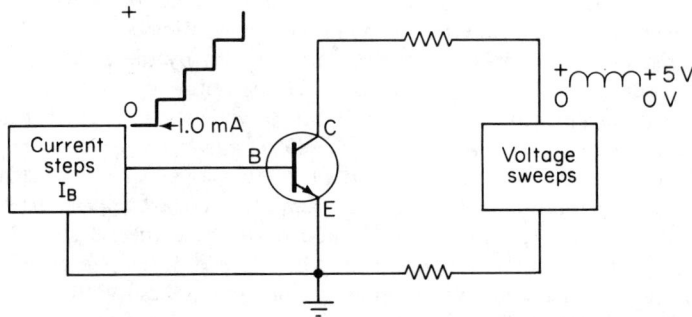

Fig. 5-23. Basic test connections and typical oscilloscope displays for transistor curve tracers (Tektronix).

transistor curve tracer, is as follows. Figure 5-23 shows the basic test connections, as well as typical oscilloscope displays.

1. The transistor is connected into a grounded-emitter circuit. Precise 1-mA current steps are applied to the base. Voltage sweeps from 0 to approximately 5 V are applied to the collector. The oscilloscope vertical deflection is obtained from the collector voltage sweeps, while horizontal deflection is obtained from the resultant collector current. All of this is accomplished by setting of switches on the curve tracer.
2. On some curve tracers, each sweep must be initiated individually, while other instruments will produce a series of ten (or more) curves in sequence, automatically.
3. Once the curves are made, they must be interpreted, as follows.
4. Choose the vertical line corresponding to the specified collector voltage. For example, the 3-V vertical line is chosen in Figure 5-23.
5. Note (on that line) the distance between the two curves that appear above

and below the specified base current (or specified collector current) and lie adjacent to the specified current. For example, assume that the transistor of Figure 5-23 is to be operated with a collector current of 30–40 mA, with 3 V at the collector. The two most logical curves would then be the 1- and 2-mA base current curves. The distance between these two curves (on the 3-V line) would be approximately 6 small divisions.

6. Translate that distance to the difference in collector current according to the current per division of the scale. One small division on the vertical (collector current) scale equals 2 mA, so the difference in collector current would be 12 mA.
7. Divide that collector current difference by the base current difference that caused it, depending on the current per step. Since the base current per step is 1 mA and the difference in collector current is 12 mA, the gain (beta) is 12. Had the base current per step been 0.1 mA and all other conditions the same, the beta would have been 120 (12/0.1 = 120).
8. When many transistors of similar characteristics are to be tested, it is often convenient to use an alternative method for computing beta from curve traces. First divide the collector current per division by the base current per step to determine the *beta per division*. In the example of Figure 5-23, the collector current per small division is 2 mA, while the base current per step is 1 mA (throughout the scale). Therefore, the beta per division is 2. This can be multiplied by the number of divisions between sweeps, on any vertical line, to find the beta. For example, there are approximately four small divisions between the 5-mA and 6-mA base current steps (on the 3-V line). This results in a beta of 8 (4 divisions times a beta per division of 2). For a further analysis of interpreting curve traces for all types of transistors (including FET and unijunction), refer to the author's Practical Semiconductor Data Book, published by Prentice-Hall, Inc.).

5-11. RECORDERS

There are two basic types of recorders used as electronic test equipment; the strip-chart or roll-chart recorder and the *X-Y* recorder (or plotter).

5-11.1. Strip-Chart Recorder

The strip-chart recorder consists essentially of a drive motor, chart paper, stylus, servo motor, follow-up potentiometer, and amplifier. As shown in Figure 5-24, the drive motor pulls a roll of chart paper past the stylus at a uniform rate. Usually, the chart paper is marked in units of time so that the recorder will provide a permanent time-related record (or log) of the signal being measured. The stylus is moved across the chart paper (at right angles to the chart paper line of motion) by the servo, which, in turn, receives its input voltage from the amplifier.

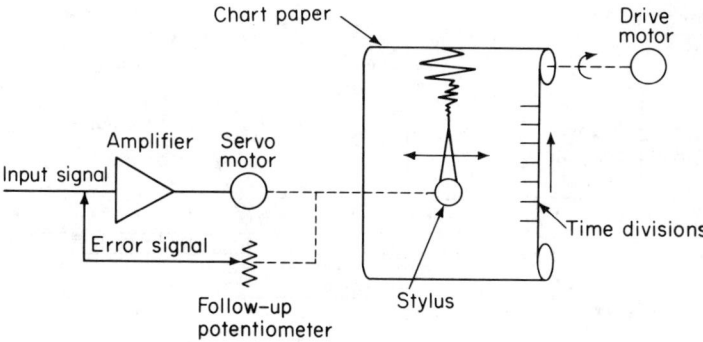

Fig. 5-24. Basic strip-chart or roll-chart recorder system.

The direction of servo motor (and stylus) motion is determined by the input signal polarity, while the amplitude of motion is determined by signal amplitude. As the stylus moves, the follow-up potentiometer produces an error signal or voltage proportional to the stylus position. The follow-up potentiometer voltage is of opposite polarity to that of the input signal. Thus, when the stylus has moved to a position corresponding to the input signal, the follow-up potentiometer output cancels the input signal, stopping the stylus. Thus the stylus constantly maintains a position that corresponds to the input signal and traces out a line on the chart paper that corresponds to the input.

Chart recorders are particularly useful in recording values over a long period of time, such as monitoring industrial processes. For example, assume that it is desired to monitor a pressure over a 24-hr period. The pressure is converted to a voltage by a *transducer* (Chapter 9). The voltage is fed to the recorder amplifier input, and the stylus moves to a position corresponding to the voltage (which represents the pressure). The drive motor is started, and the time (and/or date) is recorded on the paper or is noted. The recorder will then maintain a constant record of pressure. This record can be retained for future use.

5-11.2. X-Y Recorder (Plotter)

The X-Y recorder (described as an X-Y plotter by several manufacturers) is a two-axis instrument that records a one-shot display on a single (nonmoving) chart or graph. Both the vertical and horizontal axes are identical (or nearly so), and are similar in operation to the single axis of the strip-chart recorder. However, in an X-Y recorder the stylus is moved by two (horizontal and vertical) servo motors, which in turn receive their input from corresponding amplifiers. (See Figure 5-25.)

Often, the horizontal axis servo system will include a sweep oscillator

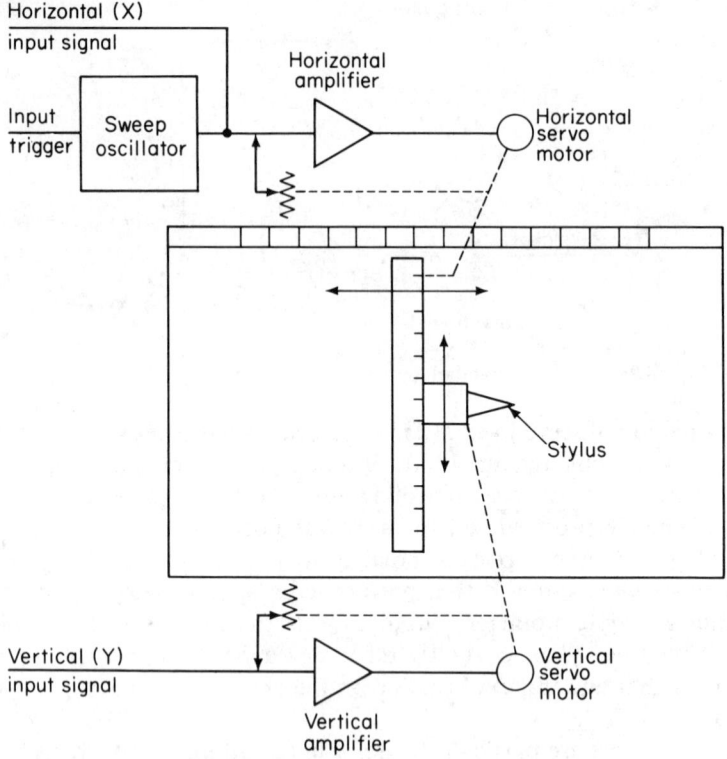

Fig. 5-25. Basic X-Y recorder or plotter system.

similar to that of an oscilloscope. The sweep oscillator produces a sawtooth voltage that moves the stylus horizontally at a fixed rate. The sawtooth sweep can be triggered internally at a fixed rate or by external signals, or it can be bypassed so that the horizontal system will respond to sine waves.

No matter what inputs or sweeps are used, the stylus constantly maintains a position that corresponds to the two input signals. Thus the display of an *X-Y* recorder is similar to that of an oscilloscope, except that the display is on paper (usually on a one-shot or single-sweep basis). In some cases, the same chart paper is used several times to superimpose displays.

5-12. OSCILLOSCOPE OPERATING PRECAUTIONS

In addition to the General Safety Precautions described in the Introduction, the following specific precautions should be observed when operating any type of oscilloscope. Most of the precautions also apply to recorders.

1. Even if you have had considerable experience with oscilloscopes, always study the instruction manual of any oscilloscope with which you are not familiar.
2. Use the procedures of Section 5-13 to place the oscilloscope in operation. It is good practice to go through the procedures each time that the oscilloscope is used. This is especially true when the oscilloscope is used by other persons. The operator cannot be certain that position, focus, and (especially) intensity controls are at safe positions, and the oscilloscope CRT could be damaged by switching it on immediately.
3. As in the case of any cathode-ray-tube device (such as a television receiver), the CRT spot should be *kept moving* on the screen. If the spot must remain in one position, keep the intensity control as low as possible.
4. Always use the *minimum intensity* necessary for good viewing.
5. If at all possible, avoid using an oscilloscope in direct sunlight or in a brightly lighted room. This will permit a low-intensity setting. When the oscilloscope must be used in a bright light, use the viewing hood.
6. Make all measurements in the center area of the screen. Even if the CRT is flat, there is a chance of reading errors caused by distortion at the edges.
7. Use only shielded probes. Never allow your fingers to slip down to the metal probe tip when the probe is in contact with a "hot" circuit.
8. Avoid operating an oscilloscope in strong magnetic fields. Such fields can cause distortion of the display. Most quality oscilloscopes are well shielded against magnetic interference. However, the face of the CRT is still exposed and is subject to magnetic interference.
9. Most oscilloscopes and their probes have some maximum input voltage specified in the instruction manual. Do not exceed this maximum. Also, do not exceed the maximum line voltage or use a different power frequency.
10. Avoid operating the oscilloscope with the shield or case removed. Besides the danger of exposing high-voltage circuits (several thousand volts are used on the CRT), there is the hazard of the CRT's imploding and scattering glass at high velocity.
11. Avoid vibration and mechanical shock. As is most electronic equipment, an oscilloscope is a delicate instrument.
12. If an internal fan or blower is used, make sure that it is operating. Keep ventilation air filters clean.
13. Do not attempt repair of an oscilloscope unless you are a qualified instrument technician. If you must adjust any internal circuits, follow the instruction manual.
14. Study the circuit under test before making any test connections. Try to match the capabilities of the oscilloscope to the circuit under test. For

example, if the circuit has a range of measurements to be made (a-c, d-c, r-f, pulse), you must use a wideband d-c oscilloscope with a low-capacitance probe (Chapter 9) and possibly a demodulator probe. Do not try to measure 3-MHz signals with a 100-kHz bandwidth oscilloscope. On the other hand, it is wasteful to use a dual-trace 50-MHz laboratory oscilloscope to check out the audio sections of transistor radios.

15. The most important oscilloscope operating precautions are summarized in Figure 5-26.

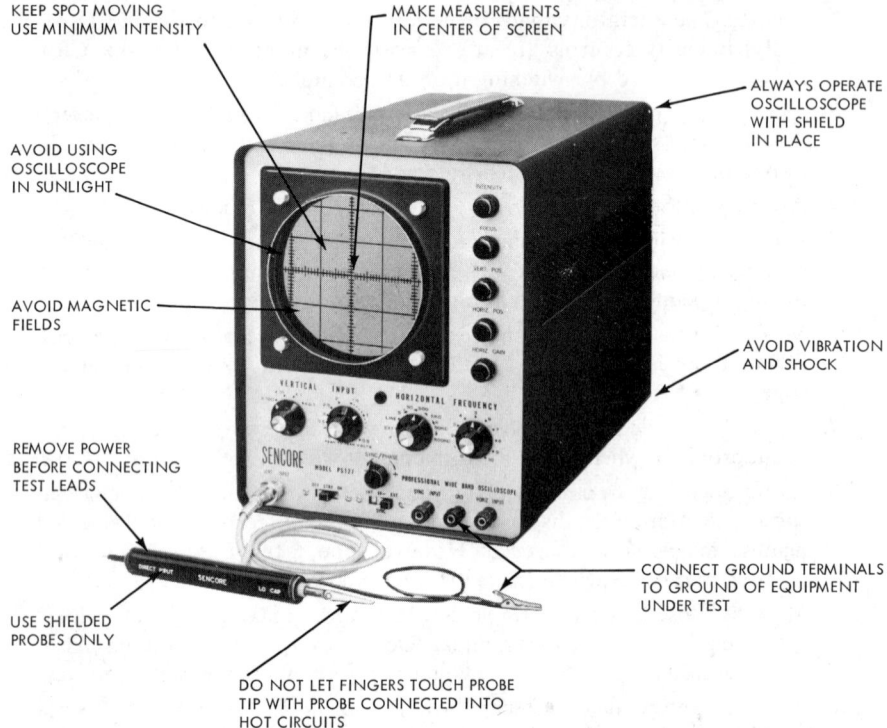

Fig. 5-26. Summary of oscilloscope operating precautions (Sencore).

5-13. PLACING AN OSCILLOSCOPE IN OPERATION

After the oscilloscope manual's setup instructions have been digested, they can be compared with the following *general or typical* procedures. Always follow the manual's procedures in case of conflict.

1. Set the power switch to OFF.
2. Set the internal recurrent sweep to OFF.

3. Set the focus, gain, intensity, and sync controls to their lowest position (usually full counterclockwise).
4. Set the sweep selector to external.
5. Set the vertical and horizontal position controls to their approximate midpoint.
6. Set the power switch to ON. It is assumed that the power cord has been connected. This is always a good idea.
7. After a suitable warm-up period (as recommended by the manual) adjust the intensity control until the trace spot appears on the screen. If a spot is not visible at any setting of the intensity control, the spot is probably off-screen (unless the oscilloscope is defective). If necessary, use the vertical and horizontal position controls to bring the spot into view. Always use the lowest setting of the intensity control needed to see the spot. This will prevent burning the oscilloscope screen.

It should be noted that d-c oscilloscopes need longer warm-up time than a-c oscilloscopes because of the drift problems associated with d-c amplifiers (as discussed in Chapter 7).

8. Set the focus control for a sharp, fine dot.
9. Set the vertical and horizontal position controls to center the spot on-screen.
10. Set the sweep selector to internal. This should be the linear internal sweep, if more than one internal sweep is available.
11. Set the internal recurrent sweep to ON. Set the sweep frequency to any frequency or recurrent rate higher than 100 Hz.
12. Adjust the horizontal gain control and check that the spot is expanded into a horizontal trace or line. The line length should be controllable by adjusting the horizontal gain control.
13. Return the horizontal gain control to zero (or its lowest setting). Set the internal recurrent sweep to OFF.
14. Set the vertical gain control to the approximate midpoint. Touch the vertical input with your finger. The stray signal pickup should cause the spot to be deflected vertically into a trace or line. Check that the line length is controllable by adjustment of the vertical gain control.
15. Return the vertical gain control to zero (or its lowest setting).
16. Set the internal recurrent sweep to ON. Advance the horizontal gain control to expand the spot into a horizontal line.
17. If required, connect a probe to the vertical input.
18. The oscilloscope should now be ready for immediate use. Depending on the test to be performed, the oscilloscope may require calibration. Voltage and current calibration procedures are described in later sections of this chapter.

5-14. RECORDING AN OSCILLOSCOPE TRACE

In many applications, an oscilloscope trace must be recorded and not merely viewed. It is obviously much easier to measure and study a permanent record than an oscilloscope trace. The X-Y recorder fills this need. However, the X-Y recorder is limited to recording of very low-frequency signals. Most high-speed recording is done with an oscilloscope and Polaroid Land camera combination. The camera is equipped with a special lens and a mounting frame attached to the oscilloscope. It is also possible to record with a conventional camera, with a moving-film camera, or even by hand in some situations.

Polaroid camera systems are the most popular for oscilloscope recording because they provide an immediate record with no wait for processing. If desired, negative-type Polaroid film can be used to provide additional prints of the recorded trace.

Figure 5-27 is a photograph of a typical oscilloscope camera system. Figure 5-28 shows the basic operating principles of the system.

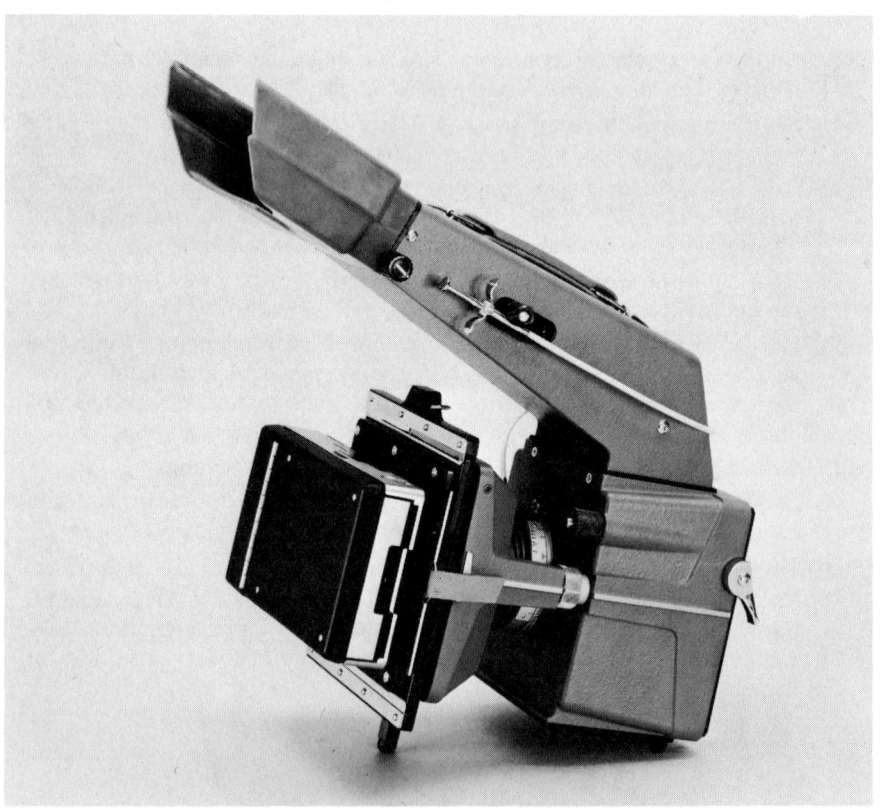

Fig. 5-27. Typical oscilloscope camera (Tektronix).

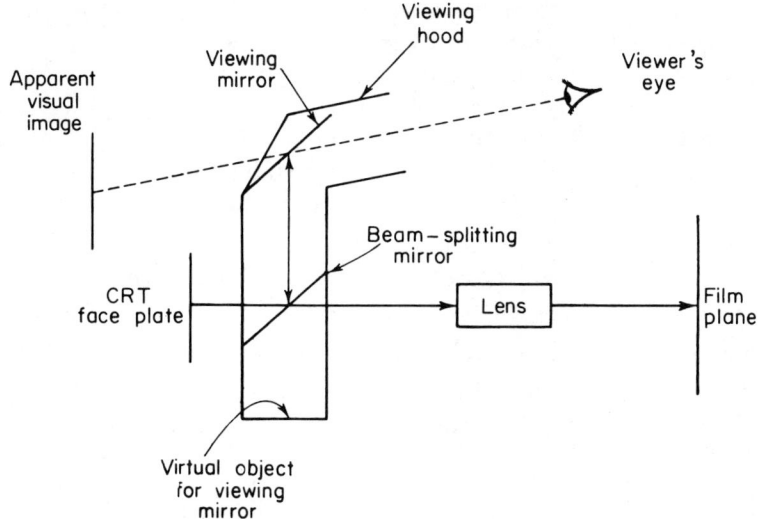

Fig. 5-28. Basic operating principles of typical oscilloscope camera system (Tektronix).

The viewing system consists of a viewing hood and two mirrors. Light from the oscilloscope screen strikes the beam-splitting mirror, where a portion of the light is transmitted to the camera lens and another portion is reflected to the second mirror. A virtual image acts as the object for the second-mirror surface. The second mirror then forms a virtual image that is viewed by the observer. Owing to the 45° arrangement of the beam-splitting mirror, the observer views the oscilloscope display as though he were looking directly toward the oscilloscope screen on a line perpendicular to the screen. This view is full size, but the image appears approximately 20 in. away. In all cases, the lens is considerably closer to the oscilloscope screen. The difference in the two distances produces a small amount of parallax between the viewed and photographed images. The small amount of parallax can usually be ignored.

A special mount (known as a bezel) is used to attach the camera to the oscilloscope. A lift-off mounting is used so that the camera can easily be mounted or removed. Swing-away hinges allow the camera to be swung out of the way when not in use.

5-15. MEASURING VOLTAGE AND CURRENT WITH AN OSCILLOSCOPE

The oscilloscope has both advantages and disadvantages when used to measure voltage and current. The most obvious advantage is that the oscilloscope shows wave form, frequency, and phase simultaneously with the amplitude of the voltage or current being measured. The volt-ohmmeter or electronic voltmeter described in Chapters 1 and 2 shows only amplitude. If the only value of interest is voltage or current amplitude, use the meter because of its simplicity in readout. Use the oscilloscope where wave-shape characteristics are of equal importance to amplitude.

The following sections describe *basic* procedures for voltage and current measurements, as well as *basic* calibration procedures.

5-15.1. Peak-to-Peak Measurements with Alternating-Current Laboratory Oscilloscopes

1. Connect the equipment as shown in Figure 5-29.
2. Place the oscilloscope in operation (Section 5-13).
3. Set the vertical step-attenuator to a deflection factor that will allow the

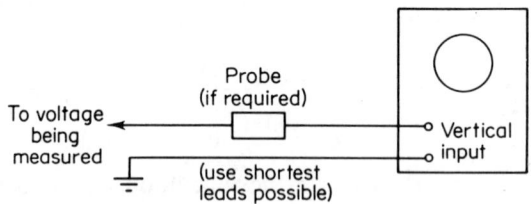

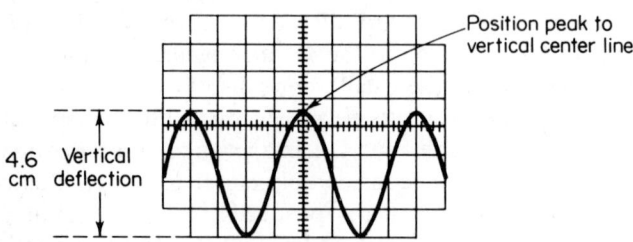

Fig. 5-29. Measuring peak-to-peak voltages.

expected signal to be displayed without overdriving the vertical amplifier.

4. Set the input selector to measure ac. Connect the probe to the signal being measured.
5. Switch on the oscilloscope internal recurrent sweep.
6. Adjust the sweep frequency for several cycles on the screen.
7. Adjust the horizontal gain control to spread the pattern over as much of the screen as desired.
8. Adjust the vertical position control so that the downward excursion of the

wave-form coincides with one of the screen lines below the centerline, as shown in Figure 5-29.

9. Adjust the horizontal position control so that one of the upper peaks of the signal lies near the vertical centerline.
10. Measure the peak-to-peak vertical deflection in divisions (usually centimeters).
11. Multiply the distance measured in step 10 by the vertical attenuator switch setting. Also include the attenuation factor of the probe, if any.

For example, assume a peak-to-peak vertical deflection of 4.6 divisions (Figure 5-29), using a $\times 10$ attenuator probe and a vertical deflection factor of 0.5 V/division:

$$\text{Volts (peak-to-peak)} = 4.6 \times 0.5 \times 10 = 23 \text{ V}$$

If the voltage being measured is a *sine wave*, the peak-to-peak value can be converted to peak, RMS, or average, as shown in Figure 1-13, and vice versa.

5-15.2. Peak-to-Peak Measurements with Alternating-Current Shop Oscilloscopes

1. Connect the equipment as shown in Figure 5-29.
2. Place the oscilloscope in operation (Section 5-13).
3. Set the vertical gain control to the calibrate-set position, as determined during the calibration procedure (refer to Section 5-15.4).
4. Set the input selector (if any) to measure ac. Connect the probe to the signal being measured.
5. Switch on the oscilloscope internal recurrent sweep.
6. Adjust the sweep frequency for several cycles on the screen.
7. Adjust the horizontal control to spread the pattern over as much of the screen as desired.
8. Adjust the vertical position control so that the downward excursion of the wave-form coincides with one of the screen lines below the centerline, as shown in Figure 5-29.

   Do not move the vertical gain control from the calibrate-set position. Use the vertical position control only.
9. Adjust the horizontal position control so that one of the upper peaks of the signal lies near the vertical centerline.
10. Measure the peak-to-peak vertical deflection in divisions.
11. Multiply the distance measured in step 10 by the calibration factor (volts per division) established during calibration. Also, include the attenuation factor of the probe (if any) and the setting of the attenuator-step switch (if any).

    For example, assume a peak-to-peak vertical deflection of 4.6 divisions (Figure 5-29) using a ×10 attenuator probe, a ×10 position of the step-attenuator, and a calibrating factor of 0.5 V/division.

$$\text{Volts (peak-to-peak)} = 4.6 \times 0.5 \times 10 \times 10 = 230 \text{ V}$$

If the voltage being measured is a *sine wave*, the peak-to-peak value can be converted to peak, RMS, or average, as shown by Figure 1-13, and vice versa.

5-15.3. Instantaneous Voltage Measurements with Direct-Current Oscilloscopes

1. Connect the equipment as shown in Figure 5-30.
2. Place the oscilloscope in operation (Section 5-13).
3. Set the vertical step-attenuator to a deflection factor that will allow the

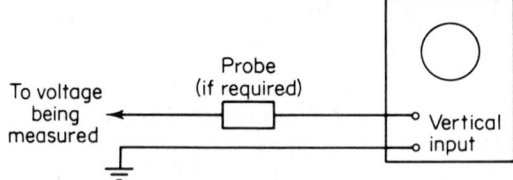

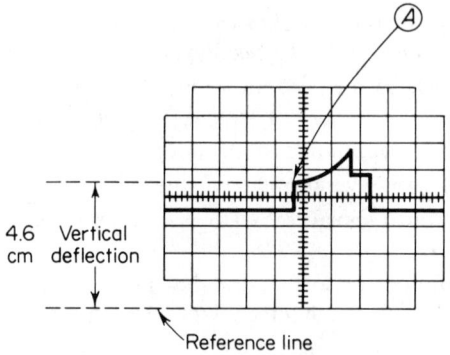

Fig. 5-30. Measuring instantaneous (or d-c) voltages.

expected signal plus any dc to be displayed without overdriving the vertical amplifier.

On a shop oscilloscope, set the vertical gain control to the calibrate-set position, as determined during the calibration procedure (refer to Section 5-15.4).

4. Set the input selector to ground. On most laboratory oscilloscopes, the input switch that selects either a-c or d-c measurements also has a position that connects both vertical input terminals to ground (or shorts them together). If no such switch position is provided, short the vertical input terminals by connecting the probe (or other lead) to ground.

5. Switch on the oscilloscope internal recurrent sweep. Adjust the horizontal gain control to spread the trace over as much of the screen as desired.
6. Using the vertical position control, position the trace to a line of the screen below the centerline, as shown in Figure 5-30. This establishes the reference line. If the average signal (ac plus dc) is negative with respect to ground, position the trace to a reference line above the screen centerline. Do not move the vertical position control after this reference line has been established.
7. Set the input selector switch to dc. (Note that not all shop oscilloscopes have a d-c input.) The ground reference line can be checked at any time by switching the input selector to the ground position. Connect the probe to the signal being measured.
8. If the wave-form is outside the viewing area, set the vertical step-attenuator so that the wave-form is visible. Not all shop oscilloscopes have a vertical step-attenuator. In any event, do not move the vertical position or vertical gain controls to bring the wave-form into view.
9. Adjust the sweep frequency and horizontal gain controls to display the desired wave-form.
10. Measure the distance in divisions between the reference line and the point on the wave-form at which the d-c level is to be measured. For example, in Figure 5-30, the measurement is made between the reference line and point A.
11. Establish polarity of the signal. If the wave-form is above the reference line, the voltage is positive; below the line, it is negative. (Any signal-inverting switches on the oscilloscope must be in the "normal" position to properly establish polarity of signals.)
12. Multiply the distance measured in step 10 by the vertical attenuator switch setting. Also include the attentuation factor of the probe, if any. On shop oscilloscopes not provided with a vertical step-attenuator, multiply the distance measured in step 10 by the calibration factor (volts per division, and so on) established during calibration.

For example, assume that the vertical distance measured is three divisions (between reference line and point A on Figure 5-30), using a $\times 10$ attenuator probe and a vertical deflection factor of 5 V/division.

$$\text{Instantaneous voltage} = 3 \times 5 \times 10 = 150 \text{ V}$$

5-15.4. Calibrating the Vertical Amplifier for Voltage Measurements

On those laboratory oscilloscopes that have a vertical step-attenuator related to some specific deflection factor (such as 5 V/cm), the calibration procedure is an internal adjustment accomplished as part of routine maintenance.

On other oscilloscopes, the vertical amplifier must be calibrated for voltage measurements. The basic procedure consists of applying a reference voltage of known amplitude to the vertical input and adjusting the vertical gain control for specific deflection. Then the reference voltage is removed and the test voltages are measured, *without changing the vertical gain setting*. The calibration will remain accurate so long as the vertical gain control is at this calibrate-set position.

5-15.4.1. Voltage Calibration with External Direct Current

On oscilloscopes that do not have internal voltage reference sources, it is necessary to use an external calibrator. Any d-c source of known accuracy can be used. It is best to select an approximate calibrating voltage value that will produce at least half-scale deflection with the vertical gain control near mid-scale and the step-attenuator (if any) in the ×1 position. (This permits the step-attenuator multiplier function to be used.) Note that accuracy of the oscilloscope voltage measurements will be no greater than the accuracy of the calibrating voltage.

1. Connect the equipment as shown in Figure 5-31.
2. Place the oscilloscope in operation (Section 5-13).
3. Using the vertical position control, move the trace to the horizontal centerline. Switch the internal recurrent sweep on (for a line trace) or off (for a dot), whichever is most convenient for calibration.
4. Turn on the calibrator and set the calibrating voltage to the desired calibrating value. The exact value of the calibration voltage depends on the scale. For example, if there are ten vertical divisions (5 above and 5 below the horizontal centerline) a value of 0.5, 5, or 50 V would be convenient.
5. Without touching the vertical position control, set the vertical gain control to move the dot or line trace vertically up to the desired number of screen divisions. For example, assuming a calibrating voltage of 5 V and a scale as shown in Figure 5-31, set the vertical gain control so that the trace is moved up to the top line or five divisions from the centerline. Thus each division will equal 1 V. In the example of Figure 5-31, this would give the oscilloscope a vertical deflection factor of 1 V/division, with the step-attenuator set to ×1. If the step-attenuator were then moved to ×10, the factor would be 10 V/division.

If the external calibrating source is fixed, the process must be reversed, and a scale factor must be selected to match the voltage. For example, assume that the only calibrating source is a 1.5-V battery of known accuracy. The vertical gain control could then be set to provide a deflection of three divisions (from the horizontal centerline up three divisions). This would give a vertical deflection factor of 0.5 V/division, with the step-attenuator set to ×1. If the attenuator were moved to ×10, the factor would be 5 V/division.

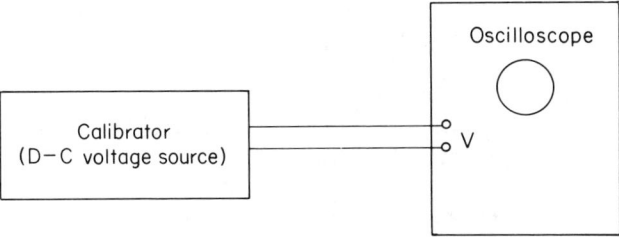

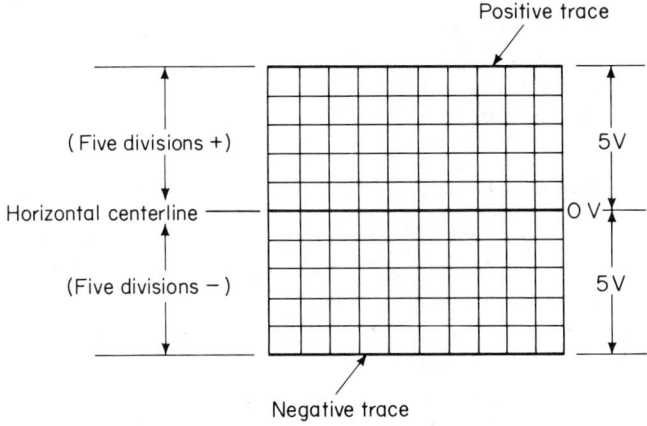

Fig. 5-31. Voltage calibration with external dc.

6. Remove the calibrating voltage and check that the trace is returned to the horizontal centerline.

7. Reverse the calibrating voltage leads. Reapply the voltage and check that the trace is moved *below* the horizontal centerline by the same number of divisions obtained in step 5.

Thus far, the procedures have provided calibration for measurement of both positive and negative d-c voltages (positive voltages are measured above the horizontal centerline, negative voltages below the centerline). Where the voltages to be measured are known to be all positive (or all negative), it may prove convenient to use the entire graticule scale. In that event, use the vertical position control to move the trace to the bottom horizontal line (for all positive voltages) or the top horizontal line (for all negative voltages). Then apply the calibrating voltage and set the vertical gain control to move the trace vertically up (or down) the desired number of screen divisions.

8. If the calibrating source voltage is variable, check the accuracy of the calibration by applying various voltages at various settings of the step-attenuator (if any).

9. If desired, the position of the vertical gain control should be noted and recorded as the calibrate-set position. Use this same position for all future voltage measurements. It is recommended that the calibration be checked at frequent intervals.

Once the calibrate-set position has been established, the reference line (calibration voltage removed, 0 V) can be moved up or down as required by the vertical position control, without affecting the V/division factor. However, any voltage measurements must be made from the reference line, wherever it is moved.

For example, assume that the horizontal centerline is used as the reference line during calibration and the deflection factor is established as 1 V/division. Then, during actual voltage measurement, the reference line is moved down two divisions below the centerline, and a voltage is measured three divisions above the centerline. Since the voltage measured is five divisions *from the reference line*, the correct reading would be 5 V.

5-15.4.2. Voltage Calibration with External Alternating Current

On those oscilloscopes that do not have internal voltage reference sources it is necessary to use an external calibrator. Any a-c source of known accuracy can be used. It is best to select an approximate calibrating voltage value that will produce near full-scale deflection with the vertical gain control near mid-scale, and the step-attenuator (if any) in the ×1 position. (This will permit the step-attenuator multiplier function to be used.)

Accuracy of the oscilloscope voltage measurements will be no greater than the accuracy of the calibrating voltage. Also, the oscilloscope display is usually calibrated for peak-to-peak voltage, whereas the meter or other device indicating the calibrating voltage will probably be in RMS. If the calibrating voltage is a sine wave, the RMS value can be converted to peak-to-peak, as shown in Figure 1-13. If the external calibrating voltage is a square wave or pulse, its value will be peak-to-peak.

1. Connect the equipment as shown in Figure 5-32.
2. Place the oscilloscope in operation (Section 5-13).
3. Using the vertical position control, position the trace to the horizontal centerline. Switch the internal recurrent sweep on (for a normal line trace) or off (for a dot), whichever is most convenient for calibration. If the internal sweep is not on, the dot trace will appear as a vertical line when the calibrating voltage is applied.
4. Turn on the calibrator and set the calibrating voltage to the desired value. The exact value of the calibrating voltage depends on the scale. For example, if there are ten vertical divisions (five above and five below the horizontal centerline), a value of 1, 10, or 100 V would be convenient.
5. Without touching the vertical position control, adjust the vertical gain control to align tips of positive half-cycles and tips of negative half-cycles

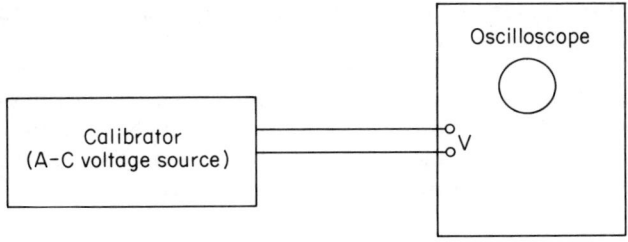

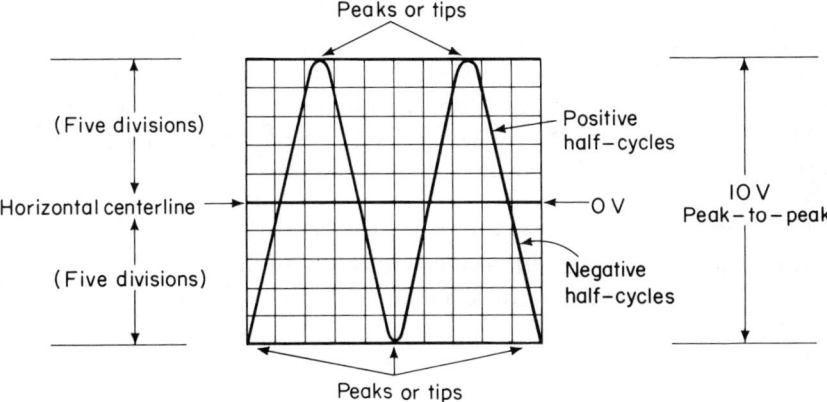

Fig. 5-32. Voltage calibration with external ac.

with the desired scale divisions. For example, assuming a calibrating voltage of 10 V peak-to-peak and a scale as shown in Figure 5-32, set the vertical gain control so that the trace is spread from the top line (five divisions up from the centerline) to the bottom line (five divisions below the centerline). Thus each division will equal 1 V peak-to-peak. In the example of Figure 5-32, this would give the oscilloscope a vertical deflection factor of 1 V/division, with the step-attenuator set to ×10. If the step-attenuator were then moved to ×10, the factor would be 10 V/division.

If the external source is fixed, then the process must be reversed, and a scale factor must be selected to match the voltage. For example, assume that the only calibrating source is a 1-V ac (RMS) of known accuracy. This is equal to 2.828 V peak-to-peak. The vertical gain control could then be set to provide a spread (positive peak to negative peak) of slightly less than three divisions (2.828 divisions). This would give a vertical deflection factor of 1 V/division peak-to-peak, with the step attenuator set to ×1. If the attenuator were moved to ×10, the factor would be 10 V/division.

6. If the calibrating source voltage is variable, check the accuracy of the calibration by applying various voltages, at various settings of the step-attenuator (if any).
7. If desired, the position of the vertical gain control should be noted and

recorded as the *calibrate-set* position. Use this same position for all future voltage measurements. It is recommended that the calibration be checked at frequent intervals. Once the calibrate-set position has been established the trace can be moved up or down as required by the vertical position control without affecting the volts per division factor. No reference line need be considered, as was the case with a d-c calibrating source. This is one of the advantages of using an a-c calibrating source.

5-15.4.3. Voltage Calibration with Internal Alternating Current

Most modern oscilloscopes have an internal voltage source of known amplitude and accuracy available for calibration. On some oscilloscopes, this calibrating voltage is available from terminals or a jack on the front panel. On other oscilloscopes, the calibrating voltage is applied to the vertical input when one of the controls (usually the vertical input selector) is set to calibrate or CAL position.

1. Connect the equipment as shown in Figure 5-33.
2. Place the oscilloscope in operation (Section 5-13).
3. With the calibrate voltage removed (front-panel connection temporarily

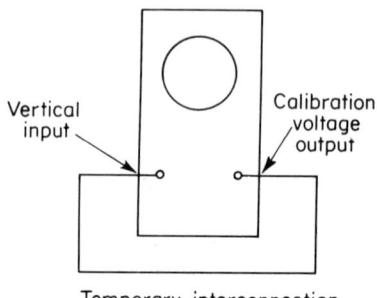

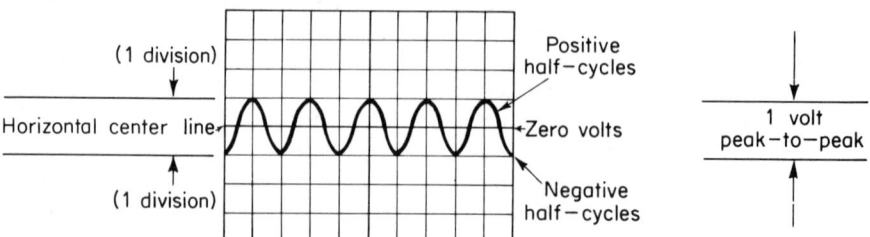

Fig. 5-33. Voltage calibration with internal ac.

Measuring Voltage and Current   215

removed or vertical input control set to normal), use the vertical position control and move the trace to the horizontal centerline. Switch the internal recurrent sweep on (for a normal line trace) or off (for a dot), whichever trace is most convenient for calibration. If the internal sweep is not on, the dot trace will appear as a vertical line when the calibrating voltage is applied.

4. Apply the calibrating voltage.
5. Without touching the vertical position control, adjust the vertical gain control to align tips of positive half-cycles and tips of negative half-cycles with the desired scale divisions. For example, assume a calibrating voltage of 1 V peak-to-peak (which is typical) and set the vertical gain control so that the trace is spread from one division above the centerline to one division below the centerline. Thus each division will equal 0.5 V peak-to-peak. In the example shown in Figure 5-33, this would give the oscilloscope a vertical deflection factor of 0.5 V/division, with the step attenuator set to ×1. If the step attenuator were then moved to ×10, the factor would be 5 V/division.
6. If desired, the position of the vertical gain control should be noted and recorded as the *calibrate-set* position. Use this same position for all future voltage masurements. It is recommended that the calibration be checked at frequent intervals. Once the calibrate-set position has been established the trace can be moved up or down as required by the vertical position control.

5-15.4.4. Voltage Calibration with Internal Square Waves

Some oscilloscopes have an internal square-wave source available for calibration. Usually, this source is variable in amplitude and is adjusted by a front-panel control. The square-wave amplitude is read from a scale on the amplitude adjustment control. The square waves are applied to the vertical input when one of the controls (usually the vertical input selector) is set to calibrate position.

1. Place the oscilloscope in operation (Section 5-13).
2. With the calibrate square waves removed (vertical input control set to normal, not at calibrate), use the vertical position control and move the trace to the horizontal centerline.
3. Apply the calibrating square waves.
4. Set the horizontal sweep frequency to some frequency *lower* than that of the internal calibrating square waves. Several cycles of square waves should appear when the sweep frequency is lower than the internal square-wave calibrating frequency. For example, four square waves should appear if the sweep frequency is one-fourth of the calibrating frequency. One square wave should appear if the sweep frequency and calibrating frequency are the same. If the sweep frequency is considerably lower than the calibrat-

ing frequency, the flat peaks of the square waves will blend and appear as two horizontal lines. If the internal sweep is not on, the trace will appear as a vertical line when the calibrating square waves are applied.

5. Without touching the vertical position control, adjust the vertical gain control and align the flat peaks of the square waves with the desired scale divisions. For example, assume square waves with a peak-to-peak amplitude of 6 V and a scale as shown in Figure 5-34. Set the vertical gain control so that the square waves are spread a total of six divisions (three divisions up from the centerline and three divisions below the centerline). Thus each division will equal 1 V peak-to-peak. In the example shown in Figure 5-34, this would give the oscilloscope a vertical deflection factor of 1 V/division, with the step-attenuator set to ×1. If the step-attenuator were then moved to ×10, the factor would be 10 V/division.

When the centerline is used as a reference, as described in step 2 of this procedure, the square-wave peaks above the centerline indicate posi-

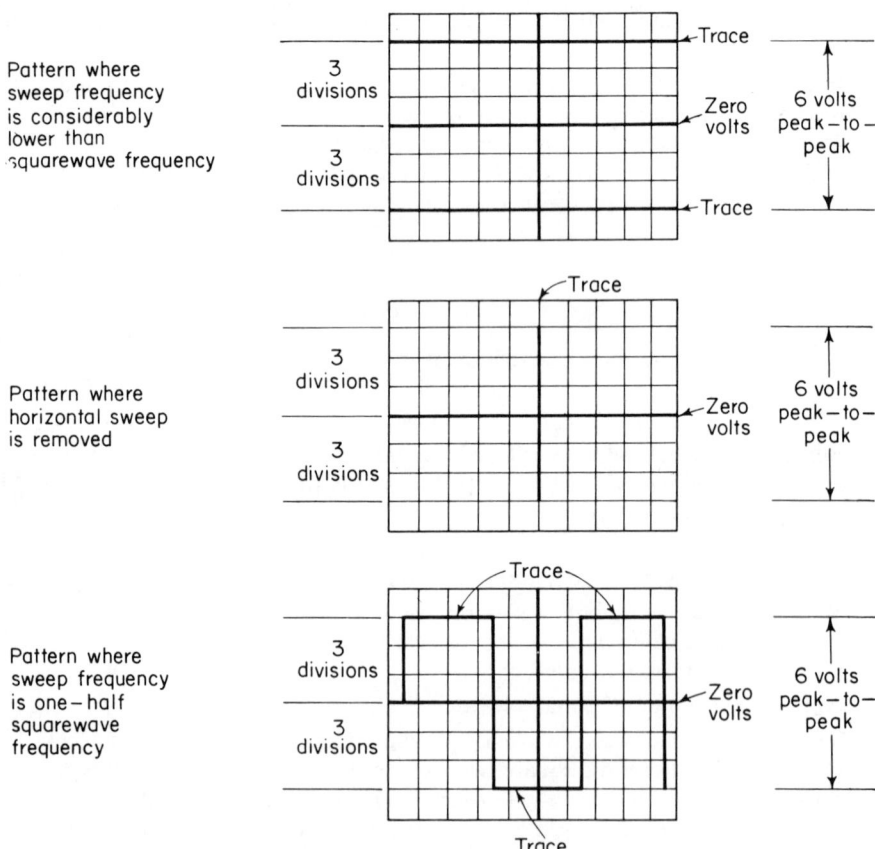

Fig. 5-34. Voltage calibration with internal square waves.

tive voltage, whereas peaks below the centerline indicate negative voltage. The amplitude of the square waves above (and below) the centerline is equal to peak a-c voltage. This is equivalent to one-half the peak-to-peak voltage.

6. If desired, the position of the vertical gain control should be noted and recorded as the *calibrate-set* position. Use the same position for all future voltage measurements. Check calibration at frequent intervals. Once the calibrate-set position has been established the trace can be moved up or down as required by the vertical position control.

5-15.5. Composite and Pulsating Voltage Measurements

In practice, most voltages measured are composites of ac and dc or are pulsating dc. For example, a transistor amplifier used to amplify an a-c signal will have both ac and dc on its collector. Such composite and pulsating voltages can be measured quite easily on an oscilloscope capable of measuring dc (an a-c oscilloscope will show only the a-c portion of the composite voltage).

The procedures for measuring a composite voltage are essentially a combination of peak-to-peak measurements and instantaneous d-c measurements, previously described.

1. Connect the equipment as shown in Figure 5-35.
2. Place the oscilloscope in operation (Section 5-13).
3. Set the vertical step-attenuator to a deflection factor that will allow the expected signal, *plus any dc*, to be displayed without overdriving the vertical amplifier.
4. Set the input selector to ground.
5. Switch on the oscilloscope internal recurrent sweep. Adjust the horizontal gain control to spread the trace over as much of the screen as desired.
6. Using the vertical position control, move the trace to a convenient location on the screen. If the voltage to be measured is a composite and the average signal (ac plus dc) is positive, move the trace below the centerline (as shown in Figure 5-35). If the average is negative, move the trace above the centerline. Do not move the vertical position control after this reference has been established.
7. Set the input selector to dc. The ground reference line can be checked at any time by switching the input selector to the ground position. Connect the probe to the signal being measured.
8. If the wave-form is outside the viewing area, set the vertical step-attenuator so the wave-form is visible.
9. Adjust the sweep frequency and horizontal gain controls to display the desired wave-form.

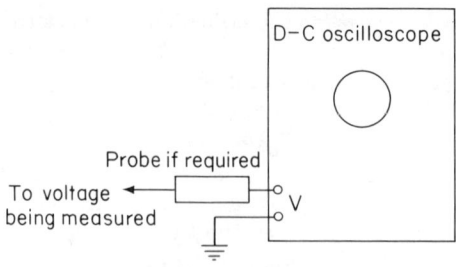

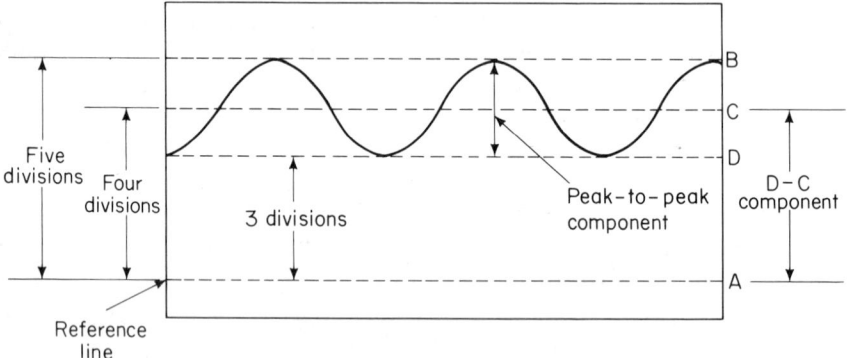

Fig. 5-35. Measurement of composite voltages.

10. Establish polarity of the signal. Any signal-inverting switches on the oscilloscope must be in the normal position. If the wave-form is above the reference line, the voltage is positive; below the line, it is negative.
11. Measure the distance in divisions between the reference line and the point on the wave-form at which the level is to be measured. With a composite of ac and dc, the trace may be on either side of the reference line, or it may possibly cross over the reference, but it usually remains on one side, as shown in Figure 5-35.
12. Multiply the distance measured in step 11 by the vertical attenuator switch setting. Also include the attenuator factor of the probe, if any.

    For example, assume that the vertical distance measured is three divisions from the reference line to point D, four divisions to point C, and five divisions to point B. Also assume that the attenuator probe is $\times 10$ and there is a vertical deflection factor of 2 V/division.

    Since the wave-form is above the reference line, the d-c component must be positive. The d-c component (reference line to point C) = $4 \times 2 \times 10 = +80$ V.

    The peak-to-peak value of the a-c component (point B to D) = $2 \times 2 \times 10 = 40$ V (peak-to-peak).

5-15.6. Oscilloscope Current Measurements

Oscilloscopes can be used to measure current. However, the current must be converted to a voltage for a practical oscilloscope display. There are two methods in general use for measuring current with an oscilloscope.

The *current probes* used with meters (Section 1-3.1) can also be used with oscilloscopes. The probe output voltage is applied to the oscilloscope vertical input and is measured in the normal manner. Since there is a direct relationship between voltage and current, the current can be calculated from the voltage indicated on the oscilloscope. If the vertical amplifier is calibrated for a given value of reference deflection, the current may be read directly from the oscilloscope screen. For example, if the probe output is 1 mV/mA of current and the oscilloscope vertical amplifier is calibrated from 1 mV/cm, the current may be read directly from the scale (1 mA/cm).

Very often, current probes are used with an amplifier, since the probe output is quite low in relation to the average oscilloscope deflection factor. A typical probe has an output of 1 mV/mA, whereas the average laboratory oscilloscope may have a vertical sensitivity of 5 mV/cm. Thus it would require 50 mA to produce a deflection of 1 cm.

5-15.6.1. Oscilloscope Current Measurement with a Test Resistance

A common method of measuring an unknown current is passing it through a resistance of known value, and then measuring the resultant voltage. (This is the basic principle of most voltmeters. See Chapter 1.) Since the oscilloscope can be used as a voltmeter, it can also be adapted to measure current. A resistor of known value and accuracy is the only other component required for the procedure. Once the voltage has been measured, the current can be calculated using the basic Ohm's law equation $I = E/R$.

1. Connect the equipment as shown in Figure 5-36.
2. Place the oscilloscope in opration (Section 5-13).
3. Apply the current through the resistor. If a 1-Ω resistor is used, the calculations will be simplified. The unknown current will be equal to the measured voltage. The wattage of the resistor must be at least *double the square* of the maximum current (in amperes). For example, if the maximum anticipated current is 10 A, the minimum wattage of the resistor should be $10^2 \times 2 = 200$ W.
4. Measure the voltage drop across the resistor in the normal manner. Calculate the current. If peak-to-peak current is measured, the resultant current value will also be peak-to-peak. To determine the RMS or average value, convert the measured voltage to RMS or average, using Figure 1-13.

   For example, assume that the voltage drop across the 1-Ω test resistor is 10 V peak-to-peak.

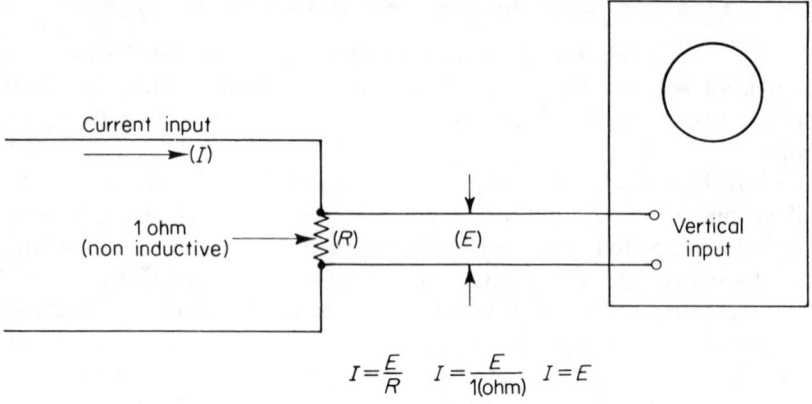

$$I=\frac{E}{R} \quad I=\frac{E}{1(\text{ohm})} \quad I=E$$

Fig. 5-36. Current measurements with a test resistor.

$$10 \text{ V peak-to-peak} = 3.54 \text{ V RMS}$$
$$I = \frac{3.54}{1} = \text{RMS current } 3.54 \text{ A}$$

5. If the current being measured is the result of a composite voltage (ac plus dc), both the a-c and d-c voltages should be measured separately, as described in Section 5-15.5. The a-c voltage (peak-to-peak) should then be converted to RMS (using Figure 1-13). The d-c and a-c currents (which are equivalent to the corresponding voltages) should be combined to find the composite or total current as follows:

$$\text{Total current} = \sqrt{\text{a-c current (RMS)}^2 + \text{d-c current}^2}$$

5-16. MEASURING TIME AND FREQUENCY WITH AN OSCILLOSCOPE

An oscilloscope is the ideal tool for measuring time and frequency of voltages and currents. If the horizontal sweep is calibrated directly in relation to time, such as 5 msec/division, the time duration of voltage waveforms may be measured directly on the screen without calculation. If the time duration of *one complete cycle* is measured, frequency can be calculated by simple division since frequency is the reciprocal of the time duration of 1 cycle. If the oscilloscope is of the shop type, where the horizontal axis is not calibrated directly in relation to time, it is still possible to obtain accurate frequency and time measurements using Lissajous figures.

5-16.1. Time-Duration Measurements with a Laboratory Oscilloscope

The horizontal sweep circuit of a laboratory oscilloscope is usually provided with a selector control that is direct-reading in relation to time; that is, each horizontal division on the oscilloscope screen has a definite relation to time, at a given position of the horizontal sweep rate switch (such as milliseconds per centimeter and microseconds per centimeter). With such oscilloscopes, the wave-form can be displayed, and the time duration of the complete wave-form (or any portion) can be measured directly.

1. Connect the equipment as shown in Figure 5-37.
2. Place the oscilloscope in operation (Section 5-13).
3. Set the vertical step-attenuator to a deflection factor that will allow the expected signal to be displayed without overdriving the vertical amplifier.
4. Connect the probe (if any) to the signal being measured.
5. Switch on the internal recurrent sweep. Set the horizontal sweep control to the *fastest* sweep rate that will display a *convenient* number of divisions between the time measurement points (Figure 5-37). On most oscilloscopes, it is recommended that the extreme sides of the screen not be used for time-duration measurements. There may be some nonlinearity at the beginning and end of the sweep.
6. Adjust the vertical position control to move the points between which the time measurement is made to the horizontal centerline.
7. Adjust the horizontal position control to move the starting point of the time measurement area to the first (vertical) screen line. If the horizontal sweep is provided with a variable control, make certain it is off or in the calibrate position.
8. Measure the horizontal distance between the time measurement points.

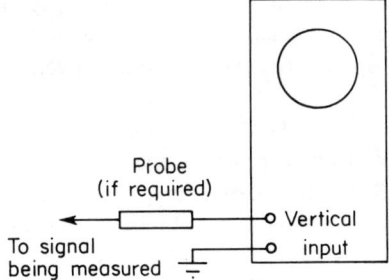

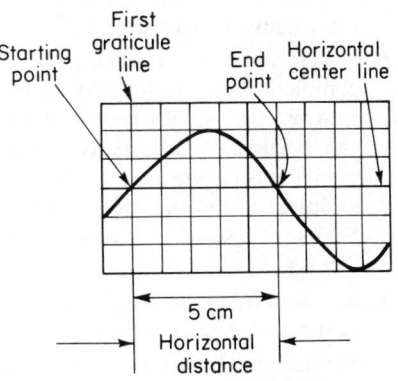

(Time measurement points)

Fig. 5-37. Measuring time duration.

9. Multiply the distance measured in step 8 by the setting of the horizontal sweep control. If sweep magnification is used, divide the answer by the magnification multiplication factor.

For example, assume that the distance between the time measurement points is five divisions (Figure 5-37), the horizontal sweep control is set to 0.1 msec/division, and there is no sweep magnification.

$$\text{Time duration} = 5 \times 0.1 = 0.5 \text{ msec}$$

5-16.2. Frequency Measurement with a Laboratory Oscilloscope

The frequency measurement of a periodically recurrent waveform is essentially the same as a time-duration measurement, except that an additional calculation must be performed. In effect, a time-duration measurement is made, then the time duration is divided into 1, since frequency of a signal is the reciprocal of 1 cycle.

1. Connect the equipment as shown in Figure 5-38.
2. Place the oscillocope in operation (Section 5-13).
3. Set the vertical step-attenuator to a deflection factor that will allow the expected signal to be displayed without overdriving the vertical amplifier.
4. Connect the probe (if any) to the signal being measured.
5. Switch on the internal recurrent sweep. Set the horizontal sweep control to a sweep rate that will display at least one complete cycle of the incoming signal. Avoid using the extreme sides of the screen.
6. Adjust the vertical position so that the beginning and end points of one complete cycle are located on the horizontal centerline.
    Any two points representing one complete cycle of the wave-form can be used. It is usually convenient to measure one complete cycle at points where the wave-form swings from negative to positive (or vice versa) or where the wave-form starts its positive (or negative) rise. Figure 5-38 shows some typical examples of a complete cycle for various waveforms.
7. Adjust the horizontal position control to move the selected starting point of the complete cycle to the first vertical line.
8. Measure the horizontal distance between the beginning and end of a complete cycle. If the horizontal sweep is provided with a variable control, make certain it is off or in the calibrate position.
9. Multiply the distance measured in step 8 by the setting of the horizontal sweep control. If sweep magnification is used, divide the answer by the magnification multiplication factor. The divide the measured time into 1 to find the frequency.
    For example, assume that the distance between the beginning and

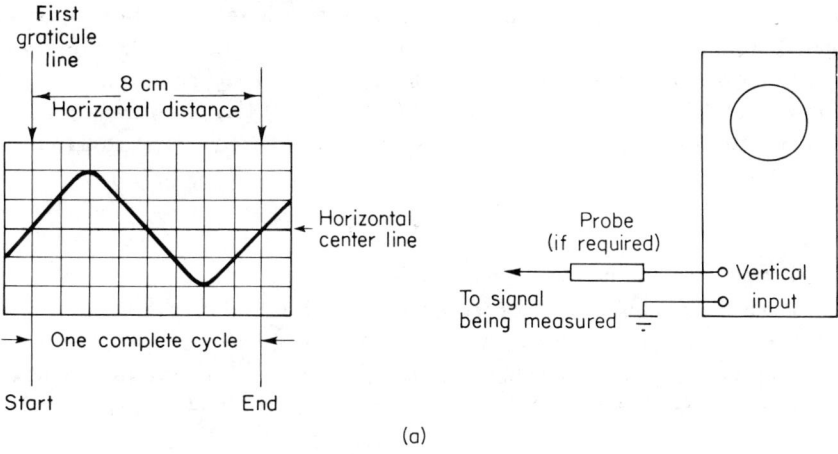

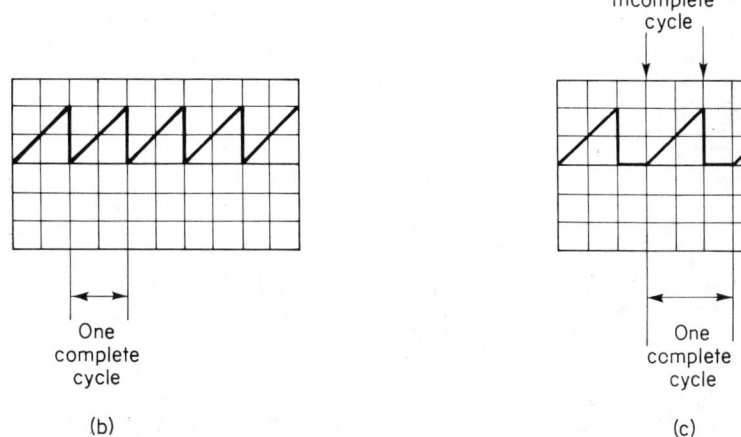

Fig. 5-38. Measuring frequency where horizontal sweep is calibrated in relation to time.

end of a complete cycle is eight divisions, the horizontal sweep control is set to 0.1 msec/division, and there is no sweep magnification.

$$\text{Time duration} = 8 \times 0.1 = 0.8 \text{ msec}$$
$$\text{Frequency} = 1/0.8 = 1250 \text{ Hz}$$

5-16.3. *Frequency and Time Measurements with Shop Oscilloscopes*

The horizontal sweep circuit of most shop oscilloscopes is provided with controls that are direct-reading in relation to frequency. Usually, there are two controls: a step selector and a vernier or variable. The sweep frequency of the horizontal trace is equal to the scale settings of the controls. Therefore, when a signal is applied to the vertical input and the horizontal sweep controls are adjusted until *one complete cycle occupies the entire length of the trace*, the vertical signal is equal in frequency to the horizontal sweep control scale settings. If desired, the frequency can then be converted to time:

$$\text{Time} = \frac{1}{\text{frequency}}$$

1. Connect the equipment as shown in Figure 5-39.
2. Place the oscilloscope in operation (Section 5-13).
3. Set the vertical step-attenuator to a deflection factor that will allow the expected signal to be displayed without overdriving the vertical amplifier.
4. Connect the probe (if any) to the signal being measured.

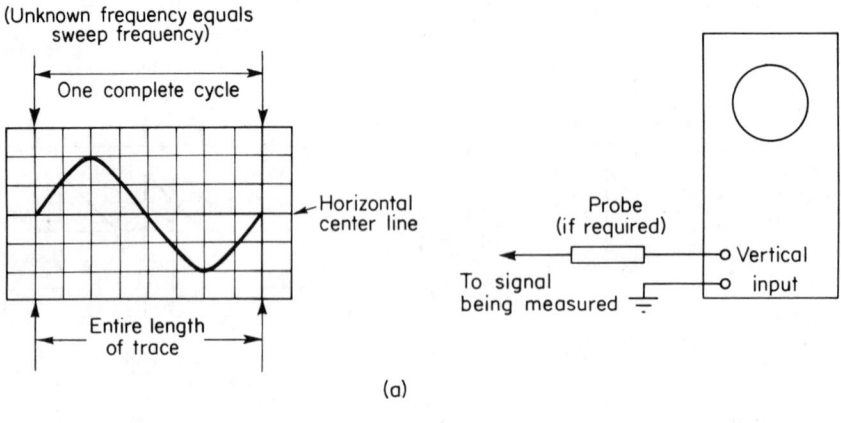

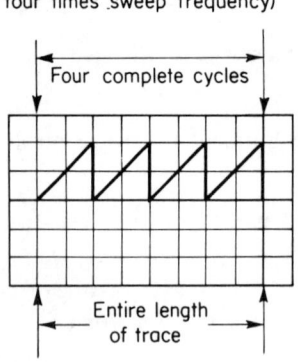

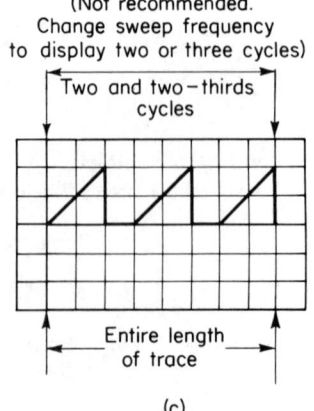

Fig. 5-39. Measuring frequency where horizontal sweep is calibrated directly in units of frequency.

5. Switch on the oscilloscope internal recurrent sweep. Set the horizontal sweep controls (step and vernier) so that one complete cycle occupies the *entire length* of the trace.
6. Read the unknown vertical signal frequency directly from the horizontal sweep frequency control settings.

    For example, assume that the step horizontal sweep control is set to the 10-kHz position and that the vernier sweep control indicates 5 (on a total scale of 10). This indicates that the horizontal sweep frequency is 5 kHz. If one complete cycle of vertical signal occupies the entire length of the trace, the vertical signal is also at a frequency of 5 kHz.
7. If it is not practical to display only 1 cycle on the trace, more than 1 cycle can be displayed, and the resultant horizontal sweep frequency indication multiplied by the *number of cycles*.

    Two important points must be remembered. First, multiply the indicated sweep frequency by the number of cycles appearing on the trace.

    For example, assume that the horizontal sweep frequency is 5 kHz, as before, but the vertical signal is *three times* that amount (three complete cycles of vertical signal occupy the entire trace). Then the signal frequency is 15 kHz.

    Next, it is absolutely essential that an *exact number of cycles occupies the entire length* of the trace.

    For example, assume that the horizontal sweep frequency is again 5 kHz and that $3\frac{1}{3}$ cycles of vertical occupy the entire length of the trace. This would indicate a frequency of 16.5 kHz. However, the exact percentage of the incomplete cycle (one-third) is quite difficult to measure. It is far simpler and more accurate to increase or decrease the horizontal sweep frequency until an exact number of cycles appears.

5-16.4. Frequency Measurements with Lissajous Figures

Lissajous figures or patterns can be used with almost any oscilloscope (shop- or laboratory-type) and will provide accurate frequency measurements. However, it must be possible to apply an external signal to the horizontal amplifier with the internal sweep disabled. Also, an accurately calibrated variable frequency signal source must be available to provide a standard frequency. Refer to Chapter 8 for a discussion of frequency standards.

The use of Lissajous figures to measure frequency involves comparing a signal of unknown frequency (usually applied to the vertical amplifier) against a standard signal of known frequency (usually applied to the horizontal amplifier). The standard frequency is then adjusted until the pattern appears as a circle or ellipse, indicating that both signals are at the same frequency. Where it is not possible to adjust the standard signal frequency to the exact frequency of the unknown signal, the standard is adjusted to a multiple or submultiple. The pattern then appears as a number of stationary loops. The ratio of horizontal loops to vertical loops provides a measure of frequency.

226   Oscilloscopes and Recorders

1. Connect the equipment as shown in Figure 5-40.
2. Place the oscilloscope in operation (Section 5-13).
3. Set the vertical step-attenuator to a deflection factor that will allow the expected signal to be displayed without overdriving the vertical amplifier.
4. Switch off the internal recurrent sweep.
5. Set the gain controls (horizontal and vertical) to spread the pattern over as much of the screen as desired.
6. Set the position controls (horizontal and vertical) until the pattern is centered on the screen.

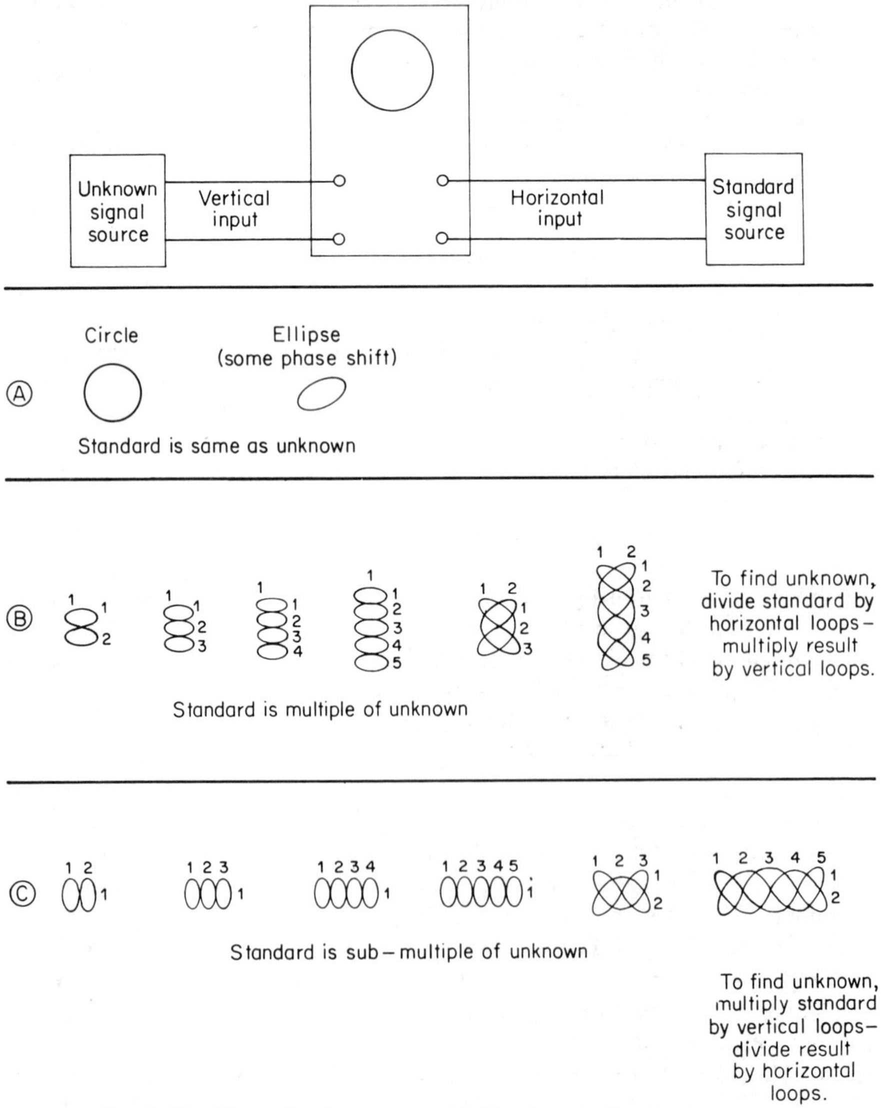

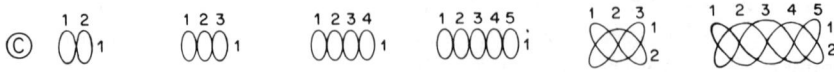

Fig. 5-40. Measuring frequency with Lissajous patterns.

7. Adjust the standard signal frequency until the pattern stands still. This indicates that the standard signal is at the same frequency as the unknown frequency (if the pattern is a circle or ellipse) or that the standard signal is at a multiple or submultiple of the unknown frequency (if the pattern is composed of stationary loops). If the pattern is still moving (usually spinning) this indicates that the standard signal is not at the same frequency (or multiple) of the unknown frequency. The pattern must be stationary before the frequency can be determined.
8. Note the standard signal frequency. Using this frequency as a basis, observe the Lissajous pattern and compare it against those shown in Figure 5-40 to determine the unknown frequency.

    If both signals are sine waves (even if not pure sine waves), are at the same frequency, and are of the same amplitude, the pattern will be a circle (or an ellipse, when the two signals are not exactly in phase), as shown in Figure 5-40.

    If the standard signal (horizontal) is a multiple of the unknown signal (vertical), the pattern will show more horizontal loops than vertical loops (Figure 5-40B). For example, if the standard signal frequency is three times that of the unknown signal frequency, there will be three horizontal loops and one vertical loop. If the standard signal frequency were 300 Hz, the unknown signal frequency would be 100 Hz. If there are two vertical loops and three horizontal loops, the unknown signal frequency would be two-thirds of the standard signal frequency.

    If the standard signal (horizontal) is a submultiple of the unknown signal (vertical), the pattern will show more vertical loops (Figure 5-40C). For example, if the standard signal frequency is one-fourth that of the unknown signal frequency, there will be four vertical loops and one horizontal loop. If the standard frequency were 100 Hz, the unknown signal frequency would be 400 Hz.
9. If two signals are to be matched without regard to frequency, it is necessary to adjust only one frequency until the circle (or ellipse) pattern is obtained.

    It is recommended that the circle pattern be used for all frequency measurements whenever possible. If this is not practical, use the *minimum* number of loops possible. Note, too, that the use of Lissajous patterns in actual practice requires considerable skill and practice to make accurate measurements.

5-17. MEASURING PHASE WITH AN OSCILLOSCOPE

An oscilloscope can be used to measure phase between two signals. There are two basic methods: *X-Y* method and dual-trace method.

5-17.1. X-Y Method of Phase Measurement

The *X-Y* phase measurement method can be used to measure the phase difference between two sine-wave signals of the *same frequency*,

228   Oscilloscopes and Recorders

up to about 100 kHz. In the *X-Y* method, one of the sine-wave signals provides horizontal deflection (*X*), and the other provides the vertical deflection (*Y*). The phase angle between the two signals can be determined from the resulting Lissajous pattern.

1. Connect the equipment as shown in Figure 5-41A.

    Figure 5-41A shows the test connections necessary to determine the inherent phase shift (if any) between the horizontal and vertical deflection systems of the oscilloscope. Even the most expensive laboratory oscilloscopes with identical horizontal and vertical amplifiers will have some inherent phase shift, particularly at the higher frequencies. Therefore, all oscilloscopes should be checked and the inherent phase shift recorded before any phase measurements are made.

    Inherent phase shift should be checked periodically. If there is excessive phase shift (in relation to the anticipated phase shift of signals to be measured), the oscilloscope should not be used. A possible exception exists when the signals to be measured are of sufficient amplitude to be applied directly to the oscilloscope deflection plates and thus bypass the horizontal and vertical amplifiers.

2. Place the oscilloscope in operation (Section 5-13).
3. Set the step-attenuators to deflection factors that will allow the expected signals to be displayed without overdriving the amplifiers.

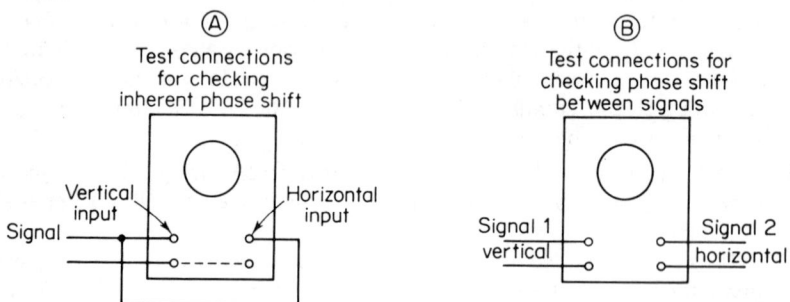

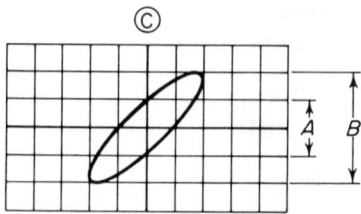

Fig. 5-41. Measuring phase difference with X-Y method.

Measuring Phase with an Oscilloscope

4. Switch off the oscilloscope internal recurrent sweep.
5. Set the gain controls (horizontal and vertical) to spread the pattern over as much of the screen as desired.
6. Set the position controls (horizontal and vertical) until the pattern is centered on the screen. Center the display in relation to the vertical centerline. Measure distances A and B, as shown in Figure 5-41C. Distance A is the vertical measurement between the two points where the trace crosses the vertical centerline. Distance B is the maximum vertical height of the display.
7. Divide A by B to obtain the sine of the phase angle between the two signals. The angle can then be obtained from Table 5-1. The resultant angle is the inherent phase shift.
8. Once the inherent phase shift has been determined, connect the equipment as shown in Figure 5-41B. Repeat steps 3–7 to find the phase angle between the two signals.
9. Subtract the inherent phase difference from the phase angle to determine the true phase difference.

For example, assume an inherent phase difference of 3°, with a display as shown in Figure 5-41C, where A is two divisions and B is four divisions.

$$\text{Sine of phase angle} = 2/4 = 0.5 = 30°$$
$$\text{Actual phase difference } 30° - 3° = 27°$$

TABLE 5-1. TABLE OF SINES

Table of Sines

Sine	Angle	Sine	Angle	Sine	Angle	Sine	Angle	Sine	Angle
0.0000	0	0.3584	21	0.6691	42	0.8910	63	0.9945	84
0.0175	1	0.3746	22	0.6820	43	0.8988	64	0.9962	85
0.0349	2	0.3907	23	0.6947	44	0.9063	65	0.9976	86
0.0523	3	0.4067	24	0.7071	45	0.9135	66	0.9986	87
0.0689	4	0.4226	25	0.7193	46	0.9205	67	0.9994	88
0.0872	5	0.4384	26	0.7314	47	0.9272	68	0.9998	89
0.1045	6	0.4540	27	0.7431	48	0.9336	69	1.0000	90
0.1219	7	0.4695	28	0.7547	49	0.9397	70		
0.1392	8	0.4848	29	0.7660	50	0.9455	71		
0.1564	9	0.5000	30	0.7771	51	0.9511	72		
0.1736	10	0.5150	31	0.7880	52	0.9563	73		
0.1908	11	0.5299	32	0.7986	53	0.9613	74		
0.2079	12	0.5446	33	0.8090	54	0.9659	75		
0.2250	13	0.5592	34	0.8192	55	0.9703	76		
0.2419	14	0.5736	35	0.8290	56	0.9744	77		
0.2588	15	0.5878	36	0.8387	57	0.9781	78		
0.2756	16	0.6018	37	0.8480	58	0.9816	79		
0.2924	17	0.6157	38	0.8572	59	0.9848	80		
0.3090	18	0.6293	39	0.8660	60	0.9877	81		
0.3256	19	0.6428	40	0.8746	61	0.9903	82		
0.3420	20	0.6561	41	0.8829	62	0.9925	83		

5-17.2. Dual-Trace Method of Phase Measurement

The dual-trace method of phase measurement provides a high degree of accuracy at all frequencies but is especially useful at frequencies above 100 kHz, where X-Y phase measurement may prove inaccurate because of inherent internal phase shift.

The dual-trace method also has the advantage of measuring phase difference between signals of different amplitudes, frequency, and wave shape. The method can be applied directly to those oscilloscopes having a built-in dual-trace feature, or to a conventional single-trace oscilloscope, using an electronic switch or "chopper" (Section 5-5.7). Either way, the procedure consists essentially of displaying both traces on the oscilloscope screen simultaneously, measuring the distance (in scale divisions) between related points on the two traces, and then converting this distance to phase.

1. Connect the equipment as shown in Fig. 5-42.
2. Place the oscilloscope in operation (Section 5-13). For the most accurate results, the cables connecting the two signals to the oscilloscope should be of the same length and characteristics. At higher frequencies, a difference in cable length or characteristics could introduce a phase shift.

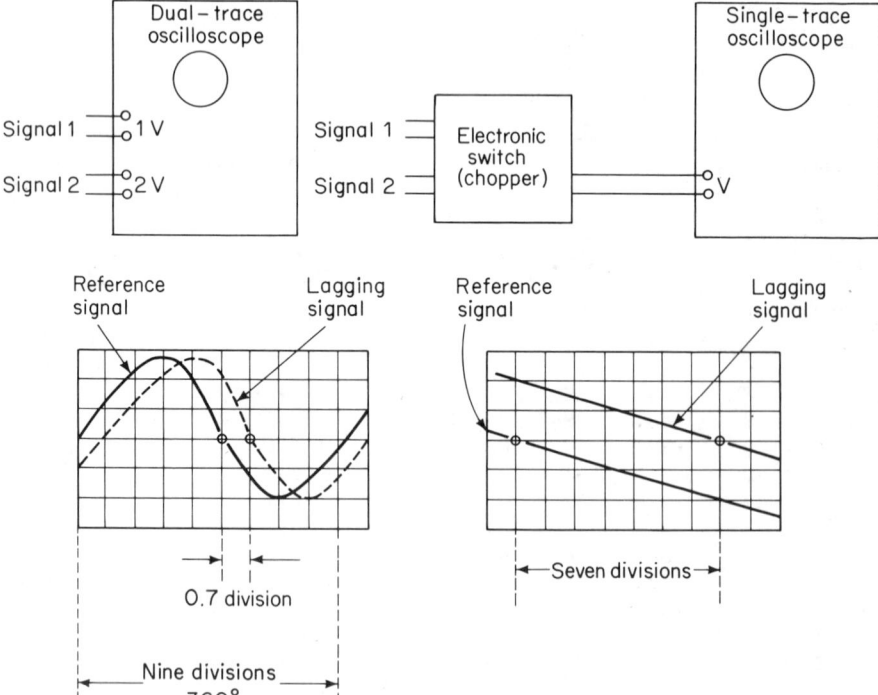

Fig. 5-42. Measuring phase difference with dual traces.

3. Set the step-attenuators to deflection factors that will allow the expected signals to be displayed without overdriving the amplifiers.
4. Switch on the oscilloscope internal recurrent sweep.
5. Set the position controls (horizontal and vertical) until the pattern is centered on the screen.
6. Set the gain controls (horizontal and vertical) to spread the patterns over as much of the screen as desired.
7. Switch on the dual-trace function of the oscilloscope, or switch on the electronic chopper.
8. Adjust the sweep controls until 1 cycle of the reference signal occupies exactly nine divisions (nine divisions horizontally) of the screen. Either of the two signals can be used as the reference signal, unless otherwise specified by requirements of the particular test. It is usually simpler if the signal of the lowest frequency is used as the reference signal.

   Note that some technicians prefer to adjust the sweep controls so that 1 cycle occupies ten divisions. However, one division will then equal 36° of phase shift (360/10 = 36). If nine divisions are used, one division will then equal 40° (360/9 = 40).
9. Determine the phase factor of the reference signal. For example, if nine divisions represent one complete cycle, or 360°, the one division represents 40°.
10. Measure the horizontal distance between corresponding points on the wave-form. Multiply the measured distance (in divisions) by 40° (or whatever phase factor has been selected) to obtain the exact amount of phase difference.

    For example, assume a horizontal difference of 0.7 divisions with a phase factor of 40°, as shown in Figure 5-42.

    $$\text{Phase difference} = 0.7 \times 40° = 28°$$

11. If the oscilloscope is provided with a sweep magnification control where the sweep rate is increased by some fixed amount ($5\times$, $10\times$, and so on) and only a portion of 1 cycle can be displayed, more accurate phase measurements can be made. In this case, the phase factor is determined as described in step 9. Then the approximate phase difference is determined as described in step 10. Without changing any other controls, the sweep rate is increased (by the sweep magnification control or the sweep rate control) and a new horizontal distance measurement is made, as shown in Figure 5-42B.

    For example, if the sweep rate were increased 10 times (with the magnifier or sweep rate control), the adjusted phase factor would be 40°/10 = 4°/division. Figure 5-42B shows the same signal as used in Figure 5-42A, but the sweep rate is set to $\times 10$. With a horizontal difference of seven divisions, the phase difference would be

    $$\text{Phase difference} = 7 \times 4° = 28°$$

6 ELECTRONIC COUNTERS

To understand the operation of electronic counters, it is necessary to have a full understanding of digital logic circuits, especially decades, BCD decoders, and numerical readouts. Information on the operation of such circuits is contained in the Appendix. Also, these circuits are discussed (relative to counters) in this chapter.

6-1. BASIC ELECTRONIC COUNTER

The electronic counter is an instrument that compares an unknown frequency or time interval to a known frequency or known time interval. The counter's logic is designed to present this information in an easy-to-read, unambiguous numerical display or readout. The readout of most electronic counters is similar to that of a digital meter (Chapter 2). In fact, a digital meter is essentially an electronic counter plus a conversion circuit for converting voltage to a series of pulses.

The accuracy of an electronic counter's measurement depends primarily on the stability of the known frequency, which is usually derived from the counter's internal oscillator or, in the case of low-cost counters, from the a-c power-line frequency. The accuracy of the internal oscillator can be checked against broadcast standards, as described in Chapter 8. Thus the electronic counter can become a shop frequency and time standard.

Counters are often used with accessories and in systems. For example, there are electronic counters that will retain their counts for automatic recording of measurements, digital clocks (Chapter 8) that control measurement intervals and supply time information for simultaneous recording, digital-to-analog converters for high-resolution analog records of digital measurements, and scanners that can receive the outputs from several electronic counters for entry into a single recording device. Also, there are magnetic and optical tachometers for revolutions-per-second measurement (such as described in

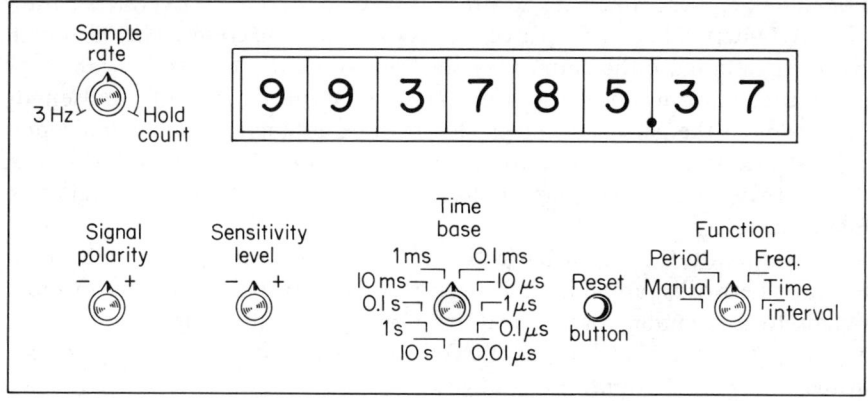

Fig. 6-1. Typical electronic counter panel controls and indicators.

Chapter 9) that are designed to provide inputs to low-frequency counters.

Figure 6-1 shows the operating controls and readout of a typical electronic counter. Note the simplicity of controls. Once the power is turned on, the operator has only to select the function, the level of the input signal to be measured, and the time base or rate at which the counting operation is to be made. Operation of these controls is discussed in the following sections.

6-2. COUNTER CIRCUIT ELEMENTS

Although there are many types of electronic counters, all counters have several basic functional sections in common. These sections are interconnected in a variety of ways to perform the different counter functions.

All electronic counters have some form of *counter and readout assemblies*. Early instruments used binary counters and readout tubes that converted the binary count to a decade readout. Such instruments have (generally) been replaced by decade counters that convert the count to binary-coded-decimal (BCD) form, decoders that convert the BCD data to decade form, and readout tubes that display the decade information directly. One readout tube is provided for each digit. For example, eight readout tubes will provide for count (or time interval) up to 99,999,999.

All electronic counters have some form of main *gate* that controls the count start and stop with respect to time. Usually, the main gate is some form of AND gate.

All electronic counters have some form of *time base* that supplies the precise increment of time to control the gate for a frequency or pulse train measurement. Usually, the gate is a crystal-controlled oscillator. The accuracy of the counter is dependent on the accuracy of the time base plus or minus one count.

For example, if the time base accuracy is 0.005 per cent, the overall accuracy of the counter will be 0.005 per cent plus or minus one count. The one-count error arises because the count may start and stop in the middle of an input pulse, thus omitting the pulse from the total count. Or part of a pulse may pass through the gate before it is closed, thus adding a pulse to the count.

Most electronic counters have *dividers that permit variation of gate time*, that is, these dividers convert the fixed-frequency time base to several other frequencies.

In addition to these four basic sections, most electronic counters have attenuator networks, amplifier and trigger circuits to shape a variety of input signals to a common form, and logic circuits to control operation of the instrument. The various modes of electronic counter operation are described in the following sections.

6-3. TOTALIZING OPERATION

Electronic counters can be operated in a totalizing mode with the main gate controlled by a manual START/STOP switch, as shown in Figure 6-2. With the switch in START (gate open), the counter assemblies totalize the input pulses until the main gate is closed by the switch being changed to STOP. The counter display then represents the input pulses received during the interval between manual START and manual STOP. Generally, totalizing can also be remotely controlled.

In this mode of operation, the main gate is a form of AND gate. As described in the Appendix, an AND gate requires two (or more) true inputs to produce an output. For example, a positive true AND gate requires two

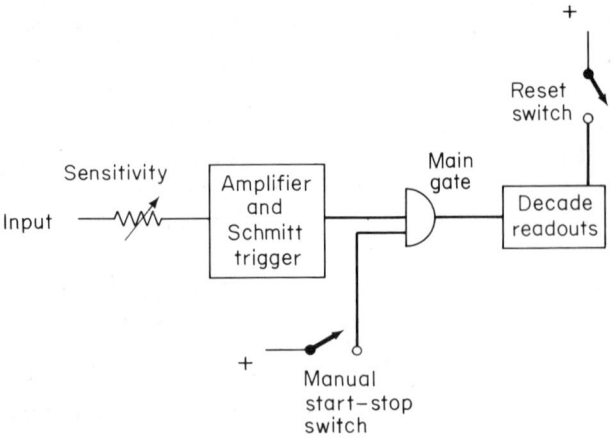

Fig. 6-2. Basic counter circuit for totalizing operation.

positive inputs to produce one positive output. When the manual control switch is set to START, a positive voltage is applied to one input of the AND gate. The input pulses (also positive) are applied to the other AND gate input. Therefore, there is one output pulse to the counter/readout for each input pulse applied. When the manual control switch is set to STOP, the positive voltage is removed from the AND gate input. Further input pulses provide only one input to the AND gate, resulting in no output pulses to the counter/readout. In some counters, the counter/readouts have the ability to totalize in either a positive or negative direction.

Note that input signals are applied to the AND gate through an attenuator network, an amplifier, and a Schmitt trigger. The attenuator network (sometimes known as a *sensitivity* control) determines the level of input signals that will produce a count. (All input signals above a certain level will be counted; all signals below the level will not be counted, for a setting of the sensitivity control.) The amplifier (not found in all counters) raises the level of input signals to a point where they can be counted.

As discussed in the Appendix, the Schmitt trigger is a wave-shaping and level-sensing element with two operational states. The Schmitt trigger is, in effect, a one-shot multivibrator that will produce a square-wave output of fixed amplitude and duration for each input signal (above a given level). The output pulse will remain fixed, regardless of the input signal shape. There fore, Schmitt triggers are useful in counter circuits for squaring signals with poor rise and fall times (such as sine waves).

Note that the counter/readout assemblies must be reset. If not, the new count would be added to the previous count each time the gate was opened. Manual reset is accomplished by pushing a RESET button. This applies a d-c voltage to all of the counters simultaneously, setting them to a position that will produce a zero readout on each readout tube. For other modes of operation (such as frequency measurement, time interval or period measurement, and so on), reset is accomplished by a pulse applied to all counters at regular intervals. This pulse is developed by a low-frequency oscillator (usually 2 or 3 Hz). Therefore, two or three counts or "samples" are taken each second. The pulse oscillator (often known as the *sample rate* oscillator) is adjustable in frequency on some counters. On such equipment, the sample rate oscillator is adjusted by a SAMPLE RATE control. One position of the SAMPLE RATE control will disable the oscillator so that the count and readout will be held.

6-4. FREQUENCY MEASUREMENT OPERATION

For direct-frequency measurements (Figure 6-3) the input signal is first converted to uniform pulses by the Schmitt trigger. These pulses are then routed through the main gate and into the counter/readout assem-

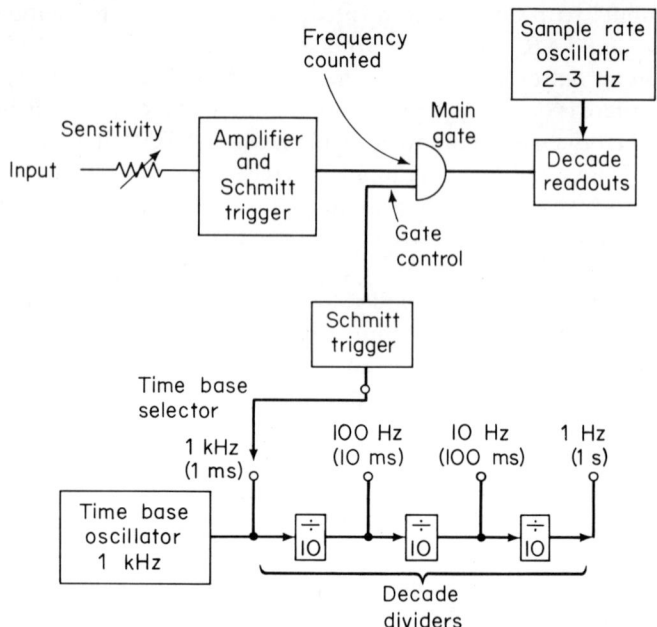

Fig. 6-3. Basic counter circuit for frequency measurement operation.

blies, where the pulses are totalized. The number of pulses totalized during the "gate-open" interval is a measure of the *average input frequency* for that interval. (For example, assume that the gate is held open for 1 sec and the count is 333. This indicates a frequency of 333 Hz.) The count obtained, with the correct decimal point, is then displayed and retained until a new sample is ready to be shown. The SAMPLE RATE control determines the time between samples (not the interval of gate opening and gate closing), resets the counter, and thus initiates the next measurement cycle.

The TIME BASE selector switch selects the gating interval, thus positioning the decimal point and selecting the appropriate measurement units. As shown in Figure 6-3, the TIME BASE selector selects one of the frequencies from the time base oscillator. If the 1-kHz signal (directly from the time base) is selected, the time interval (gate-open to gate-close) will be 1 msec. If the 100-Hz signal (from the first decade divider) is chosen, the measurement time interval will be $\frac{1}{100}$ sec (10 msec). The 10-Hz signal (from the second decade divider) will produce a $\frac{1}{10}$-sec interval (100 msec). The 1-Hz signal from the third decade divider will produce a 1-sec interval.

6-5. PERIOD MEASUREMENT OPERATION

Period is the inverse of frequency (period = 1/frequency). Therefore, period measurements are made with the input and time base

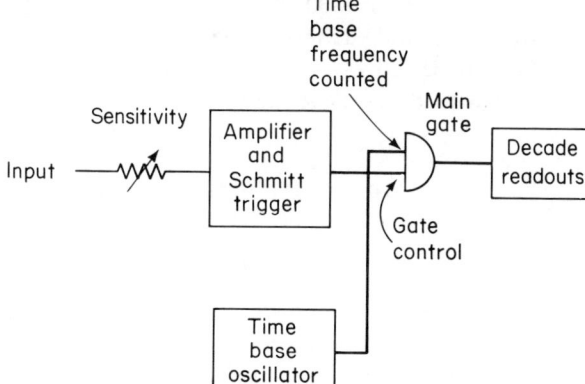

Fig. 6-4. Basic counter circuit for period measurement operation.

connections reversed from those for frequency measurement. This is shown in Figure 6-4. The unknown input signal controls the main gate time, and the time base frequency is counted and read out. For example, if the time base frequency is 1 MHz, the indicated count will be in microseconds (a count of 70 indicates that the gate has been held open for 70 μsec).

Usually, the input-shaping circuit selects the positive-going zero crossing of successive cycles of the unknown signal as trigger points for opening and closing of the gate.

The accuracy and resolution of period measurements can be increased by *period averaging*. The connections are shown in Figure 6-5. These are the same connections as for regular period measurements, except that the signal to be measured is lowered in frequency by dividers, thus extending the gate-open period. For example, if the input signal is 1 kHz, the period will be 1 msec with a regular period measurement. If the time base were 1 MHz, the

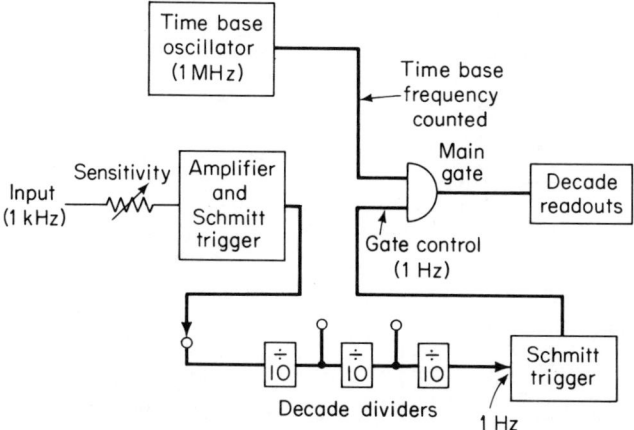

Fig. 6-5. Basic counter circuit for period averaging measurement operation.

count would be 00001000 on an 8-digit readout. If the period average method were used and the input frequency were reduced to 1 Hz, as shown in Figure 6-5, the period would be 1 sec, and the count would be 01000000 on the same 8-digit readout. Thus the resolution is increased by 10^3.

6-6. TIME-INTERVAL MEASUREMENT OPERATION

Time-interval measurements are essentially the same as period measurements. However, the time-interval mode is more concerned with time between two events, rather than the repetition rate of signals. Counters vary greatly in their time-interval-measuring capability. Some counters measure only the duration of an electrical event; others measure the interval between the start of two pulses. The most versatile models, known as universal counters, have separate inputs for the start and stop commands and have separate controls that permit setting the trigger level amplitude, polarity, slope, and type of input coupling (ac or dc) for the start and stop channel. Since stop and start commands can originate from common or separate sources, this type of instrument can measure the interval from one point on a wave-form to another point on the same wave-form.

The basic time-interval measurement circuit is shown in Figure 6-6. Note that the time base signals are counted and read out when the gate is open. Control of the gate is accomplished by two trigger circuits that receive their inputs from the signal being measured. With switch S_1 in the SEPARATE position, the two triggers receive inputs from separate lines. Assume that the START (gate-open) trigger received its input from a signal applied to an

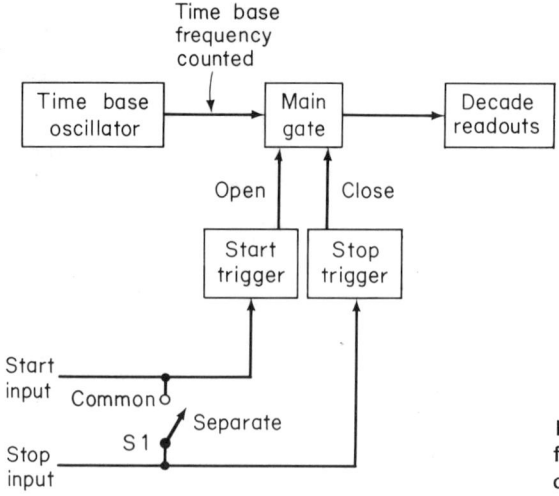

Fig. 6-6. Basic counter circuit for time-interval measurement operation.

amplifier under test, while the amplifier's output was applied to the STOP (gate-closed) trigger. Under these circumstances, the gate would be open for an interval of time equal to the delay through the amplifier under test. If the time base were 1 MHz and the count gave a readout of 33, the delay would be 33 μsec.

With switch S_1 in the COMMON position, the two triggers receive inputs from the same lines. With this arrangement, each trigger is adjusted so that it will respond to a different portion of the same wave-form. Assume that the START trigger will open the gate when the input signal rises to $+10$ V and the STOP trigger closes the gate when the input signal reaches $+15$ V. Therefore, the count will represent the time inverval between the two points.

6-7. RATIO MEASUREMENT OPERATION

The ratio of two frequencies is determined by using one signal for the gate control while the other signal is counted, as shown in Figure 6-7. With proper transducers (Chapter 9), ratio measurements may be applied to any phenomenon that may be represented by pulses or sine waves. Gear ratios and clutch slippage, as well as frequency divider or multiplier operation, are some of the measurements that can be made using this technique.

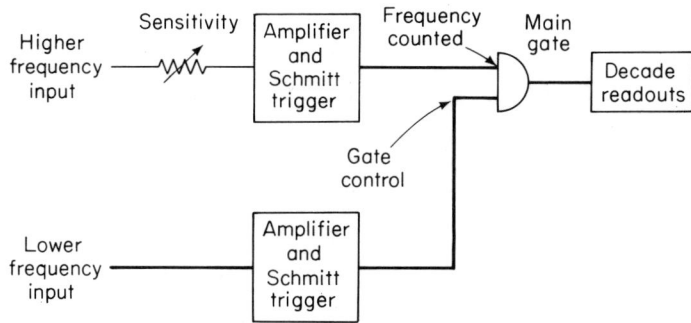

Fig. 6-7. Basic counter circuit for ratio measurement operation.

6-8. COUNTER/READOUT AND DIVIDER CIRCUIT OPERATION

The following is a description of the counter/readout circuits for a typical electronic counter. (Since these circuits involve digital logic, the reader should make reference to the Appendix for a discussion of digital circuits.)

Most counters now use decade counters (made up of four binary counters) that convert the count to a BCD code, decoders for conversion of the code to decade form, and readout tubes that display the decade information directly. The same circuits used for decade counters (four binary counters) are also used as dividers. Therefore, it is necessary to understand the operation of the basic decade before going into how the decade is used in logic circuits.

6-8.1. Decade Circuits

Decade circuits serve two purposes in counters. First, the decade will divide frequencies by ten; that is, the decade will produce one output for each ten input pulses or signals. This permits several frequencies to be obtained from one basic frequency. For example, a 1-MHz time base can be divided to 100 kHz by one decade divider, to 10 kHz by two decade dividers, to 1 kHz by three decade dividers, and so on. When decades are used for division they are often referred to as *scalers*, although dividers is a better term. The second purpose of a decade is to convert a count to a BCD logic code. The division function of a decade will be discussed first.

The basic unit of a decade divider is a 2:1 scaler, called a *binary counter*. This unit uses a *bistable multivibrator* or *flip-flop* circuit. A solid-state flip-flop (or FF) is shown in Figure 6-8. Although this circuit uses discrete components, decade flip-flops are often found in integrated circuit form, with

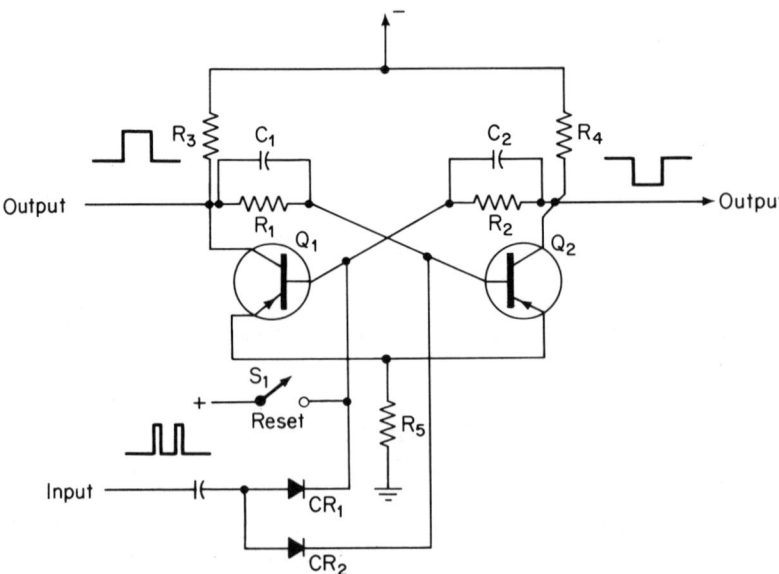

Fig. 6-8. Solid-state flip-flop (FF) circuit.

all four FFs in one package. (Sometimes several decades are found in one package.)

A flip-flop has two stable states, with Q_1 conductive and Q_2 nonconductive, and vice versa. The first input pulse flips the circuit from one state to the other. The second input pulse flops the circuit back to its original state, hence the name flip-flop. Each time the circuit is flipped from one state to the other and back again (requiring two input pulses) a single (complete) output pulse is produced.

There are several methods for distinguishing between the two stable states of the flip-flop. Generally, we consider the start to be with Q_1 cut off and Q_2 conducting. This is called the zero state. (As discussed in the Appendix, this may also be called the *false* state.) The second state, with Q_1 conducting and Q_2 cut off, is called the *one* state (or the *true* state).

Assume that in the circuit of Figure 6-8, transistor Q_2 is conducting. The voltage drop across R_4 reduces the negative voltage at the collector of Q_2. This drop in negative voltage, coupled to the base of Q_1 through R_2 and C_2, reduces the forward base-emitter bias and drives Q_1 toward cutoff. (Both Q_1 and Q_2 are PNP To be conductive, the base-emitter junction must be forward-biased; that is, the base must be negative relative to the emitter.) The negative voltage at the collector of Q_1 rises. This voltage rise, coupled to the base of Q_2 through R_1 and C_1, increases the forward baseemitter bias and drives Q_2 toward saturation. The process is cumulative and quickly results in Q_1 cut off and Q_2 conducting at saturation, the *zero* state of the FF.

The FF may be flipped to its *one* state by driving Q_2 to cutoff. This may be done by applying a positive pulse of sufficient amplitude to its base. The chain of events thus started culminates with Q_2 cut off and Q_1 at saturation. The circuit may be flopped back to its *zero* state by means of a second positive pulse applied to the base of Q_1.

The reset switch S_1 is a momentary switch that, when closed, places a positive bias on the base of Q_1. This bias is sufficient to cut off Q_1, if it is conducting. If Q_1 is already cut off, the bias has no effect. Thus closing the reset switch momentarily ensures that the cycle will start with the flip-flop in its zero state. Some electronic counters use a manual reset, while others use a repetitive pulse. It should be noted that the reset function is used primarily when the decade is used in the counter/readout circuit, rather than as a divider.

Output may be taken from the collector of Q_2. At the *zero* state when Q_2 is at saturation, the collector voltage is some low negative value. At the *one* state Q_2 is cut off and its collector voltage becomes more negative. Thus the output for the entire cycle is a negative-going square pulse. If the output were taken from the collector of Q_1, the output would be a positive-going square pulse.

The output pulses of one FF may be applied to the input of another similar FF for further frequency division. This is called *cascading*. A basic binary counter uses a cascaded chain of four FFs, as shown in Figure 6-9. The count of this chain would be 16 (2^4).

Assume that all four FFs are in their *zero* state, that a positive-going input is required to change FF states, and that the FF outputs are positive-going only when they shift from a *one* state to a *zero* state. Pulse 1 causes FF_1 to move from the *zero* to *one* state. The output of FF_1 is then negative-going and has no effect on FF_2. Pulse 2 causes FF_1 to move from the *one* to *zero* state. The output of FF_1 is then positive-going and moves FF_2 from the *zero* to *one* state. Pulse 3 causes FF_1 to move from the *zero* to *one* state (negative output with no effect on FF_2). Pulse 4 causes FF_1 to go from *one* to *zero*. This moves FF_2 from *one* back to *zero*. Thus four counts are required for FF_2 to go through a *complete* cycle. This process is repeated, requiring 16 pulses before FF_4 will go through a complete cycle.

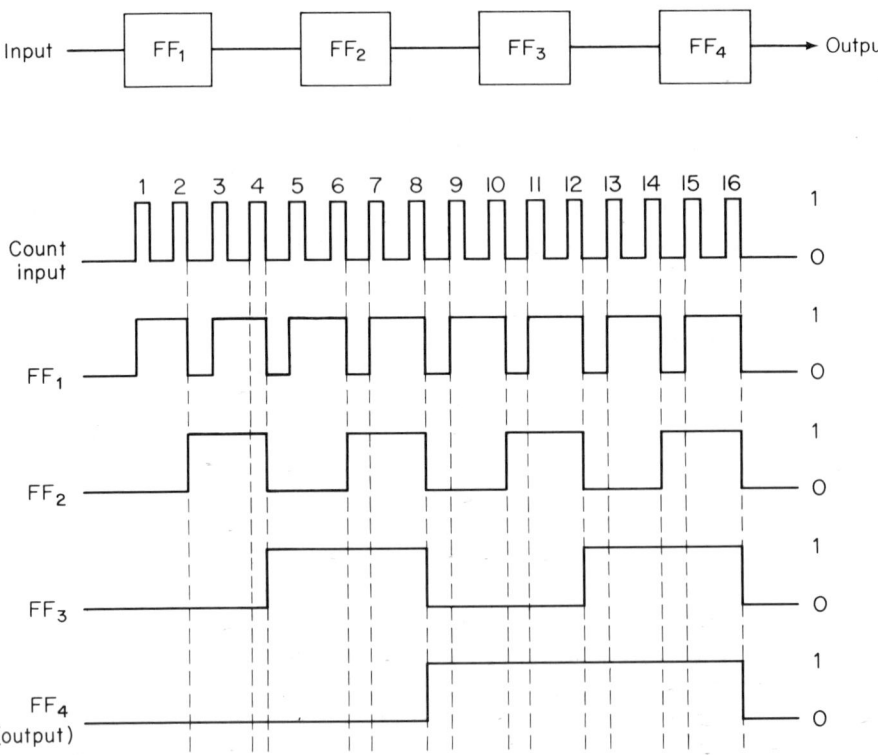

Fig. 6-9. Basic binary counter with four flip-flops in cascade.

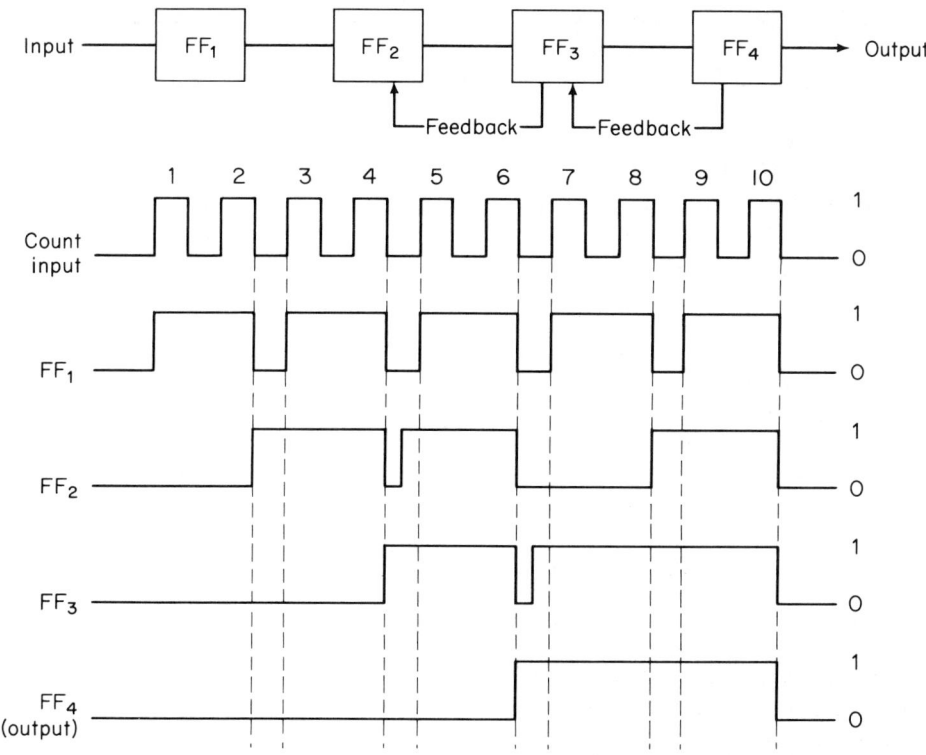

Fig. 6-10. Binary decade counter with four flip-flops in cascade, and feedback from FF_3 to FF_2, FF_4 to FF_3.

A *binary decade* counter uses feedback of the pulses to produce an output for ten input pulses, as shown in Figure 6-10.

6-8.1.1. Decade-to-Binary-Coded-Decimal Conversion

Figure 6-11 shows a decade circuit capable of converting a series of pulses to an 8421 binary code. One FF is used for each of the digits. Input pulses are fed into the 1-FF. The 2-, 4-, and 8-FFs are cascaded and receive pulses after the 1-FF.

For each state of an FF, one of the collectors is more positive than the opposite collector. When the states change, the polarities reverse. Some logic systems use *positive true*, while others use *negative true*. In the following discussions we will assume that positive true is used for all circuits; that is, when an input or output is positive, that input or output is true. In the case

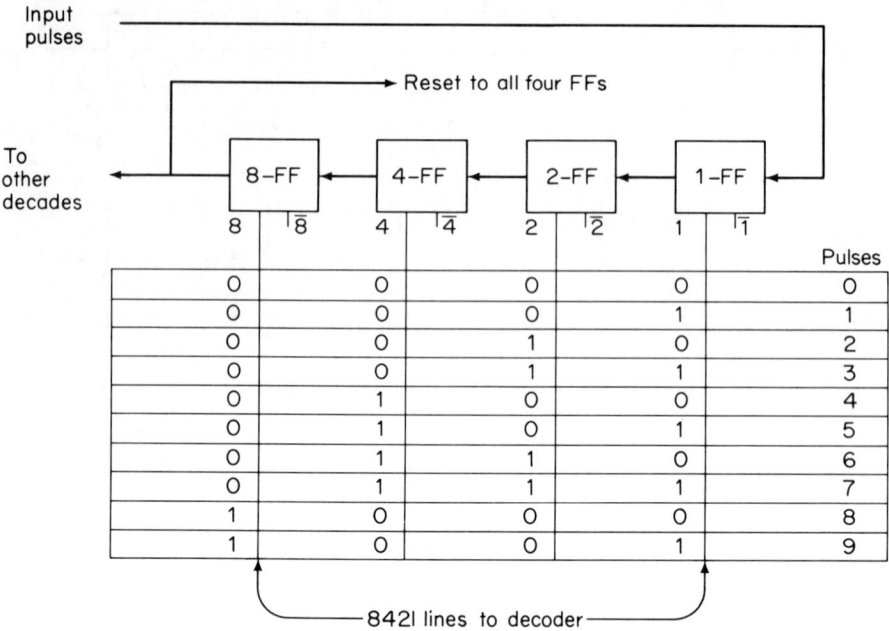

Fig. 6-11. Binary decade circuit for converting a series of pulses into an 8421 binary code.

of an FF, when one output (or collector) is true, the opposite collector is false.

For the FFs shown in Figure 6-11, when the 8421 lines are positive, the 8421 lines are true. With the FFs in the *same state*, the $\overline{8421}$ lines are negative (false). When the FFs change states, the 8421 lines are negative (false) and the $\overline{8421}$ lines are true.

At the beginning of a count, the 8421 lines are at negative (false). This is represented by 0. The $\overline{8421}$ lines are positive (true). This is represented by 1. (In a counter circuit, the decades are set (or reset) to this condition by the application of a voltage or pulse produced by a reset button, or by the sample rate generator, or by pulses from the decades at the end of a ten count.)

When the first pulse in the count is applied, the 1-FF changes states. The 1 line becomes positive, represented by a 1; the $\overline{1}$ line goes negative. (Note that the $\overline{8421}$ lines will always be at the *opposite* of the 8421 lines. The $\overline{8421}$ 0 and 1 states are omitted from Figure 6-11 for clarity.)

When the second pulse is applied, the 1-FF changes states. The 1 line goes false (0) and the 2 line goes true (1).

With the third pulse applied, the 1-FF goes true (1) but the 2-FF remains true. Remember that the 2-FF will change states for each *complete cycle* of the 1-FF.

When the fourth pulse is applied, the 1-FF goes to 0, as does the 2-FF. This causes the 4-FF to change states (the 4 line goes to positive or 1).

This process is repeated until a nine count is reached. At that point, the 8-FF moves from 1 to 0 and produces an output. This output is returned to the reset line and serves to reset all of the FFs to the false state (8421 lines at 0). The output from the 8-FF can also be applied to the 1-FF of another decade. Any number of decades can be so connected. One decade is required for each readout tube.

6-8.1.2. Binary-Coded-Decimal-to-Readout Conversion

Figure 6-12 shows the output of a decade connected to a decoder that converts the BCD code to 10-line form. Each of the 10 outputs (0–9) is connected to the corresponding cathodes of a readout tube. One readout tube (together with one decoder and one decade) is required for each digit of the counter. In practice, a *register* and/or *storage circuit* would be placed between the decades and their corresponding decoders. These registers will hold the BCD count data until cleared by the operator or will sample the count at some given rate (usually 2 or 3 times/sec). Operation of registers is discussed in the Appendix.

Note that the decoder is essentially a group of AND gates connected to provide nine inputs (8421, $\bar{8}\bar{4}\bar{2}\bar{1}$, and an enable input) and 10 outputs (0–9). As discussed in the Appendix, an AND gate requires that all inputs be true (positive in this case) to produce an output. All of the AND gates in Figure 6-12 have three inputs that must be true before they will produce an output.

The output of gates 0–9 are connected to the cathodes of a readout tube. Each readout tube consists of ten cold cathodes and a common anode, all enclosed in a gas-filled envelope. These cathodes are shaped as numbers from 0 to 9 and are stacked one above the other. When one of the decoder gates (0–9) produces an output, the corresponding cathode circuit is completed. The gas between that particular number-shaped cathode and the common anode is ionized, causing the cathode to glow, thereby revealing its number at the front panel. Only one decoder gate produces an output at any given time. Thus the remaining cathodes are nonionized and remain unlit (not visible from the front panel).

Assume that the circuit of Figure 6-12 is part of an electronic counter and the counter's gate is held open for a count of seven input pulses. Under these conditions, the $\bar{8}$, 4, 2, and 1 lines would be true (1). Likewise, the 8, $\bar{4}$, $\bar{2}$, and $\bar{1}$ lines would be false (0). The *B* AND gate of the decoder would have three true inputs ($\bar{8}$, 4, and enable). The enable pulse or voltage is always true and is applied (manually or by the sample rate generator) whenever a readout is desired. Both the *A* and *C* gates will have at least one false input. Therefore, only the 4, 5, 6, or 7 gates could produce an output. The 7 gate

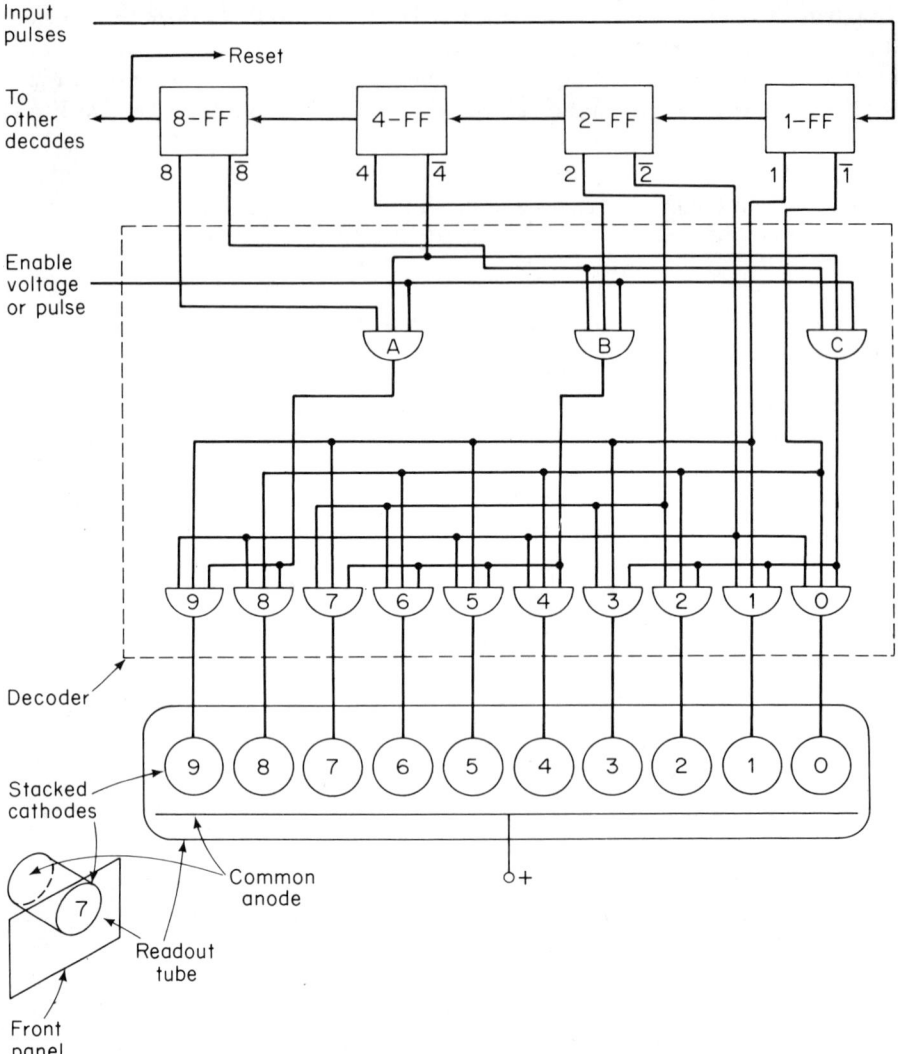

Fig. 6-12. Basic circuit for conversion of BCD code to decade readout.

would have three true inputs (B gate, 1 line, and 2 line). Gates 4, 5, and 6 would have at least one false input. Therefore, only the 7 gate will produce an output, and the numeral 7 cathode of the readout tube will glow.

6-9. COUNTER DISPLAY PROBLEMS

One problem often overlooked by inexperienced counter operators is making a count that exceeds the *readout capacity*. If a long gate time is used when a high frequency is counted, the entire answer will not be seen if the readout capacity is exceeded. To find what part of the answer will be visible, one must realize that counting starts with the rightmost digit in the readout, progresses to the next digit at the left after a count of nine has been reached, and so forth until all digits read nine. Next, account for the effect of gate time. For example, if 0.9 MHz is counted for 1 sec, a total of 900,000 counts will be gated into the counting circuits. A 6-digit counter will display 900,000, but a 5-digit counter will display 00,000.

In the most versatile 8-digit counters having gate times from 1 μsec to 10 sec, the entire answer can always he made visible by suitable gate time selection. The following table shows the maximum readout capacity (although counting rate may be much greater) for those counters having fewer than 8 digits and a limited gate time selection.

Gate	4-Digit	5-Digit	6-Digit
.01 sec	.9999 MHz	9.9999 MHz	99.9999 MHz
.1 sec	99.99 kHz	.99999 MHz	9.99999 MHz
1 sec	9.999 kHz	99.999 kHz	.999999 MHz
10 sec	.9999 kHz	9.9999 kHz	99.9999 kHz

Low-frequency count measurements can be another source of problems for the inexperienced operator. Counters can have a-c or d-c coupled inputs or both. When both inputs are available, the desired input coupling is selected by a front-panel switch. As the name implies, d-c-coupled inputs will pass input wave-forms regardless of the rise time. Alternating-current-coupled inputs discriminate against slow rise times; the frequency range specified defines sine-wave frequencies for which the sensitivity specification will be met. For example, a typical specification will be 100 mV RMS for sine-wave signals with frequencies down to 10 Hz. Most a-c-coupled counters will count sine waves below the minimum frequency specified, but a higher input amplitude will be needed. Most counters will count events of extremely low repetition rate if the input wave shape counted has a fast rise time.

When an electronic counter is used to count contact closure, be alert for spurious counts caused by *contact bounce*. Any bounce will have the same effect on the counter as a series of pulses, particularly when the counter is set for maximum sensitivity to measure a slow count.

6-10. COUNTER ACCURACY

There are three main sources of error in counter measurements: *plus or minus one count ambiguity*, *time base stability*, and *trigger error*.

6-10.1. Plus or Minus One Count Ambiguity

The plus or minus one count ambiguity is inherent in all electronic counter measurements because the input signal and the time base are

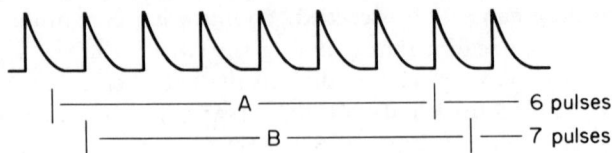

Fig. 6-13. Count registered during two identical gate times, showing ±1 count ambiguity.

normally not synchronized. As shown in Figure 6-13, the count registered during the gating time may be either six or seven, depending on the moment at which the gating time begins. Note that gate times *A* and *B* are of equal length. Thus, in any measurement, the counter's display may be incorrect by one count.

The fractional effect of the plus or minus one count ambiguity is

$$\frac{1}{\text{Total events counted}}$$

Obviously, the more events counted, the smaller this error becomes. This explains why long gate times result in better accuracy in frequency measurements.

6-10.2. Time Base Stability

The time base for most electronic counters is a crystal oscillator, mounted in a temperature-controlled oven. The frequency and stability of the crystal can be checked against broadcast standards or local standards, as described in Chapter 8. Several specifications are used by various counter manufacturers to describe the stability of the time base. The most important specifications are *crystal aging rate, short-term stability*, and error caused by *temperature change* and/or *line voltage variation*.

6-10.2.1. Crystal Aging Rate

The crystal aging rate (also specified as *long-term stability* or *drift rate* by some manufacturers) refers to slow but predictable variations in average oscillator frequency in time because of changes occurring in the quartz crystal itself. After an initial period of rapid change when the oscillator is turned on, aging in a good crystal becomes quite slow and assumes a predictable linear characteristic. The slope of this linear characteristic is the overall oscillator aging rate.

Since aging is cumulative, it is necessary to calibrate the oscillator periodically, as described in Chapter 8.

6-10.2.2. Short-Term Stability

Short-term stability specifications indicate the effects of noise generated internally in the time base oscillator on the *average frequency* over a short time, usually 1 sec.

Short-term effects are so small that the specification is listed for only the most stable time base oscillators in precision ovens. In the less stable oscillators, other errors make the short-terms specifications insignificant.

When comparing short-term stability specifications, it is important to remember that the averaging time used will determine how good the specification *appears* to be. A long averaging time will hide great frequency variations.

6-10.2.3. Line Voltage and Temperature

Line voltage and temperature specifications should be self-explanatory. However, it should be noted that the *total inaccuracy* caused by the time base is the sum of the aging, short-term line voltage and the temperature errors.

6-10.3. Trigger Error

Trigger error arises from noise on the gate-control signal. This noise causes the gate to open or close at incorrect times and results in an erroneous count.

Significant trigger error can occur only when an external signal controls the gate, that is, when period, ratio, and time-interval measurements are being made. Trigger error should not occur on frequency measurements since the trigger is supplied by a (supposedly) noise-free time base signal.

Absolute trigger error is stated in time units and the fractional effect is given by

$$\frac{\text{Error in time}}{\text{Total time gate is open}}$$

The equation explains why long period averaging is such a good method for reducing period measurement error (because the gate time is extended). As more periods are averaged, the effect of both trigger error and the plus and minus one count ambiguity are reduced proportionally.

7 AMPLIFIERS

7-1. AMPLIFIERS AS TEST EQUIPMENT

Many technicians do not consider amplifiers as test equipment. However, an amplifier can be a most useful item of test equipment for both the shop and laboratory technician. For shop work, an amplifier can serve as a signal-tracing device or as a substitute amplifier to pinpoint distortion, hum, and so on, in a suspected amplifier stage. For the laboratory technician, a differential or operational amplifier can be converted to perform integration, differentiation, amplification by a constant factor, summation, phase inversion, and so on, all by adding a simple external circuit. Thus an infinite number of tests can be performed with a basic amplifier.

7-2. TYPES AND FUNCTIONS OF AMPLIFIERS

Amplifiers have two basic functions in test equipment applications: (1) to amplify signals that are too low in level for the intended application and (2) to isolate signal sources from other circuits. In both cases, the amplifier supplies power under the control of the input signal and produces that power to the output as increased voltage and/or as increased current.

For the purpose of discussing their application as test equipment, amplifiers can be divided into two groups: (1) a-c amplifiers and (2) d-c (direct-current or direct-coupled) amplifiers. For practical purposes, all d-c amplifiers have some a-c response, but they are classified separately because of special characteristics.

Going into a further breakdown, two types of amplifiers often used in test equipment applications are the *operational amplifier* and *differential amplifier.* The following sections describe basic types of amplifiers used as test equipment.

7-2.1. Alternating-Current Amplifier

The basic a-c amplifier used in test equipment work is a conventional linear amplifier, usually with some form of *inverse feedback*. Test equipment a-c amplifiers are designed for applications requiring flat frequency response (for a-c work) and/or short rise times (for pulse work).

A simplified schematic diagram of a typical a-c amplifier is shown in Figure 7-1. This amplifier has a large inverse feedback factor, not only to reduce distortion and to broaden the frequency response, but also to ensure gain stability. A stable amplification factor (or gain) is required when an amplifier is used to measure signals, and a large amount of feedback is necessary to reduce the effects of changes in transistor characteristics. Feedback lowers the output impedance so that the amplifier performs as a constant-voltage source that is unaffected by the amount of current drawn by the output load. In the amplifier shown, the 50-Ω resistor in series with the output provides a true 50-Ω source.

Note that the input and output circuits are provided with coupling capaci-

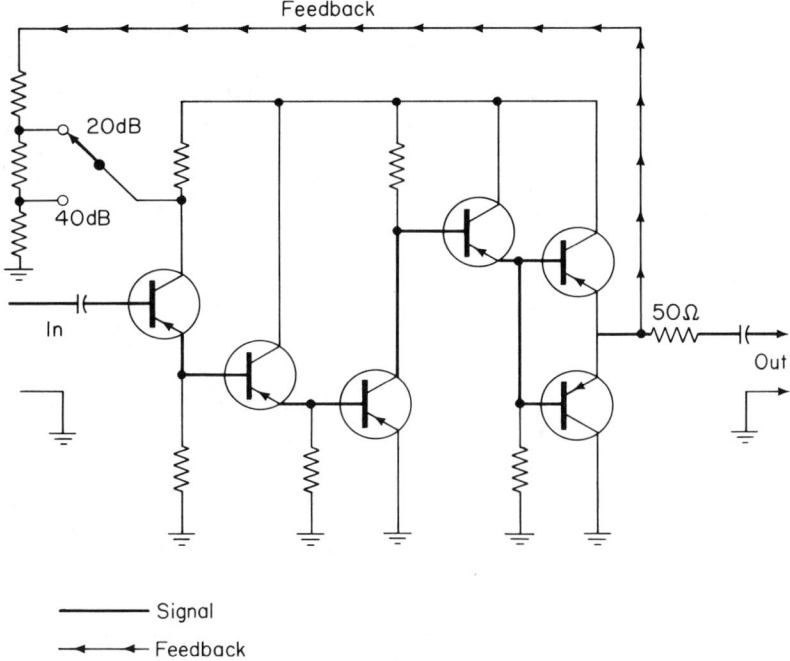

Fig. 7-1. Typical ac amplifier (Hewlett-Packard).

tors. If these were removed, the same basic circuit could be used as a d-c amplifier, since there are no interstage coupling capacitors.

7-2.2. Direct-Current Amplifier

The amplifier of Figure 7-1 suffers from a limitation that is quite serious in certain test equipment applications. It cannot amplify direct voltages and currents because the coupling capacitors are not suited for d-c operation (they will not pass direct current). And, because of the very large impedances offered by coupling capacitors at very low frequencies, such an amplifier is not suited for amplification at such frequencies.

If direct or very-low-frequency voltages or currents are to be amplified, we must use *direct-coupled* amplifiers where the signal is fed directly to the transistor without the use of a coupling capacitor.

Amplification of d-c voltage levels requires special considerations. One of the main problems is that the direct-coupled amplifier responds the same way to a change in power supply voltage (due, for example, to temperature drift) as to a change in the d-c signal level. Although high feedback can stabilize the gain of an a-c amplifier to the point where the gain is determined almost entirely by the resistors in the feedback network, the level of a d-c amplifier is not so easily stabilized.

A widely used method for circumventing the voltage change problem of direct-coupled amplifiers is to convert the d-c signal to an equivalent a-c signal (through modulation). The a-c signal is amplified in a gain-stable a-c amplifier and then reconverted to dc (through demodulation). During amplification, the signal is represented by the difference between the maximum and minimum excursions of the a-c wave-form and is not affected by drift in the absolute voltage levels within the amplifier.

One method used to convert the d-c signal to an a-c signal is to switch the amplifier input alternately to both sides of a transformer, as shown in Figure 7-2. This periodically inverts the polarity of the signal applied to the amplifier. The switches illustrated may be mechanical, transistor, or photoconductive. Another pair of contacts at the output establishes the ground level for a storage capacitor in series with the output. The output storage capacitor becomes charged to a level corresponding to the amplitude of the output square wave. Synchronous detection preserves the polarity of the input voltage and recovers both positive and negative voltages with the correct polarity. Such a direct-coupled amplifier offers drift-free amplification of low-level signals in the microvolt region.

Another modulation technique uses two photoconductors—one in series with and one parallel to the amplifier input, as shown in Figure 7-3. The photoconductor resistance is proportional to the illumination. By illuminating the photoconductors alternately, the amplifier input is connected to the

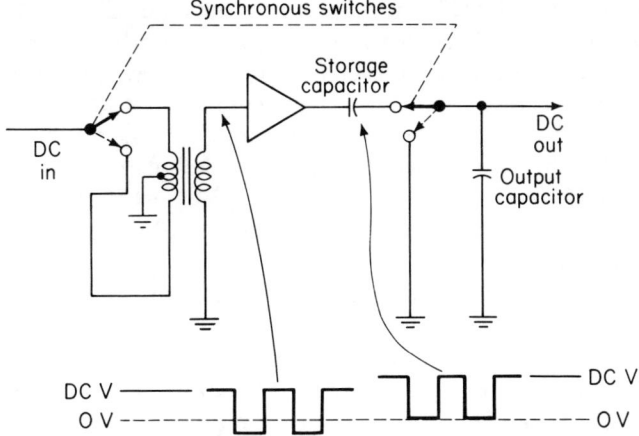

Fig. 7-2. Modulated amplifier technique for dc amplification.

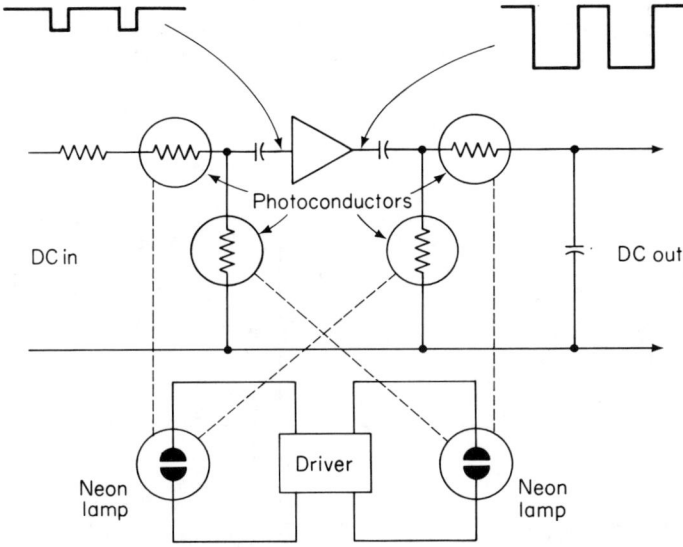

Fig. 7-3. DC amplification circuit with photoconductive modulation.

signal and to ground. By synchronizing the output photoconductors with the input photoconductors, the output storage capacitor charges to a level that corresponds to the output square-wave amplitude. This level is equivalent to the amplified d-c signal.

One problem with any type of modulated amplifier is frequency response. If the input signal is pure dc, there will be no problem using modulation. However, if the input is very-low-frequency ac, the modulating frequency must be higher than the signal frequency. If not, the input signal wave-form could be distorted or completely lost. For example, if both the input signal and modulating frequency were at 100 Hz, and if the amplifier input were shorted at the same instant as the positive swing of the input signal, the amplifier would see only the negative portion of the input.

As a rule of thumb, the modulating frequency should be 4 times that of the highest a-c input signal to be amplified. In a typical test equipment amplifier, the input signal will be limited to 100 Hz (dc–100 Hz), with 400 Hz as the modulating frequency.

Techniques other than modulation can be used where it is necessary to amplify both dc and ac (up to about 100 kHz). One such method is shown in Figure 7-4, where two parallel amplifiers are used. One amplifier is direct-coupled for dc and low-frequency ac. The other amplifier is a conventional a-c amplifier. Appropriate networks separate the two frequency bands. For example, dc is rejected from the a-c amplifier branch by coupling capacitors, but it appears at the d-c amplifier input and output as a charge across capacitors $C1$ and $C2$. High-frequency ac is attenuated at the d-c amplifier input by resistors $R1$ and $R2$. Note that feedback is used in both branches to assure uniform gain. Such an amplifier will provide stable gain (1 per cent or better) from dc up to about 100 kHz.

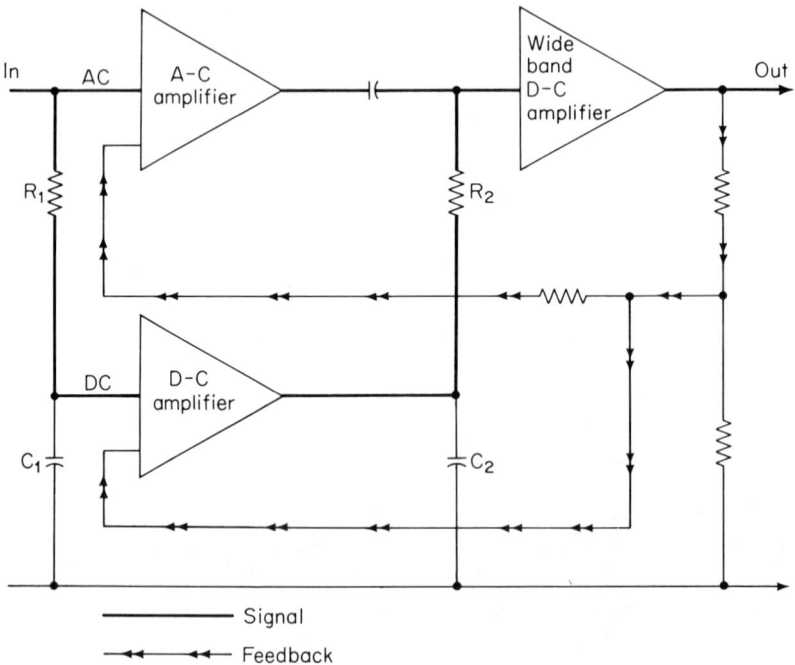

Fig. 7-4. Dual amplification circuit (Hewlett-Packard).

Another common method used in d-c amplifiers to correct for voltage change is the *chopper-stabilized* technique. This method uses a modulated amplifier to correct for voltage change or drift. The arrangement is shown in Figure 7-5. Note that the input signal (dc) is amplified through a conventional d-c amplifier. A portion of the amplified output signal is tapped off from a divider network and compared with the original input signal. The divider network reduces the output by the same amount that was amplified in the d-c amplifier. Therefore, the divider output should be equal to the input level at the summing point. Any difference at the summing point (caused by voltage change, drift, and so on) is amplified through a modulated amplifier, then applied to the main channel d-c amplifier as negative feedback to cancel the drift.

7-2.3. Differential Amplifier

Differential amplifiers have two identical input channels that function in push-pull fashion. Figure 7-6 shows a typical circuit. The basic purpose of such a circuit is to produce an output signal that is linearly proportional to the *difference between two signals applied to the input*. Such a circuit will provide an overall open-loop (no-feedback) gain of approximately 2500.

As shown, the inputs are applied to two emitter-coupled transistor amplifiers $Q1$ and $Q2$. The collectors of $Q1$ and $Q2$ are connected to a constant-current source and have the same-value load resistor. The output signals from $Q1$ and $Q2$ are equal but are 180° out of phase. Should only one input be used (with the other input grounded or returned to a bias), the resulting

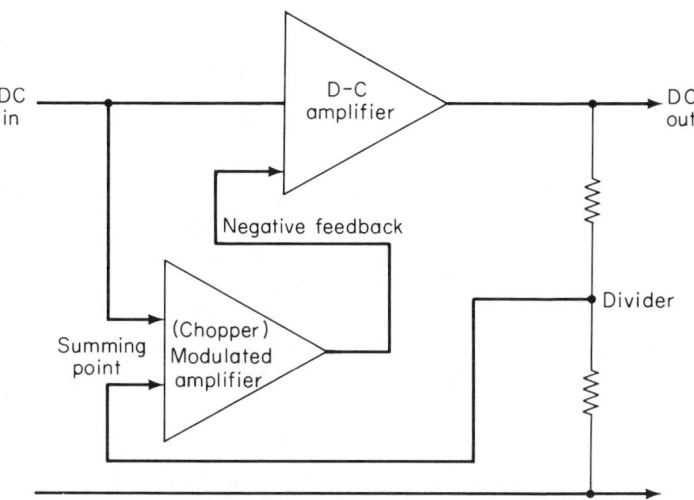

Fig. 7-5. Chopper-stabilized technique for dc amplification.

256 Amplifiers

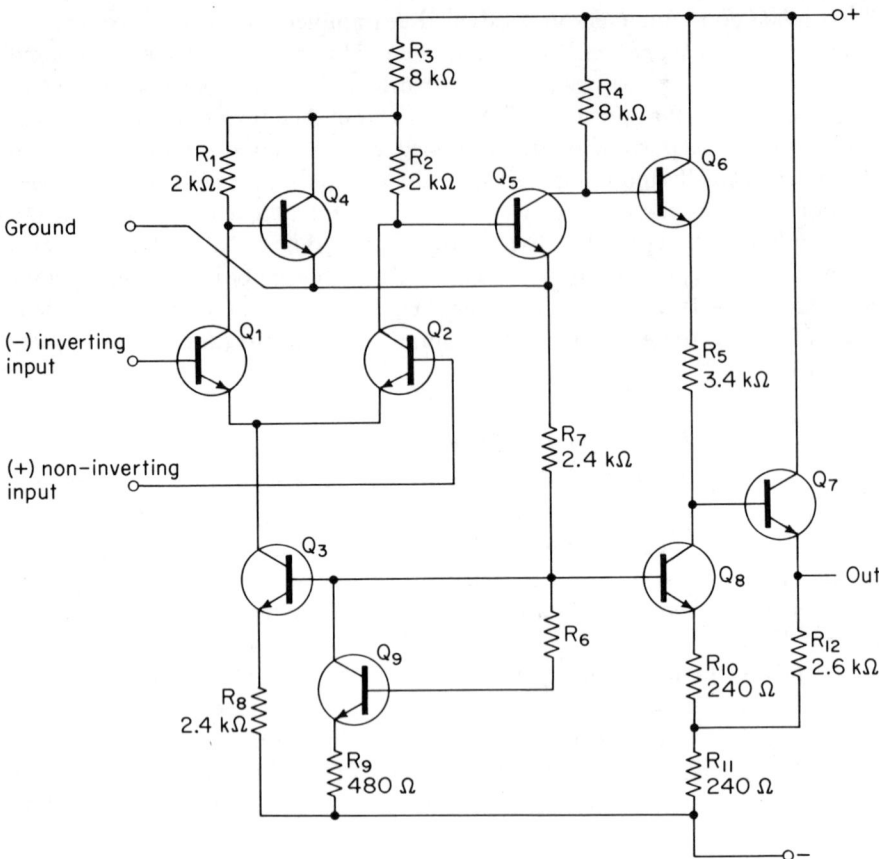

Fig. 7-6. Typical differential amplifier (Fairchild).

output would still appear at both collectors ($Q1$ and $Q2$) but the gain would be cut in half.

The output of a differential amplifier is generally single-ended and represents the amplified difference between the two input channels. This arrangement cancels hum or other interference picked up on the signal leads that appear in phase to the amplifier inputs (referred to as *common-mode signals*).

Since a differential amplifier is sensitive only to the difference between the two input signals, the signal source need not be grounded and can be *floating*. Therefore, differential amplifiers are often used in test equipment applications where the signal source is from a bridge (such as a bridge-type transducer) and the power supply is grounded.

The differential amplifier also allows injection of a fixed d-c voltage into either channel to permit establishment of a new voltage reference level at the output (some point other than 0 V). This is commonly referred to as *zero-suppression*.

When the input is floating, cable shielding between the amplifier and signal source may be connected to chassis ground rather than to signal ground. However, both a-c and d-c voltages can exist between two widely separated earth grounds, causing currents to flow. This condition is shown in Figure 7-7, where a bridge-type transducer is used with a differential amplifier.

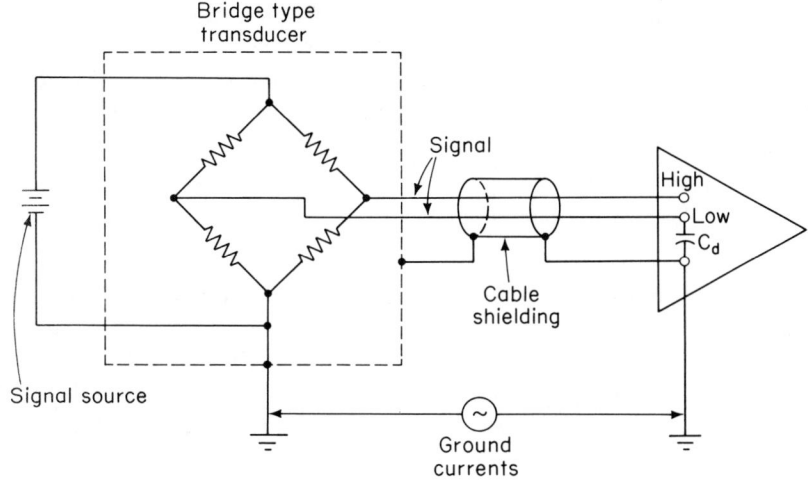

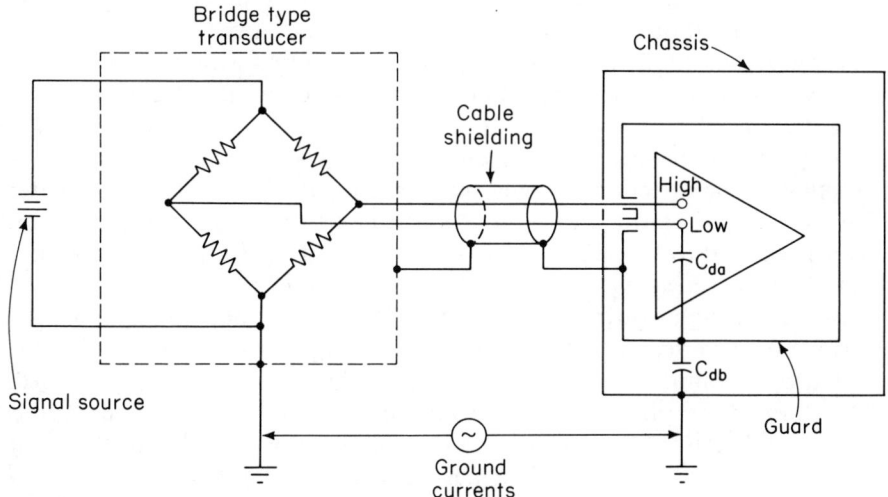

Fig. 7-7. Guard shield technique used to reduce capacitance between signal leads and ground.

Note that the signal source is connected to the transducer earth ground (local ground or physical ground, as it may sometimes be called). This ground point is connected to the amplifier ground through the cable shielding. The amplifier ground is connected to one of the differential inputs through internal capacitance (represented as C_d) of the amplifier. The same differential input is connected to the signal source through the signal leads and the transducer elements. Thus the ground currents are mixed with the signal currents.

One method to minimize this condition is shown in Figure 7-7. Here a guard shield around the input circuits of the differential amplifier provides an electrostatic shield to break the internal capacitance C_d into two series capacitances C_{da} and C_{db}. A much higher impedance is then presented to the flow of ground signals. This type of amplifier is termed a floated and guarded amplifier.

7-2.4. Operational Amplifier

An operational amplifier is a very high-gain, direct-coupled amplifier having out-of-phase input-output characteristics (180° phase shift). It is possible to convert a differential amplifier, such as shown in Figure 7-6, to an operational amplifier by the addition of a feedback network. Since the open-loop gain (without feedback) is very high, closed-loop (with feedback) characteristics can be controlled by feedback components, within the gain limits of the amplifier.

Normally, resistors and capacitors are used as input and feedback components. By selecting proper feedback networks, many operations, including linear amplification by a constant, summation of two or more signals, integration, and differentiation of voltage wave-forms with respect to time, can be performed. Such operations are discussed in later sections of this chapter.

A symbol frequently used for an amplifier is a triangle with the vertex pointing in the direction of signal flow (see Figure 7-8). Note that Figures 7-6 and 7-8 are drawn with both a positive (+) and a negative (−) input. Signals applied to the minus input are inverted at the output, while signals at the plus input arrive at the output with the same polarity.

The minus input (inverting) is normally used in an operational amplifier because the inverted output permits the use of negative feedback through the Z_f feedback components. When not in use, the plus input is usually grounded. Therefore, if the circuit of Figure 7-6 (or any similar differential circuit) is used, the output will be related to the difference between the minus input and ground, rather than a difference between two inputs.

Figure 7-8 is a *generalized or theoretical feedback circuit for a typical operational amplifier.* As shown, the *theoretical gain for* such an arrangement is

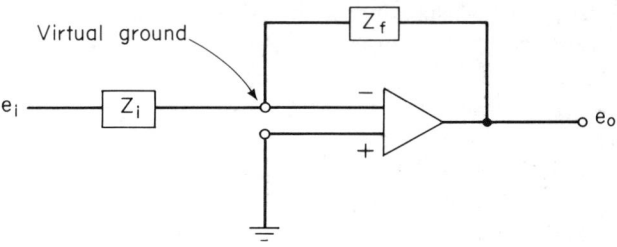

$$G = \frac{e_o}{e_i} = -\frac{Z_f}{Z_i} \left[\frac{1}{1 - \frac{1}{A}\left(1 + \frac{Z_f}{Z_i}\right)} \right]$$

G = Closed-loop gain
e_o = Output voltage
e_i = Input voltage
Z_f = Feedback impedance
Z_i = Input impedance
A = Open-loop gain

Fig. 7-8. Generalized or theoretical feedback circuit and gain equation for operational amplifier.

dependent on Z_f (feedback impedance), Z_i (input impedance), and A (open-loop amplification of the amplifier).

When the plus input is grounded, a concept of a *virtual ground* is applied to the minus input. Actually, the d-c level at the minus input of an operational amplifier is very close to ground. When an input signal is applied, the signal tends to move the base away from ground. However, the negative feedback from the output of the amplifier resists this tendency. The amount that the minus input voltage varies with a signal is dependent on the open-loop gain of the amplifier; the higher the gain, the less the minus input voltage varies. With the high open-loop gain normally encountered in an operational amplifier, the minus input voltage varies only slightly under closed-loop conditions. Therefore, it is convenient to assume that for all *practical purposes* the minus input voltage does not change with signal.

Since the minus input voltage remains essentially constant with input signal changes, it appears as though the minus input is grounded. Thus a *virtual ground* can be considered to exist at the minus input. The term "virtual" is used to indicate that, although the input of the amplifier appears to be grounded, it actually is not. Many equations for the functions performed by an operational amplifier can be most easily derived by the use of the *virtual ground concept*.

It should also be noted that since a virtual ground exists at the minus input, the input impedance of the amplifier is essentially determined by the value of the Z_i component. In a typical test equipment amplifier, the input component is a 50-Ω resistor. Thus the source sees 50 Ω, no matter what signal level is applied to the input.

Figure 7-9 is a *basic practical circuit for a typical operational amplifier*. While many forms of this basic circuit are used, the four most common test equipment applications are as follows:

1. Where an input voltage generates an output voltage (Figure 7-10)
2. Where an input voltage generates an output current (Figure 7-11)

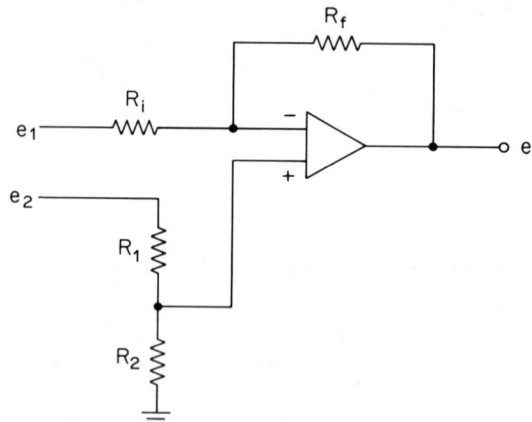

Fig. 7-9. A basic circuit of operational amplifier with both negative and positive active inputs.

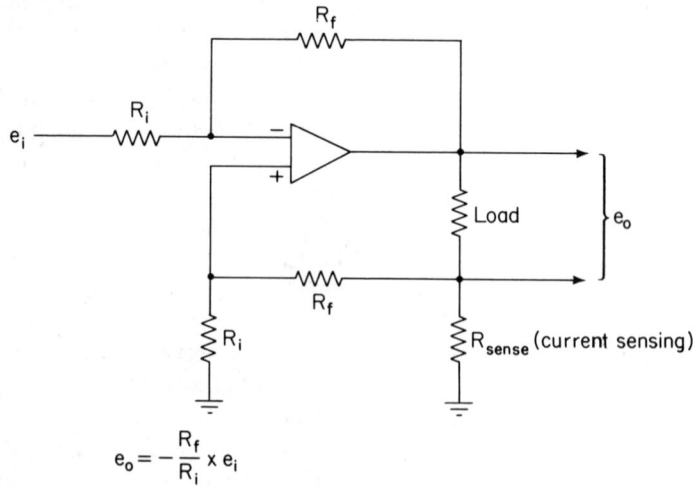

$$e_o = -\frac{R_f}{R_i} \times e_i$$

Fig. 7-10. Basic operational amplifier circuit where an input voltage generates an output voltage.

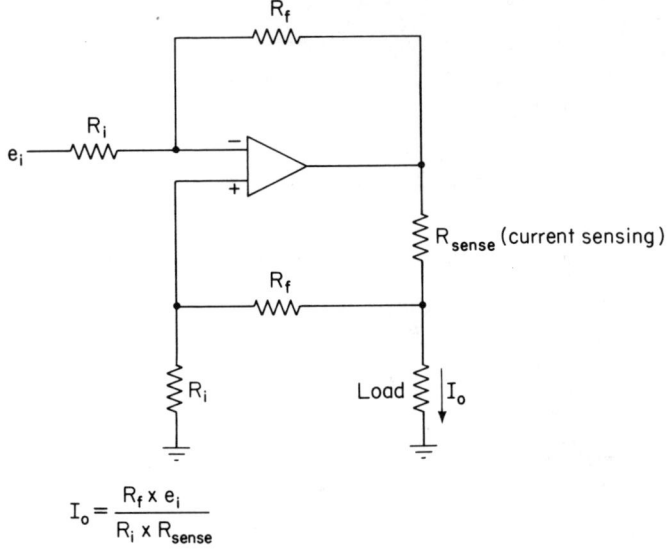

Fig. 7-11. Basic operational amplifier circuit where an input voltage generates an output current.

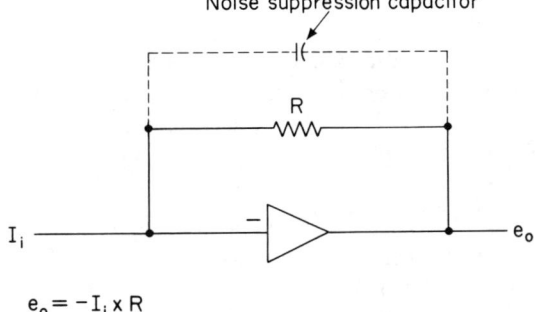

Fig. 7-12. A basic operational amplifier circuit where an input current generates an output voltage.

3. Where an input current generates an output voltage (Figure 7-12)
4. Where an input current generates an output current (Figure 7-13)

Practical variations of these four basic circuits are discussed in later sections of this chapter.

7-3. TYPICAL AMPLIFIER CHARACTERISTICS AND TEST METHODS

The major characteristics of interest in amplifiers are gain, frequency response, power output, noise and hum, distortion, and input-output impedance. Impedance is discussed in Chapter 3; distortion is discus-

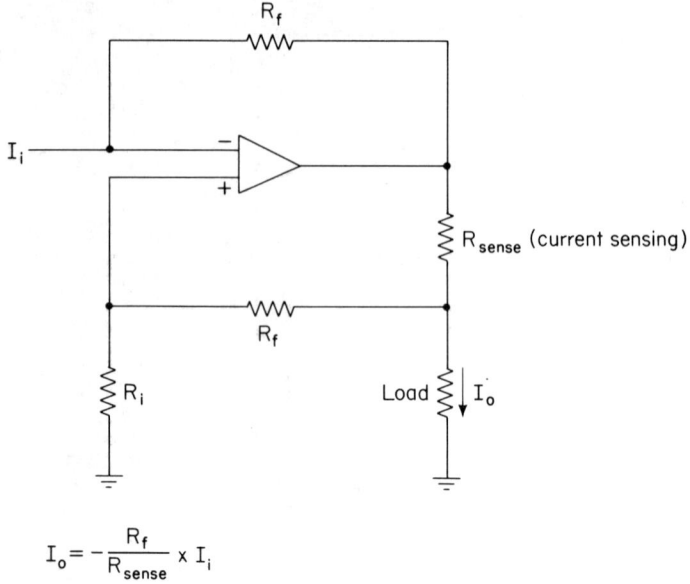

Fig. 7-13. Basic operational amplifier circuit where an input current generates an output current.

sed in Chapter 10. The following paragraphs describe basic test methods for typical test equipment amplifiers.

7-3.1. Testing Amplifier Gain

The basic connections for testing amplifier gain are shown in Figure 7-14. The basic test equipment required includes a signal source (usually a sine-wave generator) and a measuring device (meter or oscilloscope). The major problem in using meters for testing an amplifier is the limited frequency range of most meters. It is possible to use almost any meter effectively with audio amplifiers (up to about 20 kHz). However, the amplifiers used in industrial test equipment applications operate at much higher frequencies. This is especially true of amplifiers used in pulse test applications, where frequencies can be in the 150-MHz region. Even with audio amplifiers, the monitoring device should be capable of measuring signals up to about 500 kHz. This will ensure that any *harmonics* or *overtones* will be properly amplified and displayed, even though bandpass characteristics are supposedly no higher than 100 kHz.

If the frequency range of the amplifier circuit is beyond that of the meter, or if it is suspected that harmonics will be produced beyond the meter's capability, use an r-f probe, as shown.

When the amplifier is a modular unit, such as an integrated circuit or a

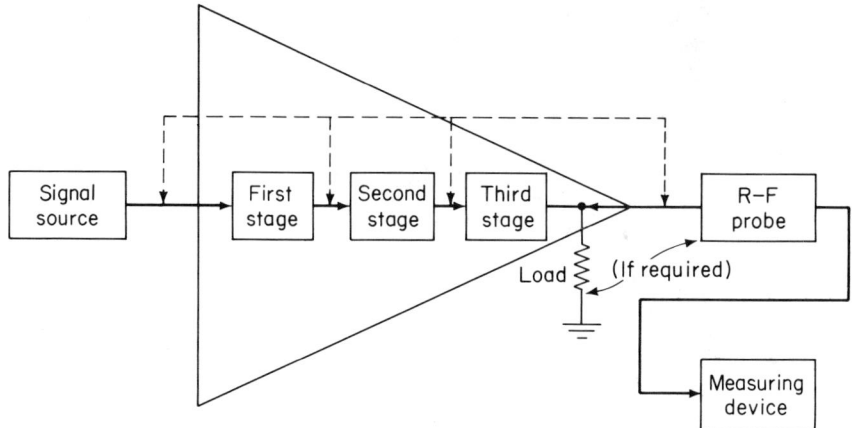

Fig. 7-14. Testing amplifier gain.

sealed solid-state amplifier, the *overall gain* is measured. On those amplifiers where the input and output of individual stages is accessible, *stage gain* can be measured, in addition to overall gain. Measuring stage gain in an amplifier is similar to *signal-tracing*. A signal of *known frequency and amplitude* is introduced into the input by means of an external generator. The signal is usually a sine wave but can be a square wave. The frequency and amplitude will depend on the manufacturer's data or other test requirements. The amplitude of the input signal is measured on the meter or oscilloscope. For practical purposes, an oscilloscope must be used for pulse or square-wave input signals. The meter or oscilloscope test lead (or probe) is then moved to the input and output of each stage in turn, until the final output is reached. In the case of a modular amplifier, only the input and final output are measured.

The gain of each stage, or overall gain, is usually measured in terms of *voltage*, rather than as a decibel value (although some meters and oscilloscopes are provided with dB scales, as previously discussed). If desired, the voltage gain can be converted to a decibel value by means of the conversion chart of Chapter 1.

One factor often overlooked in testing amplifier gain is setting the amplifier controls (such as *attenuation, gain, tone, equalizer*, and so on) to their normal operating point or to some particular point specified in the manufacturer's test data.

7-3.2. Testing Amplifier Frequency Response

The basic connections for testing amplifier frequency response are shown in Figure 7-15. The test equipment required is the same as that for testing gain. The basic method is to apply a *constant-amplitude* signal while

264   Amplifiers

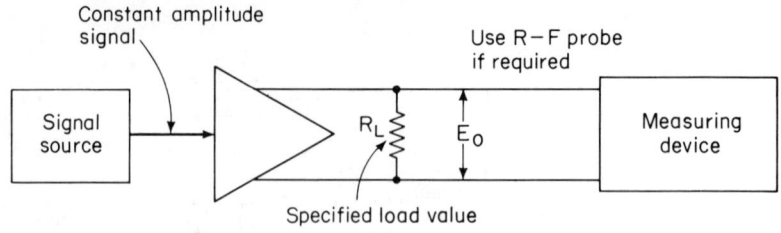

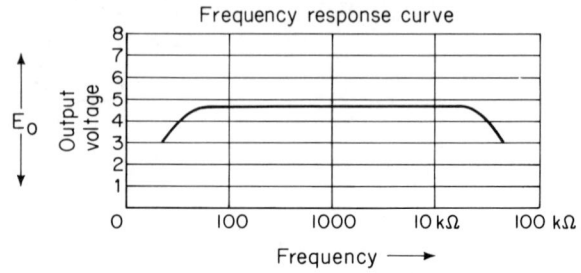

Fig. 7-15. Testing amplifier frequency response.

monitoring the amplifier final output. The input signal is varied in frequency *but not in amplitude* across the entire operating range of the amplifier. The voltage output at various frequencies across the range is plotted on a graph similar to that shown in Figure 7-15.

Usually, the first measurement is made by setting the generator output frequency to the lowest point specified in the manufacturer's data. Then the frequency is increased in steps across the entire frequency range of the amplifier. The amplifier voltage output is recorded at each step. (Steps of 100 Hz are typical for an audio amplifier.) With all of the output voltages recorded, a line is drawn on graph paper through each of the checkpoints to obtain the *frequency response curve*. A typical curve will resemble that of Figure 7-15, with a flat portion across the center and a roll-off at each end. Some amplifiers are designed to provide a high-frequency boost (where the high end of the curve increases in amplitude) or low-frequency boost (where the low end shows an amplitude increase). The manufacturer's data must be consulted for this information.

A fact often overlooked in making a frequency response test of an amplifier is that generator output may vary with changes in frequency (even with precision laboratory generators). This can result in considerable error. The problem can be prevented by monitoring generator output after each change in frequency. Then, if necessary, the generator output amplitude can be reset to the correct value. Within extremes, it is more important that the generator output remain constant rather than at some specific value, when making a frequency response check.

As in the case of gain measurements, the amplitude of the input signal, as well as settings of amplifier controls, should be as specified in the manufacturer's test data. In the absence of such data, set the controls at their normal operating positions.

If a particular input level is not specified for test, an arbitrary level that will not *overdrive* the amplifier can be used.

A simple method of determining a satisfactory input level is to monitor the amplifier output and increase the generator output until the amplifier is overdriven. This point will be indicated when further increases in generator output do not cause further increases in amplifier output. Set the generator output just below this point. Measure the generator output and keep it constant throughout the test.

Note that the load resistor R_L of Figure 7-15 is used for power amplifiers. Most manufacturers recommend that power amplifiers not be operated without a load. Also, the load serves to stabilize the amplifier output during test. The value of the load resistor should equal the normal output impedance of the amplifier.

7-3.3. Testing Amplifier Power Output

The test connections for power output measurement are identical to those for frequency response (Figure 7-15). The basic procedure is essentially the same as for gain and frequency response tests. A signal of specified frequency and amplitude is applied to the input, and the output voltage is measured across a load. The power output is then calculated by

$$\text{Power output} = \frac{(\text{Maximum voltage})^2}{\text{Value of load resistor (or load impedance)}}$$

7-3.4. Testing Amplifier Noise and Hum

The basic connections for testing amplifier noise and hum are shown in Figure 7-16. The major concern in such a test is that the monitor-

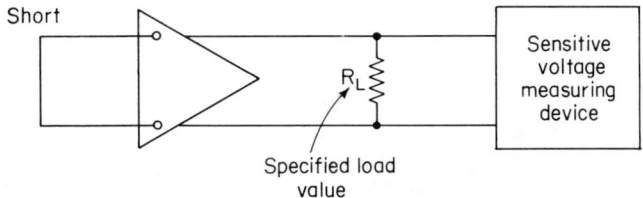

Fig. 7-16. Testing amplifier noise and hum.

ing device be sufficiently sensitive to measure possible noise and hum signals present within the amplifier. Such signals are usually in the order of a few microvolts and are almost always less than 1 mV for laboratory-type amplifiers.

The basic procedure consists of measuring amplifier output with the gain control at maximum but without an input signal. An oscilloscope is superior to a meter for noise-level measurement since the frequency and nature of noise (or other signal) will be displayed visually on the oscilloscope. The meter will show only the fact that a voltage is present and its amplitude.

After noting the voltage level at the amplifier output with the input short in place, the short is disconnected. Then any change is noted in the output. If the voltage level increases with the short removed, the voltage is probably a result of pickup (external to the amplifier). If the voltage indication remains constant with or without the input short, the voltage is the result of background noise, oscillation, and so on (within the amplifier).

7-3.5. Testing Amplifier Gain-Bandwidth Product

In some laboratory amplifiers used as test equipment a *gain-bandwidth product* is specified. This product is a combination of gain and frequency response and is so tested. The gain-bandwidth product is that *high frequency* at which the amplifier gain drops to unity, or one. For example, the gain-bandwidth product of the amplifier shown in Figure 7-6 is 15 MHz, while the normal open-loop gain is -2500. (The minus figure is applied because it is assumed that the minus input will be used and the output will be inverted. If a $+1$ mV signal were applied, the output would be -2500 mV, or -2.5 V.)

This output should remain constant from dc to about 10 MHz and then begin to drop off. At 15 MHz, the amplifier output should be 1 mV, indicating that there is unity gain at that frequency.

7-4. OPERATING PRECAUTIONS FOR AMPLIFIERS

Following are a few precautions to be observed when using amplifiers as test equipment.

7-4.1. Connecting Input and Output Signals

To assure good results, certain precautions should be taken in connecting signals to an amplifier. For low-level signals, it may be necessary to use shielded leads to minimize stray pickup. This is especially important when an operational amplifier is used for differentiation. High-frequency noise is particularly a problem with differentiation since the output of the differentiator is proportional to frequency.

Shielded input cables are recommended when signals to be amplified are obtained from a high-impedance source or when leads are long.

When shielded leads are used, the shields must be securely grounded to the chassis of both the signal source and amplifier input, as well as the amplifier output and the load.

It may be necessary to terminate signal cables to prevent standing waves in the cable. Such terminations are generally placed at the load end of the cable, although some sources also require a termination at the source end. When practical, simulate actual operating conditions in the equipment being tested by permitting it to work into a normal load.

7-4.2. Using Probes with an Amplifier

Any of the probes described in Chapter 9 can be used with an amplifier, but certain problems must be considered.

If an *attenuator* or *divider probe* is used with an amplifier, the overall gain will be reduced. For example, if a 10X divider probe were used with an amplifier having a gain of 2500, the net gain for the probe-amplifier combination would be 250 ($\frac{2500}{10} = 250$).

If an *r-f probe* is used, the amplifier must be of the d-c type, since the output of an r-f probe is a direct current corresponding to the r-f signal strength. If a demodulator probe is used, the probe output will be a direct current proportional to the modulation present on the r-f signal.

No matter what type of probe is used, it should be compensated, as described in Chapter 9.

7-5. DIFFERENTIAL AND OPERATIONAL AMPLIFIERS IN TEST APPLICATIONS

The following sections describe some typical test equipment applications for differential and operational amplifiers.

Note that all of the following applications require an operational amplifier. Some, but not all, require a differential input. If a differential amplifier is used for any of the following applications and only one input is shown, the signal should be applied to the minus input, with the plus input grounded.

7-5.1. Amplification by a Constant

When it is desired to amplify a signal by a constant (that is, to provide a fixed gain of 100, 1000, and so on), the circuit of Figure 7-17 is used.

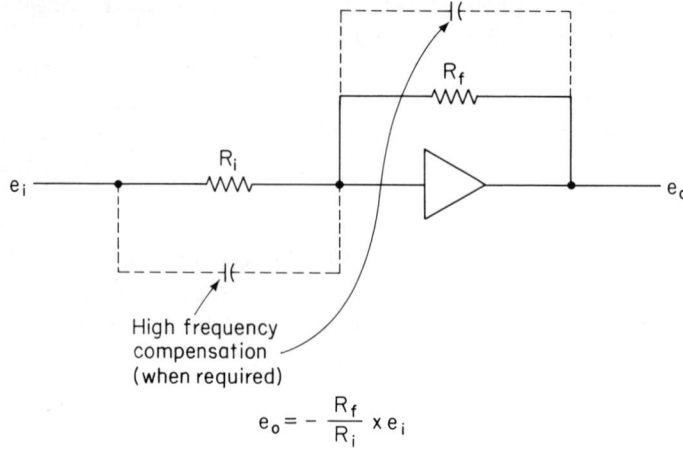

Fig. 7-17. Amplification by a constant.

$$e_o = -\frac{R_f}{R_i} \times e_i$$

Any desired gain can be obtained with such a circuit, provided the gain is within the open-loop gain and frequency limits of the amplifier. The closed-loop gain is determined by the feedback resistance values In effect, the input voltage will be multiplied by the ratio of the feedback resistance R_f to input resistance R_i. For example, if R_i is 1000 and R_f is 10,000, then a gain of 10 will be obtained. If a 0.5-V input were applied, the output would be 5 V. If the output were capable of a 10-V swing, an input signal up to 1 V could be applied without overdriving the amplifier.

The output of this circuit will be inverted. Therefore, the circuit can serve as a sign changer, with or without amplification. For example, if the circuit is used to amplify a 1-mV positive pulse and the resistance values are selected to provide a gain of 10, the output will be negative (−) 10-mV pulses.

The circuit can be made to provide variable gain instead of gain by a fixed constant, if the feedback resistor R_f is a potentiometer. As the potentiometer setting is changed, the gain of the amplifier will also change. The use of a potentiometer for R_f will allow the input impedance of the amplifier to remain essentially constant, while permitting the gain to be varied.

Feedback resistance R_f can be replaced by thermistors, photoresistors, or other variable-resistance elements. The gain will then be a function of the temperature, light level, or other variable, depending on the test equipment application.

7-5.2. Integration of Signals

When it is desired to integrate signals, the circuit of Figure 7-18 is used.

Note that the input circuit is formed by a series resistance R_i, while the

Differential and Operational Amplifiers 269

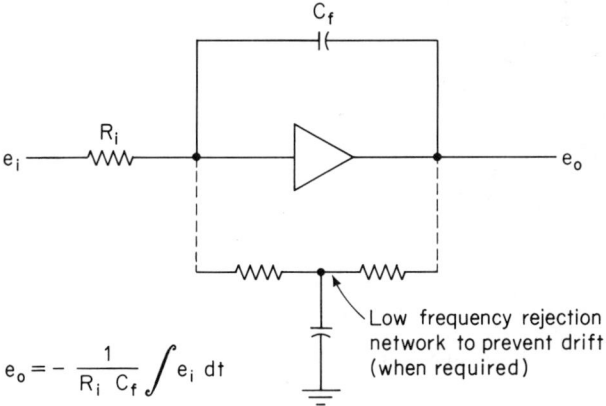

Fig. 7-18. Integration of signals with an amplifier.

feedback is accomplished with a fixed capacitor C_f. With such an arrangement, the output voltage is directly proportional to the integral of the input voltage and inversely proportional to the time constant of the feedback network (R_iC_f). In such an application, the R_iC_f time constant is approximately equal to the period of the signal to be integrated.

In some test equipment applications, the integrator circuit of Figure 7-18 is used as a precision 90° *phase shifter*. Sine waves applied to the input of the integrator are shifted in phase by exactly $+90°$. Sine waves of any frequency can be applied, provided that they are within the frequency limits of the amplifier.

One problem often encountered in practical applications of the circuit is that it will integrate *all signals* present at the input; that is, the output will represent the integral of all inputs (ac, dc, drift, and so on). This could cause a problem where it is desired to integrate only the a-c signals. For example, if the output were displayed on an oscilloscope and dc were present with the a-c signals at the input, the output would gradually increase, resulting in drift of the oscilloscope display. Some test equipment amplifiers incorporate a *low-frequency rejection* network (similar to that of Figure 7-18) to prevent such output drift.

7-5.3. Differentiation of Signals

When it is desired to differentiate signals, the circuit of Figure 7-19 is used.

Note that the input circuit is formed by a series capacitor C_i, while the feedback is accomplished with a fixed resistor R_f. With such an arrangement, the output voltage is driectly proportional to the time rate of change (or frequency) of the input voltage and is inversely proportional to the feed-

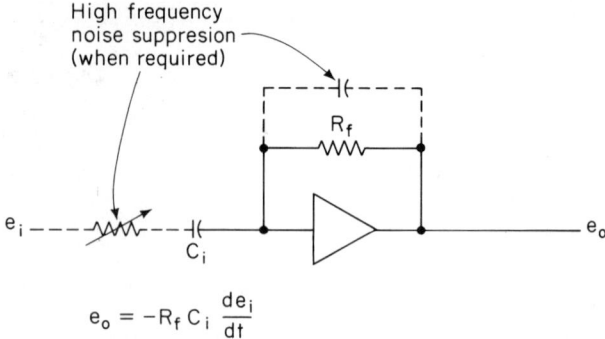

Fig. 7-19. Differentiation of signals with an amplifier.

back time constant ($R_f C_i$). In such an application, the $R_f C_i$ time constant is approximately equal to the rise time of the signal to be differentiated.

In some test equipment applications, the differentiator circuit of Figure 7-19 is used as a precision 90° *phase shifter*. Sine waves applied to the input of the differentiator are shifted in phase by exactly −90° (as opposed to a +90° shift for an integrator). Sine waves of any frequency can be applied, provided that they are within the frequency limits of the amplifier.

The circuit of Figure 7-19 permits slight changes in input slope to produce very significant changes in the output. This can be put to use in determining the linearity of supposedly linear waves, such as a sweep-sawtooth wave-form. In that application, the output would be displayed on an oscilloscope. Nonlinearity results from *change in the slope* of the wave-form. Therefore, if any nonlinearity is present, the differentiated wave-form indicates the points of nonlinearity quite clearly.

Since the output of a differentiator is directly proportional to frequency (within the limits of the amplifier) the circuit of Figure 7-19 can be used as a frequency-to-voltage converter. The frequency of an unknown signal can be determined by comparing the amplitude of the output voltage to that obtained using a known input frequency. If the circuit is used with an oscilloscope, the oscilloscope graticule can be calibrated for frequency-per-division. If the circuit output is applied to a meter, the output can be related on a frequency-per-volts basis. Either way, a constant-amplitude input signal must be used in the application to prevent changes in amplitude from disturbing the measurement.

One problem often encountered in practical applications of the circuit is that a differentiator will accentuate high-frequency noise. Some test equipment amplifiers incorporate a noise suppression circuit to minimize this condition. High-frequency noise suppression in a differentiation circuit can be accomplished by means of a capacitor across the feedback resistor or by a resistor in series with the input capacitor, as shown in Figure 7-19.

In other test equipment applications, the differentiation circuit is used to detect the presence of high-frequency noise.

7-5.4. Summation of Signals

When it is desired to sum a number of voltages, the circuit of Figure 7-20 is used.

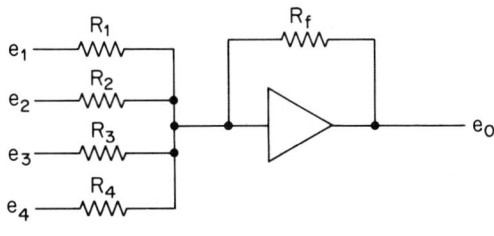

$$e_o = -\left(\frac{R_f}{R_1}e_1 + \frac{R_f}{R_2}e_2 + \frac{R_f}{R_3}e_3 + \frac{R_f}{R_4}e_4\right)$$

When $R_1 = R_2 = R_3 = R_4 = R_f$

Then $e_o = -(e_1 + e_2 + e_3 + e_4)$

Fig. 7-20. Summation of signals with an amplifier.

Note that the input circuit is formed by a number of parallel resistors, one for each voltage to be summed, while the feedback is accomplished by a single resistor. This arrangement is similar to the basic operational amplifier circuit of Figure 7-17.

When the values of all resistors are the same, the output of the amplifier is the sum of all the input voltages, with the sign inverted. For example, assume that all of the resistors (R_1, R_2, R_3, R_4, and R_f) are 10 Ω and that the voltage across R_1 is +10 across R_2 is −7 across R_3 is +5, and across R_4 is −1 V. Then the output would be −7 V, since

$$+10 - 7 + 5 - 1 = +7$$
$$+7 \text{ inverted} = -7$$

7-5.5. Unity Gain with High Input Impedance

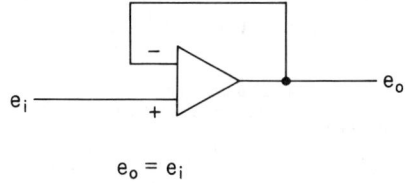

$e_o = e_i$

Fig. 7-21. Unity gain with high input impedance.

When it is desired to provide a high impedance input with unity gain, the circuit of Figure 7-21 is used.

272   Amplifiers

Note that there are no input or feedback resistors. The feedback is applied directly to the minus input to provide a gain of 1 at the amplifier output. Since the signal is applied to the plus input, the output is not inverted.

Because the signal is applied directly to the plus input and there are no resistance elements between it and ground, the input impedance of the circuit is determined primarily by the input current. Since this is quite small in most cases, the input impedance is very high. Of course, this is the d-c impedance. For a-c signals, the shunt capacitance brings the impedance down to a much lower level.

7-5.6. Amplification with High Input Impedance

When it is desired to provide a high input impedance with some gain, the circuit of Figure 7-22 is used.

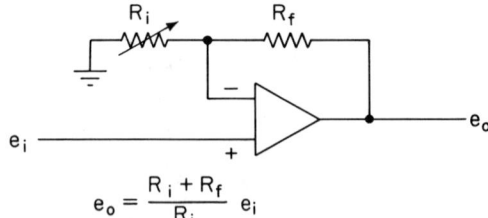

$$e_o = \frac{R_i + R_f}{R_i} e_i$$

Fig. 7-22. Amplification with high input impedance.

Note that this circuit is similar to the circuit of Figure 7-21. The gain is positive (noninverting), and is determined by the equation shown in Figure 7-22.

In this circuit, the signal is applied directly to the plus input, with feedback from the output to the minus input. The amount of feedback (and the amount of gain) is controlled by the values of R_i and R_f. Note that it is *not* possible to obtain a gain of less than 1 with this circuit.

7-5.7. Subtraction and/or Difference Amplification

Using a test equipment amplifier connected in the circuit of Figure 7-23, one signal voltage can be subtracted from another by simultaneous application of signals to both inputs of the amplifier. The signal applied to the minus input is subtracted from the signal applied to the plus input.

If the values of the resistor are not all the same, the amplifier gain for signals applied to the plus input is

$$\frac{R_1 + R_2}{R_1} \times \frac{R_4}{R_3 + R_4}$$

Gain of the amplifier for signals applied to the minus input is

$$-\frac{R_2}{R_1}$$

Differential and Operational Amplifiers 273

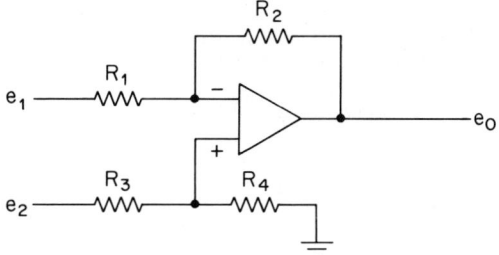

Subtract:

$$e_o = -\frac{R_2}{R_1} e_1 + \left[\frac{R_4}{R_3 + R_4}\right]\left[\frac{R_1 + R_2}{R_1}\right] e_2$$

Difference:

When $R_1 = R_2 = R_3 = R_4$

Then $e_o = e_2 - e_1$

Fig. 7-23. Subtraction and/or difference amplification.

By use of these two equations, a desired amplification can be combined with the subtractive process. This produces an output shown by the generalized expression

$$e_o = -\frac{R_2}{R_1} e_1 + \frac{R_1 + R_2}{R_1} \times \frac{R_4}{R_3 + R_4} e_2$$

In this application, the amplifier is used essentially as a difference amplifier.

A difference amplifier may be operated with compensated frequency response by adding small variable capacitors across R_3 and R_4 of the preceding circuit. This permits balancing the time constants, thus extending the usable bandpass of the difference amplifier. For example, in cases where all resistors are not equal, compensation for high frequencies may be accomplished by making all time constants equal.

7-5.8. Voltage-to-Current Converter (Transadmittance Amplifier)

Using a test equipment amplifier connected in the circuit of Figure 7-24, it is possible to supply a current to a load, with the current proportional to the amplifier input voltage. The current supplied to the load is relatively independent of the load characteristics. This circuit is essentially a *current feedback amplifier*.

A current-sampling resistor r is used to provide the feedback to the plus input. When $R_i = R_f = R_3 = R_4$, the feedback maintains the voltage across current sampling resistor r at a value of $-R_f/R_i \times e_i$, regardless of the load.

If a constant input voltage is applied, the voltage across r also remains con-

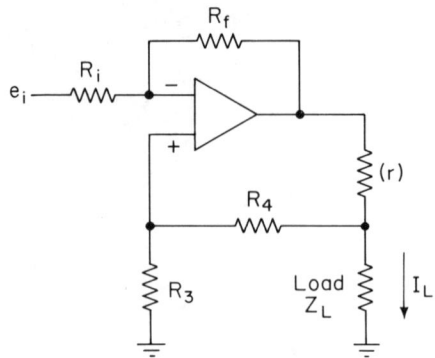

When $R_i = R_f$ $R_3 = R_4$
And $R_3 + R_4 \gg Z_L$

Then $I_L = -\dfrac{e_i}{(r)}$

Fig. 7-24. Voltage to current converter (transadmittance amplifier).

stant, regardless of the load. If the voltage across r remains constant, the current through r must also remain constant. With R_3 and R_4 normally much higher than the load impedance, the current through the load must remain nearly constant, regardless of the impedance.

The values of R_i, R_f, R_3 and R_4 should normally be the same. The current sampling resistor r is then selected for the desired load currents. Of course, the value of r should be selected so that the maximum rated output current is not exceeded.

With r expressed in kilohms, the current through the load $I_L = e_i/r$, as milliamperes per volt of input signal.

7-5.9. Voltage-to-Voltage Amplifier (Voltage Gain Amplifier)

Using a test equipment amplifier connected in the circuit of Figure 7-25, it is possible to supply a voltage across a load that will remain constant, regardless of changes in the load. This circuit is similar to that of Figure 7-24 (voltage-to-current converter) except that the load and current-sensing resistor r are transposed. With $R_i = R_f = R_3 = R_4$, the feedback to the minus input maintains the voltage across the load equal to minus the input voltage, regardless of load (within the current limitations of the amplifier). Operation of the circuit is essentially the same as that of the voltage-to-current converter.

7-5.10. Bandpass Amplifier

An operational amplifier can be used as a bandpass amplifier for test equipment purposes. The basic circuit is shown in Figure 7-26.

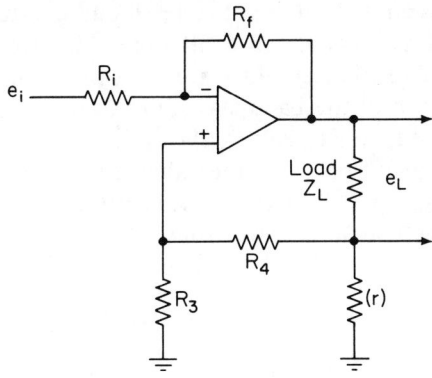

Fig. 7-25. Voltage to voltage amplifier (voltage gain amplifier).

When $R_i = R_f$ and $R_3 = R_4$ then $e_L = -e_i$

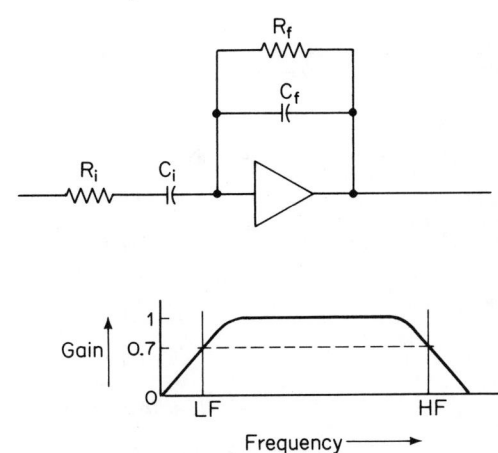

If both R_i and $R_f = 159\ k\Omega$ then:

LF	C_f
1 Hz	1 µF
10 Hz	0.1 µF
100 Hz	0.01 µF
1 kHz	0.001 µF
10 kHz	0.0001 µF
100 kHz	10 pF

HF	C_f
100 kHz	10 pF
10 kHz	0.0001 µF
1 kHz	0.001 µF
100 Hz	0.01 µF
10 Hz	0.1 µF
1 Hz	1 µF

Fig. 7-26. Basic bandpass amplifier.

In this circuit, the input series R and C attenuate low frequencies, and the feedback parallel R and C attenuate high frequencies. If both the input and feedback resistance values are the same, there will be unity gain across the flat portion of the bandpass curve shown in Figure 7-26.

The values for C_i and C_f shown in Figure 7-26 will provide an *approximate* 3-dB/octave drop when the values of both resistors are 159 kΩ.

For example, if 0.1 μF is used for C_i and 0.0001 μF is used for C_f, then the output will be 3 dB down from unity gain at 10 Hz and at 10 kHz.

7-5.11. Frequency-to-Voltage Conversion

An operational amplifier can be used to provide a direct-current output that is proportional to input frequency. The basic circuit is shown in Figure 7-27. Note that this is similar to the differentiator circuit of Figure 7-19. However, the output of Figure 7-27 is rectified and used to charge a capacitor. Since the output of a differentiator is proportional to frequency, the capacitor charge (and the d-c voltage thus obtained) is proprotional to frequency.

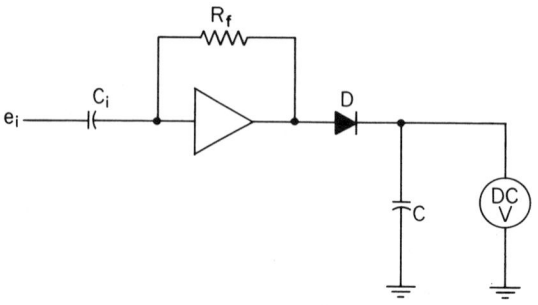

Fig. 7-27. Frequency to voltage conversion.

7-5.12. Peak-Reading Amplifier

An operational amplifier can be used to measure peak voltage of various wave-forms. The basic circuit is shown in Figure 7-28. In this circuit, advantage is taken of the high input impedance feature of the plus input. When a positive pulse is applied, the diode conducts, charging the capacitor to the peak voltage. Because of the high input impedance, the capacitor charge is retained for a relatively long time. Under these conditions unity gain is obtained from the amplifier. Therefore, the output is equal to the peak voltage of the input pulse.

In order for the circuit to operate properly, the time constant of the source impedance and the capacitor to be charged must be short enough so that the capacitor can charge to the peak voltage in the time that the pulse remains at the peak. Thus the capacitor value should be as small as possible. The capacitor cannot be too small, however, or it will discharge too rapidly.

A capacitor with a very low leakage should be selected to prevent rapid loss of the charge. Also, the diode reverse current should be very low to prevent the capacitor charge from being lost too rapidly. The forward drop across some silicon diodes is great enough to prevent the capacitor from charging to the peak voltage.

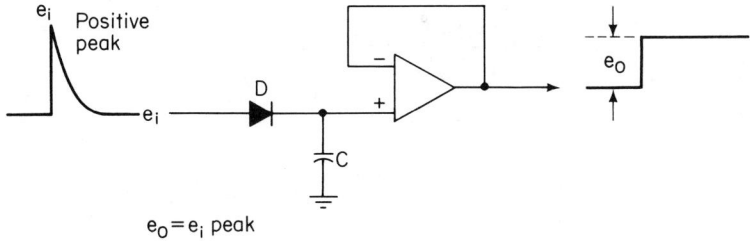

Fig. 7-28. Peak reading amplifier.

8 FREQUENCY AND TIME STANDARDS

Frequency standards are required for shop and laboratory work. Both frequency and time standards are used in most electronic laboratories. Many shops and all laboratories have at least one standard, usually a signal generator, digital counter, frequency meter, oscillator, or similar instrument. This standard is compared against government time and frequency standard broadcasts at regular intervals. All other time and frequency measuring equipment in the shop or laboratory is then compared against the standard. In some laboratories, there are one *primary standard* and several *secondary standards*. The primary standard is calibrated or checked against the government broadcasts. The secondary standards are then checked against the primary. The secondary standards are not used for routine work but serve as standards for all frequency and time measuring instruments.

In many shops and laboratories, the "standard" is simply a conventional signal generator, or counter, such as described in other chapters. In laboratories where more precise frequency and time measurements are required, special primary and secondary standards must be used. The two most popular primary standards are the *hydrogen maser* and the *cesium beam standard*. Secondary standards include the *rubidium vapor standard* and the temperature-controlled quartz oscillator.

To understand operation of such standards, as well as the methods used to calibrate and check the standards, it is first necessary to understand how time is kept, the standards on which time and frequency are based, and the schedule of time and frequency standard broadcasts. This information is discussed in the following sections. The last sections of this chapter discuss operational principles of specific standards, as well as check and calibration procedures.

8-1. TIMEKEEPING METHODS

Accurate timekeeping requires a reference standard for a uniform time scale, together with a means to interpolate the standard from a small interval or an extended period. Throughout the years, many reference standards have been used for time. Even today, more than one reference is used for different timekeeping purposes. The following is a summary of the various timekeeping methods.

Apparent solar time is based on rotation of the earth about its axis with respect to the sun. A unit of time derived from observations of the apparent movement of the sun will obviously be a constant value only if the sun reappears over a fixed point of observation at uniform invervals. As man has increased the precision with which astronomical observations can be made, it has been found that the earth's rotation does not represent a uniform time scale. Even after all possible corrections are made for the known regular variations in the measurement conditions, there still remain irregular variations in the earth's rotational speed.

Mean solar time is simply apparent time, averaged to eliminate variations due to orbital eccentricity and the tilt of the earth's axis. A *mean solar day* is the average of all the apparent days in the year, and a *mean solar second* is equal to a mean solar day divided by 86,400. A solar year is presently equal, in mean solar time, to 365 days, 5 hr, 48 min, and 45.5 sec, or in decimal form, 365.24219879 mean solar days. Because the solar year is 365 days plus a fraction (approximately 6 hr or $\frac{1}{4}$ day), corrections must be made to our calendar at various times (1 day is added every 4th yr, leap year) in order to make it correspond with the sun.

Universal time (UT) is also based on the earth's rotation. The units of UT were chosen so that, on the average, local noon would occur when the sun was on the local meridian. UT, thus defined, makes the assumption that the earth's rotation is constant and equivalent to mean solar time, when uncorrected for variations in earth's rotation, Uncorrected UT, also known as Greenwich Mean Time (GMT), is identified as a time scale by the designation UT_0. Correction to UT_0 has led to two subsequent universal time scales: UT_1 and UT_2.

UT_1 recognizes that the earth is subject to polar motion. The effect of this polar motion is to give an error to any uncorrected measurement of the earth's angular rotation. Therefore, UT_1 is a time scale based on the true angular rotation of the earth about its axis.

UT_2 time is equivalent to UT_1, but with an additional correction for seasonal variations in the rotation of the earth. UT_2 is the time scale in widest use today and represents the mean angular motion of the earth. UT_2 time is free of periodic variations, but it is still affected by irregular variation and requires a yearly correction, or offset, by an amount determined through international agreement.

Sidereal time is a time scale that takes its reference from the relative position of the stars with respect to the earth's rotation. A sidereal day is strictly defined as the interval between two successive transits of the first point of Aries (a northern constellation) over the upper meridian of any place. A sidereal day contains 24 sidereal hr, each having 60 sidereal min of 60 sidereal sec. In mean solar time, a sidereal day is about 23 hr, 56 min, and 4.09 sec. A sidereal year is a measure of the exact period of revolution of the earth around the sun and is about 20 min longer than a solar year. Sidereal time is used primarily by astronomers.

Ephemeris time (ET) is astronomical time based on the motion of the earth about the sun. A second of ET is defined as the fraction 1/31,556,925.9747 of the tropical year for January 0, 1900 at 12 hr ephemeris time. The tropical year for this moment is the length the tropical year would be if the sun continued at its apparent instantaneous rate, corrected for all variations. The second of ET, thus defined, appears to fulfill the requirements for an invariable unit of time. However, universal time (UT) is the time by which we live. Most time and frequency standard broadcasts are in agreement with UT.

The *atomic time* scale is derived from an atomic frequency standard (the cesium beam) and a unit of time based on the natural frequencies of atoms and molecules.

The atomic second is defined by international agreement to be 9,192,-631,770 cycles of the resonance-frequency transition of the cesium atom under zero-field conditions (resonant frequency of the cesium atom measured without any external fields).

The international designation for the atomic time scale is A-1 (sometimes A.1). In the United States, the National Bureau of Standards maintains an atomic time scale designated as NBS-A. The NBS-A time scale is based on the average count of several precision oscillators, compared against a cesium beam standard. In turn, the cesium beam standard (maintained by the U.S. Naval Observatory) is compared against other standards throughout the world (such as the TA_1 atomic time scale maintained by the Observatory of Neuchatel, Switzerland.

Clocks and other time/frequency standards based on the atomic time scale establish a perfectly uniform time, since the resonance frequency of atoms does not change under a given set of conditions. Therefore, there is always some disagreement between atomic time (A-1 or NBS-A) and universal time, which is based on astronomical observations. As a result, a correction must be applied when converting from one time scale to the other. By international agreement, a correction factor or offset is published each year. Many nations have adopted a "coordinated universal time" scale (designated UTC) that is supervised by the Bureau International de l'Heure, Observatoire de Paris, Paris, France. A typical UTC offset from atomic time is -300×10^{-10}. The predicted average rate of UTC offset (for this century) is 150×10^{-10} lower than atomic time.

U.S. standard time differs from nominal UT_2 by an integral number of hours, and is kept by the U.S. Naval Observatory's master clock (consisting of an atomic resonator, a quartz crystal oscillator, and a clock movement). The atomic resonator monitors the frequency of a 2.5-MHz quartz oscillator, kept offset from the atomic frequency to yield the best approximations to UT_2. Oscillator output, divided down to 100,000 Hz, is fed to a Hewlett-Packard clock consisting of a divider, clock movement, and seconds pulser. The 1000-Hz output drives a synchronous motor geared to indicate hours, minutes, and seconds. The seconds pulses serve as the precise reference.

The master clock is compared with observations for universal time made by the U.S. Naval Observatory on each clear night with photographic zenith tubes, or PZT. (A PZT is a specialized telescope fitted for extremely accurate photographic observations of stars that transit near the zenith). The U.S. Naval Observatory is responsible for maintaining the reference for time interval and epoch time.

The continental United States is divided into four standard *time zones* with the time in each an integral number of hours different from (earlier than) Greenwich Mean Time (which is the same as Universal Time). The U.S. time zones are: Eastern, 5 hr earlier; Central, 6 hr, Mountain, 7 hr, and Pacific, 8 hr.

There are 24 *world time zones* based on longitude. By international agreement, the central cross hair of an historic instrument called the Airy transit circle, in Greenwich, England, marks the prime meridian at 0° longitude. This prime meridian is an imaginary great circle crossing both geographic poles. Standard time zones are established at intervals of 15° of longitude east and west of the prime meridian. Time zones in the continental United States are roughly centered along the meridians of 75°, 90°, 105°, and 120°. The exact boundaries of the time zones are modified by political and geographic considerations, but they are approximately a belt 7.5° on either side of the zone reference meridians.

8-2. TIME AND FREQUENCY BROADCASTS

Many nations broadcast time and frequency information from various radio stations. In the United States, time and frequency broadcasts are made from four stations: WWV, WWVB, and WWVL (Fort Collins, Colorado), and WWVH (Maui, Hawaii). The broadcast schedules from these stations are subject to change. For complete information on the schedules, refer to NBS (National Bureau of Standards) *Standard Frequency and Time Services* (Miscellaneous Publication 236), available from Superintendent of Documents, U.S. Government Printing Office, Washington, D.C. 20402. The following is a summary of the broadcast schedules.

The NBS radio stations provide standard radio frequencies, standard audio frequencies, musical pitch, time intervals (1 sec, 1 min, 1 hr, and so on), time signals indicating universal time, universal time corrections (for agreement with UT_2), propagation forecasts, and geophysical alerts. The hourly broadcast schedules are shown in Figure 8-1.

Radio broadcasts from WWV on 2.5, 5, 10, 15, 20, and 25 MHz are continuous night and day, except for silent periods of approximately 4 min beginning 45 min after each hour. Broadcast frequencies are held constant to within 5 parts in 10^{11}. WWVH transmits on the four lower frequencies with 4-min silent periods beginning 15 min after each hour. WWVB and WWVL broadcast continuously on 60 and 20 kHz, respectively, with a nominal stability of 2 parts in 10^{11}.

Standard audio frequencies of 440 and 600 Hz are transmitted alternately in 5-min intervals beginning with 600 Hz on the hour. The first tone transmission from WWV is 3 min long; all others are 2 min long. At WWVH, all tone transmissions are of 3-min duration. WWVB and WWVL do not transmit standard audio tones.

For time signals, the audio frequencies are interrupted 3 min before each hour at WWV and 2 min before each hour at WWVH. The tones are restored on the hour and at 5-min intervals throughout the hour.

Time signals from WWVB are exactly 1 sec (international atomic interval) apart. Those from WWV and WWVH are slightly greater than 1 sec apart, since they are referenced to UT_2. WWVL does not presently broadcast time signals.

Time signals are kept close to UT_2 by (1) maintaining the broadcast *frequency* offset from the U.S. frequency standard (in accordance with the Bureau International de l'Heure offset announced for the year, as described in Section 8-1), and by (2) making 100-msec step adjustments in phase, as needed, on the first of the month. Since WWV and WWVH time signals are kept locked to transmission frequency, they may continuoulsy depart from UT_2. Differences are determined and published by the U.S. Naval Observatory. Fractional frequency deviations of the NBS stations are published monthly in *Proceedings of the IEEE* under the heading "Standard Frequency and Time Notices."

WWVB time signals are the international second (atomic) and are not offset. However, WWVB time signals are locked to transmission frequency (60 kHz), and are kept close to UT_2 by 200-msec step adjustments in phase, made on the first of the month as needed. Their difference from UT_2 is coded on the broadcasts.

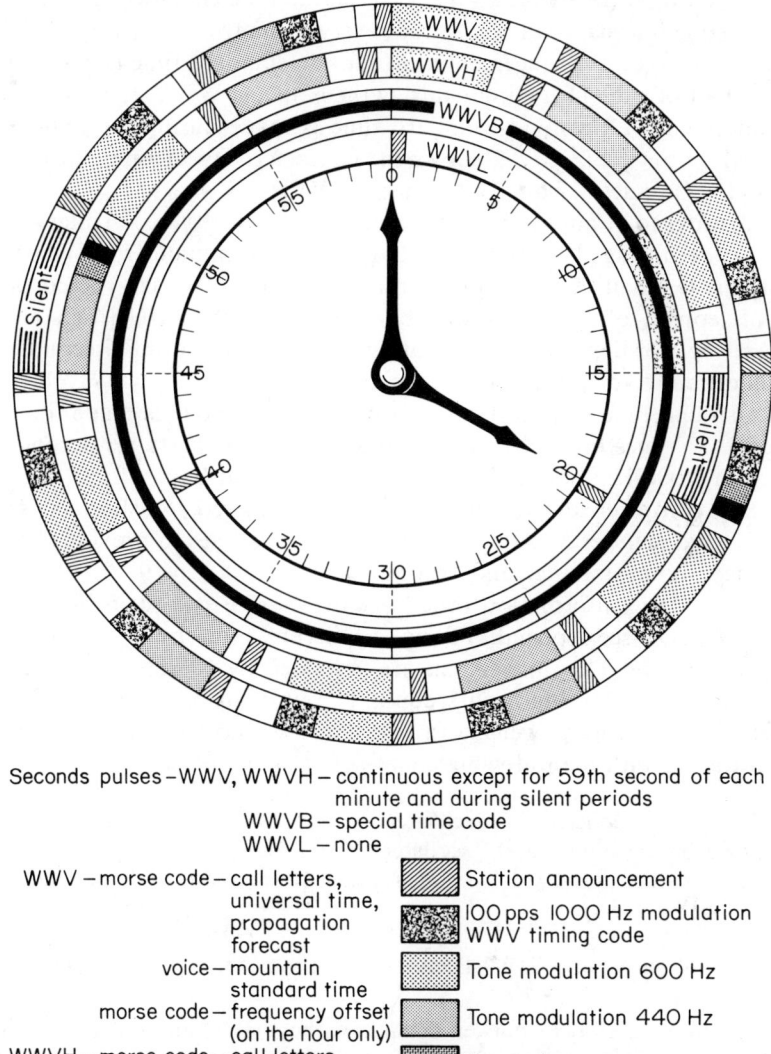

Seconds pulses – WWV, WWVH – continuous except for 59th second of each minute and during silent periods
WWVB – special time code
WWVL – none

WWV – morse code – call letters, universal time, propagation forecast
voice – mountain standard time
morse code – frequency offset (on the hour only)

WWVH – morse code – call letters, universal time,
voice – Hawaiian standard time
morse code – frequency offset (on the hour only)

WWVL – morse code – call letters, frequency offset

▨ Station announcement
▦ 100 pps 1000 Hz modulation WWV timing code
▢ Tone modulation 600 Hz
▧ Tone modulation 440 Hz
▩ Geoalerts
▦ Identification phase shift
■ UT-2 time correction
▬ Special time code

Fig. 8-1. Hourly broadcast schedules of WWV, WWVH, WWVB, and WWVL.

Universal time (reference to the zero meridian at Greenwich, England) is announced in international Morse code every 5 min from WWV and WWVH. At WWV, a voice announcement in Mountain standard time is made during the last half of each fifth minute during the hour. At WWVH, a similar voice announcement of Hawaiian standard time is made during the first half of each fifth minute. Figure 8-2 shows the characteristics of time pulses (1-sec time ticks) broadcast from WWV and WWVH.

Corrections to be applied to universal time signals are given in international Morse code during the last half of the nineteenth minute from WWV and during the last half of the forty-ninth minute from WWVH. The correction symbols consist of UT_2, followed by AD (for add) or SU (for subtract), followed by a 3-digit number indicating the correction in milliseconds.

In addition to voice and Morse code announcements of time, WWV broadcasts a special time code in binary form. This time code is made for 1 min out of 5, 10 times/hr, and is produced at a 100-pps (parts per second) rate carried on 1000-Hz modulation. Since the code is in binary form, a digital-to-analog converter and/or digital readout is required. As shown in Figure 8-3, the code requires 36 binary digits and requires 1 sec to provide full clock data (day, hour, minute, and second).

The time code information broadcast by WWVB is also in binary-code-decimal (BCD—see Appendix) form, and is presented by means of a *level-shift-carrier code*. The time signals are indicated by 60 drops in power level per minute, one marking each second. The amount of drop is 10 dB. The length of time before power is restored indicates more detailed information in accordance with the following code:

1. For a binary code *zero*, 0.2 sec later
2. For a binary code *one*, 0.5 sec later
3. For a marker pulse, which designates each 10 sec and the minute reference, 0.8 sec later

Figure 8-4 shows the WWVB code, with the zero, one, and marker pulse encircled.

The minute reference marker occurs in the first second of each minute. The 10-sec reference markers occur each 10 sec, including the sixtieth. Thus a double marker pulse starts each minute and gives an easily identified origin point.

The BCD code presents time-of-year information each minute: the minute, the hour, the day of the year, and the millisecond *difference* between the time of broadcast signal and the *best approximation* to UT_2.

An example of a typical minute is shown in Figure 8-4. Note that the first bit of information presented after time zero is the minute reference marker, a 10-dB reduction 0.8 sec in length. Proceeding from left to right (advancing in time) are the presentations for minutes, hours, day of the year, and the

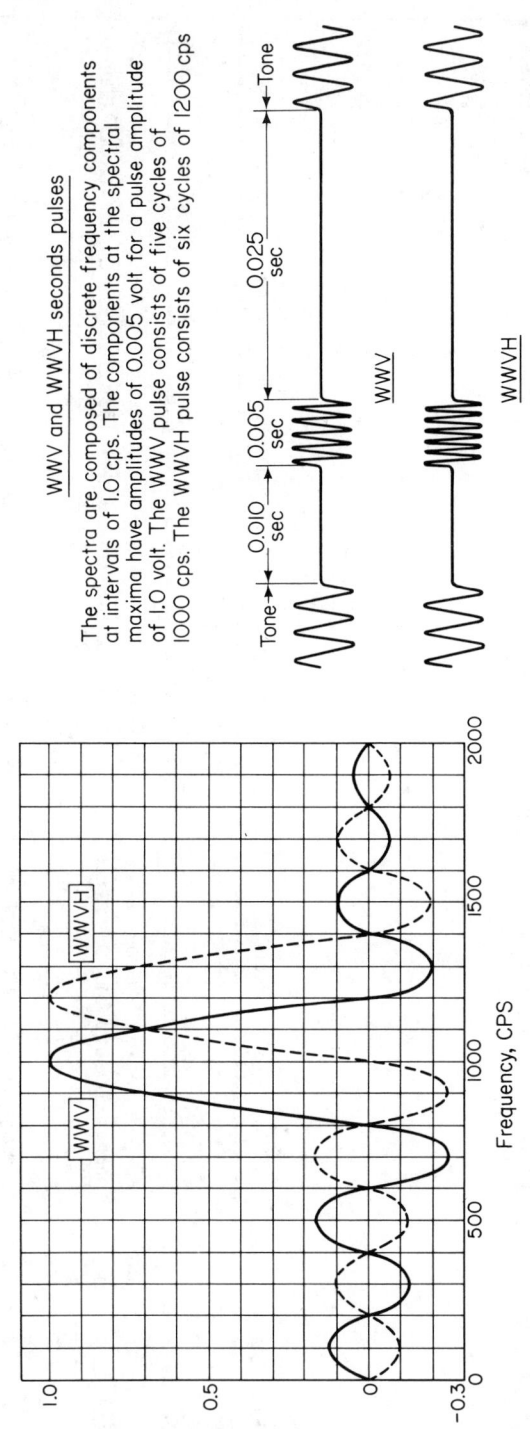

Fig. 8-2. Characteristics of time pulses broadcast from WWV and WWVH.

WWV and WWVH seconds pulses

The spectra are composed of discrete frequency components at intervals of 1.0 cps. The components at the spectral maxima have amplitudes of 0.005 volt for a pulse amplitude of 1.0 volt. The WWV pulse consists of five cycles of 1000 cps. The WWVH pulse consists of six cycles of 1200 cps.

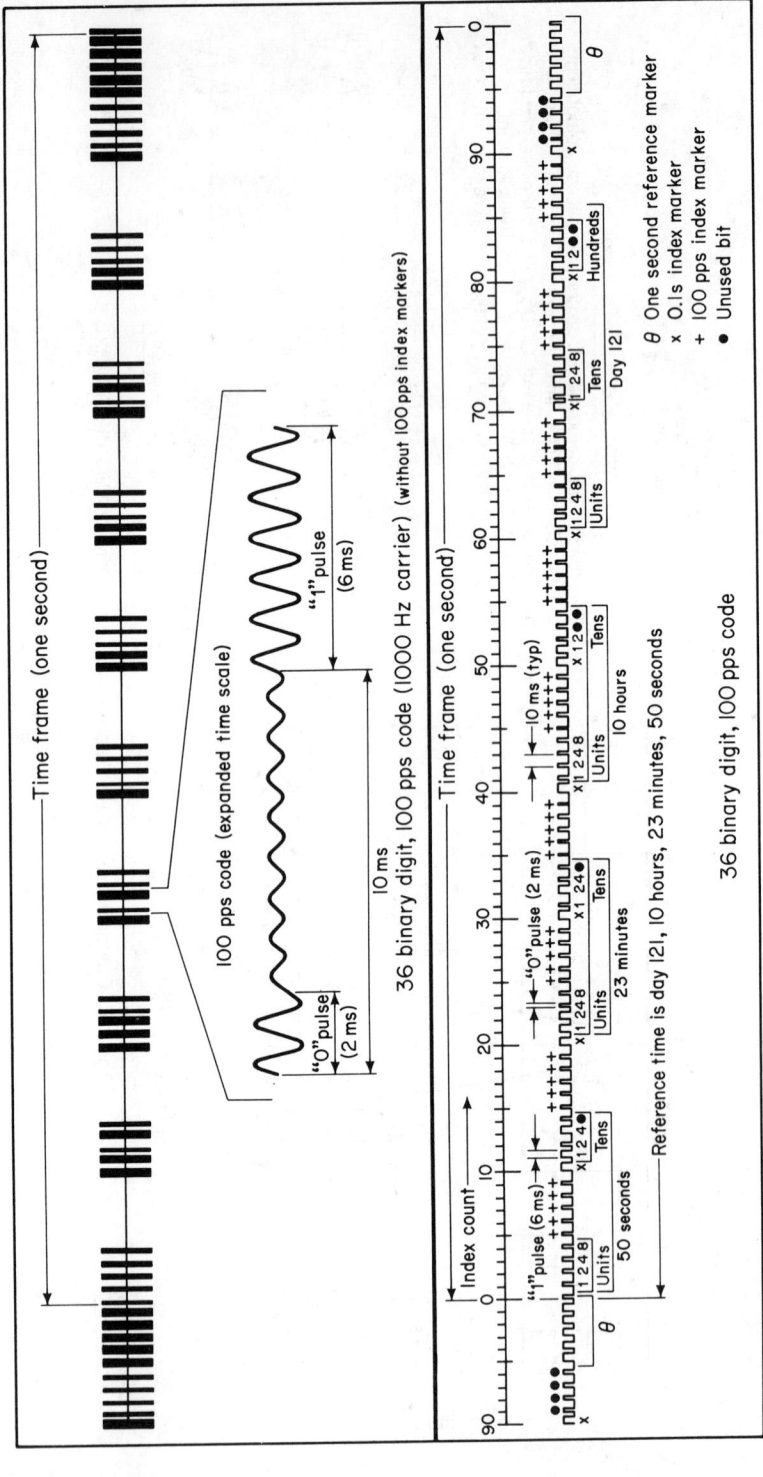

Fig. 8-3. Time code transmissions from WWV.

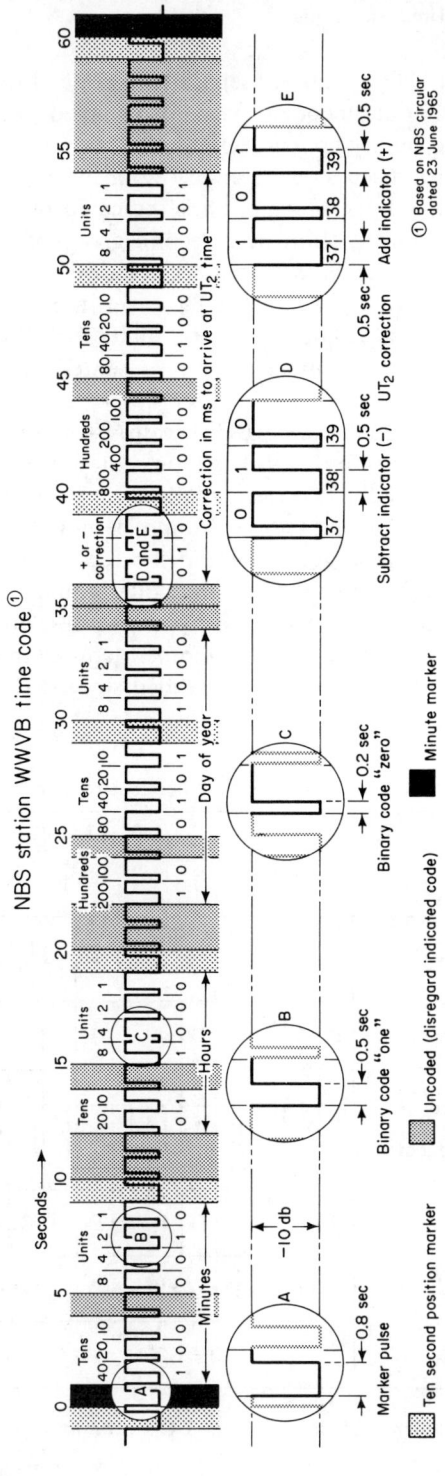

Fig. 8-4. Time code transmissions from WWVB.

correction to arrive at UT_2. Areas are separated by one of the 10-sec markers (shaded for emphasis) or an uncoded space (also shaded). Each area presents its information in the binary-coded-decimal (BCD) notation. (Refer to the Appendix for a further discussion of binary notation.)

In BCD notation, the units information is indicated by a group of binary digits or bits, the tens information by a second group of bits, the hundreds information by a third group, and so on. As described in the Appendix, the normal BCD group consists of four bits. However, a group of two bits can serve to represent digits 0–3, and three bits, digits 4–7. Shorter groups are often used where the largest number to be represented does not require the full set.

The decimal digit represented by a BCD group is obtained by addition of the bit values present. The number 461 is represented in Figure 8-5 as it appears in the WWVB level-shift-carrier code and in BCD notation, to show the correspondence between the two. It should be noted that the level-shift code wave-forms shown in Figures 8-4 and 8-5 represent those that would appear on an oscilloscope. In practice, the code would be converted by a digital-to-analog converter and displayed by means of a ditial readout.

Returning to Figure 8-4, note that the number denoted in the minutes area is 42 (the fourth bit in the tens set is not needed since the largest number to be represented is 60):

Tens set	Units set	Number represented
4 × 10 = 40	2 × 1 = 2	40 + 2 = 42

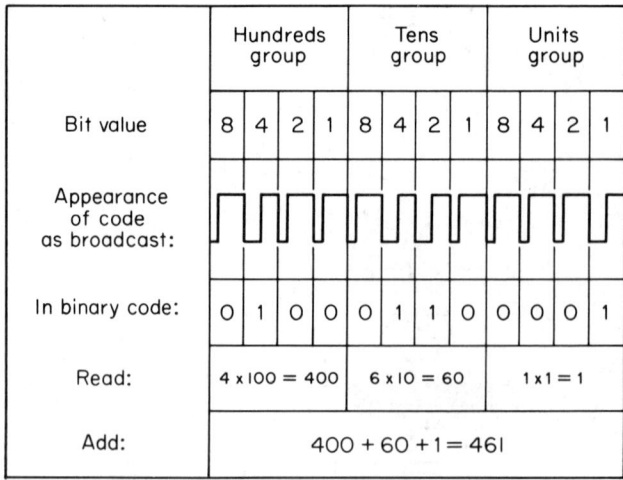

Fig. 8-5. Relationship of time code to BCD notation.

By a similar process, the hours area is found to read 18 and the days are to read 258. The code thus far (up through second 34) reads as the forty-second minute of the eighteenth hour of the two-hundred-fifty-eight day of the year.

The remainder of the code shows the correction in milliseconds to be applied to the time as broadcast, in order to arrive at the best approximation of UT_2. First, the sign of the correction is indicated: a 1 in seconds 37 and 39 designates a plus sign; the correction is to be added. A 1 in second 38 designates a minus sign; the correction is to be subtracted.

The numerical correction itself is indicated in the next area. The example shown is read: "Subtract 41 (msec)."

WWVB is identified by an advance of the 60-kHz carrier phase by 45° at 10 min after the hour, returned to normal phase at 15 min after the hour. The shift is initiated with a time accuracy of 1 msec and spaced with a precision of 1 μsec.

In addition to time and frequency information, WWV and WWVH broadcast *propagation* and *geophysical* data. Propagation forecast announcements in Morse code indicating ionospheric conditions and the quality of radio reception that can be expected within the next 6 hr are made from WWV at 0500, 1200 (1100 in summer), 1700, and 2300 UT. The forecast consists of a letter and a number. The letters N, U, and W indicate normal, unsettled, and disturbed conditions, repectively. Nos. 1–4 (disturbed) indicate useless, very poor, poor, and poor-to-fair; 5 (unsettled) is fair; 6 to 9 (normal) indicate fair-to-good, good, very good, and excellent, respectively.

A letter symbol indicating the current geophysical alert is broadcast in slow Morse code from WWV during the first half of the nineteenth minute of each hour and from WWVH during the first half of each forty-ninth minute. The geophysical alert identifies days in which outstanding solar or geophysical events are expected or have occurred during the preceding 24-hr period.

The geophysical announcement is identified by the letters GEO followed by the letter symbol repeated 5 times. The letter M means magnetic storm, N is magnetic quiet, C is cosmic-ray event, E is no geoalert issued, S indicates the presence of solar activity, Q is solar quiet, and W is stratospheric warning.

8-3. TIME AND FREQUENCY STANDARD EQUIPMENT

Time standards and frequency standards have no basic differences—they are based on dual aspects of the same phenomenon. The reciprocal of time interval is frequency.

As a practical matter, a frequency standard can serve as the basis for time measurement, and vice-versa, with certain limitations.

To avoid errors when a frequency standard is used to maintain time, either interval or epoch, care must be taken to reference the frequency to the time scale of interest (atomic, UT_2, and so on).

The first requirement for a time standard is an ultrastable oscillator capable of producing an accurate, precise frequency. Such an oscillator can be made the basis of a clock.

When two independent oscillators are made to drive clocks, then the time kept is only as accurate as the frequency. For example, assume that two quartz oscillators with a known *relative* stability of about 10 parts in 10^{11} are used to drive clocks. In a day, the two clocks could accumulate a *relative* error of nearly 10 μsec. Even when the same frequency is used to drive two identical clocks, they might show two different times unless one had been set against the other to high precision.

Therefore, to maintain a practical time standard, an additional requirement is placed on top of those associated with maintaining a frequency standard.

8-3.1. Types of Frequency Standards

At the present time, four types of frequency standards are in common use:

1. The atomic hydrogen maser (primary standard)
2. The cesium atomic beam controlled oscillator (primary standard)
3. The rubidium gas cell controlled oscillator (secondary standard)
4. The quartz crystal oscillator (secondary standard)

The difference between a primary standard and a secondary standard is that the primary standard does not require any other reference for calibration. A secondary standard requires calibrations both during manufacture and during use, as well as at certain intervals depending on the stability desired.

8-3.2. The Hydrogen Maser

Figure 8-6 is a schematic diagram of the *master oscillator* portion of a typical hydrogen maser. A beam of atomic hydrogen is directed through a magnetic field that acts to select atoms in states of higher energy from those in states of lower energy and allows the atoms to enter the quartz bulb. The bulb is enclosed in a tuned microwave cavity (Chapter 11) set to the transition frequency of hydrogen atoms. The resonant cavity acts to absorb some of the energy given up by the atoms moving about within the quartz crystal. Since the cavity and the energy are at the same frequency, the cavity becomes an r-f source at the resonant frequency. In effect, the cavity becomes an oscillator that produces an r-f signal at the transition

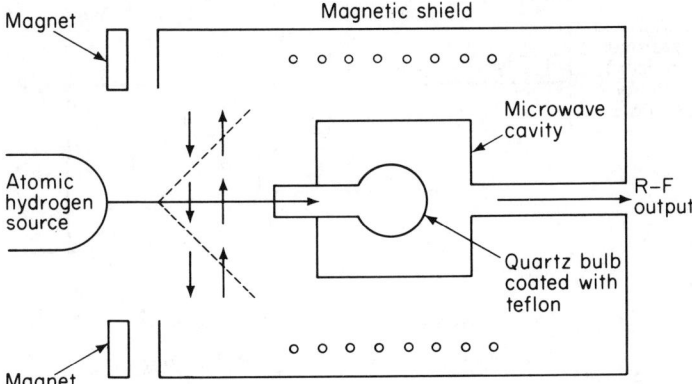

Fig. 8-6. Master oscillator portion of typical hydrogen maser (Hewlett-Packard).

frequency of hydrogen atoms (approximately 1.42 GHz). Since the atomic frequency does not vary (for a given set of conditions), the r-f output from the cavity can be used as a primary standard. Once the cavity is adjusted and checked against NBS standards, the frequency should remain constant indefinitely.

In most practical applications, the master oscillator portion of a hydrogen maser is not used by itself. Instead, the master oscillator is used to control a slave oscillator operating at a lower frequency (such as 1 MHz, 5 MHz, and so on). The block diagram of a typical hydrogen maser frequency standard is shown in Figure 8-7. In this arrangement, the crystal-controlled slave oscillator produces a standard 5-MHz signal. The slave oscillator is locked in frequency by a signal from the phase detector. There are two inputs applied to the phase detector. Both are of the same frequency (20.405752 MHz) and, if both signals are in phase, there will be no output from the phase detector. If there is any phase difference, the detector output shifts the slave oscillator phase (and consequently the frequency) until the difference is removed. One of the phase detector inputs is from a frequency synthesizer (Chapter 4) locked to the 5-MHz standard signal from the slave oscillator. The other phase detector input is obtained by mixing the hydrogen maser master oscillator signal (1420.405752 MHz) with a 1400-MHz signal from the frequency multiplier, also locked to the 5-MHz standard signal. Should the slave oscillator frequency vary for any reason, a correction is applied to offset the variation. For example, if the slave oscillator frequency should increase, the frequency multiplier output will also increase. The mixer output would then decrease because the master oscillator frequency remains constant. The frequency synthesizer output would increase in frequency, causing one phase detector input (from synthesizer) to be high and one to be low (from mixer). This difference in frequency (or phase) in the phase detector results in a cor-

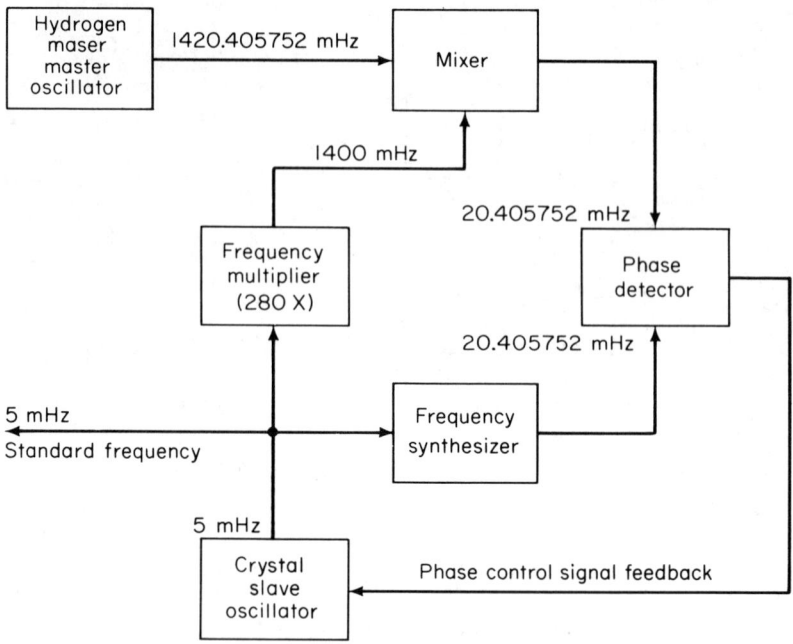

Fig. 8-7. Typical/hydrogen maser block diagram.

rective feedback signal being applied to the slave oscillator. Since the phase detector will produce an output when there is a slight difference in phase between the two input signals, it is possible to maintain the slave oscillator at the desired frequency within a fraction of 1 Hz.

8-3.3. The Cesium Beam Frequency Standard

The cesium beam frequency standard is similar to the hydrogen maser in that both depend on the natural, unvarying resonant frequency of atoms passing from one state to another. Also, like the hydrogen maser, the cesium beam resonator is used to control the frequency of a crystal-controlled slave oscillator, rather than being used as a direct-frequency standard. However, operation of the cesium beam resonator is somewhat different.

Figure 8-8 is a schematic diagram of the *resonator* portion of a typical cesium beam frequency standard. As shown, cesium atoms are sorted and focused by passing a beam of them through a magnetic field (magnet *A*). The atoms pass through a vacuum chamber that has a low magnetic field (field *C*), and they are subjected to an r-f injection signal. This signal is obtained from the slave oscillator, after multiplication and frequency synthesis. The r-f injection signal is at the transition frequency of cesium-133 atoms (approximately

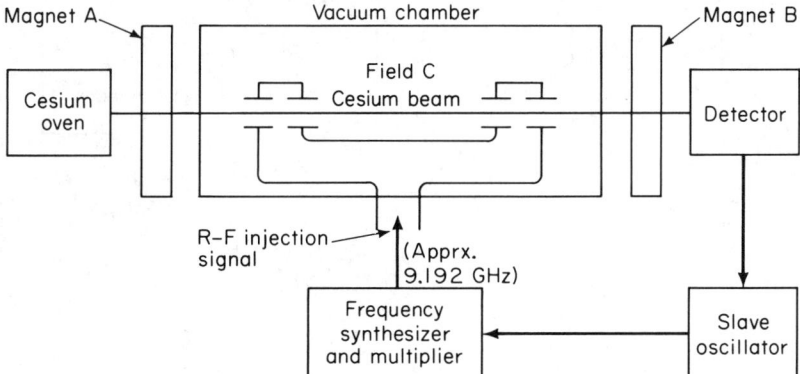

Fig. 8-8. Resonator portion of typical cesium beam frequency standard (Hewlett-Packard).

9.192 GHz). After leaving the vacuum chamber, the atoms are directed onto a detector by a second magnet (magnet B). The output of the detector is used to control frequency of the slave oscillator.

8-3.4. Rubidium Vapor Frequency Standard

Rubidium standards are similar to cesium beam standards in that an atomic resonant element prevents drift of a standard frequency slave oscillator through a feedback loop. However, the resonant frequency of a rubidium gas or vapor cell is dependent on gas mixture and gas pressure in the cell and is therefore subject to change (or frequency drift). For this reason, rubidium cells are used as *secondary frequency* standards.

Figure 8-9 is a schematic diagram of the resonator portion of a typical rubidium vapor frequency standard. In operation, a rubidium spectral lamp is driven by an r-f oscillator. The r-f signal energizes rubidium atoms within the lamp, causing them to produce light waves at two different wavelengths. One wavelength is removed by the filter, while the other is absorbed in a microwave cavity. The cavity is subjected to an r-f injection signal obtained from the slave oscillator (after multiplication and frequency synthesis). The r-f injection signal is at the transition frequency of rubidium atoms (approximately 6.834 GHz).

Both the light waves and r-f signals act on the atoms within the cavity. The light waves act to deplete the atoms so that they can absorb less light, while the r-f signals restore the atoms to the full light absorption capability. If the r-f signals are at the exact atomic frequency, the atoms will remain in the maximum absorption condition, and a minimum of light will pass to the solar cell detector. The output of this detector controls the frequency

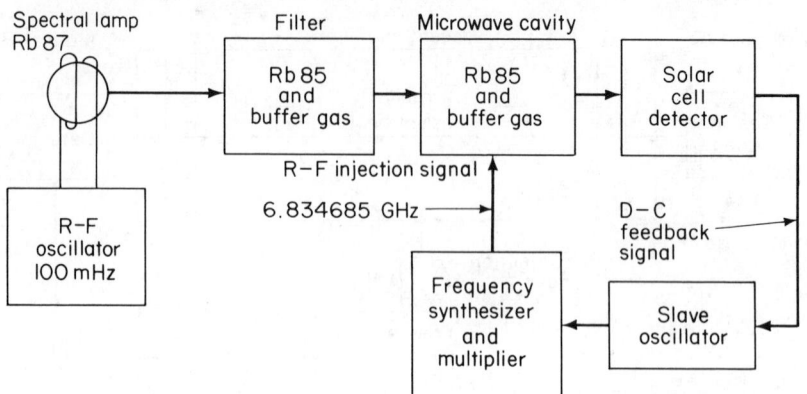

Fig. 8-9. Resonator portion of typical cesium beam frequency standards (Hewlett-Packard).

of the slave oscillator. If the r-f signals are not at the desired frequency, the atoms absorb less light, and more light will pass to the detector. The increase in detector output will shift slave oscillator frequency until light is again at a minimum. In a typical system arrangment, detector output reaches a minimum when the r-f signal frequency corresponds to the rubidium atom transition frequency (6,834,685 MHz).

8-3.5. Quartz Crystal Oscillator

The quartz crystal oscillator is the most common type of *secondary* frequency and time standard. Almost any crystal oscillator can be used as a secondary frquency standard (or the time base for a time standard), provided that the crystal temperature is maintained constant and the amount of frequency drift is checked at regular intervals. One of the characteristics of crystal oscillators is that their resonant frequency changes slightly as they age. This aging rate or drift of temperature-controlled crystal oscillators is almost constant. After the initial aging period (a few days to a month) the rate can be taken to be constant with but slight error. Once the rate is measured, it is usually easy to correct data to remove its effect. Over a long period, the accumulated error drift could amount to a serious error. Thus periodic frequency checks are needed to maintain a quartz crystal frequency standard.

The quartz crystal oscillator is not usually used as a standard by itself. Instead, the oscillator serves as the basic reference frequency (or time base) signal source that is multiplied and/or divided as necessary. This is the principle of the *frequency synthesizers* described in Chapter 4.

Figure 8-10 is the block diagram of a typical working (secondary) frequency

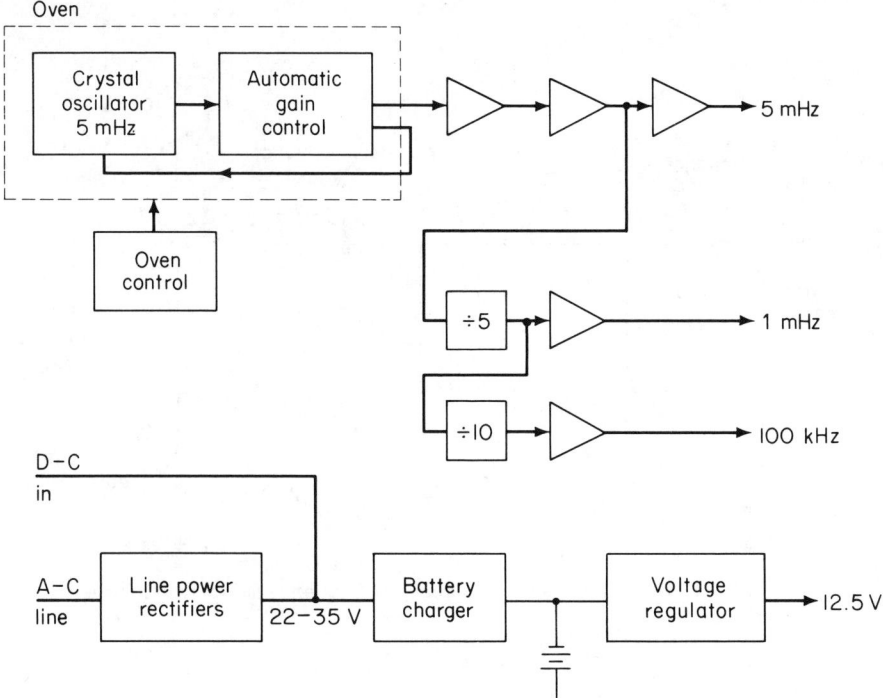

Fig. 8-10. Typical working frequency or timing standard (General Radio).

(or timing) standard. The instrument contains a 5-MHz crystal-controlled oscillator in a proportional-control oven. Amplifiers provide isolation and power for the 5-MHz output. The 5-MHz output is also divided by regenerative frequency dividers to produce outputs at 1 MHz and 100 kHz.

The frequency can be adjusted (within limits) electrically by a potentiometer whose dial is direct-reading in parts in 10^{10}. This adjustment permits the instrument to be calibrated against a primary standard.

The power supply section has a line power rectifier, battery charger, and voltage regulator. In case of power-line failure, operation switches automatically to a self-contained battery and remains in operation for several hours. The battery is recharged rapidly after power failure and is then maintained at optimum charge.

The basic oscillator circuit for the frequency standard is shown in Figure 8-11. As in the case of any frequency standard, both the *output frequency* and *output amplitude* must remain constant. Output amplitude is maintained at a fixed level by an *automatic gain control* (AGC) circuit, represented by resistance R. The complete circuit for gain control is shown in Figure 8-12. The

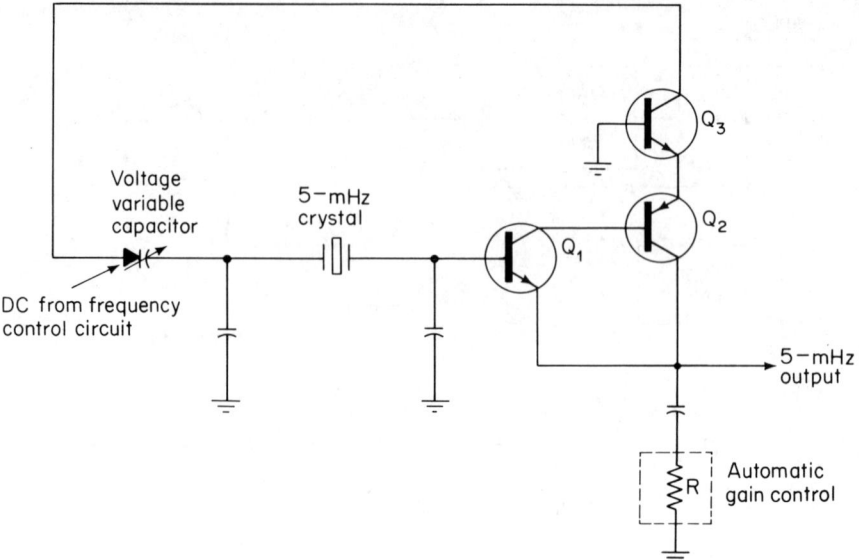

Fig. 8-11. Simplified oscillator circuit for frequency standard (General Radio).

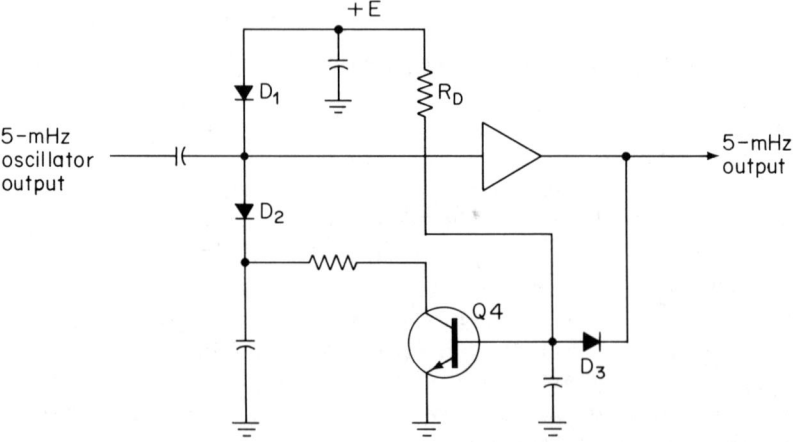

Fig. 8-12. Simplified AGC circuit for frequency standard (General Radio).

AGC circuit varies the bias current through diodes D_1 and D_2 to change their forward resistance. The r-f output from the oscillator is amplified by a two-stage amplifier and rectified by D_3. As long as there is no r-f voltage, Q_4 is biased on to pass a maximum current through the AGC diodes D_1 and D_2.

This results in maximum gain in the oscillator to start oscillations. As the amplitude increases, D_3 reduces the drive to Q_4 until Q_4 goes out of saturation. Any further increase in r-f amplitude reduces the current through the diodes D_1 and D_2 and reduces the gain of the oscillator. Resistance R_D determines the r-f amplitude at which Q_4 turns off and thus sets the r-f level.

The oscillator frequency is normally at 5 MHz, as determined by the quartz crystal. However, the voltage-variable capacitor (VVC) in series with the crystal permits the frequency to be adjusted over a narrow range. This provides for calibration against a primary standard. The VVC is controlled by a d-c bias voltage that, in turn, is set by a front-panel control. The frequency control dial is coupled to a mechanical digital readout. As the bias is changed, the VVC capacitance varies, producing a change in oscillator frequency. The digital readout indicates frequency increments of 1×10^{-10} per digit. The total range of frequency tuning is 2700×10^{-10}. The complete bias network is shown in Figure 8-13. Resistors R609 and R608 set the total range of the frequency dial, while resistors R606 and R607 affect linearity. Note that R607 and R609 values are selected at manufacture to provide the desired frequency range and linearity. Typical linearity is about $\pm 7 \times 10^{-10}$ (out of 2700×10^{-10}), or about 0.25 per cent.

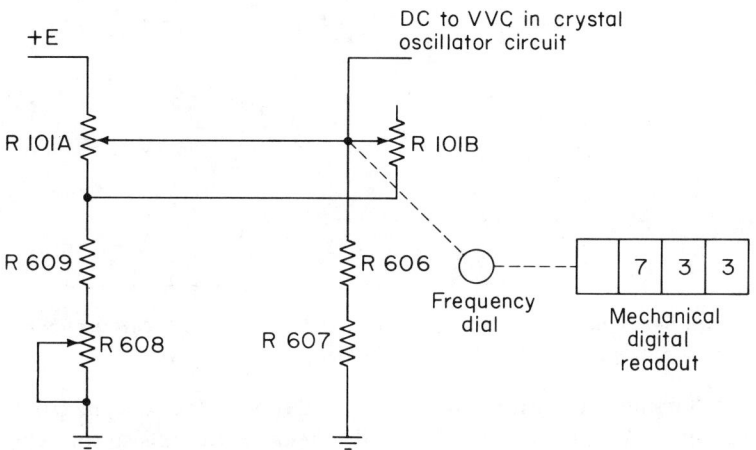

Fig. 8-13. Simplified frequency control circuit for frequency standard (General Radio).

Three stages of amplification are used between the oscillator and the 5-MHz output. As shown in Figure 8-14, the first two stages are of the cascade type (that is, each stage consists of a grounded emitter driving a grounded base). The third stage is the output power amplifier. A crystal filter is used between the second isolation amplifier and the output amplifier.

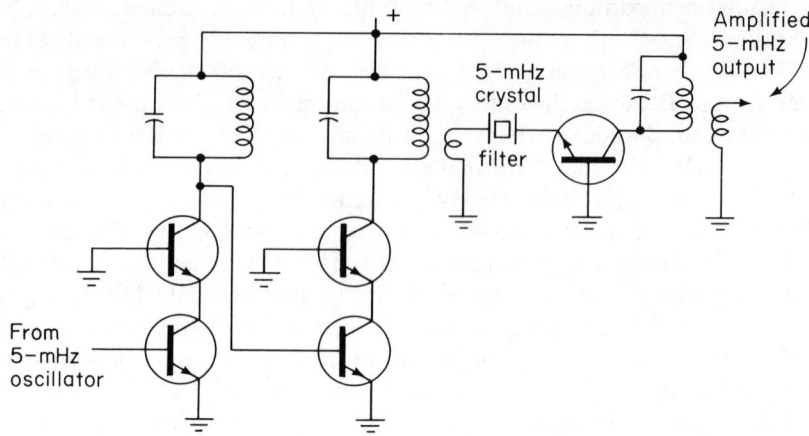

Fig. 8-14. Simplified 5 MHz amplifier circuit for frequency standard (General Radio).

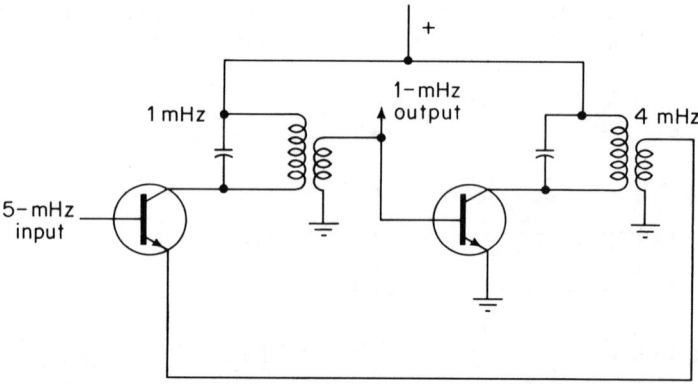

Fig. 8-15. Simplified regenerative divider (5 MHz to 1 MHz) circuit for frequency standard (General Radio).

A self-starting regenerative divider provides a 1-MHz output from the 5-MHz input. The basic circuit shown in Figure 8-15 consists of two tuned stages (1-MHz and 4-MHz). The emitter of the 1-MHz stage is coupled to ground through a pickup coil on the 4-MHz tank coil. The base of the 4-MHz stage is coupled to ground through a pickup coil on the 1-MHz tank. The combination of a 4-MHz signal and the 5-MHz input produces both a 1- and a 9-MHz signal. Since there is a 1-MHz tank, only that signal frequency appears at the output. The 1-MHz signal from the divider is amplified by an output amplifier similar to the one used in the 5-MHz section.

Another self-starting regenerative divider provides a 100-kHz output from the 1-MHz input. The basic circuit shown in Figure 8-16 is similar to that for the 5:1-MHz divider, except that one tank is tuned to 900 kHz

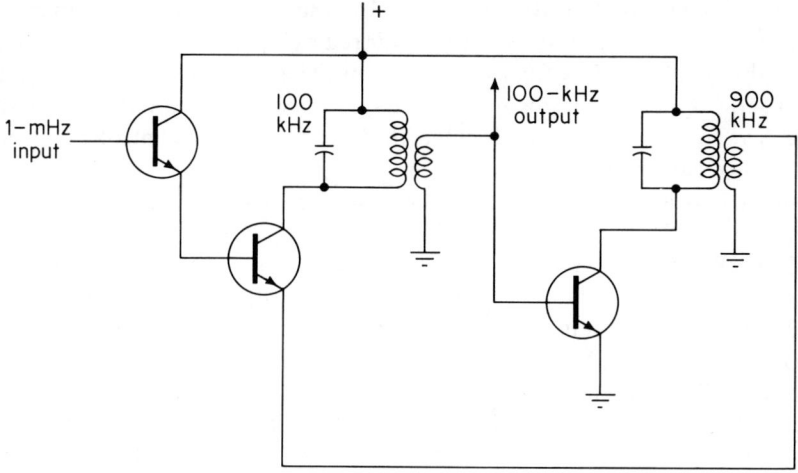

Fig. 8-16. Simplified regenerative divider (1 MHz to 100 kHz) circuit for frequency standard (General Radio).

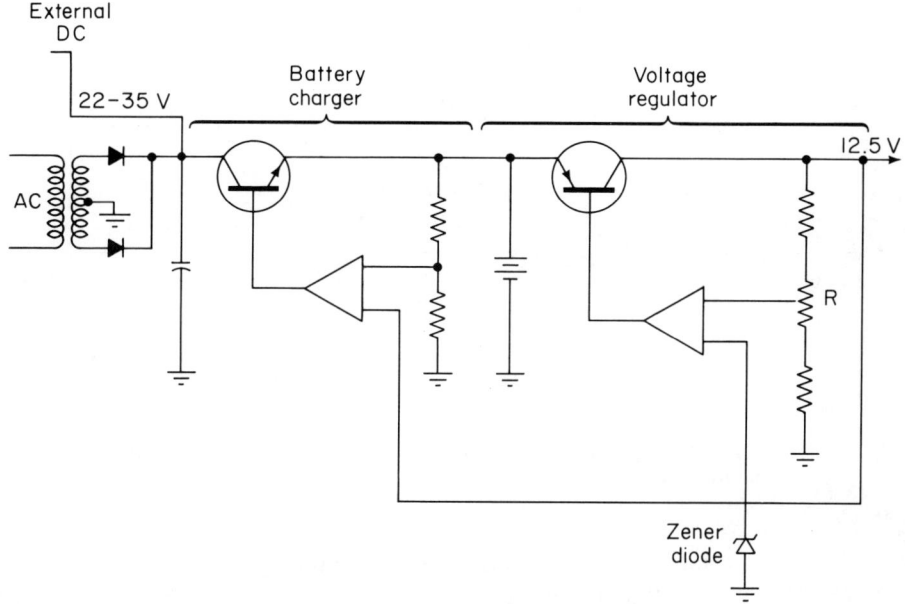

Fig. 8-17. Simplified power supply circuit for frequency standard (General Radio).

and the other (output) tank is tuned to 100 kHz. The 100-kHz divider also has an emitter follower at the input.

The power supply and battery circuit are shown in Figure 8-17. Line power is rectified, filtered, and applied to the input of the battery charger, which provides a current-limited voltage source for the battery and instrument. The limit voltage for the battery can be adjusted by potentiometer R, which controls the regulated $B+$ for the instrument. The battery voltage is regulated by a series-type voltage regulator. The reference voltage for this regulator is supplied by a Zener diode, located in the temperature-controlled oven for best temperature stability.

A nickel-cadmium battery powers the instrument upon line failure. Typically, 35 hr of operation can be expected at 25°C ambient temperature.

The quartz (oscillator) crystal is operated in a temperature-controlled oven to assure frequency stability. The temperature-control circuit is shown in Figure 8-18. As long as the oven temperature is low, positive feedback is applied around amplifier A_1, and the oven-control circuit oscillates at about 1 kHz. (The frequency is determined by the tuned transformer.) This a-c signal is rectified and amplified by the d-c amplifier A_2 and is then applied as heater power. As the temperature increases, less signal is fed back around A_1 and the amplitude decreases, reducing the heater power so that stable operation results near the balance of the bridge (consisting of ratio transformer T, the thermistor, and fixed resistor R).

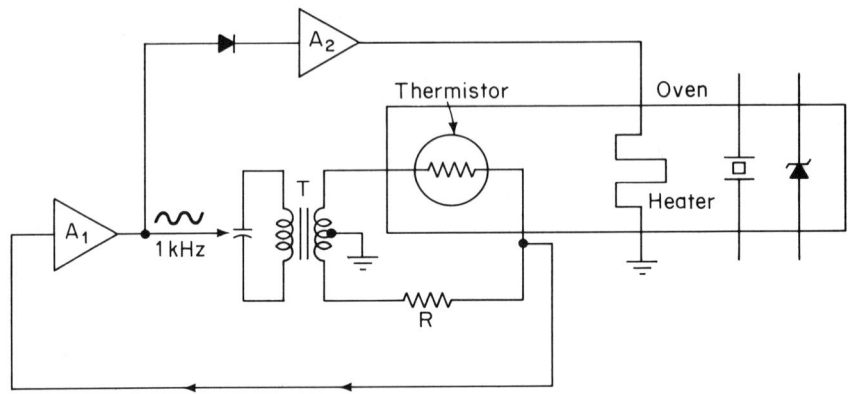

Fig. 8-18. Simplified oven temperature-control circuit for frequency standard (General Radio).

8-4. PRACTICAL FREQUENCY CALIBRATION TECHNIQUES

There are many methods for calibrating a "house" or "local" standard against a frequency standard broadcast. Two types of information can be obtained from such comparisons: (1) the frequency offset (or difference), and (2) the aging or frequency drift characteristics of the local standard. With these data, the local standard can be adjusted to agree with the

Practical Frequency Calibration Techniques   301

broadcast, or the difference can be noted and recorded for future use. The latter method is usually preferred, unless there is a very large difference between the broadcast and the standard.

Either way, it is usually impractical to maintain a large number of frequency standards (crystal oscillators in synthesizers, counter time bases, and so on) directly to a standard broadcast. It is often desirable to maintain a house standard to the broadcast, then use the house instrument as a transfer standard for local frequency offset and drift comparison work. Also, it is generally impossible to use a received frequency standard broadcast signal for short-term stability (fractional frequency deviations or fractional phase deviations). Instead, a stable, spectrally pure (see definitions in Section 8-7) local standard is used.

To be useful, methods for comparing frequency standard oscillators against broadcasts or against other standard instruments must be capable of resolving extremely small differences. The oscilloscope (Chapter 10) and the electronic counter (Chapter 6) are the most common instruments for making such comparisons. If a comparison is to be made against standard broadcasts, a good communications-type receiver is required.

8-4.1. Shop-Level Frequency Calibration

Most shop-level frequency calibration procedures can be performed with a radio receiver, using the *zero-beat* method. The basic components required for any form of zero-beat frequency measurement are a signal or voltage source of known accuracy and a detector that will show when the two signals (standard and unknown) are at the same frequency or some exact multiple of each other. A radio receiver is a common detector for zero-beat frequency measurement. In operation, the two signals (known standard broadcast and unknown local signal source) are applied to the radio receiver input. As the unknown signal frequency is adjusted close to that of the known signal (so that the difference in frequency is within the audio range) a tone, whistle, or "beat note" is heard on the receiver.

For example, assume that the known signal is at 5 MHz (WWV or WWVH broadcast) and the unknown signal (say from a signal generator to be calibrated) is adjusted to 5.001 MHz (or 10-kHz difference) so that the 10-kHz-difference signal can be heard. As the unknown signal frequency is brought closer to the known, say to 5.0001 or 100-Hz difference, a tone is still heard. When the unknown signal is adjusted to exactly the same frequency as the known, there is no difference signal, and the tone can no longer be heard. In effect, the tone drops to zero and the two signals are at zero beat.

Figure 8-19 shows the basic circuit for shop-level frequency calibration against standard broadcasts. In operation, the receiver is tuned to one (or more) of the standard broadcast frequencies. (Note that the receiver fre-

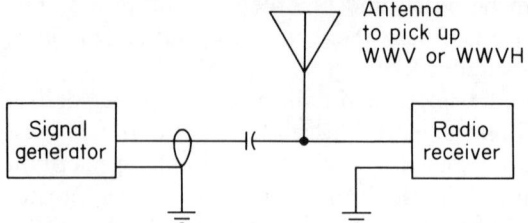

Fig. 8-19. The basic circuit for shop-level frequency calibration against standard broadcast.

quency dial need not be highly accurate, but it is necessary to identify the stations correctly.) The signal generator (or other shop standard being calibrated) is adjusted to obtain a zero beat against the station carrier (not station modulation). When a zero beat is heard in the receiver loudspeaker, the signal generator has an output frequency equal to that of the station carrier frequency.

It should be noted that as the signal generator is tuned to one-half (or other multiples) of that station frequency, a whistle or zero beat will be heard. This is because most signal generators have some harmonic output. However, the *fundamental* zero-beat note is the loudest. Good use can be made of harmonic beats, since one station of known frequency can be used to calibrate several generator bands. It should be noted that if the signal generator remains in the zero-beat condition over a long period of time (especially with the harmonic zero beats), the thermal stability of the generator is good.

If it is desired to calibrate an audio generator against the *audio tones* transmitted by WWV or WWVH, connect the equipment as shown in Figure

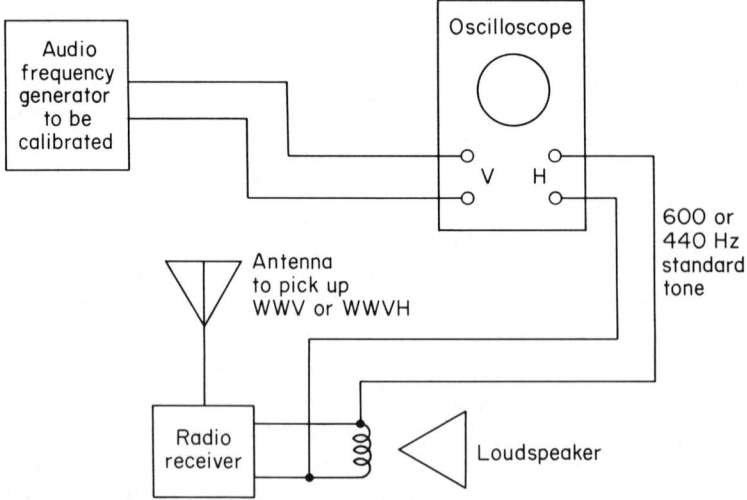

Fig. 8-20. Basic circuit for calibration of audio generator against the audio tones of standard broadcast.

8-20. Tune the receiver to one of the government standard station frequencies. Under these conditions, the oscilloscope horizontal sweep will be obtained from the 440- or 600-Hz tone from WWV or WWVH, while vertical deflection will be obtained from the audio generator to be calibrated. Adjust the audio generator until the pattern stands still. This indicates that the audio generator is at the same frequency as the standard tone (if the pattern is a circle or ellipse) or that the audio generator is at a multiple or submultiple of the standard tone (if the pattern is composed of stationary loops). If the pattern is still moving (usually spinning), this indicates that the audio generator is not at the same frequency (or multiple) of the standard tone. The pattern must be stationary before the frequency can be determined. Note the standard tone frequency (440 or 600 Hz). Using this frequency as a basis, observe the Lissajous pattern (Chapter 5) to determine the audio generator frequency.

8-4.2. Basic Laboratory Frequency Calibration Using Oscillopscope Lissajous Patterns

The Lissajous patterns described in Chapter 5 are a form of zero-beat frequency measurement. In this case, the oscilloscope is used as the detector, and the resultant patterns are used to indicate relationship of one signal frequency to the other. When an oscilloscope is used as the detector, the signal source is usually an audio generator or low-frequency r-f generator. When high-frequency r-f signals are to be measured, an oscilloscope is not satisfactory as a detector since the passband of most oscilloscopes does not extend into the higher r-f ranges. However, there are special purpose laboratory oscilloscopes that have high-frequency amplifiers.

A 1:1 relationship of comparison frequencies using an oscilloscope produces an ellipse with opening and inclination dependent on amplitude and phase relationships of the applied signals. Phase-shift determinations are often made from such an ellipse, as discussed in Chapter 10. Slight frequency differences cause the ellipse to pass repeatedly through all the orientations from 0° to 360°. It is possible to time the completion of a 360° sequence and to find frequency difference, which is the reciprocal of this time (in seconds). To match the frequency of a local standard closely to that of a standard broadcast, the local standard oscillator is adjusted until the ellipse is stationary.

The practical limit for use of this technique for frequency offset adjustments and comparisons is about one part in 10^9. As an example, if the time required for a Lissajous figure to "rotate" through 360° is 100 sec, and the two signals used as inputs are 1 MHz, the offset will be

$$\Delta f = f \text{ offset} \frac{1}{T \text{ offset}} = \frac{1}{100} = 1 \times 10^{-2} \text{ Hz}$$

or related to the 1-MHz signal,

$$\frac{\Delta f}{f} = \frac{1 \times 10^{-2}}{10^6} = 1 \times 10^{-8} \text{ (offset)}$$

The technique of timing oscilloscope pattern rotations is not practical for quantitive *fractional* frequency deviation or calibration measurements and is not in common use.

8-4.3. Laboratory Frequency Calibration Using Oscilloscope Pattern Drift

An oscillator can be compared against a standard by externally triggering the oscilloscope from the standard while a pattern of several cycles of the oscillator is displayed. The ratio of drift of the oscilloscope pattern is related to the frequency error of the oscillator under test.

For example, suppose an oscilloscope is being used to check the oscillator frequency against a house standard. (It is assumed that the house standard has been checked against a standard broadcast and the error has been corrected, or recorded.) The equipment is connected as shown in Figure 8-21. The output of the oscillator being tested is connected to the oscilloscope's vertical input. The standard output is used to trigger the oscilloscope's horizontal sweep. The oscilloscope must be operated in the external sync condition (Chapter 5).

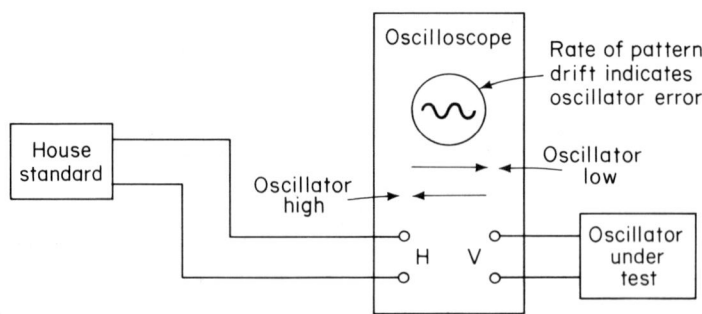

Fig. 8-21. Frequency calibration using oscilloscope pattern drift (Hewlett-Packard).

If the oscilloscope pattern apparently moves to the right, the oscillator frequency is low compared to that of the standard; if the pattern moves to the left, the oscillator frequency is high. *Rate of movement* can be interpreted in terms of *frequency error*. For example, with a 1-MHz signal from the standard used to trigger the display of a 1-MHz signal from the oscillator, the time required for the pattern to apparently drift the *width of 1 cycle* of the display is noted. Suppose that the pattern drifts left the width of 1 cycle in a time of 10 sec. This is equivalent to a frequency difference of 0.1 cycle/sec. Therefore, frequency error is 1 part in 10^7 (high):

$$\frac{\Delta f}{f} = \frac{0.1}{10^6} = 1 \times 10^{-7}$$

If it takes 100 sec for the pattern to drift the width of 1 cycle, the error is 1 part in 10^8:

$$\frac{\Delta f}{f} = \frac{0.01}{10^6} = 1 \times 10^{-8}$$

With the oscilloscope pattern drift method, as in the Lissajous method (Section 8-4.2), the largest error will come from the method used for timing. The best technique for both of these frequency difference measurements is to adjust the oscillator being tested until there is little or no apparent movement on the oscilloscope. This will allow a longer observation time for the drift of 1 cycle and will reduce the visual timing error.

8-4.4. Laboratory Frequency Calibration Using Direct-Frequency Comparison with a Counter

A frequency counter (Chapter 6) can serve to compare two oscillators against each other. The standard oscillator, substituted for the counter's internal time base, as shown in Figure 8-22, established the interval; that is, the interval selected at the counter is locked to the standard frequency source. The counter is set to measure frequency. Then cycles from the oscillator under test are counted during the interval.

Assume that two 1-MHz oscillators are being compared. The plus or minus one count error (common to electronic counters, as described in Chapter 6) limits precision to 1 part in 10^6 for a 1-sec interval. A 10-sec interval would increase precision to 1 part in 10^7, but to intercompare two oscillators to a precision of parts in 10^{11} would require an interval an entire day in length.

The time needed per measurement can be reduced by multiplying the frequency of the oscillator being checked *before it is counted*. Suppose the signal being measured is multiplied by 1000 to 1 GHz. Then a 1-sec counting interval will give a comparison of 1 part in 10^9, and a 100-sec interval will give 1 part in 10^{11}. However, if the multiplication factor is very large, the possible gain in precision is cancelled by the lack of coherence in the measurement. For that reason, the direct counter readout of frequency error, described in Section 8-4.5, is usually preferred.

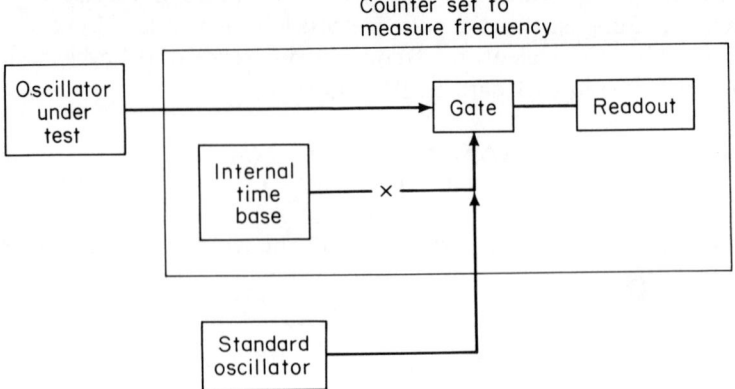

Fig. 8-22. Frequency calibration using direct frequency comparison with a counter (Hewlett-Packard).

8-4.5. Laboratory Frequency Calibration Using Direct Counter Readout of Frequency Error

Where a number of comparisons of precision oscillators expected to agree in frequency within parts in 10^6 (or better) are to be made against a house frequency standard, it is convenient to arrange equipment so that a counter's readout can be interpreted directly in terms of frequency error to an accuracy of parts in 10^6, 10^7, 10^8, and so on. The equipment arrangement to accomplish this direct readout uses a precision oscillator that has been offset by a predetermined amount. The offset frequency and the frequency from the oscillator under test are mixed, and the *period of their difference frequency* is measured. Such a comparison constitutes a short-term stability measurement; the changes in period of the difference frequency indicate the instabilities of the test oscillator (the reference frequency is considered to be stable). The period displayed on the counter's readout can easily be interpreted (digit by digit from left to right) as frequency error (for example, parts in 10^6, in 10^7, and so on).

To illustrate the method, assume an oscillator's 1-MHz output is to be compared against a reference oscillator. This reference oscillator need not be a standard, but it must have at least short-term stability and be adjustable over a small range. The test connection diagram is shown in Figure 8-23. First, the house standard and the reference oscillator are connected to the mixer. The counter is set to measure period. The combination of 1-MHz signals (reference and standard) produces a zero output to the gate trigger, producing a zero reading on the counter. The reference oscillator is then offset as necessary so that there is a 1-Hz difference between standard and reference. This 1 Hz triggers the counter gate open for 1 sec. The counter

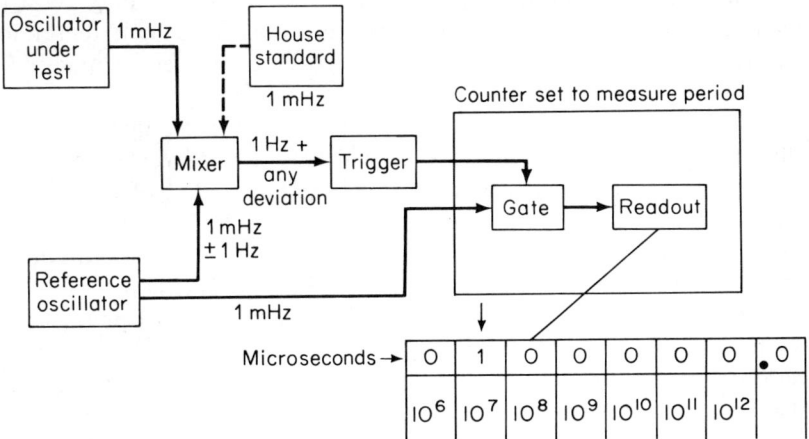

Fig. 8-23. Frequency calibration using direct counter readout of frequency error (Hewlett-Packard).

readout is 1,000,000.0 μsec (or 1 sec), since the external time base is 1 MHz.

With the reference oscillator adjusted to produce a 1,000,000.0-μsec readout, the house standard is removed from the mixer and replaced by the oscillator under test. If the oscillator under test is at the same frequency as the house standard, the counter readout will remain 1,000,000.0 μsec (indicating an offset of 1 part in 10^6). If the oscillator under test is at the same exact frequency as the reference oscillator, the counter readout will drop to zero (indicating no offset).

If the oscillator under test is between these extremes, which is likely the case, the reading in microseconds indicates the frequency error between the reference oscillator and the oscillator under test. For example, assume that the reference oscillator was set to 999,999.0 Hz during initial adjustment against the house standard. Then assume that the oscillator under test is at 999,999.1 Hz. The resultant signal at the mixer would be 0.1 Hz, and the gate would be triggered open for 0.1 sec. This would produce a count of 100,000 μsec. The actual counter readout would be 0,100,000.0. Since the 1 is the second digit from the left, this indicates that the test oscillator is in agreement with the reference oscillator by 1 part in 10^7 (as shown in Figure 8-23).

8-5. PRACTICAL TIME CALIBRATION TECHNIQUES

Several types of time information can be obtained from the standard broadcasts. The binary-coded-decimal time information can be used to check local clocks. However, this requires a digital-to-analog converter and a digital readout. The BCD time codes are not in general use. Instead, the voice and Morse code time information is used to set local clocks, then the 1-sec time ticks are used to check continued accuracy of the clocks.

A typical electronic clock (Figure 8-24) consists of a highly stable crystal oscillator time base (the frequency of which is divided down to produce 1 pulse / sec pulses) and a digital readout similar to that of an electronic counter. Usually, the readout is 6 digits (2 for seconds, 2 for minutes, and 2 for hours). Once set, the clock operates continuously with the count increasing by 1 sec for each pulse. The clock is set by introducing a series of fast pulses (usually from the dividers at some rate much faster at 1 pulse/sec) into the counter readout until the correct time is indicated. Then the counter input is returned to 1-pulse/sec. In addition to the visual readout, some clocks provide a BCD readout that can be applied to other equipment.

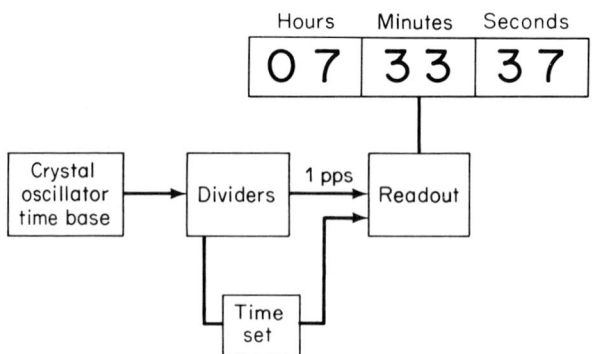

Fig. 8-24. Basic circuit for electronic clock.

Many electronic clocks incorporate a synchronizer or comparator feature (such as the Hewlett-Packard frequency divider and clock, or the General Radio digital synchronizer) that permits the clock to be compared against the standard broadcasts. Generally, an oscilloscope is used to compare the 1 pulse/sec master tick from the clock against the standard broadcast ticks.

In some clocks the master tick can be synchronized so that the master tick occurs at the same instant as the broadcast tick. Usually, the local clock must be synchronized at regular intervals (each day, each hour, or whatever is required) since the local standard may drift.

In other clocks, the offset or difference in clock tick and broadcast tick is measured (in microseconds). Then the difference is compared or measured at regular intervals to observe for drift.

It sould be noted that these same time comparisons or calibrations can be used to check a *local frequency standard* (provided the clock can be triggered by an external source). With such an arrangement, the local frequency standard is substituted for the clock time base oscillator, and the resultant clock tick is compared with the broadcast tick. Any difference in time between the two ticks is the result of frequency drift by the local frequency standard.

8-5.1. Basic Time Comparison Methods

Figure 8-25 shows one method of comparing local time ticks (from clock internal time base standards or from local frequency standards) against standard broadcast time ticks. Note that the local tick drives the oscilloscope horizontal sweep, while the vertical deflection is provided by the broadcast tick. If a dual-trace oscilloscope (Chapter 5) is available, the broadcast tick can be applied to one trace, while the local tick is applied to the other trace.

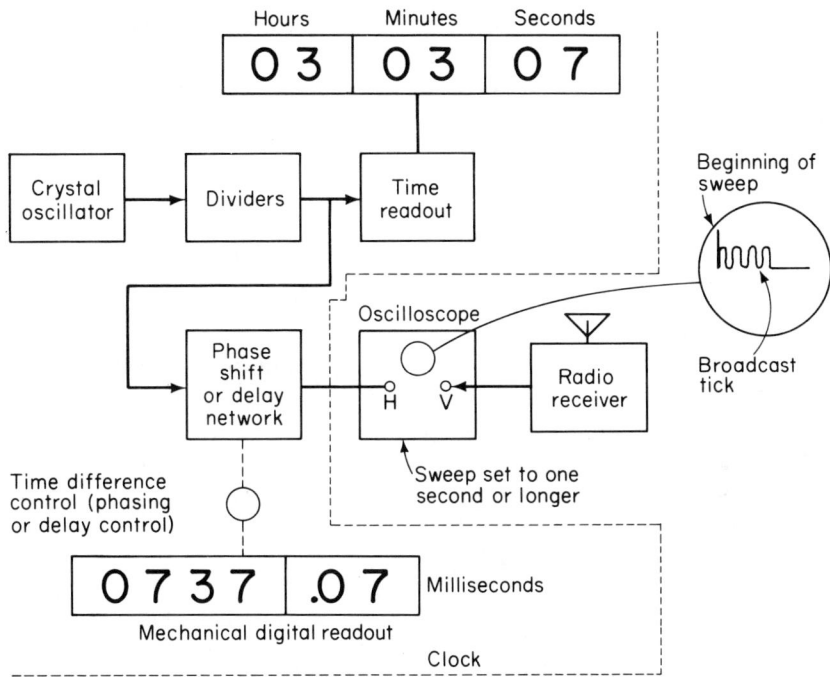

Fig. 8-25. Basic time comparison circuit.

Either way, the local tick passes through a phase-shift network (usually a capacitor and variable resistor) that changes the phase but not the frequency of the local tick. In some clock comparator systems, a *variable delay* is used instead of a phase shift. The phase relationship (or delay) is set by a front-panel control coupled to a mechanical digital readout. As the phase (or delay) is changed, the digital readout indicates the time difference (usually in milliseconds or microseconds).

Upon initial observation, the local tick and the broadcast tick may be apart as much as half a second. With the oscilloscope sweep set at 1 sec (or longer), the broadcast tick may be located with reference to the local tick. Adjustment of the phasing (or delay) control is then made to bring the broadcast tick toward the beginning of the oscilloscope trace. Successive adjustments of the phasing (or delay) control and oscilloscope sweep speed are made until the two ticks are brought into coincidence (or near coincidence).

Once the two ticks have been brought into coincidence, the mechanical digital readout geared to the phasing (or delay) control gives the initial *time reference* between local time and the standard broadcast time. At this point, the mechanical readout is logged. As the local standard drifts with respect to the broadcast standard, the phasing (or delay) control is readjusted to again establish coincidence of the two ticks. The amount of this readjustment (which indicates time drift of the local standard) is again logged.

This information, taken over a period of time and plotted, will allow accurate determination of drift rate and time (or frequency) error. Time comparisons made over several days can yield comparison accuracies of a few parts in 10^8 or better.

8-5.2. Precautions for High-Frequency Time Comparisons

High-frequency signal propagations (such as those from WWV and WWVH) are subject to erratic variations. Erratic phase delays are common. Therefore, it is important that certain precautions be observed to reduce the effect of these variations. The following is a summary of these precautions:

1. Schedule observation for an all-daylight or all-night transmission path between transmitter and receiver. Avoid twilight hours.
2. Choose the highest reception frequency that provides consistent reception.
3. Observe tick transmissions for a few minutes to judge propagation conditions. The best measurements are made on days when signals show little jitter or fading. If erratic conditions seem to exist, indicated by considerable fading and jitter in tick timing, postpone the measurement. Ionospheric disturbances causing erratic reception sometimes last less than an hour but may last several days.
4. Make time comparison measurements using the ticks with the earliest consistent arrival time.
5. A good-communications-quality receiver that can be tuned to the needed frequencies is the basic requirement. It is preferable that the antenna be of the directional type, oriented to favor that transmission mode which consistently provides the shortest propagation path.

8-6. PRACTICAL PHASE COMPARISON TECHNIQUES

A local frequency standard can be maintained to within 1 part in 10^{10}, or better, by comparison of its *relative phase* to that of a received *very-low-frequency* (VLF) carrier, such as that transmitted by WWVB at 60 kHz. Any one of a number of monitoring systems may be chosen to make this comparison possible, depending on the degree of precision required of the relative phase measurement. For greatest precision, the local standard must have a low drift that is predictable to within a few parts in 10^{10} over several days.

If no better than 1 part in 10^8 is wanted, a nearly instantaneous direct comparison for a short time may be used. If one part in 10^9 is wanted, comparison must be continued for a long enough time to reveal any ionospheric disturbance. Although VLF signals are relatively immune to propagation variations, best results usually are obtained when the total propagation path is in sunlight and conditions are stable. Near sunrise and sunset there are noticeable shifts in both amplitude and phase.

Figure 8-26 shows a typical VLF comparator system. As shown, the basic system consists of a 60-kHz receiver with directional antenna, divider and multiplier circuits to convert the local standard frequency to 60 kHz, a phase detector to compare phase of the two signals and produce an output with an amplitude corresponding to relative phase, and a strip-chart recorder to

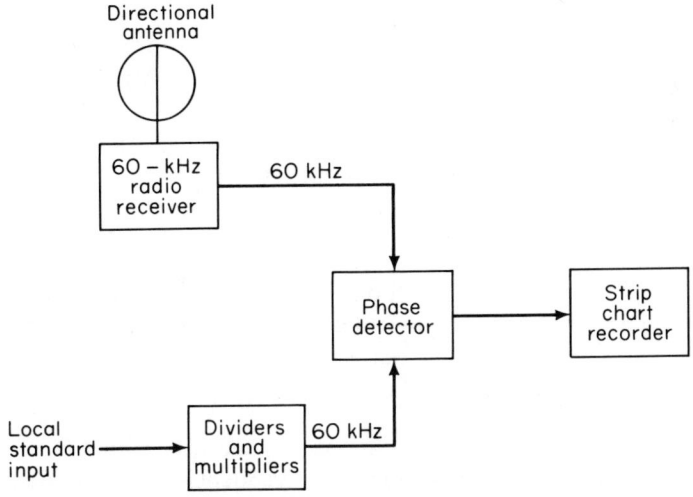

Fig. 8-26. Basic VLF Phase Comparator circuit (Hewlett-Packard).

monitor the phase detector output. A chart recorder is used because it is usually most practical to measure relative phase over a long period of time.

If the relative phase of the two signals remains constant, the chart trace will appear as a straight line. If the relative phase shifts, the chart trace will appear to slope or slant. The slope of the trace made by the chart recorder is, at a given instant, frequency offset between the local standard and the broadcast standard. So that this slope may be read at a glance, a set of transparent templates relating slope to frequency offset is provided with a typical comparator system (such as the Hewlett-Packard VLF comparator).

In use, the template is overlayed on the trace, the matching slope is selected, and the frequency offset (together with its sign, such as $+3 \times 10^{10}$ or -3×10^7) is then read from the template. The template is oriented so that it is aligned with the chart (with the long lines on the template parallel to the chart edges) and it is moved back and forth along the chart until one of the template lines is found to have the same slope as the chart trace. The offset is then read directly from the template.

To establish the drift rate of a local standard, two determinations of the offset at separate times are required. The rate of change of the slope is frequency drift rate, or aging rate, of the local standard. Figure 8-27 shows a template superimposed on a typical recorder chart at two separate chart times so that a drift determination can be made. At around 1 P.M. (point t_1), offset is read to be about $+1.5$ parts in 10^{-9} ($+1.5 \times 10^{-9}$). The following day, again about 1 P.M. (point t_2), offset is read to be about $+1.9 \times 10^{-9}$. Therefore, drift is

$$\begin{array}{r} +1.9 \times 10^{-9} \text{ (at } t_2) \\ -\ +1.5 \times 10^{-9} \text{ (at } t_1) \\ \hline +0.4 \times 10^{-9} \text{ (drift for 24 hr)} \end{array}$$

8-7. DEFINITIONS OF TERMS USED IN FREQUENCY AND TIME STANDARD INSTRUMENTS

The following definitions provide a basis for understanding frequency and time standard equipment and techniques. Reference to publications of the engineering societies, the National Bureau of Standards, and so on, should be made for more definitive statements.

A-1 or A.1. Atomic time scale maintained by the U.S. Naval Observatory, presently based on weighted averages of frequencies from cesium beam devices operated at a number of laboratories.

Accuracy. For a measured or calculated value, the degree to which it conforms to the accepted standard or rule.

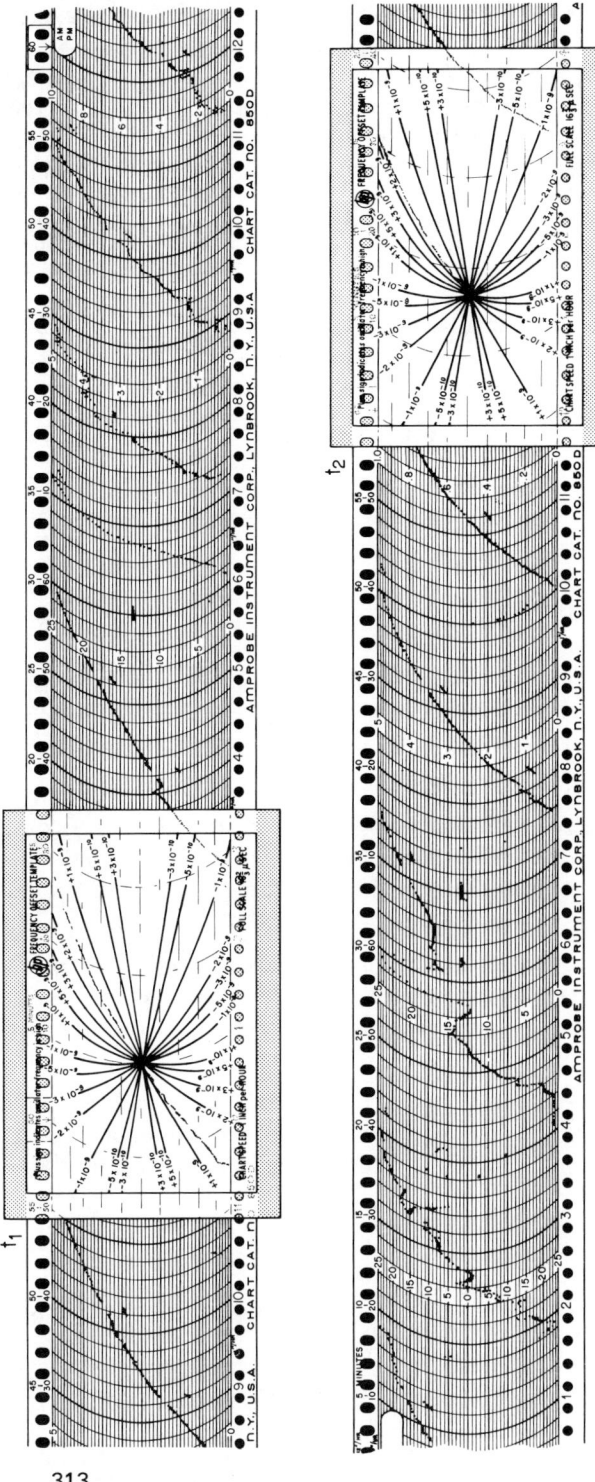

Fig. 8-27. Typical recorder chart traces with template to convert trace slope into frequency shift (Hewlett-Packard).

BCD. In the *binary-coded-decimal* system, individual decimal digits are represented by some binary code. For example, 16 might be represented in an 8–4–2–1 BCD notation as 0001 (for 1) and 0110 (for 6). In pure binary, 16 is 10,000. Refer to the Appendix for additional data on binary coding.

ET. Ephemeris time is astronomical time based on the motion of the earth about the sun during the tropical year 1900. The second of ET is *apparently* not variable.

EPOCH. Usually denotes the interval elapsed from an arbitrary origin; this origin is customarily stated in terms of a natural phenomenon common to all observers. Epoch may also denote a particular instant of time.

ERROR. The difference between the true value and an observed or calculated one.

FRACTIONAL FREQUENCY DEVIATION. Statements of stability (hence quality) for frequency standards usually are made in terms of RMS fractional frequency deviation based on a statistical approach.

FREQUENCY OFFSET. Fractional frequency offset is the amount by which a frequency lies above or below reference frequency. For example, if a frequency measures 1.000,001 MHz when compared against a reference frequency of 1.000,000 MHz, then its fractional frequency offset is 1 Hz/1 MHz, or 1 part in 10^6.

NBS-A. Atomic time scale kept by NBS, presently based on four quartz oscillators and one rubidium standard that are frequency compared daily against the U.S. frequency standard (cesium beam). The cycles are counted (an accumulated total is kept) and a suitable weighted average is taken (a computer program is used).

NBS-SA. "Stepped Atomic" time signals are broadcast on NBS station WWVB. These signals are in accordance with NBS-A, which is based on the USFS; an 0.2-sec step in phase is made on the first of the month as necessary to keep ticks within 0.1 sec of UT_2.

NBS-UA. The universal atomic time scale of NBS is referenced to the USFS and incorporates the frequency offsets and shifts in epoch as announced by the Bureau International de l'Heure; thus NBS-UA is an approximation of UT_2.

PHASE ANOMALY. A sudden irregularity in the phase of an LF or VLF transmission.

PHASE NOISE. A measure of the random phase instability of a signal.

PRECISION, REPRODUCIBILITY, RESETTABILITY. All of these imply the existence of an observation consisting of a series of readings taken in a prescribed manner.

STABILITY, SHORT-TERM AND LONG-TERM. The stability of a frequency or time standard is one of its most important characteristics.

LONG-TERM stability (or instability) refers to the low changes in average frequency with time arising from changes in the resonator or other element of an oscillator. Statements of long-term stability for quartz oscillators often term this characteistic "aging rate" and specify it as "parts per day" (fractional frequency change over 24 hr). For cesium standards, this term commonly refers to total fractional frequency drift for the life of the cesium beam tube.

SHORT-TERM stability (or instability) is an expression of the change in average frequency over a time sufficiently short (but exceeding some minimum time) that long-term effects are of small significance. Since at present there is no general agreement fixing the averaging time, one must be specified for each statement of short-term stability. A typical statement for an ultraprecise quartz oscillator is "fractional frequency deviation is 1 part in 10^{11} for an 0.1-sec average time and is 5 parts in 10^{11} per day."

SPECTRAL PURITY. A frequency standard must provide a stable, spectrally pure signal if it is to yield a narrow spectrum after multiplication to the microwave region. A standard that was spectrally pure would have no sidebands. This is not practical, as discussed in Chapter 10. However, the higher the degree of spectral purity, the narrower the sidebands (before and after frequency multiplication).

STANDARDS. Frequency and time standards are designated in a variety of ways:

HOUSE STANDARD—A house or company frequency or time standard is the best (most nearly accurate and stable, lowest drift, and so on) frequency source on the premises. House standards are normally kept referenced to national frequency standards via radio broadcasts.

LOCAL STANDARD—A local standard is local in the sense that it is not a national standard; hence it may be a house standard, a reference standard, or a working standard.

REFERENCE STANDARD, TRANSFER STANDARD—Often, the standard used to convey frequency or time from the house standard to working standards (the house standard may itself be considered a reference standard).

WORKING STANDARD—The standard against which instruments are compared for purposes of test or repair.

TIME CODE. WWV broadcasts (since 1961) a binary-coded-decimal time code (pulse-coded signals) at a 100-pps rate carried on 1-kHz modulation. WWVB broadcasts a binary-coded-decimal time code presented by means of a level shift, with markers generated by a 10-dB drop each second. These codes present time-of-year and time-of-day information.

USFS. United States frequency standard, maintained and operated by the Radio Standards Laboratory of the National Bureau of Standards. USFS realizes the idealized cesium atomic resonance frequency to 1 part in 10^{11}.

UT_2. Universal time (UT_0, UT_1, UT_2) is established by astronomical measurement of the rotation of the earth. Universal time is often called Greenwich Mean Time. For the United States, the U.S. Naval Observatory determines universal time. UT_2 means UT corrected for polar motion and for seasonal variations.

UNCERTAINTY. Once error has been corrected for, what remains is uncertainty.

9 PROBES AND TRANSDUCERS

Practically all meters and oscilloscopes operate with some type of probe. In addition to providing for electrical contact to the circuit under test, probes serve to modify the voltage being measured to a condition suitable for display on an oscilloscope or readout on a meter. For example, assume that a very high voltage must be measured and this voltage is beyond the maximum input limits of the meter or oscilloscope. A voltage-divider probe can be used to reduce the voltage to a safe level for measurement. Under these circumstances, the voltage will be reduced by a fixed amount (known as the *attenuation factor*), usually in the order of 10:1 or 100:1.

Transducers serve a similar function. However, a transducer operates to convert a physical property to an electrical signal suitable for meter or oscilloscope inputs. Although a variety of probes are found in both shop and laboratory, the use of transducers is generally limited to the laboratory or industrial applications.

9-1. THE BASIC PROBE

In its simplest form, the basic probe is a test prod. In physical appearance, the probe is a thin metal rod connected to the meter or oscilloscope input through an insulated flexible lead. All but a small tip of the rod is covered with an insulating handle, so that the probe can be connected to any point of the circuit without touching nearby circuit parts. Sometimes the probe tip is provided with an alligator clip so that it is not necessary to hold the probe at the circuit point.

Such probes work well on circuits carrying d-c and audio-frequency signals. If, however, the ac is at a high frequency or if the gain of the meter (such as an electronic meter) or oscilloscope amplifier is high, it may be necessary to use a special low-capacitance probe. Hand capacitance in a simple probe can cause hum pickup, particularly if the amplifier gain is high. This condition can be offset by shielding in low-capacitance probes. In a more important problem, the input impedance of the meter or oscilloscope is connected directly to the circuit under test by a simple probe. Such input impedance could disturb circuit conditions (as discussed in Chapter 1). The low-capacitance probe contains a series capacitor and resistor that increase the meter impedance.

9-2. LOW-CAPACITANCE PROBES

The basic circuit of a low-capacitance probe is shown in Figure 9-1. The series resistance R_1 and capacitance C_1, as well as the parallel or shunt resistance R_2, are surrounded by a shielded handle. The values of R_1 and C_1 are preset at the factory by screwdriver adjustment and should not be disturbed unless recalibration is required, as discussed in later sections of this chapter.

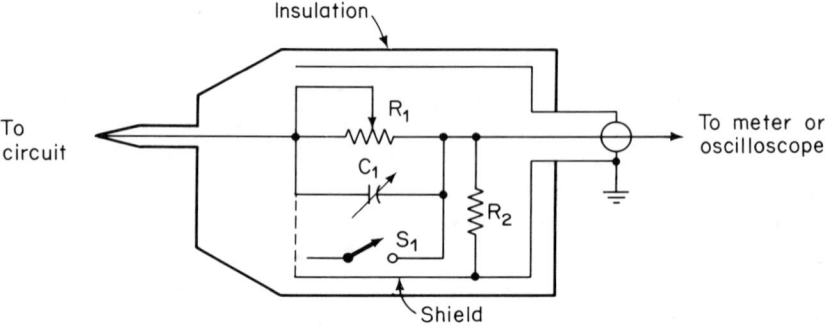

For a 10:1 voltage division:
$R_1 = 9$ times R_2
$R_2 =$ input impedance of meter or oscilloscope

Fig. 9-1. Typical low-capacitance probe circuit.

In most low-capacitance probes, the values of R_1 and R_2 are selected to form a 10:1 voltage divider between the circuit under test and the meter input. Thus the probes serve the dual purpose of capacitance reduction and voltage reduction. The operator should remember that voltage indications will be one-tenth (or whatever value of attenuation is used) of the actual value when such probes are connected at the inputs of meters or oscilloscopes.

The capacitance value of C^1 in combination with the values of R_1 and R_2 also provides a capacitance reduction (usually in the range of 3:1–11:1).

There are probes that combine the features of low-capacitance probes and the simple probes described in Section 9-1. In such probes, a switch (shown as S_1 in dotted form in Figure 9-1) is used to short both C_1 and R_1 when a direct input is required. With S_1 open, both C_1 and R_1 are connected in series with the input, and the probe provides the low-capacitance and voltage-division features.

9-3. RESISTANCE-TYPE VOLTAGE-DIVIDER PROBES

A resistance-type voltage-divider probe is used when the primary concern is reduction of voltage. The resistance-type probe, shown in Figure 9-2, is similar to the low-capacitance probe described in Section 9-2, except that the frequency-compensating capacitor is omitted.

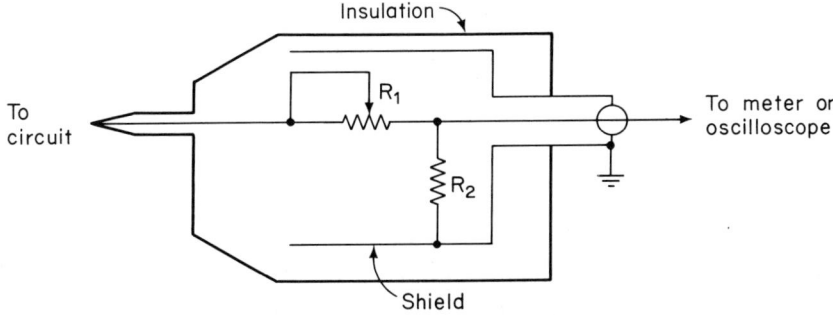

For a 10:1 voltage division:
R_1 = 9 times R_2
R_2 = input impedance of meter or oscilloscope

Fig. 9-2. Typical resistance-type voltage-divider probe circuit.

Usually, the straight resistance-type probe is used when a voltage reduction of 100:1, or greater, is required, and when a flat frequency response is of no particular concern.

As shown in Figure 9-2, the values of R_1 and R_2 are selected to provide the necessary voltage division and to match the input impedance of the meter or oscilloscope. Resistor R_1 is usually made variable so that an exact voltage division can be obtained.

Because of their voltage-reduction capabilities, resistance-type probes are often known as *high-voltage* probes. Some resistance-type probes are capable of measuring potentials at or near 40 kV (with a 1000:1 voltage reduction).

9-4. CAPACITANCE-TYPE VOLTAGE-DIVIDER PROBES

In certain isolated cases, the resistance-type voltage-divider probes described in Section 9-3 are not suitable for measurement of high voltages because stray conduction paths are set up by the resistors. A capacitance-type probe, shown in Figure 9-3, can be used in those cases.

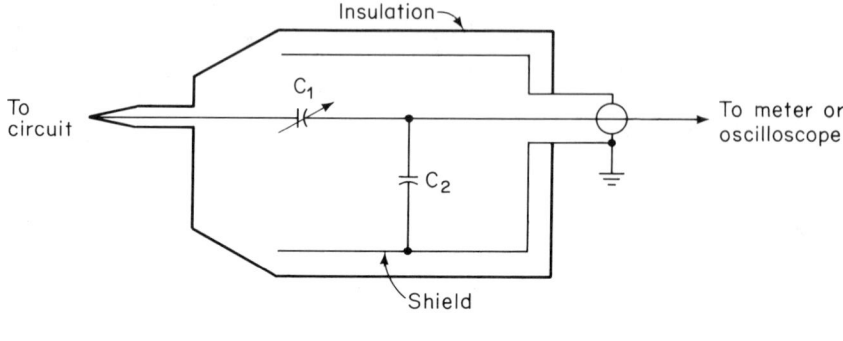

For a 10:1 voltage reduction:
$C_1 = \frac{1}{9}$ of C_2
C_2 = input capacitance of meter or oscilloscope

Fig. 9-3. Typical capacitance-type voltage-divider probe circuit.

In such capacitance probes, the values of C_1 and C_2 are selected to provide the necessary voltage division and to match the input capacitance of the meter or oscilloscope. Capacitor C_1 is usually made variable so that an exact voltage division can be obtained.

9-5. RADIO-FREQUENCY PROBES

When the signals to be measured are at radio frequencies and are beyond the frequency capabilities of the meter or oscilloscope circuits, a radio-frequency (r-f) probe is required. Such probes convert (rectify) the r-f signals to a d-c output voltage that is equal to the peak r-f voltage. The d-c output of the probe is then applied to the meter or oscilloscope input and is displayed as a voltage readout in the normal manner. In some r-f probes, the d-c output is equivalent to peak r-f voltage, whereas in other probes the readout is equal to RMS voltage.

The basic circuit of a radio-frequency probe is shown in Figure 9-4. This circuit can be used to provide either peak output or RMS output. Capacitor C_1 is a high-capacitance d-c blocking capacitor used to protect diode CR_1. Usually, a germanium diode is used for CR_1, which rectifies the r-f voltage and

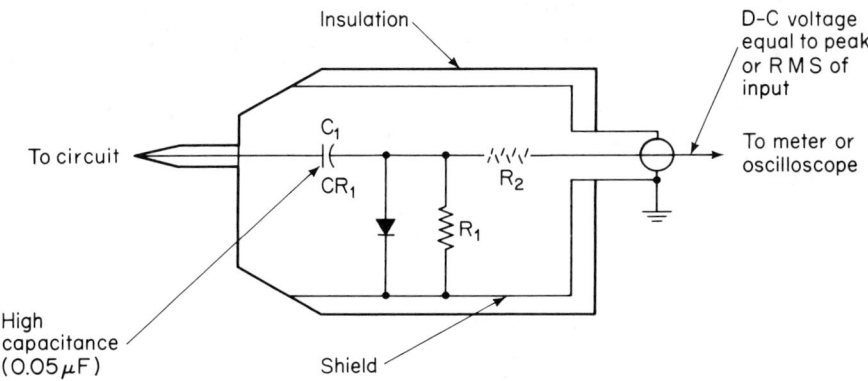

Fig. 9-4. Typical half-wave radio-frequency probe circuit.

produces a d-c output across R_1. In some probes, R_1 is omitted so that the d-c voltage is developed directly across the input circuit of the meter or oscilloscope. This d-c voltage is equal to the peak r-f voltage, less whatever forward drop exists across the diode CR_1.

When it is desired to produce a d-c output voltage equal to the RMS of the r-f voltage, a series-dropping resistor (shown in dotted form in Figure 9-4 as R_2) is added to the circuit. Resistor R_2 drops the d-c output voltage to a level that equals 0.707 of the peak r-f value.

The r-f probe of Figure 9-4 is a half-wave probe. A full-wave probe will provide an output to the meter or oscilloscope that is (approximately) equal to the peak-to-peak value of the voltage being measured. This is particularly important when measuring pulses, square waves, and any other complex wave-form.

A full-wave probe circuit is shown in Figure 9-5. Such probes are usually found with meters rather than oscilloscopes. Since most meters are calibrated

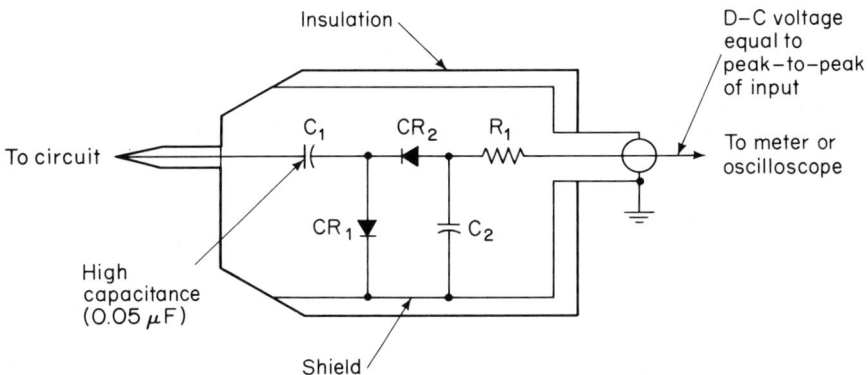

Fig. 9-5. Typical full-wave radio-frequency probe circuit.

to read in RMS values, the probe output is reduced to 0.3535 of the peak-to-peak value. This is accomplished by selecting or adjusting the value of resistor R_1. A few VTVM and electronic meters are provided with peak-to-peak scales. These meters are used in certain laboratory applications and in television service to measure such values as horizontal-oscillator or deflection-coil voltages, input to video amplifier, output of the vertical amplifier, and so on. On such meters, resistor R_1 is omitted so that the d-c output will be equivalent to peak-to-peak input.

9-6. DEMODULATOR PROBES

The circuit of a demodulator probe is essentially like that of the r-f probe described in Section 9-5. The circuit values and the basic functions are somewhat different.

Both the half-wave and full-wave probes are used to convert high-frequency signals (usually an r-f carrier) to a d-c voltage that can be measured on a meter or oscilloscope. When the high-frequency signals contain a-c or pulsating d-c modulation (such as a modulated r-f carrier), a demodulator probe will be more effective for signal tracing.

The basic circuit of the demodulator probe is shown in Figure 9-6. Here, capacitor C_1 is a low-capacitance, d-c blocking capacitor. (In the r-f probe a high-capacitance value is required for C_1 to ensure that the diode operates at the peak of the r-f signal. This is not required for a demodulator probe.) Germaninum diode CR_1 demodulates (or detects) the amplitude-modulated signal and produces voltages across load resistor R_1. (C_1 and R_1 also act as a filter.)

The demodulator probe produces both an a-c and d-c output. The r-f carrier frequency is converted to a d-c voltage equal to the peak of the r-f car-

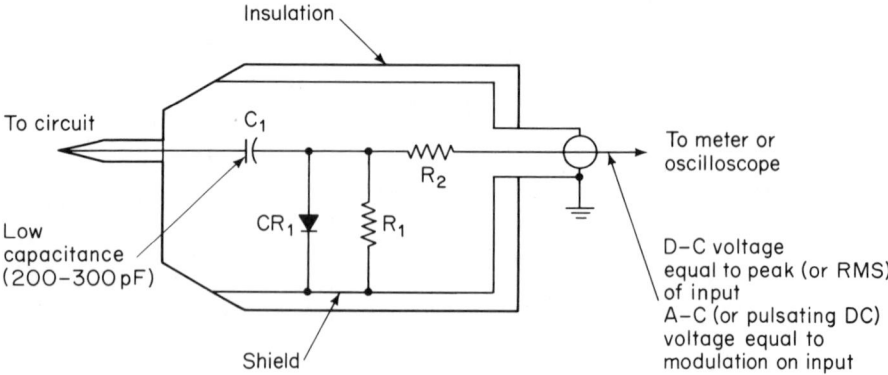

Fig. 9-6. Typical demodulator probe circuit.

rier. The low-frequency modulating voltage appears as ac (or pulsating dc) at the probe output.

In use, the meter or oscilloscope is set to measure dc, and the r-f carrier is measured. Then the meter or oscilloscope is set to measure ac, and the modulating voltage is measured. Resistor R_2 is used primarily for isolation between the circuit under test and the meter or oscilloscope input. Resistor R_2 can also serve as a calibrating resistance so that the output will equal value (RMS, peak, and so on). However, in general, demodulator probes are used primarily for *signal tracing*, and their output is not calibrated to any particular value.

9-7. SPECIAL PURPOSE PROBES

The probes described thus far are in common use. There are also a number of special probes for limited applications. The transistorized signal-tracing probe and the cathode follower probes are typical examples.

9-7.1. Cathode Follower Probes

The cathode follower probe provides a means of connecting into a circuit without disturbing circuit operation. A cathode follower probe provides a high input impedance (as does an electronic meter or VTVM, Chapter 1) that will not disturb circuit conditions and a low-impedance output to match the input of an amplifier (such as some amplifiers used in oscilloscopes).

The basic circuit of a cathode follower probe is shown in Figure 9-7. The grid input resistor R_1 is of high resistance (usually several megohms); the cathode load resistor R_2 is of the same value as the amplifier input impedance.

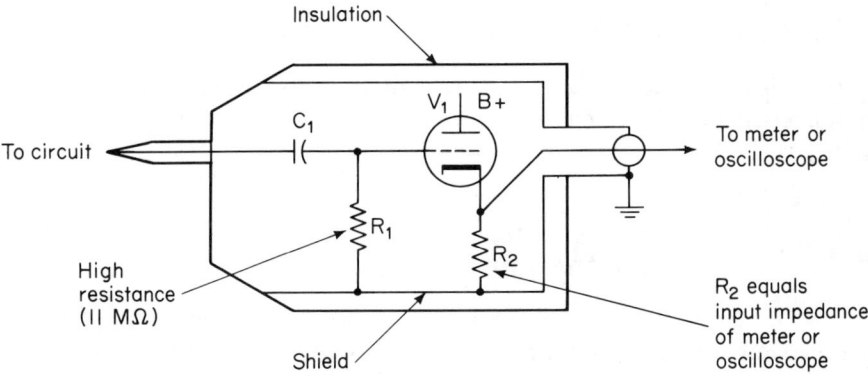

Fig. 9-7. Typical cathode follower probe circuit.

Therefore, the circuit under test sees a high impedance that does not disturb the circuit, whereas the amplifier (such as an oscilloscope vertical amplifier) sees a matched impedance. The cathode follower probe has one disadvantage in that $B+$ and filament power are required.

9-7.2. Transistorized Signal-Tracing Probes

It is possible to increase the sensitivity of a probe with a transistor amplifier. Such an arrangement is particularly useful with a VOM (Chapter 1) for measuring small-signal voltages. An amplifier is not usually required for a VTVM or electronic voltmeter or with an oscilloscope, since such instruments contain built-in amplifiers.

A transistor probe and amplifier circuit is shown in Figure 9-8. This circuit will increase the sensitivity of the probe by at least 10:1 and should provide good response up to about 500 MHz. The circuit is not normally calibrated to provide a specific voltage indication but is used to increase the sensitivity of the probe for signal-tracing purposes.

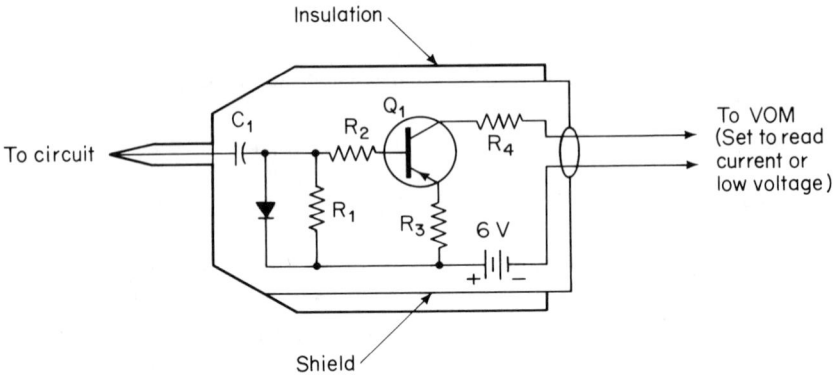

Fig. 9-8. Typical transistorized signal-tracing probe circuit.

It is also possible to use a half-wave or full-wave probe with one of the amplifier circuits described in Chapter 7 to provide an effective signal-tracing tool.

9-8. PROBE COMPENSATION AND CALIBRATION

Probes must be calibrated to provide a proper output to the meter or oscilloscope with which they are to be used. Probe compensation and calibration is done at the factory and must be accomplished with the proper test equipment. The following paragraphs describe the *general* procedures for compensating and calibrating probes. *Never* attempt to adjust a probe unless you follow the instruction manual and have the proper test equipment. An improperly adjusted probe will provide erroneous readings and may possibly cause undesired circuit loading.

9-8.1. Probe Compensation

The capacitors that compensate for excessive attenuation of high-frequency signal components (through the probe resistance dividers) affect the entire frequency range from some midband point upward. (Capacitor C_1 in Figure 9-1 is an example of such a compensating capacitor.)

Compensating capacitors must be adjusted so that the higher-frequency components will be attenuated by the same amount as low frequencies and dc. It is possible to check adjustment of the probe-compensating capacitors using a square-wave signal source.

This is done by applying the square-wave signals directly to an oscilloscope input, then applying the same signals through the probe and noting any change in pattern. There should be no change (except for a reduction of the amplitude) in a properly compensated probe.

Figure 9-9 shows typical square-wave displays with the probe properly compensated, undercompensated (high frequencies underemphasized), and overcompensated (high frequencies overemphasized).

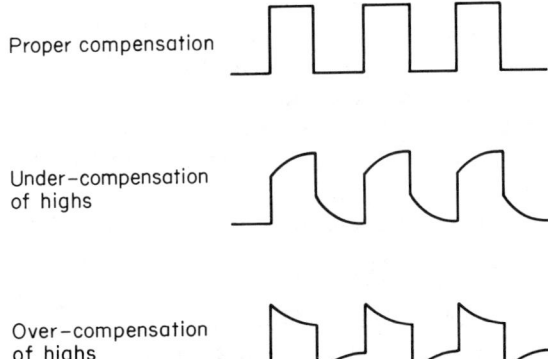

Fig. 9-9. Typical square-wave displays showing frequency compensation of probes.

Proper compensation of probes is often neglected, especially when probes are used interchangeably with meters or oscilloscopes having *different* input characteristics. It is recommended that any probe be checked with square-wave signals before use.

Another problem related to probe compensation is that *input capacitance* of the meter or oscilloscope may change with age. Also, with vacuum-tube meters and oscilloscopes, the input capacitance can change when tubes are changed. Either way, all of the compensated dividers will be improperly adjusted. Readjustment of the probe *will not correct* for the change needed by the meter or oscilloscope input circuits.

9-8.2. Probe Calibration

The main purpose of probe calibration is to provide a specific output for a given input. For example, the value of resistor R_1 in Figure 9-2 is adjusted (or selected) to provide a specific amount of voltage division. During calibration, a voltage of known value and accuracy is applied to the input. The output is monitored, and R_1 is adjusted to provide a given value. For example, on a 10:1 divider probe, 10 V could be applied to the input and R_1 adjusted for a 1-V output.

As another example, the value of resistor R_2 in Figure 9-4 is adjusted (or selected) to provide a d-c output that equals 0.707 of the peak r-f input value. For example, with a 10-V peak r-f input, R_2 would be adjusted to provide a 7.07-V d-c output.

9-9. PROBE OPERATING TECHNIQUES

Although a probe is a simple instrument and does not require specific operating procedures, several points should be considered to use a probe effectively.

Circuit loading is a major problem in many measurements. Connection of a meter or oscilloscope to a circuit may alter the amplitude or wave-form at the point of connection. To prevent this, the impedance of the circuit being measured must be a small fraction of the input impedance of the meter or oscilloscope. When a probe is used, the *probe's impedance* determines the amount of circuit loading.

The ratio of the two impedances (probe and circuit) represents the amount of probable error. For example, a ratio of 100:1 (perhaps a 100-mΩ probe to measure the voltage across a 1-mΩ circuit) will account for about a 1 per cent error. A ratio of 10:1 would produce a possible 9 per cent error.

Remember that input impedance is not the same at all frequencies but continues to get smaller at higher frequencies because of the input capacitance. (Capacitive reactance and impedance decrease with an increase in frequency.) All probes will have some input capacitance. Even an increase at audio frequencies may produce a significant change in impedance.

When using a shielded cable with a probe to minimize pickup of stray signals and hum, the additional capacitance of the cable should be recognized. The capacitance effects of a shielded cable can be minimized by terminating the cable at one end in its characteristic impedance. Unfortunately, this is not always possible with most meter and oscillosocope input circuits.

The reduction of resistive loading due to probes may not be as much as the attenuation ratio of the probe, but capacitive loading will not be reduced to the same extent because of the additional capacitance of the probe cable. For example, a typical 5:1 attenuator probe may be able to reduce capacitive loading somewhat better than 2:1. A 50:1 attenuator probe may reduce capacitive loading by about 10:1. Beyond this point, little improvement can be expected because of stray capacitance at the probe tip.

Whether a particular probe connection is disturbing a circuit can be judged by attaching and detaching another connection of similar kind (such as another probe) and observing any difference in meter reading. If there is little or no change when the additional probe is touched to the circuit, it is safe to assume that the probe has little effect on the circuit.

Long probes should be restricted to the measurement of relatively slow-changing signals (d-c and low-frequency a-c). The same is true for long ground leads.

The ground lead should be connected where no hum or high-frequency signal components exist in the ground path between that point and the signal pickoff point. (Refer to Chapters 1 and 2.)

Avoid applying more than the rated voltage to a probe. Using a high-voltage coupling capacitor between a probe tip and a very high d-c level *may not always prevent probe burnout*. This is because the capacitor must charge and discharge through the probe. If care is taken to charge and discharge the blocking capacitor through a path that shunts the probe, the technique can be successful. On way to accomplish this is to ground *the junction of the capacitor and the probe tip* whenever the capacitor is being charged or discharged.

For example, assume that the probe is to be used to measure 500 V and that it is rated at 250 V. By connecting a high-voltage capacitor to the probe tip, grounding the junction of the probe tip and capacitor, and then connecting the free end of the capacitor to the 500-V point, the capacitor will be charged to 500 V, and the measurement can continue as normal. At the completion of the measurement, the capacitor must be discharged by touching the probe-capacitor junction to ground. In this way, both charging and discharging of the capacitor will be accomplished outside the probe circuit.

9-10. TRANSDUCERS

Although there are many types of transducers used as test equipment, transducers can be divided into three broad categories: resistance-changing, self-generating, and inductance- or capacitance-changing. Practically any condition can be measured by one or more of these three types of transducers and can be converted to a voltage suitable for readout on a meter or display on an oscilloscope. The remaining sections of this chapter summarize the three types of transducers used as test equipment.

9-11. RESISTANCE-CHANGING TRANSDUCERS

A resistance-changing transducer has the widest application in test equipment. The resistance can be changed by mechanical means or by some external condition, such as temperature. Likewise, the resistance element can be connected as part of a voltage divider or as part of a bridge circuit.

The classic voltage-divider circuit is shown in Figure 9-10. Here a fixed-voltage source is placed across the resistance element. The element contact arm is actuated by mechanical force. This force could be a bellows (to measure pressure differences), a spring-loaded weight that slides in a direction parallel to motion (to measure acceleration), or the force could come from movement of an arm coupled to some mechanical device (to measure position). In any event, the output voltage to the meter or oscilloscope will be proportional to the mechanical force and can be so related.

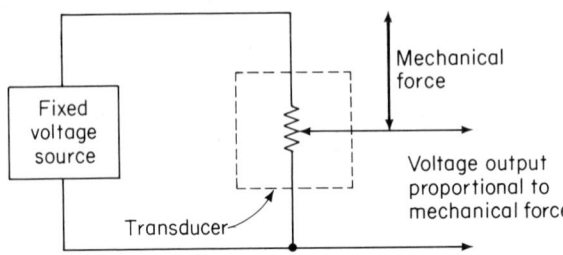

Fig. 9-10. Typical voltage-divider type of resistance-changing transducer.

For example, assume that the mechanical force is pressure that operates against a bellows that, in turn, operates the resistance wiper arm. By applying a known fixed voltage across the resistance element, it is possible to determine whether a given pressure will equal a given voltage. This relation of pressure to voltage is known as *scale factor*.

Another version of the voltage-divider circuit used in test equipment transducers is shown in Figure 9-11. Here the fixed voltage source is placed across the resistance element and a fixed resistor in series. The resistance element changes resistance value with changes in external condition. An example of this is a thermistor, which is a resistor that has a negative temperature coefficient (resistance decreases with temperature increase). Thermistors are discussed further in Chapter 11.

Since the voltage divides itself across the two resistances, the output voltage will be proportional to the value of the resistance element. This, in turn,

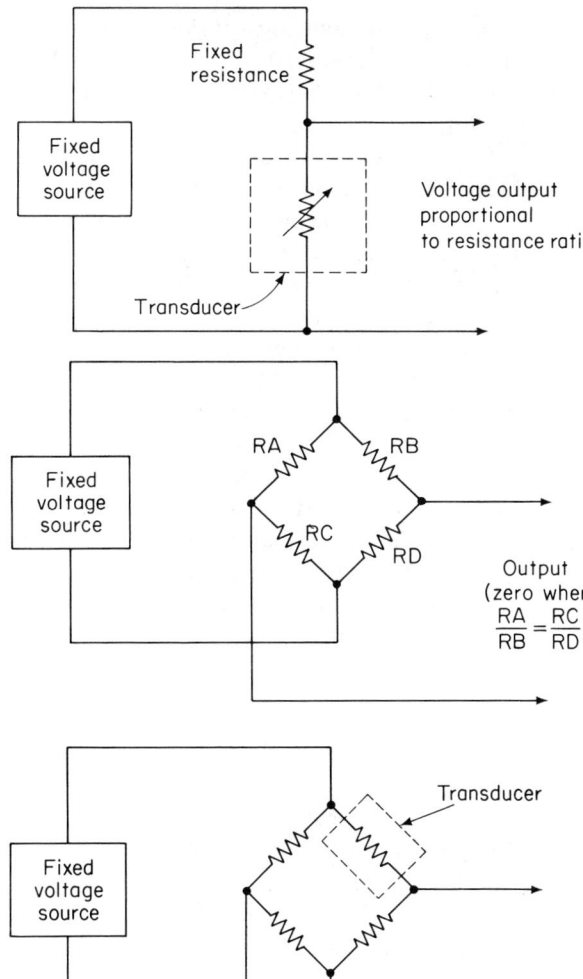

Fig. 9-11. Typical resistance-ratio type of resistance-changing transducer.

Fig. 9-12. Basic bridge circuit used in transducers.

Fig. 9-13. Basic one-sensitive-arm transducer bridge.

is in relation to the temperature. Thus the output voltage is proportional to the temperature surrounding the thermistor.

Resistance-changing transducers are often used in bridge circuits similar to those discussed in Chapter 3. Figure 9-12 shows the basic bridge circuit. The bridge will be balanced and will have a zero output when RA/RB is equal to RC/RD. When there is a change in one or more of the bridge arms, the bridge will become unbalanced and there will be an output voltage proportional to the change.

One or more arms of the bridge may be active. When only one arm is active, as shown in Figure 9-13, the transducer bridge is said to be a *one-sensitive-arm bridge*. Two active arms (Figure 9-14) make the bridge a *two-sensitive-arm bridge*.

There are a number of drawbacks to any resistance-changing type of transducer, as well as some advantages.

One of the main advantages is that a minimum of *signal conditioning* is required, except in very special applications; that is, the transducer output voltage can be applied directly to the meter or oscilloscope without modification (amplification, rectification, and so on). Likewise, this high output permits operating even with severe impedance mismatch.

As for the disadvantages, resistance-changing transducers that are operated by mechanical movement are subject to shock and vibration, since the resistance element contact arm cannot tell the difference between desired and undesired movement. The thermistor-type transducers also have a problem because current passing through the resistance element creates heat. This heat combines with the ambient heat and changes the thermistor resistance. For these and various other design considerations, a number of self-generating transducers are used both in scientific and routine laboratory test equipment applications.

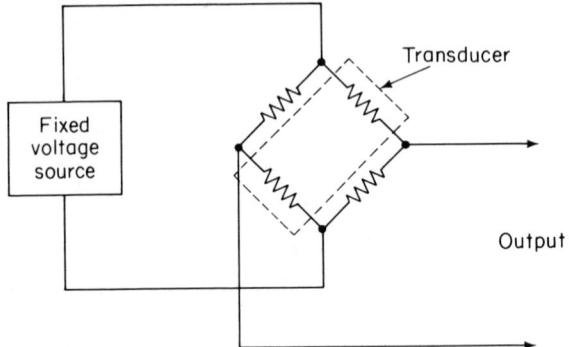

Fig. 9-14. Basic two-sensitive-arm transducer bridge.

9-12. SELF-GENERATING TRANSDUCERS

It would be impossible to describe all of the self-generating transducers in test equipment use. And in many cases the various transducers use different versions of the same basic circuit. The following circuits are typical of many commercially available transducer and signal-conditioning systems.

Thermocouples are often used to measure temperature because of their simplicity and ruggedness. Thermocouple transducers are based on the fact that a voltage will be produced when two dissimilar metals are joined and subjected to heat. Test equipment transducers often use chromel and alumel as the dissimilar metals.

One of the drawbacks to a thermocouple transducer is that the voltage output is very low. This is usually resolved by adding a d-c amplifier (Chapter 7) between the thermocouple and meter or oscilloscope. Another drawback is that the thermocouple output will vary with changes in ambient temperature and with temperature changes caused by heating the wires. This is compensated by the use of reference junctions and a bridge circuit. As shown in Figure 9-15, the reference junctions are connected in series with the measuring-junction output. The reference junctions are also placed near a thermistor element that forms one leg of a bridge circuit. The output of the bridge is also in series with the measuring-junction output. Both the reference junctions and the thermistor are subjected to identical temperature conditions. Resistance changes in RA cause changes in the bridge unbalance so that the bridge-applied voltage across RB is of the same value as the change in voltage from the reference thermocouple junctions. The accuracy of the system depends on the ability of the bridge to track changes in the reference junction output.

Turbine-type flowmeters are a typical example of self-generating transducers

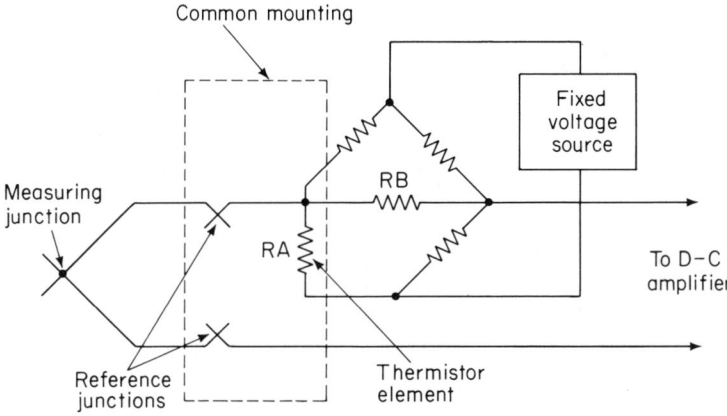

Fig. 9-15. Thermocouple transducer with reference-compensation circuit.

that measure some physical property such as liquid flow, air flow, and so on. Their operation is similar to that of a *tachometer*, except that they measure liquid flow rather than mechanical rotation.

As shown in Figure 9-16, a flowmeter transducer consists essentially of a turbine placed in the flow path, a magnet in one of the turbine blades, and a pickup coil placed near the turbine-blade tips. As the turbine turns in the flow path, the pickup coil produces an alternating current or signal. The frequency of this signal is proportional to flow rate. This alternating-current

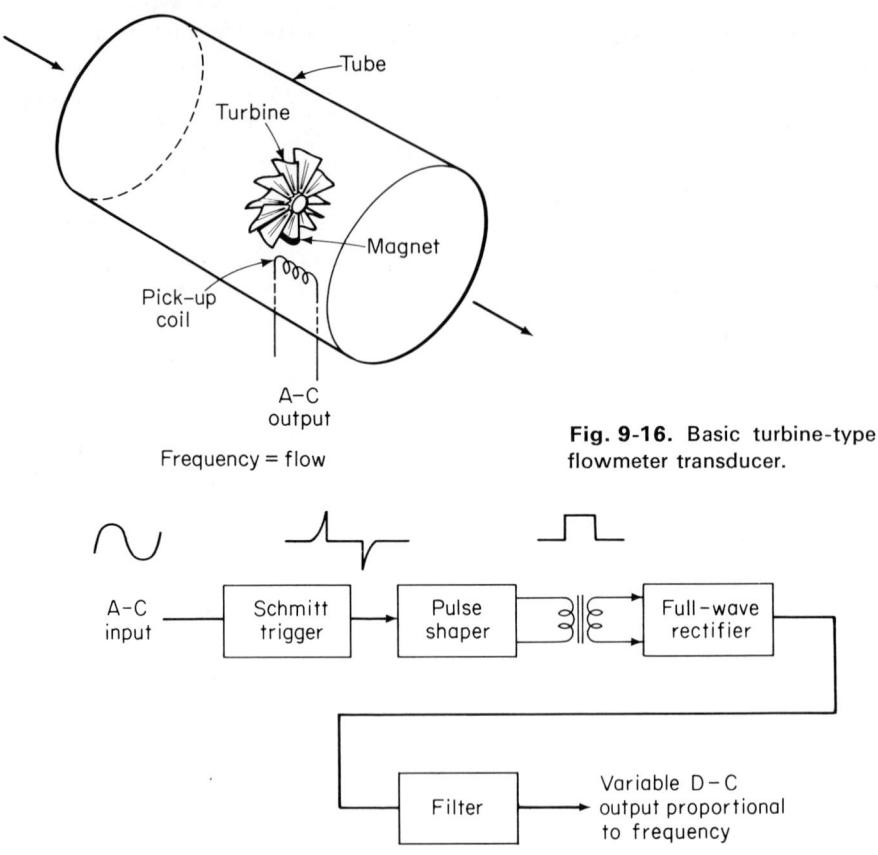

Fig. 9-16. Basic turbine-type flowmeter transducer.

Fig. 9-17. Basic frequency-to-analog conversion circuit used with tachometer-type transducers.

signal could be applied directly to a frequency-measuring device. However, it is more common to convert the a-c signal to a d-c voltage that is proportional to frequency.

There are several circuits used for this frequency-to-analog conversion. Most of the circuits involve a Schmitt trigger and a rectifier as the basic elements. Figure 9-17 shows such a circuit in simplified block form. Here the Schmitt trigger produces a pulse output for each cycle of a-c input. (Schmitt triggers are discussed further in the Appendix.) The width of the Schmitt trigger output pulses remains constant. Therefore, as frequency increases (indicating an increased flow), the on time of the trigger output increases. In turn, this increases the average d-c output from the rectifier. If this current were used with a flowmeter transducer, the net result would be a variable d-c voltage whose amplitude is proportional to the rate of flow.

Tachometer-type transducers operate in a manner similar to that of turbine flowmeters, except that their function is to measure actual *frequency* rather than to produce an output where frequency is proportional to some other value. Therefore, tachometer-type transducers are used with oscilloscopes (where the frequency can be measured as described in Chapter 5) but not with meters.

Usually, tachometer-type transducers are used to measure mechanical rotation, as are conventional tachometers. Tachometer-type transducers consist of a pickup coil and magnet arrangement similar to that of Figure 9-16. However, the magnet is driven past the coil by mechanical rotation and produces an a-c output, usually at the rate of 1 cycle/rotation.

Magnetic-induction transducers are particularly effective as self-generating devices. Their basic operating principle is the same as that of sound-powered telephones. A magnetic-induction transducer consists essentially of a high-density permanent magnet with a coil. Any movement of the magnet (in relation to the coil) will produce a proportional output from the coil. The magnet is driven by the mechanical force to be measured. One of the main advantages of the magnetic-induction type of transducers is that the magnet can be suspended, thus eliminating the problems of friction and sticking. A typical magnetic-induction type of self-generating transducer circuit is shown in Figure 9-18.

Piezoelectric-type transducers are also in common use as self-generating sensing devices. The piezoelectric principle is based on the fact that certain materials (such as quartz crystal) will produce a voltage when subjected to pressure. As shown in Figure 9-19, the piezoelectric material is placed between two plates, and the resultant voltage is applied to the meter or oscilloscope.

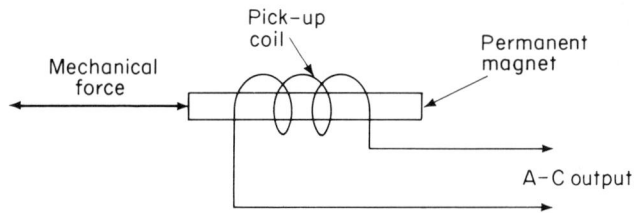

Amplitude proportional to displacement.
Frequency equal to mechanical force frequency.

Fig. 9-18. Basic magnetic-induction type transducer.

A prime example of this is the piezoelectric accelerometer, often used for vibration measurement. Because of the low voltages produced, the output of piezoelectric transducers usually requires considerable amplification to operate with a meter or oscilloscope.

Solar-cell transducers are often used with test equipment designed to mea-

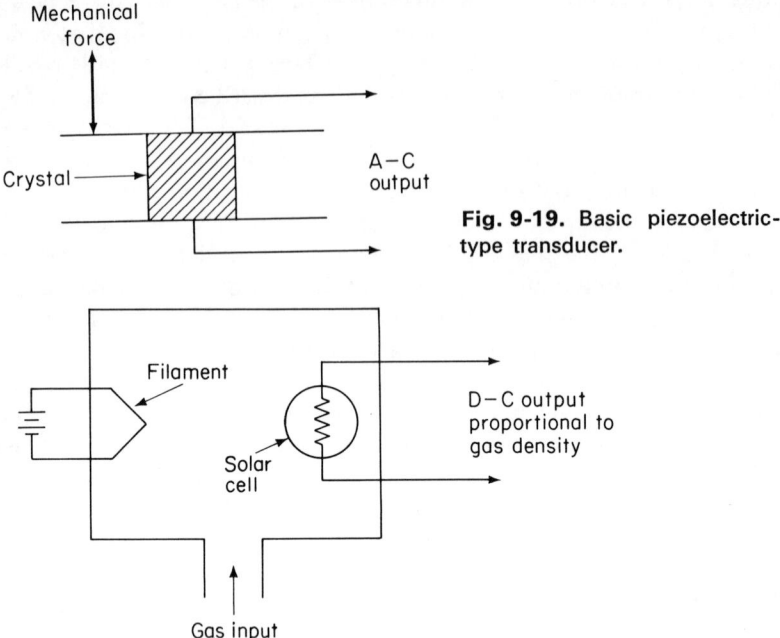

Fig. 9-19. Basic piezoelectric-type transducer.

Fig. 9-20. Gas detector; typical example of solar cell transducer.

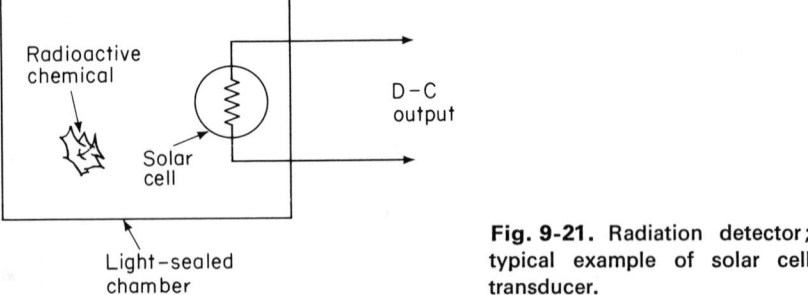

Fig. 9-21. Radiation detector; typical example of solar cell transducer.

sure the presence of nuclear radiation or gas density. In either case, the solar cell produces a voltage output proportional to the amount of light applied to the cell surface.

In the case of a gas detector, a metal filament is placed near the solar cell, and a current is passed through the filament. (See Figure 9-20). This current causes the filament to heat and glow, producing a voltage output from the solar cell. A metal that will glow with greater intensity in the presence of various gases is chosen. (One good example of this is platinum, which will glow with increased brilliance in the presence of hydrocarbons, such as those

Inductance- and Capacitance-Changing Transducers   335

of gasoline fumes.) Such gas-detector transducers are used in industrial test equipment. The output of the solar cell is applied to a d-c meter or oscilloscope (sometimes through a d-c amplifier).

In the case of a radiation detector, a chemical or mineral is placed near the solar cell (Figure 9-21). In the presence of nuclear radiation, the chemical gives off light. The intensity of the light and solar-cell output are proportional to the amount of radiation.

9-13. INDUCTANCE- AND CAPACITANCE-CHANGING TRANSDUCERS

Inductance- and capacitance-changing transducers are usually used in conjunction with a bridge circuit operating on alternating current.

The *linear transformer* is a good example of an inductance-changing transducer used to measure mechanical position (instead of the resistance-changing transducer previously described). As shown in Figure 9-22, the transducer is essentially a transformer with a secondary winding that can be rotated in relation to the primary. A reference voltage is applied to the primary, and the mechanical driving force is applied to rotate the secondary winding, whose output is applied to the meter or oscilloscope.

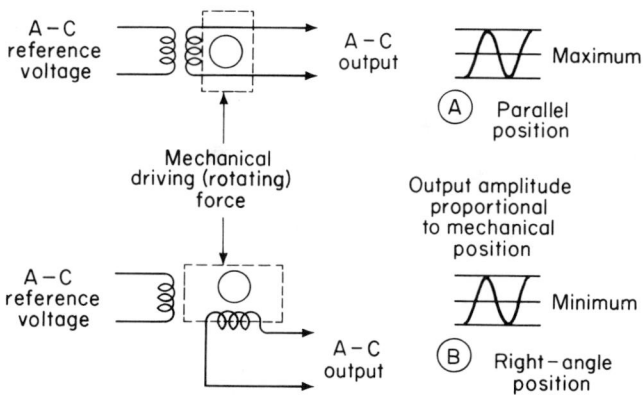

Fig. 9-22. Linear-transformer type of inductance-changing transducer.

As the secondary is rotated toward the parallel position (Figure 9-22A), the output voltage increases. Conversely, the output will decrease when the secondary winding is moved toward the right-angle position (Figure 9-22B). Therefore, the output of the transducer is proportional to mechanical position.

The *liquid-level-measuring* transducer is a good example of a capacitance-changing transducer. With such transducers, the liquid itself forms one plate of the capacitor (Figure 9-23). Since the capacitance value of any capacitor is determined both by the plate size and spacing, any variation in liquid level will cause a corresponding variation in capacitance. The capacitor formed by the liquid and the transducer plate is usually connected in a bridge circuit. The matching leg of the bridge is formed by a variable capacitor. This permits the bridge circuit to be zeroed at minimum or maximum capacity. Either way, the output of the bridge is proportional to liquid level.

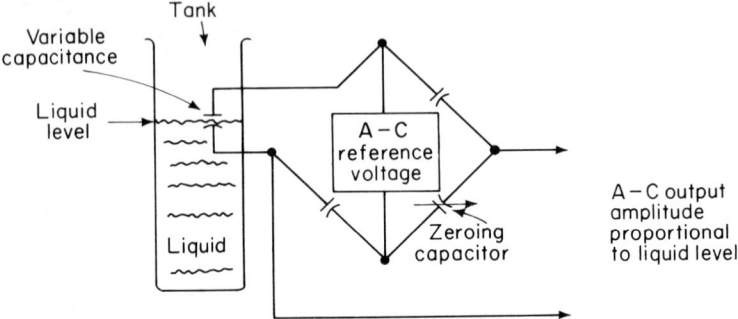

Fig. 9-23. Liquid-level type of capacitance-changing transducer.

10 WAVE ANALYSIS INSTRUMENTS

Much of today's test equipment is involved with the analysis or measurement of wave-forms. As discussed in Chapter 5, the wave-forms of square waves, pulses, or any other type of signal can be observed on the face of an oscilloscope screen. This permits the time duration, frequency, amplitude, shape, and so on, of the wave-form to be measured directly. It is also possible to use an oscilloscope to measure distortion and modulation of wave-forms.

In addition to an oscilloscope, wave-forms are measured with three test instruments: *distortion analyzers*, *wave analyzers*, and *spectrum analyzers*.

The *time domain reflectometer* uses the technique of wave-form analysis to determine characteristics of transmission lines, cables, and so on.

In this chapter, we will discuss operation of these devices and the basic techniques used for the measurement and analysis of wave-forms.

It is assumed that the readers are already familiar with basic theory and operation of oscilloscopes. If not, the readers should study Chapter 5 before starting this chapter.

10-1. DISTORTION MEASUREMENT

There are many techniques for measuring distortion of signals of any frequency. (It should be noted that distortion may occur in signals at any frequency. However, distortion of audio-frequency signals is usually of the greatest concern, since audio-frequency distortion is readily detected by the human ear.) An oscilloscope (Chapter 5) is the most effective tool for distortion measurement. There are four basic methods of distortion measurement that involve the use of an oscilloscope: analysis of sine-wave patterns, analysis of square-wave patterns, measurement of harmonic distortion (fun-

338   Wave Analysis Instruments

damental suppression method), and measurement of intermodulation distortion.

10-1.1. Measuring Distortion by Sine-Wave Analysis

The procedure for measurement of distortion using sine-waves is essentially *visual signal tracing*. Such a procedure is illustrated in Figure 10-1, which shows visual signal tracing through an audio amplifier (Chapter 7).

In such a procedure, a sine wave is introduced into the input by means of an external generator (an audio generator in this case—refer to Chapter 4). The amplitude and wave-form of the input signal are measured on the oscilloscope. The oscilloscope probe is then moved to the input and output of each stage, in turn, until the final output (usually at a loudspeaker or output transformer) is reached. The gain of each stage is measured as a voltage on the oscilloscope vertical scale. In addition, it is possible to observe *any change in wave-form* from that applied to the input by the external generator. Thus stage gain and distortion (if any) are established quickly with an oscilloscope.

When measuring distortion, the primary concern is *deviation* of the amplifier (or stage) output wave-form from the input wave-form. If there is no change (except in amplitude), there is no distortion. If there is a change in the wave-form, the nature of the change will often reveal the cause of distortion.

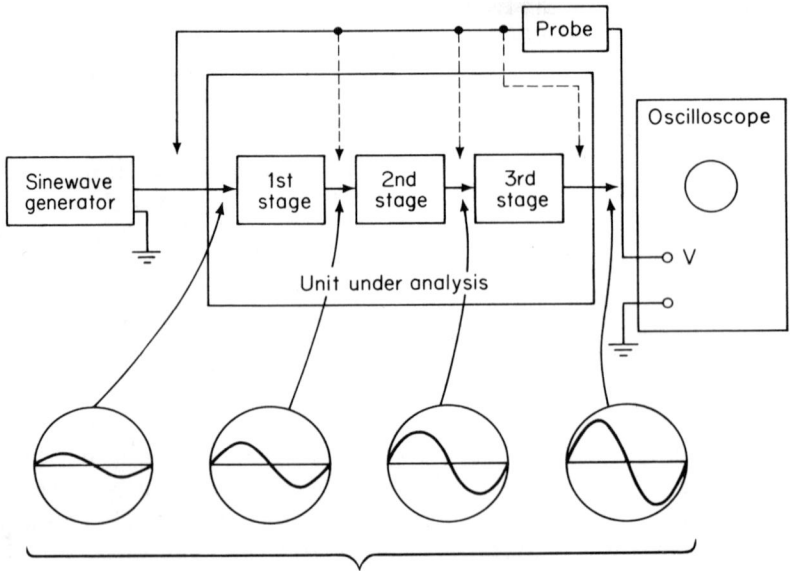

Amplitude increases with each stage; waveform remains substantially the same

Fig. 10-1. Measuring distortion by sine-wave analysis.

In practice, analyzing sine waves to pinpoint distortion is a difficult job requiring considerable experience. Unless the distortion is severe, it may pass unnoticed. Therefore, sine waves are best used where *harmonic distortion* or *intermodulation distortion* analyzers are combined with oscilloscopes for distortion analysis. If an oscilloscope is to be used alone, square waves provide the best basis for distortion analysis.

10-1.2. Measuring Distortion by Square-Wave Analysis

The procedure of measuring distortion by means of square waves is essentially the same as for sine waves. Distortion analysis is more effective with square waves because of their high odd-harmonic content, and because it is easier to see a deviation from a straight line with sharp corners than from a curving line. As in the case of sine-wave distortion measurement, square waves are introduced into the amplifier (or stage) input, while the output is monitored on an oscilloscope. The primary concern is deviation of the amplifier (or stage) output wave-form from the input wave-form (which is also monitored on the oscilloscope). If the oscilloscope has the dual-trace feature, the input and output can be monitored simultaneously.

If there is a change in wave-form, the nature of the change will often reveal the cause of distortion. For example, poor high-frequency response will round the edge of the output square wave, as shown in Figure 10-2.

The third, fifth, seventh, and ninth harmonics of a clean square wave are emphasized. Therefore, if an amplifier (or stage) passes a given audio frequency and produces a clean square-wave output, it is reasonable to assume that the frequency response is good up to at least 9 times the fundamental frequency. For example, if an amplifier passes a clean square wave at 3000 Hz, it shows a good response up to 27 kHz, which is beyond the top limit of the audio range.

In any form of square-wave analysis, the square waves must be clean. (Refer to Chapter 4 for an analysis of pulse and square-wave characteristics.) In square-wave distortion alalysis, the output wave-form can be no better than the input wave-form.

10-1.3. Measuring Harmonic Distortion

No matter what amplifier circuit is used or how well the circuit is designed, there is always the possibility of odd or even harmonics (multiples of the basic frequency) being present with the fundamental signal frequency.

The effects of even and odd harmonics on the fundamental signal are shown in Figure 10-3. For example, a second harmonic (an even harmonic) will cause the positive half-cycle of a sine-wave signal to be a different shape than the negative half-cycle. When a fundamental and its third harmonic (an odd harmonic) are 180° out of phase, the resultant wave-form will be peaked. If

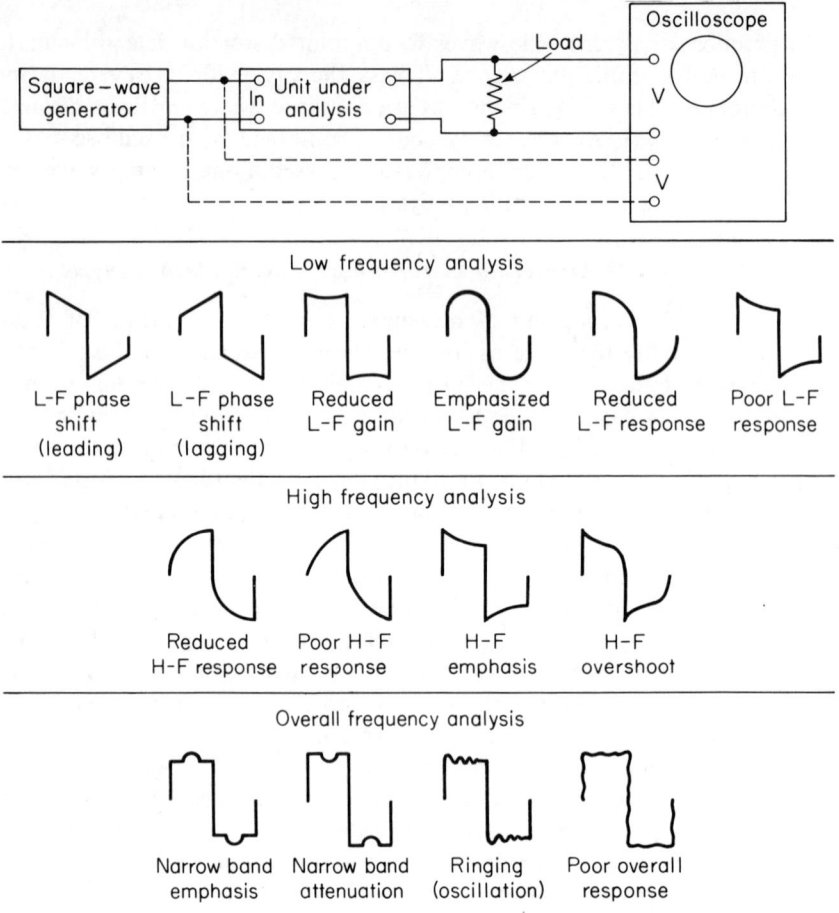

Fig. 10-2. Measuring distortion by square-wave analysis.

a fundamental and the same third harmonic are in phase, the resultant waveform will be flat on top or will sag.

Commercial harmonic distortion meters operate on the *fundamental suppression principle*. As shown in Figure 10-4, a sine wave is applied to the amplifier input, and the output is measured on a meter and/or oscilloscope. (Most harmonic distortion meters do not include an oscilloscope, but have provisions for connecting an external oscilloscope.)

After an initial measurement of the output signal, the output is applied through a filter that suppresses the fundamental frequency. Any output from the filter is then the result of harmonics. This output can also be displayed on the oscilloscope, where the signal frequency can be checked (to determine the harmonic content).

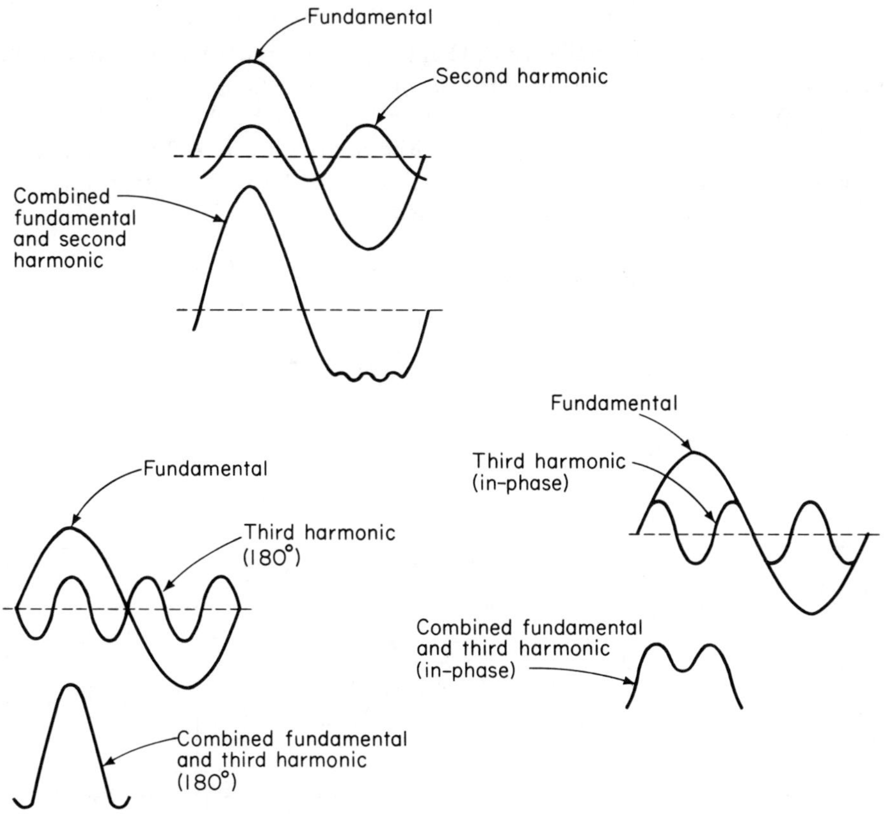

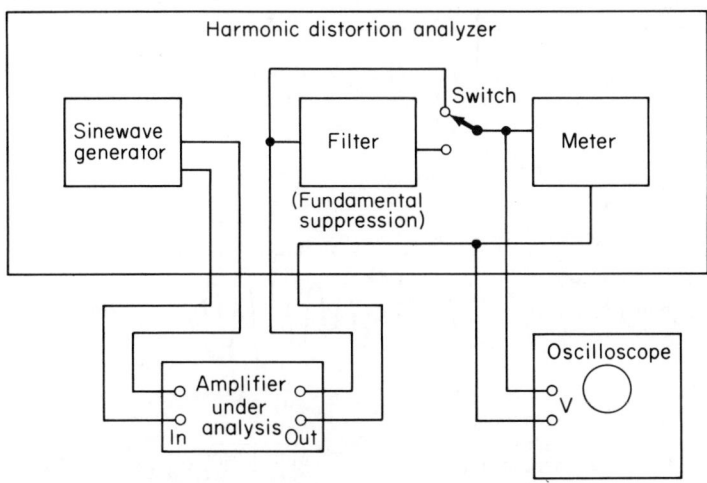

Fig. 10-3. Effects of harmonic distortion.

Fig. 10-4. Basic circuit for harmonic distortion analyzers.

For example, if the input were 1000 Hz and the output after filter were 3000 Hz, this would be a result of third harmonic distortion.

The percentage of harmonic distortion can also be determined by this method. For example, if the output without the filter is 100 mV and with the filter is 3 mV, this would indicate a 3 per cent harmonic distortion.

Total harmonic distortion is calculated by

$$\text{Harmonic distortion} = 100 \, \frac{\text{output before filtering}}{\text{output after filtering}}$$

10-1.4. Measuring Intermodulation Distortion

When two signals of different frequency are mixed in an amplifier (or other circuit), there is a possibility of the lower-frequency signal

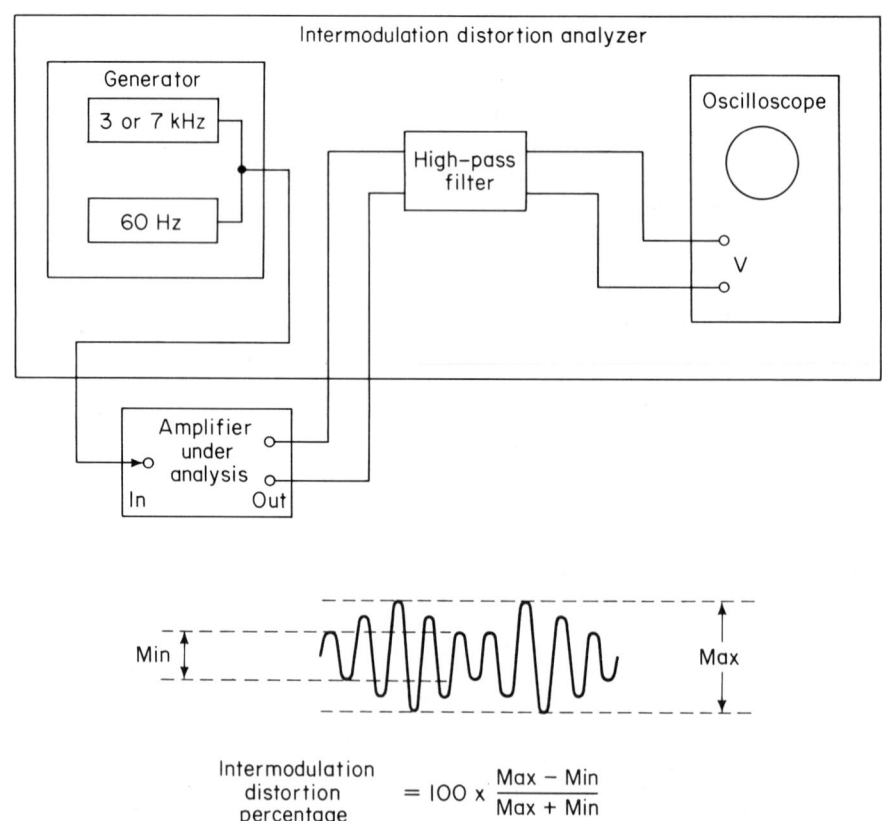

Fig. 10-5. Basic circuit for intermodulation distortion analyzers.

amplitude modulating the higher-frequency signal. This produces a form of distortion known as *intermodulation distortion*.

Commercial intermodulation distortion analyzers consist of a signal generator and high-pass filter, as shown in Figure 10-5. The signal-generator portion of the meter produces a high-frequency signal (usually in the range of 3–7 kHz) that is modulated by a low-frequency signal (usually 60-Hz).

The mixed signals are applied to the amplifier input. The amplifier output is connected through a high-pass filter to the oscilloscope vertical channel. The high-pass filter removes the low-frequency (60-Hz) signal. Therefore, the only signal appearing on the oscilloscope vertical channel should be the high-frequency (3–7-kHz). If any 60-Hz is present on the display, it is being passed through as modulation on the 3–7-kHz signal.

Intermodulation distortion is calculated by

$$\text{Intermodulation distortion} = 100 \frac{\text{MAX} - \text{MIN}}{\text{MAX} + \text{MIN}}$$

where MAX and MIN are dimensions shown in Figure 10-5.

10-2. MODULATION MEASUREMENT

As oscilloscope can be used to display the carrier of an amplitude-modulated wave. There are two basic methods: direct measurement of the modulation envelope and conversion of the envelope to a trapezoidal pattern. The trapezoidal method is the most effective since it is easier to measure straight-line dimension than curving dimensions. Also, nonlinearity in modulation can be checked easily with the straight-line trapezoid. With either method, the percentage of modulation can be calculated from the dimension of the modulation pattern.

10-2.1. Direct Measurement of the Modulation Envelope

The equipment is connected as shown in Figure 10-6 for direct measurement of the modulation envelope. If the vertical channel response of the oscilloscope is capable of handling the transmitter output frequency, the output is applied through the oscilloscope vertical amplifier. If not, the transmitter output must be applied directly to the vertical deflection plates of the oscilloscope cathode-ray tube. When the transmitter is amplitude-modulated with a sine wave, the resultant wave-form is similar to that shown in Figure 10-6. If the transmitter is operating properly, the modulation envelope should reflect the sine-wave modulation signal. By measuring the vertical dimensions MAX and MIN shown in Figure 10-6, the percentage of modulation can be calculated.

344   Wave Analysis Instruments

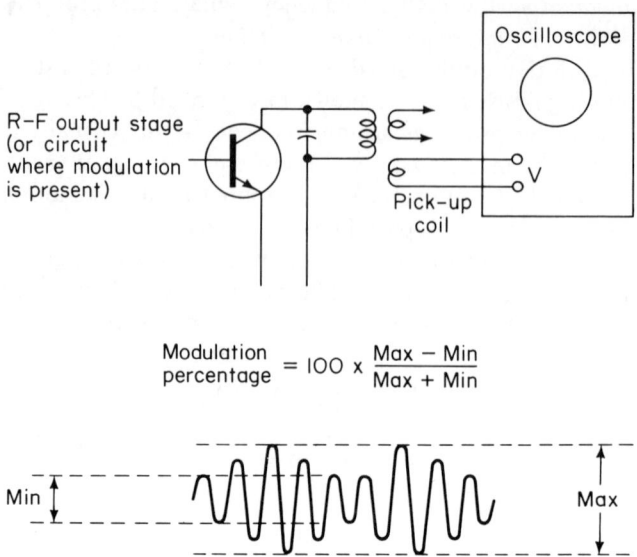

$$\text{Modulation percentage} = 100 \times \frac{\text{Max} - \text{Min}}{\text{Max} + \text{Min}}$$

Fig. 10-6. Direct measurement of modulation envelope.

10-2.2. Trapezoidal Measurement of the Modulation Envelope

The equipment is connected as shown in Figure 10-7 for trapezoidal measurement of the modulation envelope. With this method, the oscilloscope amplifiers are not used. Instead, both the horizontal and vertical connections are made directly to the oscilloscope cathode-ray tube. When the transmitter is amplitude-modulated with a sine wave, the resultant waveform is similar to that shown in Figure 10-7.

The oscilloscope display width is adjusted by means of resistor R_1. The height of the oscilloscope display is adjusted by varying the coupling between the pickup coil and the output tank of the transmitter.

Figures 10-7C, D, and E show typical patterns for 50 per cent modulation, 90–95 per cent modulation, and overmodulation (over 100 per cent modulation) respectively. By measuring the vertical dimensions MAX and MIN shown in Figure 10-7F, the percentage of modulation can be calculated.

10-3. WAVE ANALYZERS

Distortion and other characteristics of wave-forms are often measured by means of a wave analyzer, rather than a distortion analyzer. A wave analyzer can be thought of as a finite bandwidth "window" filter that can be tuned throughout a particular frequency range, as shown in Figure

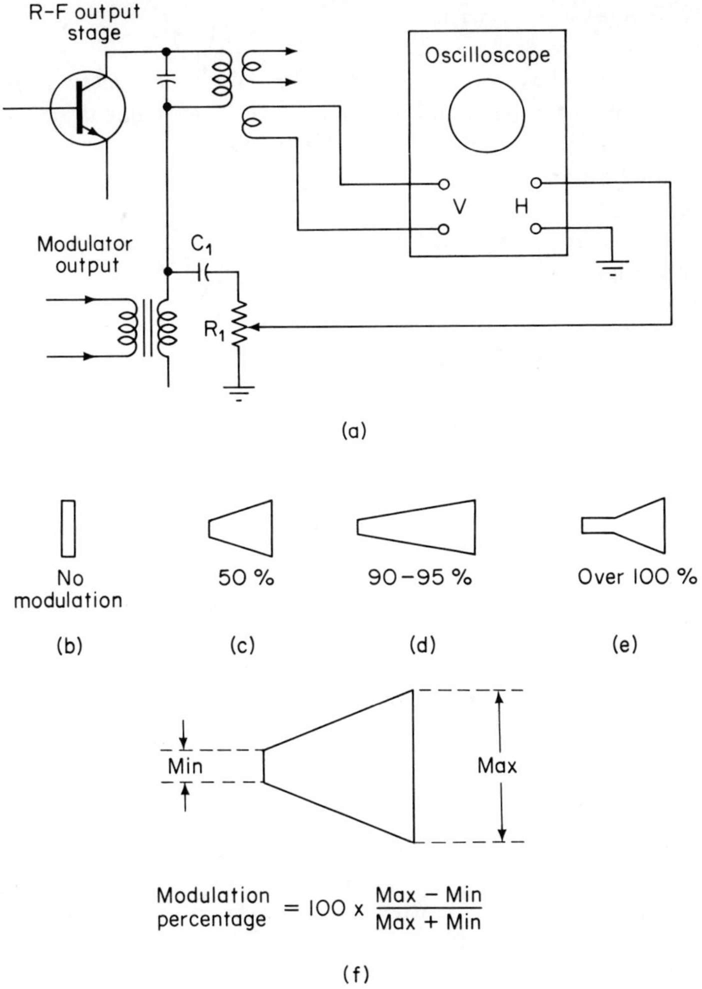

(a)

(b) No modulation

(c) 50 %

(d) 90-95 %

(e) Over 100 %

$$\text{Modulation percentage} = 100 \times \frac{\text{Max} - \text{Min}}{\text{Max} + \text{Min}}$$

(f)

Fig. 10-7. Trapezoidal measurement of modulation envelope.

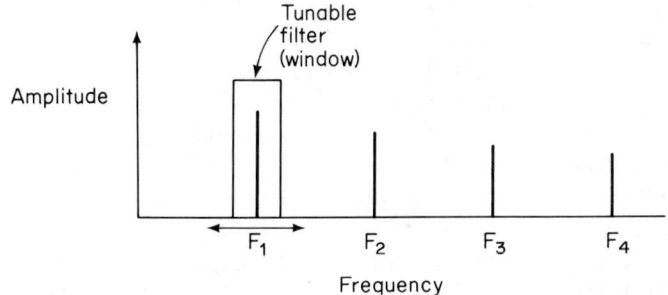

Fig. 10-8. Basic wave analyzer function (tunable filter) (Hewlett-Packard).

10-8. Signals located on the frequency spectrum will be selectively measured as they are framed by the "window." Thus, for a particular signal, the wave analyzer can indicate its frequency (window position) and amplitude. Amplitude is read on an analog meter; frequency is read on either a mechanical or electronic readout.

In its simplest form, a wave analyzer is a frequency-selective voltmeter, as shown in Figure 10-9. In operation, the filter is tuned by means of a cali-

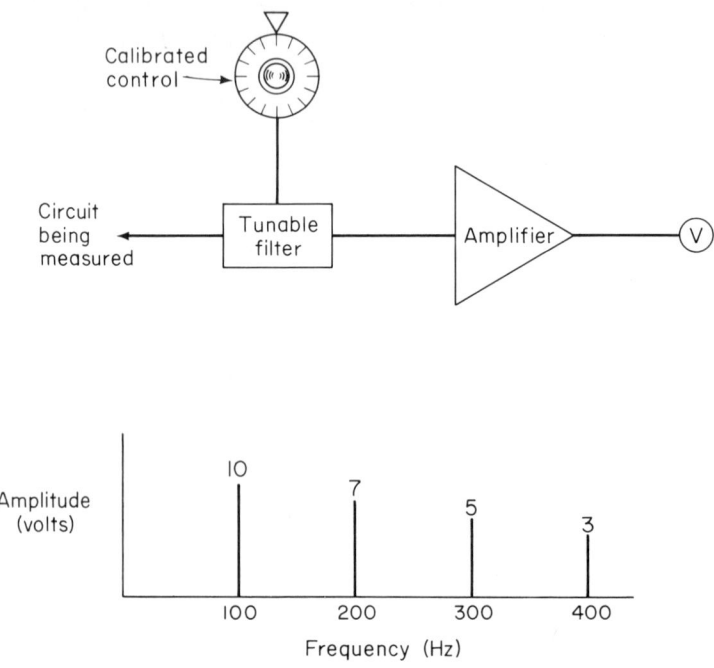

Fig. 10-9. Basic wave analyzer circuit.

brated front-panel control to the desired frequency. For example, assume that harmonic distortion of a 100-Hz wave-form is to be measured. The filter would then be tuned to 200 Hz, 300 Hz, 400 Hz, and so on, in turn. The output voltage is then measured at each harmonic frequency. If there is no output voltage measured at any frequency (except the fundamental) there is no harmonic distortion.

An improved version of the wave analyzer is shown in Figure 10-10. Note that this circuit is similar to that of a heterodyne receiver. The wave-form to be measured is mixed with a variable frequency signal. This signal source is tuned by means of a calibrated front-panel control. In effect, the input signal is heterodyned to a higher intermediate frequency by an internal local oscillator (tuned by the calibrated dial). Filtering is performed by the inter-

Wave Analyzers

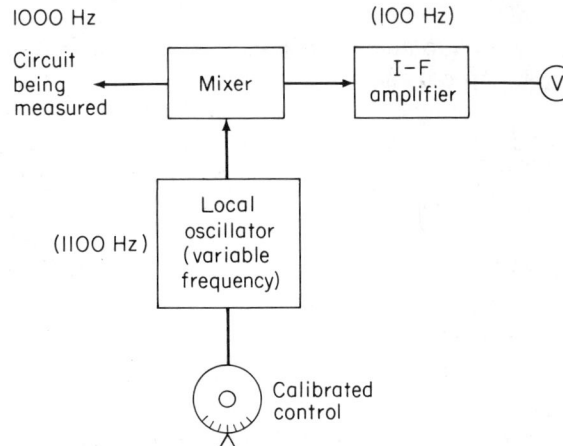

Fig. 10-10. Improved wave analyzer circuit (heterodyne type).

mediate-frequency amplifier so that the wave analyzer's passband remains constant regardless of the tuning.

Tuning the local oscillator shifts the various input wave-form frequency components into the passband of the intermediate-frequency amplifier. For example, if the intermediate-frequency were 100 Hz and the input waveform to be measured were 1000 Hz, the local oscillator would be tuned to 1100 Hz (1000 + 100 Hz). The output of the intermediate-frequency amplifier is rectified and supplied to the metering circuit.

10-3.1. Wave Analyzer Versus Distortion Analyzer

The choice between a wave analyzer and a distortion analyzer depends on the kind of measurement desired. The wave analyzer is a narrowband *pass* filter that is tuned to select and measure the strength of individual components of a wave-form, one at a time. In contrast to the frequency selection of a wave analyzer, the distortion analyzer is a narrowband *rejection* filter that, when properly tuned, removes the fundamental frequency, so that the amplitude of the remaining components can be measured all at once.

The distortion analyzer is used for fast, quantitative measurements of *total distortion*, whereas the wave analyzer provides detailed information concerning *each harmonic* and intermodulation product.

10-3.2. Typical Wave Analyzer Measurements

Wave analyzer measurements can be divided into three broad areas:

Selective measurements of signals with large differences in level. For example, distortion analysis, measurements of low-level signals very close in frequency to much larger signals, or identification of low-level signals obscured by broadband noise.

Determination of noise characteristics by use of the well-defined bandwidth of a wave analyzer. Noise power spectral density (noise signals over a given bandwidth) can also be measured over the entire frequency range of the instrument.

Frequency response testing, using the tracking output (output of the local oscillator or another signal source locked in frequency to the local oscillator) as an excitation source to make tests at ultralow threshold levels. The wave analyzer's high sensitivity eliminates harmonics, spurious responses, and ground loop effects.

10-3.3. Wave Analyzer Characteristics

There are many wave analyzers available for laboratory work. Some of these are the simple tunable filter and broadband voltmeter. Other wave analyzers incorporate such features as automatic range-changing circuits, automatic frequency control (AFC), electronic sweeping of the frequency range, digital readouts, selectable bandwidths, and outputs suitable for recorders. Operation of these circuits is the same as for other test instruments discussed elsewhere in this handbook and will not be repeated here. However, there are two characteristics common to all wave analyzers that should be discussed. These are *selectivity* and *dynamic range*.

10-3.3.1. Wave Analyzer Selectivity

The selectivity of a wave analyzer is its greatest asset and a most important specification. Selectivity is usually defined by the 3-dB bandwidth and the *shape factor* of the bandpass. Figure 10-11 shows the shape factor for different wave analyzers. The smaller the shape factor number, the more selective the instrument will be.

Note the passband (dotted line) in Figure 10-11. Specifying just the 3-dB bandwidth (bandpass C) can be misleading. However, specifying the *ratio of the two selected bandwidths* (usually -3 dB or -6 dB and the -60-dB points) provides further definition of the sharpness of the skirt (solid line in Figure 10-11).

A shape factor so defined gives

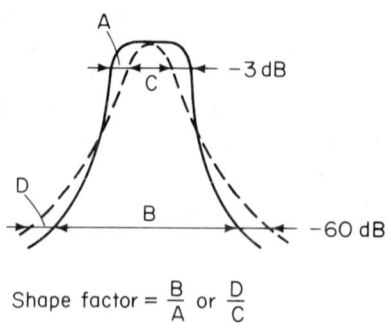

Fig. 10-11. Definition of wave analyzer shape factor (Hewlett-Packard).

a true picture of the bandpass. High-quality laboratory wave analyzers have shape factors as low as 2:1. These are especially useful in making critical frequency measurements where signal density is high.

10-3.3.2. Wave Analyzer Dynamic Range

The dynamic range of a wave analyzer defines the range of the smallest to the largest signal the instrument can accommodate simultaneously. Some wave analyzers are capable of an 80–90-dB range (voltage ratios of 10^{-4} or 10^{-5}); that is, they must be capable of measuring and displaying a 1-V signal at one frequency and then measuring a signal of a few microvolts at a slightly different frequency.

Figure 10-12 shows the relationship between attenuation and dynamic

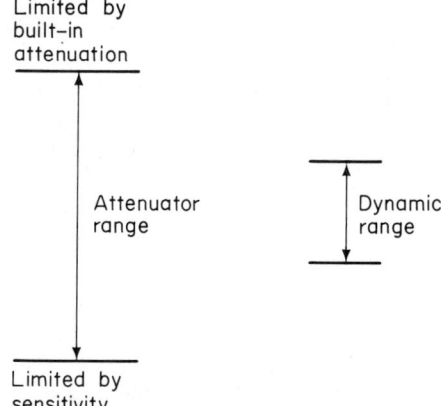

Fig. 10-12. Relationship between attenuation range and dynamic range of wave analyzers.

range. The high end of the attenuator range is limited by the amount of attenuation built in. The low end of the attenuator range is set by the instrument's sensitivity. (A highly sensitive wave-analyzer amplifier will require some minimum value of attenuation to prevent the meter circuits from being overdriven.) Dynamic range is limited by nonlinearities in the analyzer's amplifier and by background noise.

10-4. SPECTRUM ANALYZERS

Wave-forms (particularly those resulting from pulse or frequency modulation) are often measured by means of a spectrum analyzer. The basic circuit for a spectrum analyzer is shown in Figure 10-13. The spectrum analyzer is essentially a narrowband receiver, electrically tuned over a given frequency range, combined with an oscilloscope (Chapter 5). As shown, the local oscillator is swept over a given range of frequencies by the

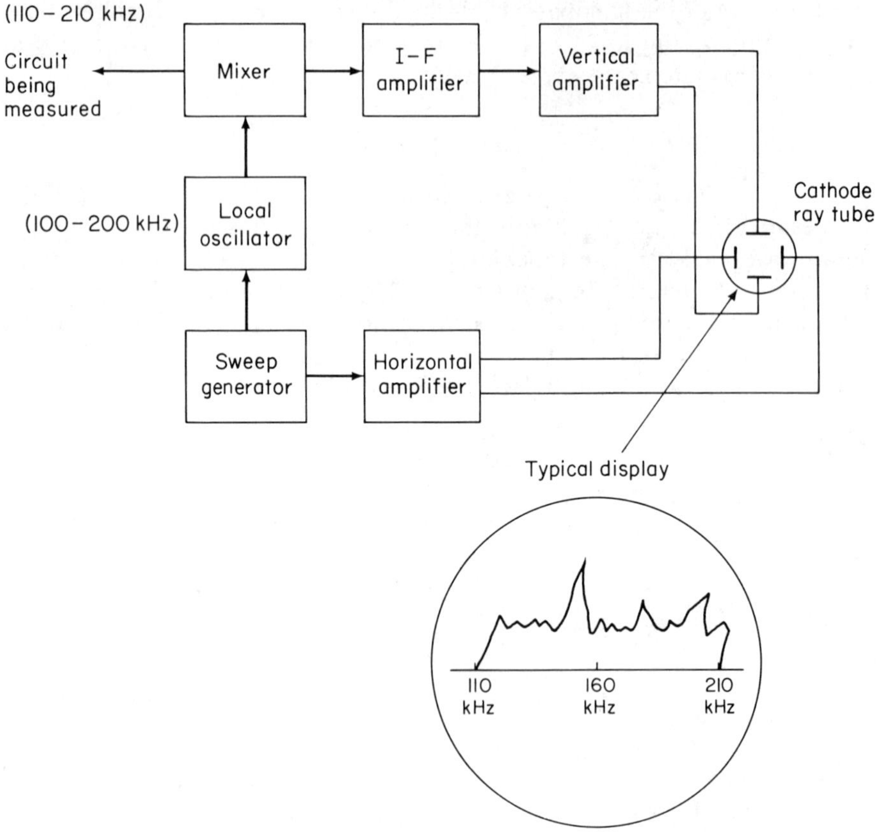

350 Wave Analysis Instruments

Fig. 10-13. Basic spectrum analyzer circuit.

sweep generator (Chapter 4). Since the intermediate-frequency amplifier passband remains fixed, the input circuits and mixer are swept over a corresponding range of frequencies.

For example, if the intermediate frequency were 10 kHz and the local oscillator swept a band of frequencies from 100 to 200 kHz, the input would be capable of receiving wave-forms in the range of 110–210 kHz. The output of the intermediate-frequency amplifier is further amplified and supplied to the vertical deflection plates of the cathode-ray tube.

The cathode-ray-tube horizontal plates obtain their signal from the same sweep generator used to tune the local oscillator. Therefore, the length of the horizontal sweep should represent the total sweep spectrum. For example, if the sweep is from 110 to 210 kHz, the left-hand end of the horizontal trace represents 110 kHz and the right-hand end represents 210 kHz. Any point along the horizontal trace will represent a corresponding frequency. For example, the midpoint on the trace would represent 160 kHz.

10-4.1. Time-Amplitude Displays Versus Frequency-Amplitude Displays

To gain a better understanding of the usefulness and application of a spectrum analyzer, it is important to understand what the spectral display is and how to interpret it.

Figure 10-14 shows the relationship of time-amplitude and frequency-

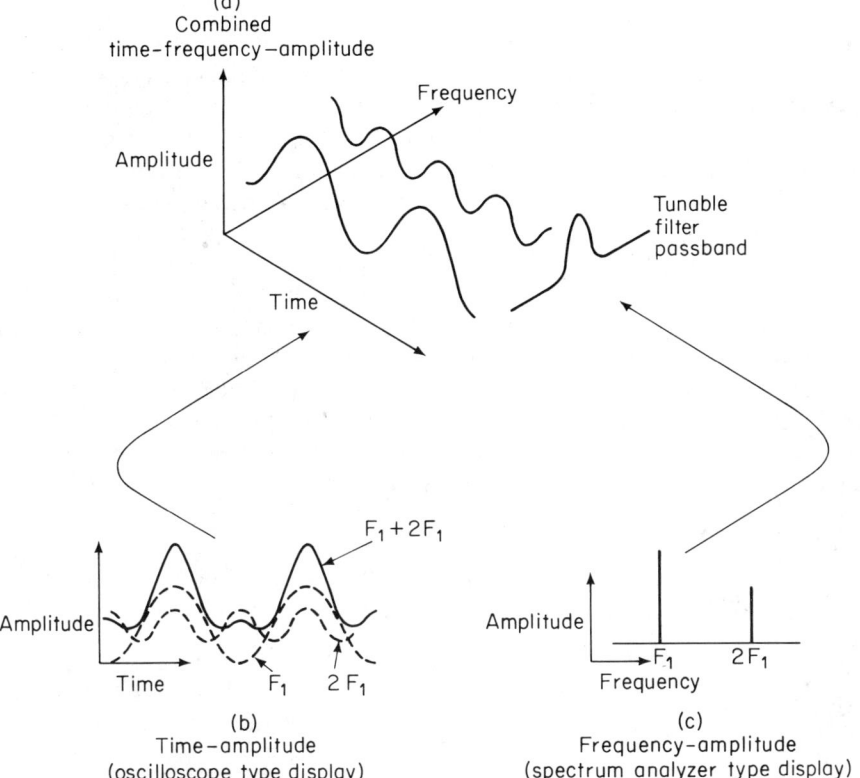

Fig. 10-14. Relationship of time-amplitude and frequency-amplitude displays (Hewlett-Packard).

amplitude displays. As discussed in Chapter 5, the conventional oscilloscope produces a time-amplitude display. For example, pulse rise time and width are read directly on the X axis of the cathode-ray tube. A spectrum analyzer produces a frequency-amplitude display where signals (unmodulated, AM, FM, or pulse) are broken down into their individual *frequency* components and displayed on the cathode-ray-tube X axis.

In Figure 10-14A, both the time-amplitude and frequency-amplitude coordinates are shown together. The example given is that showing the addition of a fundamental frequency and its second harmonic.

In Figure 10-14B, only the time-amplitude coordinates are shown. The solid line (which is the composite of fundamental f_1 and second harmonic $2f_1$) is the only display that would appear on a conventional oscilloscope.

In Figure 10-14C, only the frequency-amplitude coordinates are shown. Note how the components (f_1 and $2f_1$) of the composite signal are clearly seen here.

10-4.2. Practical Spectrum Analysis

Spectrum analyzers are often used in conjunction with Fourier analysis and transform analysis. Both of these techniques are quite complex and beyond the scope of this publication. Instead, we will concentrate on the practical aspects of spectrum analyzers; that is, what display will result from a given input signal and how the display can be interpreted.

10-4.2.1. Unmodulated Signal Displays

If the spectrum analyzer's local oscillator sweeps through an unmodulated or *continuous-wave* (CW) signal slowly, the resulting response on the analyzer screen is simply a plot of the analyzer's intermediate-frequency amplifier passband. A pure CW signal will by definition have energy at only one frequency and should therefore appear as a single spike on the analyzer screen (Figure 10-15). This will occur provided the total sweep width or so-

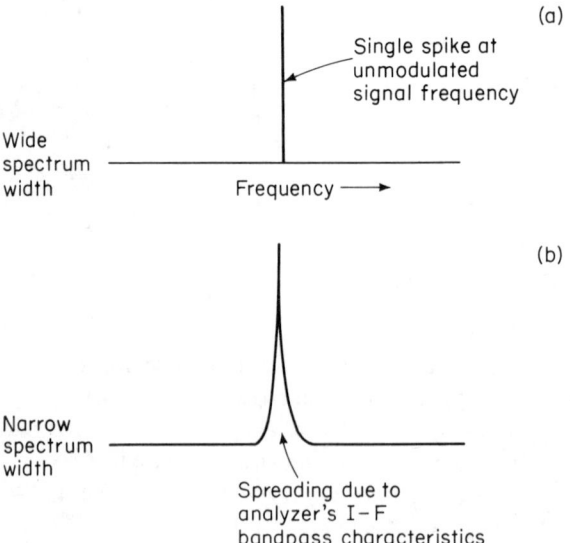

Fig. 10-15. Unmodulated (CW) spectrum-analyzer display.

called "spectrum width" is wide enough compared to the intermediate-frequency bandwidth in the analyzer. As spectrum width is reduced, the spike response begins to spread out until the intermediate-frequency bandpass characteristics begin to appear, as shown in Figure 10-15.

10-4.2.2. Amplitude-Modulated Signal Displays

A pure sine wave represents a single frequency. Its spectrum is shown in Figure 10-16A and is the same as the unmodulated signal display

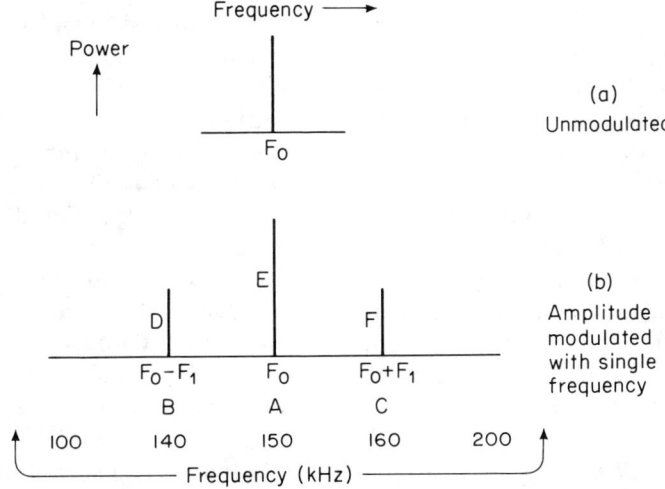

Position of A = carrier frequency (150 kHz)
Distance between B and A, or A and C = frequency dispersion or modulation frequency (10 kHz)
Ratio of D to E, or F to E = one half percentage of modulation

Fig. 10-16. Frequency spectrum for single-tone amplitude-modulated carrier, and rules for interpretation.

of Figure 10-15 (a single vertical line). The height of line F_0 represents the power contained in the single frequency. Figure 10-16B shows the spectrum for a single sine-wave frequency F_0, amplitude-modulated by a second sine wave F_1. In this case, two *sidebands* are formed, one higher than and one lower than the frequency F_0. These sidebands correspond to the sum and difference frequencies, as shown. If more than one modulating frequency is used (as is the case with most practical amplitude-modulated signals), two sidebands are added for *each frequency*.

Note that if the frequency, spectrum width, and vertical response of the analyzer are calibrated (as they are with any modern laboratory instrument) it is possible to find (1) the carrier frequency, (2) the modulation frequency,

(3) the modulation percentage, and (4) the nonlinear modulation (if any) and incidental FM (if any).

An amplitude-modulated spectrum display can be interpreted as follows:

The *carrier frequency* is determined by the position of the center vertical line F_0 on the X-axis. For example, if the total spectrum is from 100 to 200 kHz and F_0 is in the center, the carrier frequency is 150 kHz.

The *modulation frequency* is determined by the position of the sideband lines $F_0 - F_1$ or $F_0 + F_1$ on the X axis. For example, if sideband $F_0 - F_1$ is at 140 kHz and F_0 is at 150 kHz, then the modulating frequency is 10 kHz. Under these conditions, the upper sideband $F_0 + F_1$ should be 160 kHz. The distance between the carrier line F_0 and either sideband is sometimes known as the *frequency dispersion* and is equal to the modulation frequency.

The *modulation percentage* is determined by the ratio of the sideband amplitude to the carrier amplitude. The amplitude of either sideband with respect to the carrier voltage is *one half of the percentage of modulation*. For example, if the carrier amplitude is 100 mV and either sideband is 50 mV, this indicates 100 per cent modulation. If the carrier amplitude is 100 mV and either sideband is 33 mV, this indicates 66 per cent modulation.

Nonlinear modulation is indicated when the sidebands are of unequal amplitude or are not equally spaced on both sides of the carrier frequency. Unequal amplitude indicates nonlinear modulation that results from a form of distortion. Unequal frequency dispersion of the sidebands is usually a result of undesired *frequency modulation* combined with the amplitude modulation.

Incidental FM is indicated by any shift in the vertical signals along the X axis. For example, any horizontal "jitter" of the signals indicates rapid frequency modulation of the carrier.

The rules for interpreting amplitude-modulated spectrum-analyzer displays are summed up in Figure 10-16.

In practical applications, carrier signals are often amplitude-modulated at many frequencies simultaneously. This results in many sidebands (two for each modulating frequency) on the display. To resolve this complex spectrum, the operator should make sure that the analyzer bandwidth is less than the *lowest* modulating frequency or less than the *difference* between any two modulating frequencies, whichever is the smaller.

Overmodulation will also produce extra sideband frequencies. The spectrum for overmodulation is very similar to multifrequency modulation. However, overmodulation is usually distinguished from multifrequency modulation by the facts that (1) the spacing between overmodulated sidebands is *equal*, while multifrequency sidebands may be arbitrarily spaced (unless the modulating frequencies are harmonically related), and (2) the amplitude of overmodulated sidebands decreases progressively out from the carrier, but the amplitude of multifrequency-modulated signals is determined by the modulation percentage of each frequency and can be arbitrary.

10-4.2.3. Frequency-Modulated Signal Displays

The mathematical expression for a frequency-modulated wave-form is long and complicated, involving a special mathematical operator known as the *Bessel function*. However, the spectrum representation of the FM wave-form is straightforward. Such a representation is shown in Figure 10-17, which illustrates the frequency spectrum of a carrier that has been frequency-modulated by a single-tone (1-kHz) sine wave. Figure 10-17A

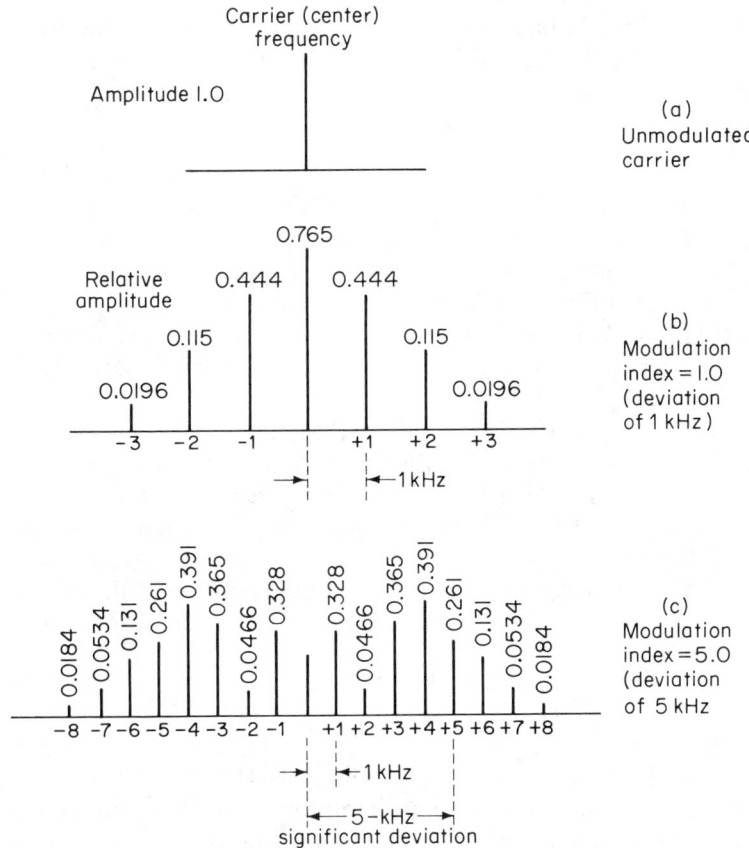

Fig. 10-17. Frequency spectrum for single-tone (1 kHz) frequency-modulated carrier.

shows the unmodulated carrier spectrum wave-form; Figure 10-17B shows the relative amplitudes of the wave-form when the carrier is frequency-modulated with a deviation of 1 kHz (modulation index of 1.0); Figure 10-17C shows the relative amplitudes of the wave-form when the carrier is frequency-modulated with a deviation of 5 kHz (modulation index of 5.0). Note that the term *modulation index* is given by

$$\text{Modulation index} = \frac{\text{Maximum frequency deviation}}{\text{Modulating frequency}}$$

The term *maximum frequency deviation* is theoretical. If a continuous-wave signal F_c is frequency-modulated at a rate f_r, it will *theoretically* produce an infinite number of sidebands. These sidebands will be located at intervals of $F_c \pm nf_r$, where $n = 1, 2, 3,$ and so on.

However, as a practical matter, only the sidebands containing *significant power* are usually considered. For a quick approximation of the bandwidth occupied by the significant sidebands, multiply the sum of the carrier deviation and the modulating frequency by two, or

$$\text{Bandwidth} = 2 \text{ (carrier deviation plus modulating frequency)}$$

As a rule of thumb, when using a spectrum analyzer to find maximum deviation of an FM signal, locate the sideband where the amplitude begins to drop and continues to drop as the frequency moves from the center. For example, in Figure 10-17C, sidebands 1, 2, 3, and 4 rise and fall, but sideband 5 falls, and all sidebands after 5 continue to fall. Since each sideband is 1 kHz from the center, this indicates a *practical or significant* deviation of 5 kHz. (It also indicates a modulation index of 5.0.)

As in the case of amplitude modulation, the center frequency and modulating frequency can be determined by the spectrum analyzer display.

The *carrier frequency* is determined by the position of the *center* vertical line on the X axis. (The centerline is not always the highest amplitude, as shown in Figure 10-17.)

The *modulating frequency* is determined by the position of the sidebands in relation to the center line or the distance between sidebands (frequency dispersion).

10-4.2.4. Pulse-Modulated Signal Displays

In the microwave region (Chapter 11), wave-form measurement by spectrum analysis has become an important aid. This is because microwave equipment (such as radar) is often pulse-modulated. Also, the increasing use of pulse time, pulse width, and pulse code modulation for telemetry and digital computer equipment has increased the applications of spectrum wave-form analysis techniques.

A perfect square wave is effectively made up of a fundamental sine wave plus an infinite number of odd-harmonic, in-phase sine waves that are progressively smaller in amplitude as the harmonic number increases. Theoretically, a 100-Hz square wave contains frequencies of 100, 300, 500, 700 Hz, and so on. In practice, it contains only a limited number of harmonics, since a perfect wave is impossible to obtain. It is the imperfect wave-forms produced by practical limitations of electronic circuitry that limit the number of harmonics produced, thereby making spectrum analysis a practical measurement process. However, a good square wave may contain frequencies up to the one-hundredth harmonic. A rectangular pulse is merely an extension of this principle and by changing the relative amplitudes and phases of harmonics, both odd and even, it is possible to plot an infinite number of wave shapes.

For example, a microwave radar transmission is modulated by short, rectangular pulses occurring at the PRF (pulse repetition frequency) of the radar set. Under these conditions, two distinct modulating components are present. One component consists of the PRF and its harmonics, and the other consists of the fundamental and odd-harmonic frequencies that make up the rectangular pulse. The spectrum analyzer effectively "unplots" square or rectangular wave-forms and presents the fundamental and each harmonic contained in the wave-form.

When a carrier is pulse-modulated, the carrier is periodically turned on and off. The on period is determined by the modulating pulse width; the off period is related to the pulse repetition rate or frequency. (In most practical applications, carriers are modulated by rectangular pulses rather than square pulses.)

Figure 10-18 illustrates a theoretical spectrum of a square-pulse, pulse-modulated oscillator (or transmitter). Figure 10-18A is the true *voltage* spectrum (where adjacent power "lobes" are 180° out of phase), while Figure 10-18B shows the same wave-form as it would appear on a spectrum analyzer (all power lobes in phase and shown above the X axis).

In Figure 10-18A, the main lobe and the side lobes are shown as groups of spectral lines extending above and below the base line. The number of these side lobes for a true square or rectangular pulse approaches infinity. Any two adjacent side lobes are separated on the frequency scale by a distance equal to the inverse of the modulating pulse width.

Although adjacent lobes are actualy 180° out of phase, the spectrum analyzer is insensitive to phase. Therefore, only the absolute value of the spectrum is displayed and appears as shown in Figure 10-18B.

A pulse-modulated spectrum display can be interpreted as follows:

The *pulse repetition frequency* (PRF) is determined by the frequency difference between two vertical lines, as shown in Figure 10-18.

The *pulse width* is determined by measuring the frequency difference (or width) of a side lobe, then finding the reciprocal of this value.

The *transmitting oscillator* stability can be checked by noting the degree of frequency shift, as in the case of amplitude modulation.

Transmitter tuning can be accomplished visually by tuning the transmitter oscillator so that the major lobe (or most of the output power) is within the frequency range of the receiver bandwidth.

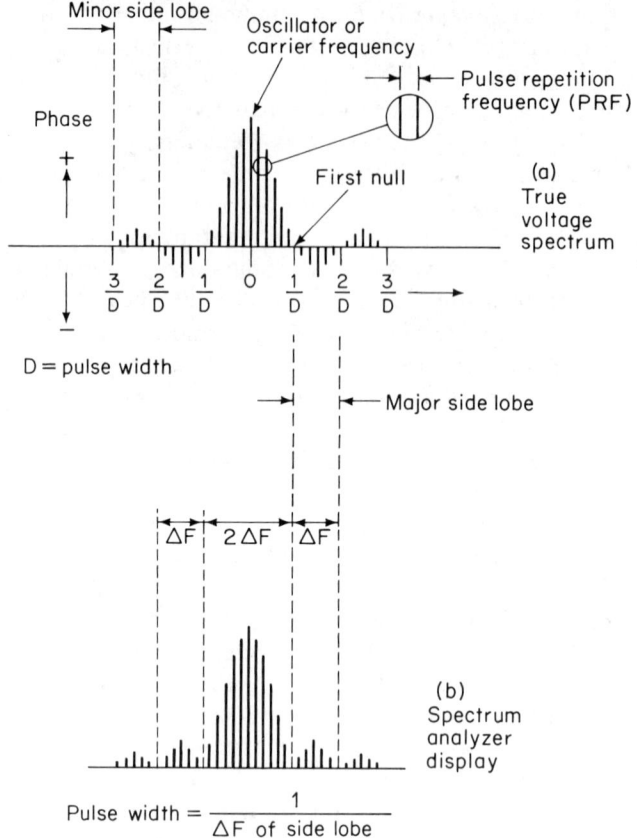

Fig. 10-18. Relationship of true voltage spectrum to corresponding display of spectrum analyzer (Tektronix).

If the spectrum has two peaks or what appear to be two major lobes, the oscillator is operating in two modes or is being pulled in frequency by some external factor. Likewise, if the spectrum is without deep minimum points on either side of the major lobe, this indicates the presence of frequency modulation (in addition to the pulse modulation). Usually, this is an undesired condition.

10-5. TIME DOMAIN REFLECTOMETERS

Time domain reflectometers are not wave analysis instruments, as such. Instead, time domain reflectometers use wave analysis to determine the characteristics of other devices, such as transmission lines, fittings, and so on.

The time domain reflectometer (originally developed by Hewlett-Packard) uses a step or pulse generator (Chapter 4) and an oscilloscope (usually a sampling oscilloscope—Chapter 5) in a system best described as a "closed-loop radar." A voltage step is propagated down a transmission line under investigation, and the incident (outgoing) and reflected voltage waves are monitored by the oscilloscope at a particular point on the line. This "echo" technique reveals at a glance the characteristic impedance of the line, as well as both the position and nature (resistive, inductive, or capacitive) of any discontinuity on the line. Time domain reflectometers also indicate whether losses in a transmission system are series losses or shunt losses.

10-5.1. Basic Time Domain Reflectometer Operation

The basic time domain reflectometer circuit is shown in Figure 10-19. In operation, the step generator produces a positive-going wave that is fed into the transmission system under test. The step wave travels down the transmission line at the velocity of propagation of the line. If the load impedance is equal to the characteristic impedance of the line, no wave is reflected, and all that will be seen on the oscilloscope is the outgoing voltage step re-

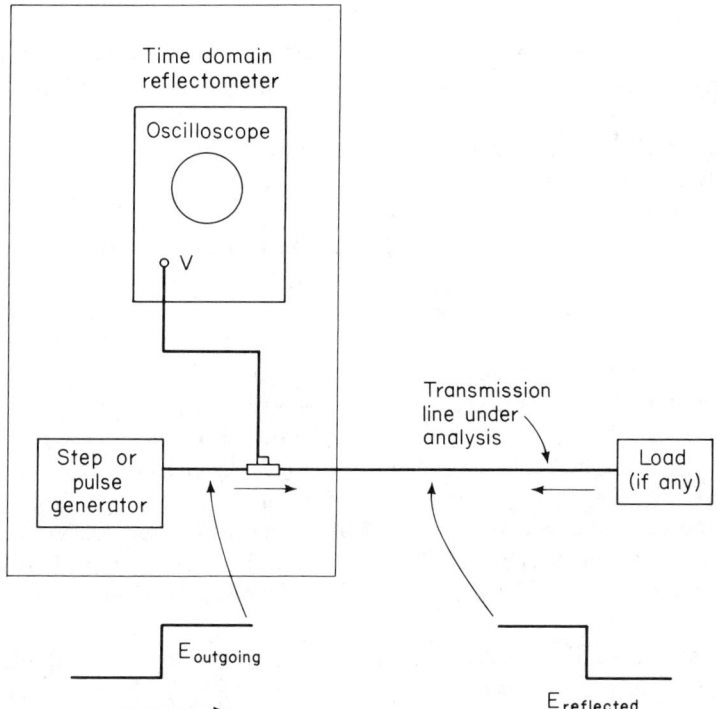

Fig. 10-19. Basic time domain reflectometer circuit (Hewlett-Packard).

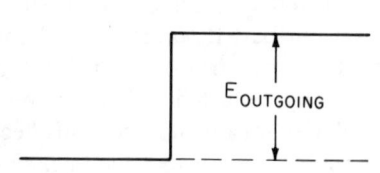

Fig. 10-20. TDR pattern when reflected signal is zero.

corded as the wave passes the point on the line monitored by the oscilloscope. (See Figure 10-20.) If a mismatch exists at the load, part of the outgoing wave is reflected. The reflected voltage will appear on the oscilloscope display, algebraically added to the outgoing wave. (See Figure 10-21.)

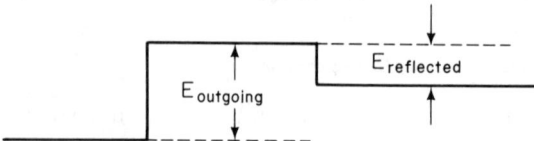

Fig. 10-21. TDR pattern when reflected signal is added algebraically to outgoing signal.

10-5.2. Analyzing Time Domain Reflectometer Displays

The shape of a wave-form on a time domain reflectometer (or TDR) requires analysis. The shape and amplitude of the reflected wave in relation to the outgoing wave reveal the nature and magnitude of mismatches, loads, discontinuities, and so on, on the line.

Figure 10-22 summarizes typical TDR displays and the corresponding load or mismatch responsible for the displays. It should be noted that these are *theoretical displays.*

In Figure 10-22A (an open-circuit termination where the load impedance is infinity), the full voltage will be reflected back and will be added to the outgoing voltage.

In Figure 10-22B (a short-circuit termination where the load impedance is zero), the voltage will drop to zero and cancel the outgoing voltage.

In Figure 10-22C (a pure resistive load where the load impedance is twice the impedance of the transmission line), one-third of the voltage will be reflected back and will be added to the outgoing voltage.

In Figure 10-22D (a pure resistive load where the load impedance is one-half the impedance of the transmission line), one-third of the voltage will be reflected back and will cancel one-third of the outgoing voltage.

Note that the reflected pulses from an *open, a short, or a pure resistive load* will be the same shape as those of the pulse generator, and they will be added algebraically to the outgoing pulse on the oscilloscope display. Unlike pure resistive load, *capacitive or inductive loads* will change the *shape* of the reflected pulse.

In Figure 10-22E (a load combination of resistance and inductance in series), the leading edge of the reflected wave is the same shape as the outgoing pulse and is added algebraically. With time, however, the pulse slopes

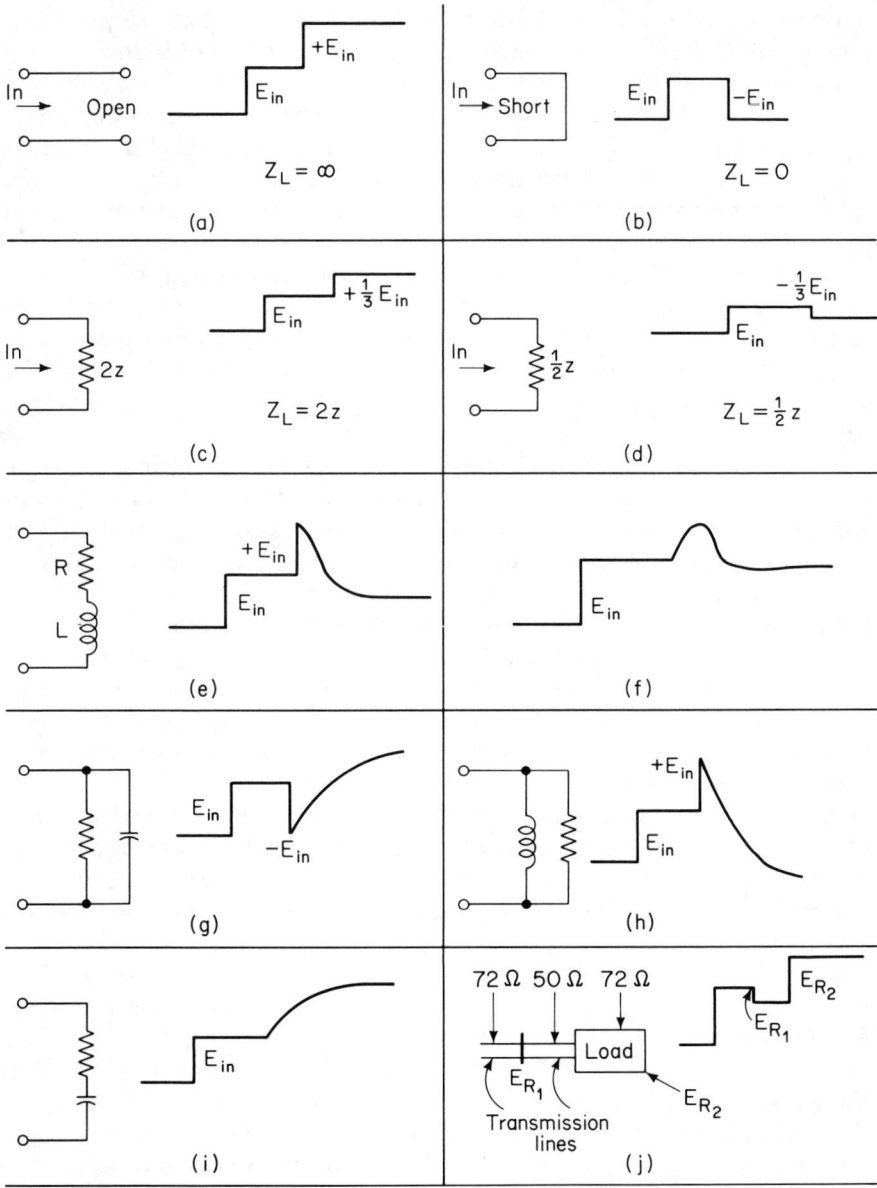

Fig. 10-22. Typical TDR displays and corresponding condition responsible for displays.

off to a value *below* that of the outgoing pulse. This slope is caused by the effect of the inductance. When the pulse wave-form is first applied, the inductance opposes current flow and apears as an infinite impedance. As current starts to flow, the impedance begins to drop, and the voltage slopes off in proportion. Current flow is limited only by the effect of the resistance.

The display of Figure 10-22E would appear only for very large values of inductance and resistance. If the values were small (particularly the inductance) the display would be more like that of Figure 10-22F. This is because the time constant of the reflected wave is so short that it decays to its final value almost before the TDR oscilloscope can rise. This problem is minimized by the use of a sampling oscilloscope (Chapter 5).

In Figure 10-22G (a load combination of capacitance and resistance in shunt), the capacitor acts like a short to the pulse wave-form. Initially, the load impedance is zero, and the voltage drops to zero and cancels the incident voltage. With time, however, the voltage builds up across the capacitor, and current flow is reduced. The rate of capacitor charge and discharge is determined by the RC circuit values.

In Figure 10-22H (a load combination of resistance and inductance in shunt), the leading edge of the reflected wave is the same shape as the outgoing pulse and is added algebraically. With time, however, the pulse slopes off to zero. This is a similar reaction to that of a resistance-inductance in series. However, since the inductance is in shunt, the current flow is not limited by the resistance. Therefore, the voltage drops to zero.

In Figure 10-22I (a load combination of capacitance and resistance in series), the capacitor initially appears as a short to the pulse leading edge, leaving only the resistance. With time, however, the voltage builds up across the capacitor, and is added algebraically. The rate of capacitor charge and discharge is determined by the RC circuit values.

In Figure 10-22J (multiple mismatches) it will be seen that multiple mismatches (or a mismatch and a load) produce reflections that can be analyzed separately. The mismatch at the junction of the two transmission lines (72 and 50 Ω) generates a reflected wave E_{R1}. Similarly, the mismatch at the load also creates a reflection E_{R2}.

By studying Figure 10-22, it will be seen that the reflected wave shape determines the type of mismatch or load present on the transmission line under test. Once these wave shapes are learned, both the type and location of the mismatch can be determined. If the amplitude and rate of slope could be measured accurately, the actual values of resistance, inductance, and capacitance causing the particular wave shape could be calculated. However, such analysis and calculations require a study of complex wave-forms and are beyond the scope of this publication.

10-5.3. Locating Mismatches or Discontinuities on Lines

In addition to determining the characteristics of mismatches or loads in transmission lines, a TDR can be used to locate mismatches (physically) on lines.

The basic procedure for locating mismatches is to measure the travel time and speed of the step or pulse as it passes down the transmission line and then back to the oscilloscope. This is accomplished by measuring the distance (in time) between the outgoing and the reflected wave. The reflected wave is readily identified, since it is separated in time from the outgoing wave. By measuring both the time and the velocity of propagation, the distance can be calculated by

$$\text{Distance} = \text{Velocity of propagation} \times \frac{T}{2}$$

where $T =$ the transit time from monitoring point to the mismatch and back again, as measured on the oscilloscope

If the velocity of propagation for a particular type of transmission line is not known, it can be found from an experiment on a known length of the same type of line. For example, the time required for the outgoing wave to travel down and the reflected wave to travel back from an open-circuit termination at the end of a 120-cm length of RG-9A coaxial cable is 11.4 nsec. This means that the velocity of propagation is 21 cm/nsec or 2.1×10^{10} cm/sec.

11 MICROWAVE TEST EQUIPMENT

Radio waves at frequencies above about 1000 MHz are known as *microwaves*, since they are quite short. A full wave at 1000 MHz (or 1 GHz) is about 1 ft. Radio waves at these frequencies behave somewhat like light waves (travel in a straight line). Also, they require special handling with respect to transmission lines, attenuation, modulation, resonant circuits, and so on.

Because of the special handling, microwave test equipment uses special components and techniques quite different from those of test equipment operating at lower frequencies. Much of today's microwave test equipment is special purpose; that is, the test instruments are used to test one particular type of equipment (such as microwave telephone relay test equipment, radar test equipment, and so on). The instruction manuals supplied with such specialized test equipment generally describe both the operation and application in great detail. Therefore, this chapter is devoted to the basic principles involved in microwave test equipment. A thorough study of this chapter will familiarize the reader with microwave test equipment techniques in common use.

It is assumed that the reader is already familiar with microwave principles, such as the operating theory of waveguides, resonant cavities, Klystrons, traveling-wave tubes, and so on. Although not intended as a basic course in microwave theory, the first sections of this chapter provide a summary for the student, as well as a refresher for the experienced technician.

11-1. MICROWAVE TRANSMISSION LINES

Although it is possible to transmit microwaves along a conventional transmission line, such as a coaxial cable, it is not practical at higher frequencies. Because of the *skin effect*, high-frequency radio waves tend to travel on the outside of a conductor. The higher the frequency, the nearer the skin they travel. This means that the conducting area decreases as the frequency increases. (See Figure 11-1.) Since the resistance increases as the conducting area decreases, high frequencies produce high resistance.

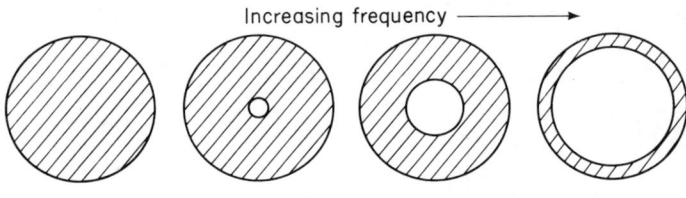

Fig. 11-1. Cross section of conductor showing "skin effect" as frequency is increased.

This problem is overcome by means of hollow transmission lines known as *waveguides*. As shown in Figure 11-2, a waveguide is made up of two parallel

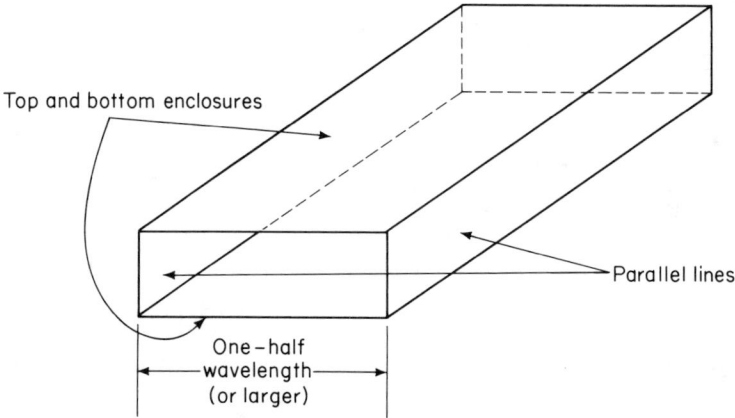

Fig. 11-2. Basic waveguide construction.

strips of metal, with two additional strips at the top and bottom to enclose the radio waves in conducting metal. Since the waveguide is hollow, all of the conducting area is at the surface or skin. Therefore, the radio waves (which travel on the skin only) have maximum conducting area and minimum resistance or loss.

Figure 11-3 shows the relationship between the electromagnetic and electrostatic fields within a rectangular waveguide. Both the electric and magnetic

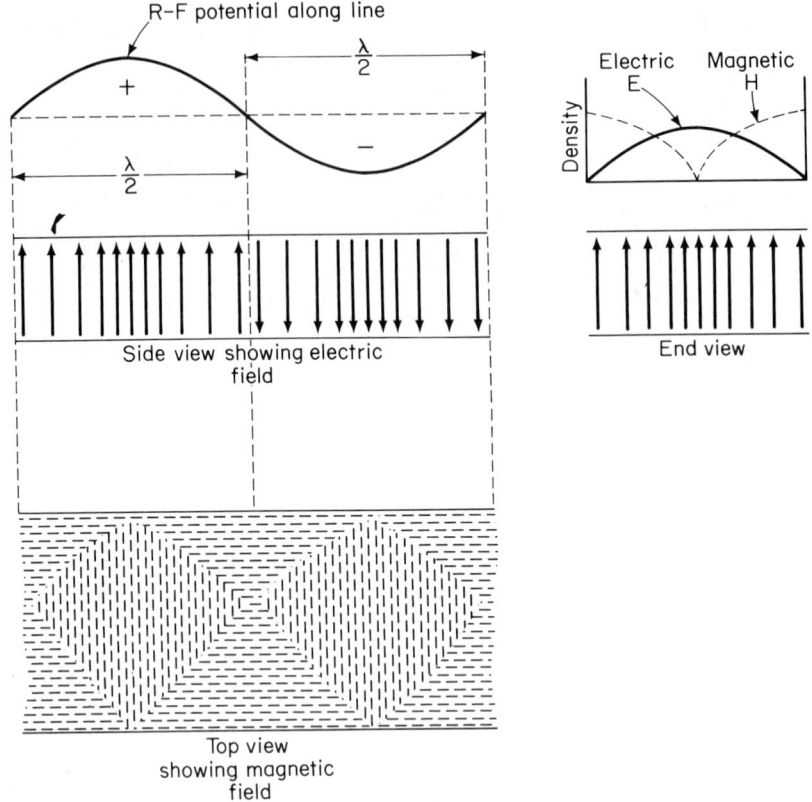

Fig. 11-3. Magnetic and electric fields in a waveguide.

waves are contained within the waveguide, although they are at 90° angles to each other and at 90° angles with respect to the path of radio waves passing along the waveguide. The density of the electric fields is maximum at the center of the waveguide, while the magnetic fields are maximum at the sides.

11-2. MICROWAVE RESONANT CIRCUITS

Conventional resonant circuits cannot be used at microwave frequencies. Instead, resonant circuits are made up of waveguide sections, or cavities, that can be connected into waveguides.

11-2.1. Waveguide Resonant Circuits

It is possible to make a waveguide act as a resonant circuit or a reactive component by placing metal plates within the waveguide. As shown

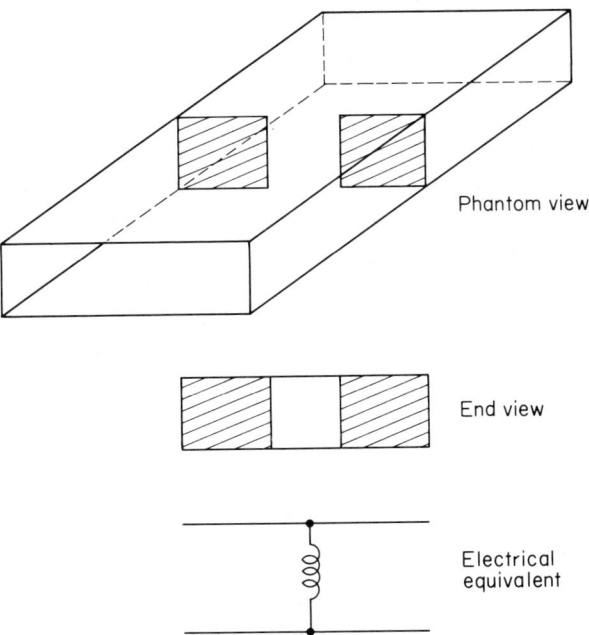

Fig. 11-4. Forming inductive reactance in waveguide by adding plates.

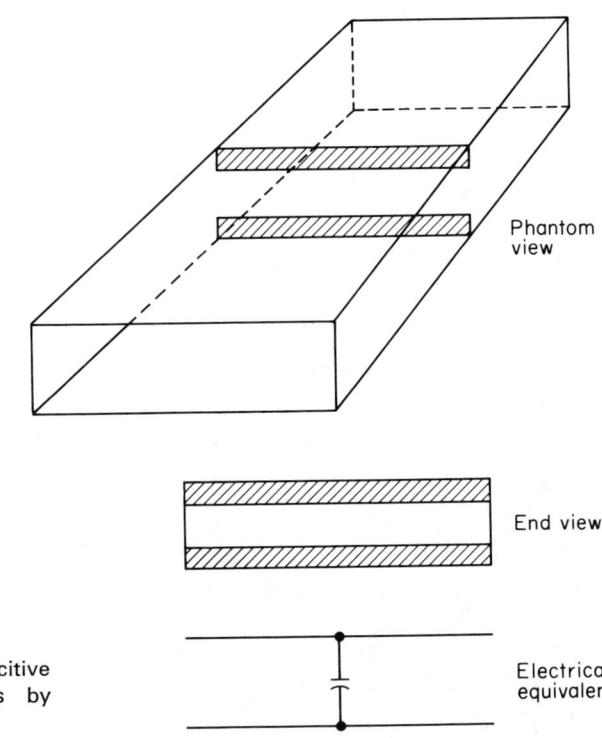

Fig. 11-5. Forming capacitive reactance in waveguides by adding plates.

in Figure 11-4, an inductive reactance is formed when plates are placed across the short side of the waveguide. When plates are placed across the long side of the waveguide (Figure 11-5), a capacitive reactance is introduced into the waveguide. Either way, the plates cause a portion of the radio waves to be reflected and set up standing waves (as is discussed in later sections of this chapter). The amount of reactance is determined by the spacing between the plates. An increased space causes an increased reactance.

As shown in Figure 11-6, a parallel-resonant circuit is formed with a combination of plates. The resonant point or frequency is determined by the size of the opening.

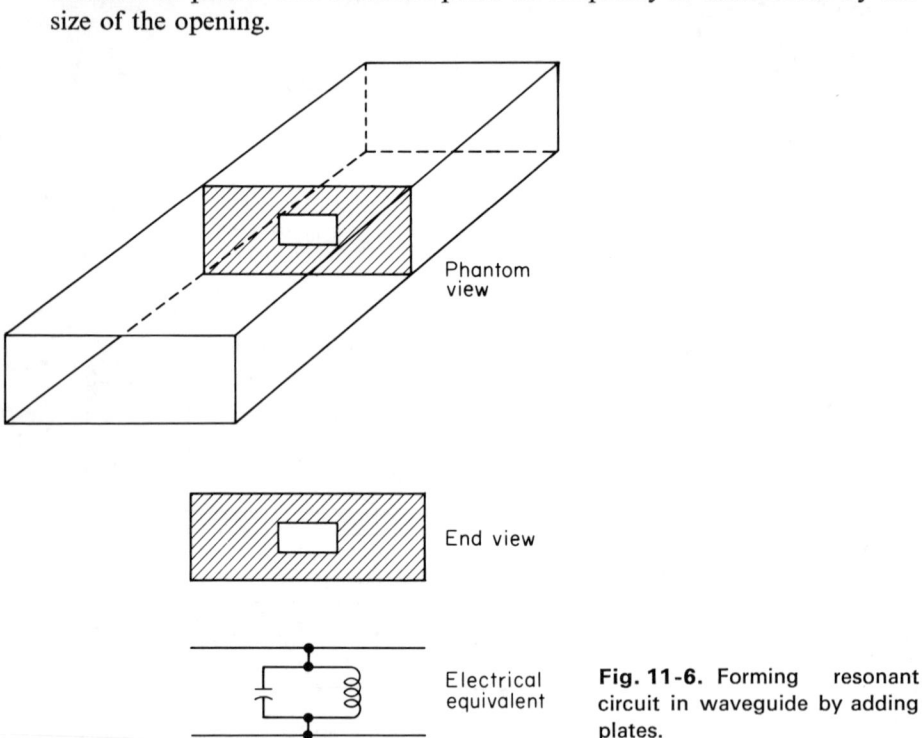

Fig. 11-6. Forming resonant circuit in waveguide by adding plates.

11-2.2. Microwave Resonant Circuits

A resonant cavity is essentially a tuned, or tunable, section of waveguide that acts like a resonant circuit. The size of the cavity determines the resonant frequency. Resonant cavities are made in various shapes. However, their operation is best understood when the cavity is considered to be rectangular.

Assume that the cavity consists of two identical sections, as shown in Figure 11-7. Both sections are one-quarter wave long at the desired frequency, and both are shorted at the ends. Any signals generated at the center point will

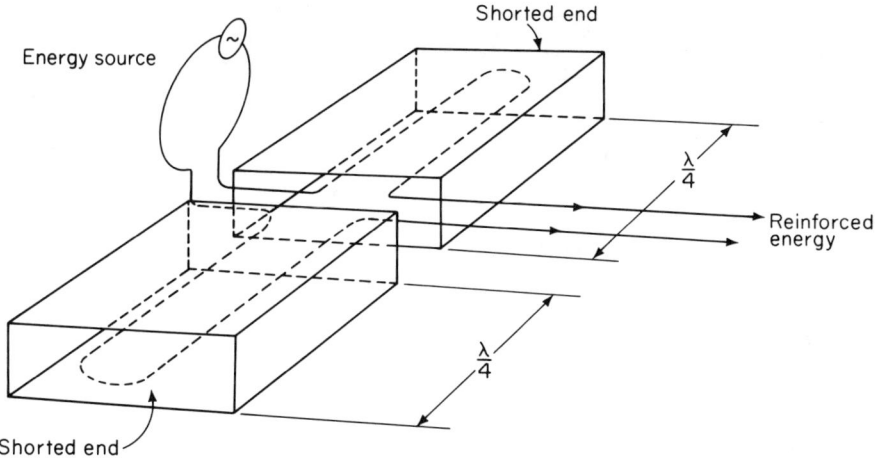

Fig. 11-7. Basic microwave resonant cavity.

travel down the rectangular sections, hit the short, and be returned. Since each section is a quarter-wave long, the returning energy will be returned at the same time from each section and will be reinforced. The cavity thus acts like a parallel-resonant circuit, with the resonant frequency being determined by the length of the rectangular sections.

Note that, although it is easier to understand operation of a resonant cavity shown in rectangular form, most resonant cavities are cylindrical.

11-2.3. Tuning Microwave Resonant Cavities

There are several methods for tuning resonant cavities. Three common methods are shown in Figure 11-8.

The resonant frequency of the cavity shown in Figure 11-8A is varied by changing the length of the cavity (frequency depends on cavity dimensions). The cavity of Figure 11-8B is tuned by varying the capacity between the cavity sides. (Since a resonant cavity is essentially a section of waveguide, or transmission line, any changes in capacity between conductors will change the frequency.) The third method of tuning a resonant cavity (Figure 11-8C) involves changing the density of electric and magnetic waves within the cavity.

The density of the magnetic fields within a waveguide, or cavity, are maximum at the sides, while density of the electric fields is maximum at the center. If a metal object is placed in the center, where the electric fields are maximum, the density of lines is changed so that the cavity frequency is lowered. If the same object is placed near the outer end or outer diameter of the cavity, where the magnetic lines are strongest, the frequency will be increased. In practice,

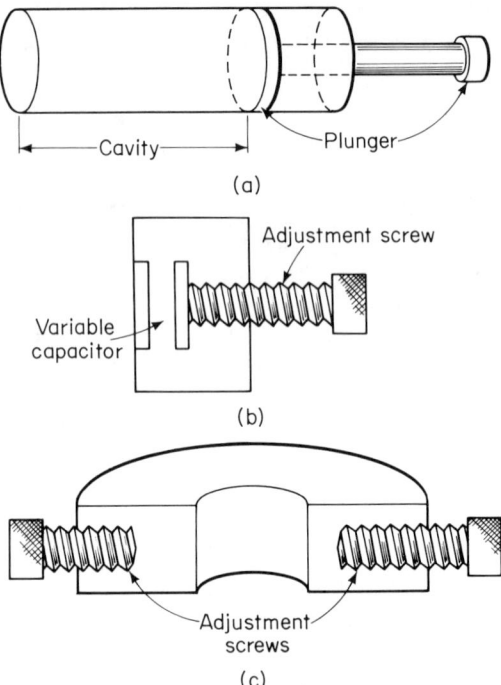

Fig. 11-8. Basic methods of tuning resonant cavities.

this changing of frequency by varying the density of radio waves is accomplished with adjustable screws or slugs.

11-3. TRANSFER OF ENERGY IN MICROWAVE CIRCUITS

There are several methods for coupling energy into waveguides and cavities. For maximum transfer of energy, such as when a microwave oscillator is to be coupled to an antenna through a waveguide, the oscillator is built into a section of waveguide. In turn, this waveguide section is coupled directly to the antenna waveguide, just as two sections of pipe might be joined.

It is often necessary to introduce power into a waveguide or cavity, or to remove power, through a coaxial cable. As shown in Figure 11-9, energy is introduced by means of a *probe*, which is a continuation of the coaxial cable inner conductor. The outer conductor of the coaxial cable is in contact with the top and bottom of the waveguide or cavity. The inner conductor acts as an antenna, radiating the signal in all directions. However, since the signal is enclosed in all directions, it passes down the waveguide in the desired direction or radiates within the cavity.

When energy is to be removed, the coaxial cable inner conductor is bent to

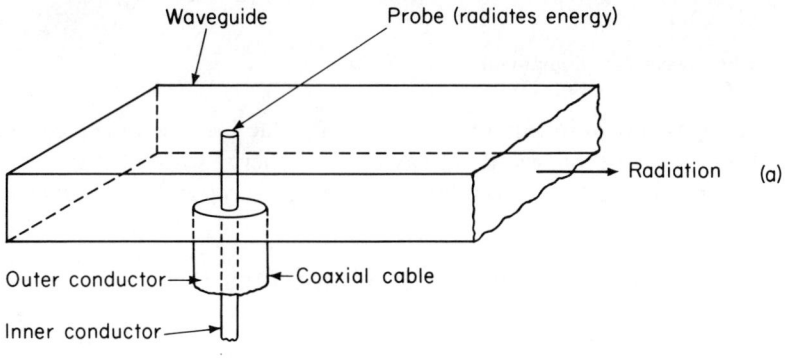

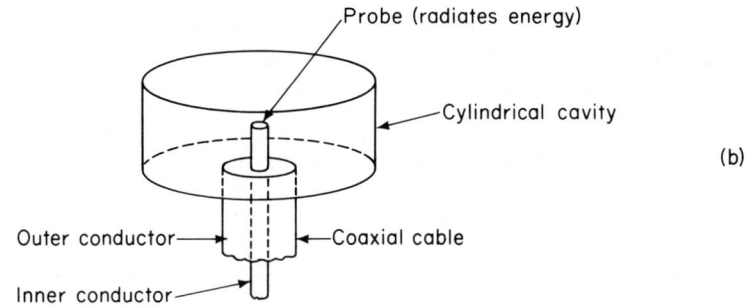

Fig. 11-9. Introducing RF power into waveguides and cavities.

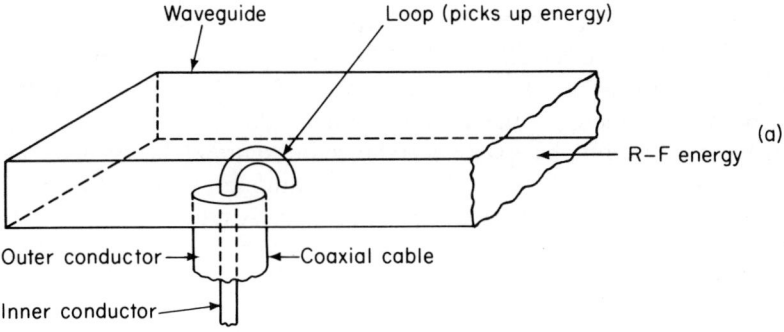

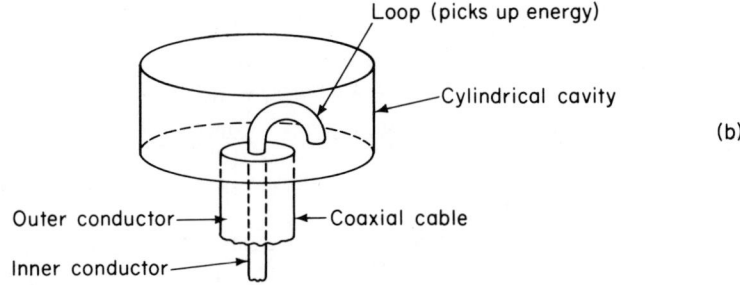

Fig. 11-10. Removing RF power from waveguides and cavities.

form a loop, as shown in Figure 11-10. Usually, the loop is connected to the top or bottom of the waveguide or cavity. Any energy within the waveguide or cavity is picked up by the loop and transmitted through the coaxial cable.

11-4. MICROWAVE SIGNAL SOURCES

There are three major sources for signals of microwave frequency: Klystrons, traveling-wave tubes, and magnetrons. The Klystron and traveling-wave tube are used frequently in microwave test equipment. Magnetrons produce high power levels not usually needed by test equipment. Magnetrons are most often found in radar equipment.

11-4.1. Klystron-Tube Operation

Operation of a Klystron is based on *velocity modulation*. One of the factors that limits the maximum frequency of radio waves is the transit time of electrons from cathode to anode of vacuum tubes. All other factors being equal, if the transit time were 1 μsec, the maximum frequency limit would be 1 MHz. Transit time is determined primarily by the spacing between cathode and anode and by acceleration of the electrons. If electrons can be speeded, the frequency can be increased. Therefore, the frequency of the r-f energy can be controlled by controlling the velocity or acceleration of electron flow. If the electrons can be made to move past a point at a given velocity, energy can be taken from that point, and this energy will be at a frequency determined by the acceleration.

The basic operation of a Klystron is shown in Figure 11-11. As in a conventional vacuum tube, the electrons pass from cathode to anode (identified as the collector plate) through an accelerating grid. A repeller or reflector ele-

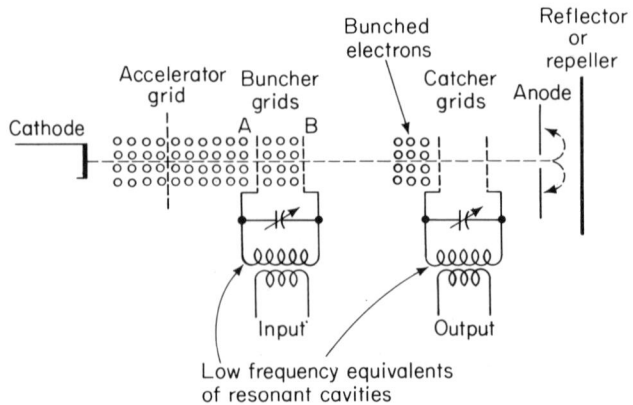

Fig. 11-11. Basic operation of klystron tube.

ment is included in a Klystron to control velocity of the electron stream. Overall control of the Klystron frequency is obtained by varying the voltage relationship of the elements. Usually, the reflector or repeller voltage is varied to provide frequency control.

Operation of the Klystron to produce a specific frequency is determined by the buncher and catcher circuits. As shown in Figure 11-11, the low-frequency equivalents of the buncher and catcher circuits are parallel-resonant circuits with grids placed in the electron stream. The buncher controls the electron stream acceleration at some given frequency, while the catcher extracts energy from the electron streams at that frequency.

As the electron stream passes the A and B grids of the buncher parallel-resonant circuit, a voltage differential is developed across the grids. The capacitor is charged and then discharges through the inductance (as in any parallel-resonant circuit) at a frequency determined by the circuit values. This makes the A and B grids alternately positive and negative. When grid A is positive, the electrons are speeded up. As these electrons approach grid B, they are slowed down. Therefore, the electrons are bunched and are at the resonant frequency of the buncher circuit.

When these bunched electrons arrive at the catcher circuit, they produce the same effect, except that they are reinforced because they have been bunched at the resonant frequency of the catcher. Consequently, the Klystron produces some amplifier action, as well as oscillation. If the catcher is coupled to an external circuit, part of the energy across the parallel-resonant circuit can be taken from the catcher, just as with the parallel circuit of a transmitter or oscillator tank.

In practice, the buncher and catcher parallel circuits are replaced with *resonant cavities*. The energy is extracted from the catcher by means of a probe or loop. In some Klystrons, it is possible to inject a signal into the buncher so that it can be amplified and taken from the catcher. In other Klystrons, part of the catcher output is fed back to the buncher to obtain oscillation.

One of the unique features offered by a Klystron is that, being a vacuum tube with built-in resonant circuits, it will function as a *one-piece microwave oscillator*. In practical applications, the frequency limits are set by cavity dimensions, while the frequency is controlled by adjustment of the control element voltage, or by cavity tuning, or both.

11-4.2. Traveling-Wave-Tube Operation

The traveling-wave-tube (TWT) is used in microwave and UHF test equipment as a wideband amplifier. As shown in Figure 11-12, a TWT is made up of three basic components: (1) the electron gun, (2) a slow-wave structure or *helix*, and (3) a collector.

A beam of electrons from the electron gun is sent to the collector through

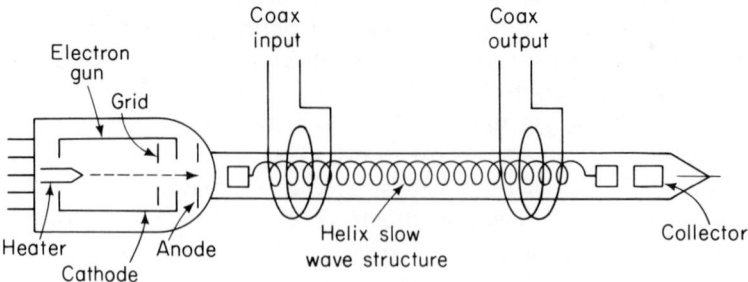

Fig. 11-12. Basic operation of traveling wave tube (TWT).

the slow-wave structure. An electromagnetic wave is propagated by the slow-wave structure. This progagation occurs at a velocity where the electron beam can be made to travel at the *same speed*. When this occurs, the electrons are continuously accelerated or decelerated by the waves (once for each turn of the slow-wave structure). This results in bunching of the electrons at radio frequencies. The slow-wave structure can then take power from the bunched electrons as they move down the circuit. The longer the slow-wave structure, the more power can be taken. Therefore, the TWT can act as an amplifier. Tube gain in decibels is *approximately proportional to the tube length*, as measured in wavelengths of the operating frequency. For example, if the slow-wave structure is three wavelengths long at the operating frequency, approximately 3-dB gain can be obtained.

The bandwidth of a TWT amplifier is usually quite large since there are no resonant circuits, as such, involved. Therefore, TWTs can be used as amplifiers for such test instruments as signal generators that operate over a wide frequency range.

11-5. MODULATION, ATTENUATION, AND LEVEL CONTROL FOR MICROWAVE SIGNAL SOURCES

Special techniques must be used for modulation, attenuation, and output-level control of Klystron oscillators used in test equipment.

11-5.1. Klystron Modulation

Any form of modulation based on varying element voltages to the Klystron can result in frequency shift or a form of frequency modulation. This problem is overcome in microwave signal generators through the use of PIN diodes. Operation of a PIN diode is similar to that of conventional diodes, except that they have a very long storage time; that is, the forward current must be applied for some time before the diode will pass current. At higher frequencies (above about 100 MHz) the direction of current flow changes so rapidly that the diodes do not pass any current. However, if a forward d-c bias is applied to the diodes, they will conduct. The stronger the bias, the more they conduct.

As modulators, the diodes are placed across the Klystron output, as shown in Figure 11-13. Without any dc applied, the diodes appear as high-resistance

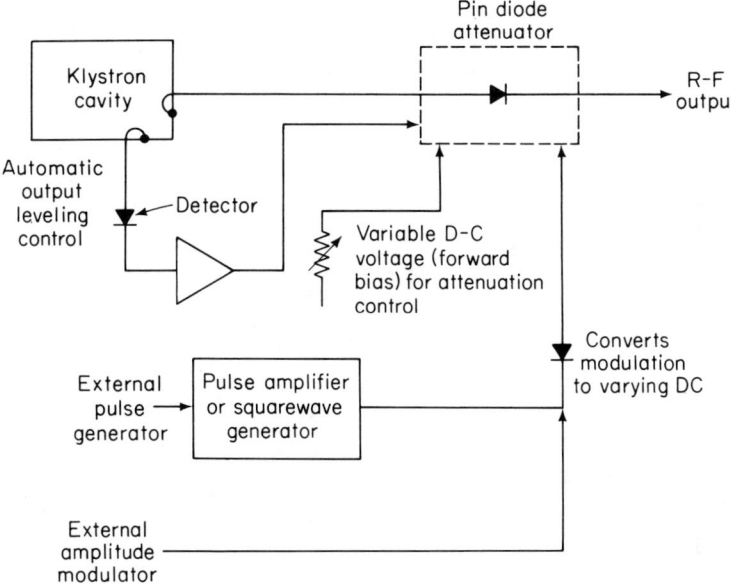

Fig. 11-13. Using PIN diodes for attenuation, modulation, and leveling of klystron output.

shunt across the output. When a varying dc is applied, the diode resistance drops and absorbs part of the output. In effect, the diodes act as a low-resistance variable resistor shunting the waveguide output. The diode resistance and the degree of attenuation of an r-f signal are functions of the applied dc. The modulating signal is converted to a varying dc before application to the diodes.

11-5.2. Klystron Output Attenuation

The output of a Klystron oscillator cannot be attenuated by the same resistive attenuators used in lower-frequency equipment, such as the signal generators discussed in Chapter 4. Conventional resistive attenuators will become reactive at microwave frequencies, possibly changing frequency of the signal source.

When a Klystron is used as a microwave signal source for test equipment, two basic attenuation methods are used: mechanical and electronic.

Mechanical attenuators are placed in the waveguide between the Klystron oscillator and the output connector. Both shutter-type and absorption-type attenuators are used, as shown in Figure 11-14. The shutter-type attenuator

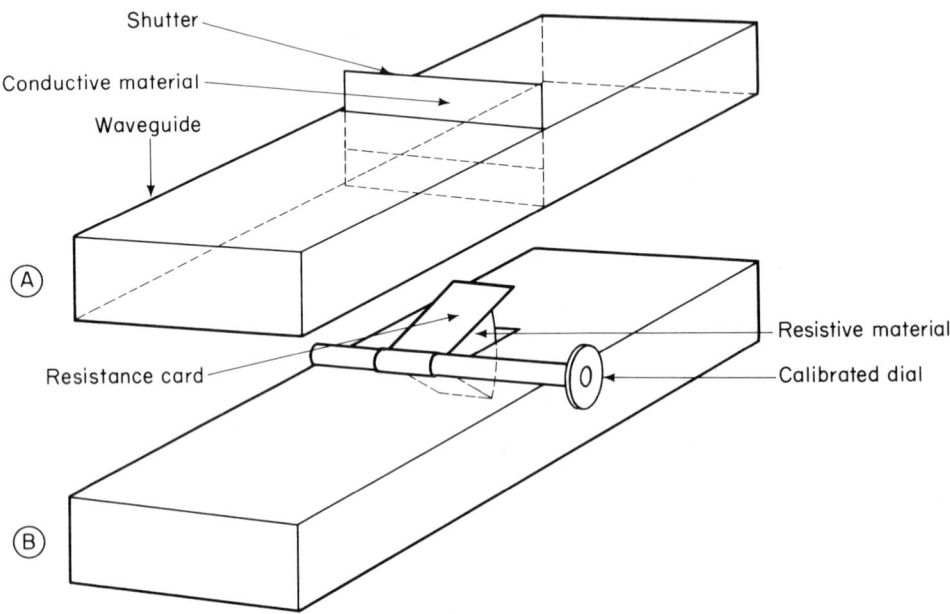

Fig. 11-14. Basic shutter-type and absorption-type waveguide attenuators.

(Figure 11-14A) is essentially a conductive material placed across the waveguide. When the waveguide is completely closed off, the signal will be prevented from passing (complete attenuation). If the waveguide is partially blocked by the conductive material, partial attenuation will result. However, this type of attenuator will also cause reflection of the signals and standing waves (discussed in later sections of this chapter.)

When a signal attenuation is wanted without reflection, an absorption-type attenuator (Figure 11-14B) is used. A calibrated dial controls a resistive card or shutter. The card is coated on the surface with a resistive material and is inserted in the waveguide parallel to the electric field. The resistive material absorbs energy without producing reflections.

It is also possible to use PIN diodes to provide attenuation. As previously discussed, if PIN diodes are placed across the output of an oscillator in the waveguide and sufficient forward bias is applied, the diodes will act as a low-resistance shunt, absorbing all of the power output (Figure 11-13.). As the forward bias is lowered, the diode shunt resistance increases and a portion of the signal is allowed to pass. This permits the diodes to act as attenuators.

11-5.3. Klystron Output Leveling

Usually, the same diodes are used as modulators, attenuators, and automatic-leveling controls. This is shown in Figure 11-13. The automatic-leveling system consists of a detector that samples the Klystron output and produces a d-c output of corresponding amplitude. This d-c output is applied to the diode attenuator. If the Klystron output increases in amplitude, the detector output also increases, raising the bias on the attenuator. This serves to reduce the output and offset the Klystron increase.

11-5.3.1. Klystron Sweep Output Leveling

Klystrons are often used in swept-frequency generators. These are the microwave version of the sweep generators described in Chapter 4. The frequency sweep is accomplished by applying a sawtooth voltage to the Klystron control element grid, thus rapidly sweeping the Klystron frequency over a wide range.

Microwave sweep generators must also have an output-leveling function, as do lower-frequency sweep generators. Generally, a PIN diode circuit is used. However, instead of sampling part of the signal directly from the Klystron, the sweep output is sampled in the waveguide by means of a *directional coupler* (discussed in later sections of this chapter). The basic circuit for automatic output leveling using a directional coupler is shown in Figure 11-15.

The directional coupler functions to pass r-f signals in one direction only and to sample a portion of the r-f signals. The detector in the directional

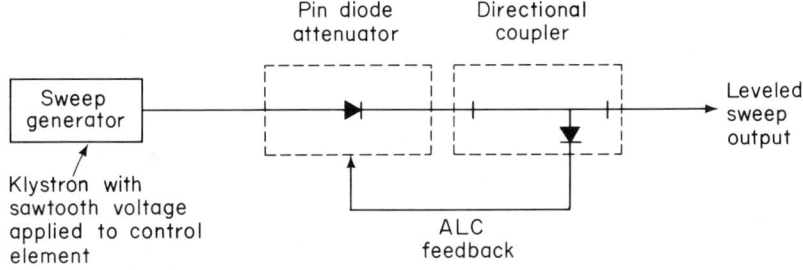

Fig. 11-15. Automatic leveling control circuit for Klystron used as microwave sweep generator.

coupler produces a d-c voltage in proportion to the output amplitude. The d-c voltage is returned to act as the forward bias for the attenuator diode. If the amplitude decreases as the frequency is swept, the detector voltage also decreases, reducing the forward bias on the attenuator diodes. This serves to increase the output. An increase in output produces the opposite effect.

11-6. MICROWAVE POWER MEASUREMENT

The current and voltage in a microwave circuit are complex in nature and difficult to evaluate in terms of their ability to do work. On the other hand, power is a real quantity that can be measured and easily related to circuit performance. Also, in a loss-free line, microwave power remains constant with position of measurement (unlike voltage and current levels along the line). Therefore, power is one of the basic measurements made at microwave frequencies.

11-6.1. Bolometric Power Measurements

Below 10 mW, power is usually measured with bolometers (temperature-sensitive resistive elements) in conjunction with a balanced bridge. This basic bolometer is a thin wire placed within the r-f energy field. Changes in temperature (caused by changes in r-f field signal strength) cause a corresponding change in wire resistance.

There are two types of bolometers: *thermistors* (Figure 11-16), whose

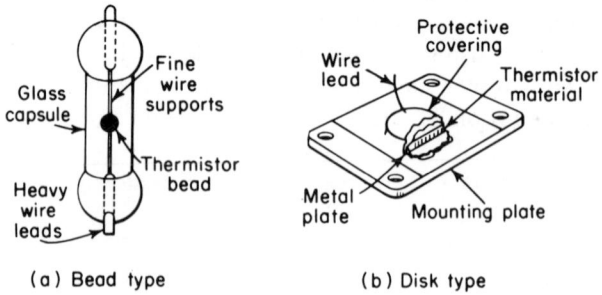

Fig. 11-16. Typical thermistor.

resistance decreases with temperature increases (negative temperature coefficient) and *barretters* (Figure 11-17), which have a positive temperature coefficient. Thermistors are more commonly used since they are more rugged, both physically and electrically, than the barretters.

Bolometer elements are mounted in devices that ideally present a perfect

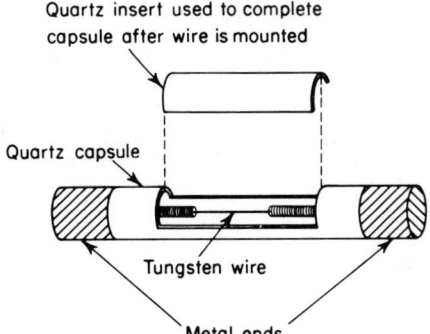

Fig. 11-17. Typical barretter.

impedance match to microwave transmission lines, either coaxial or waveguide (Figure 11-18). Such devices, appropriately termed *bolometer mounts*, allow a "bias" connection to the bolometer element, as well as a proper entry point for the r-f signal.

In use, the bolometer is connected as one leg of a Wheatstone bridge (or some modification thereof) through the bias connection, and an excitation voltage is applied to the bridge (Figure 11-19). The d-c or low-frequency a-c bridge excitation serves as the bolometer-element bias power that affects

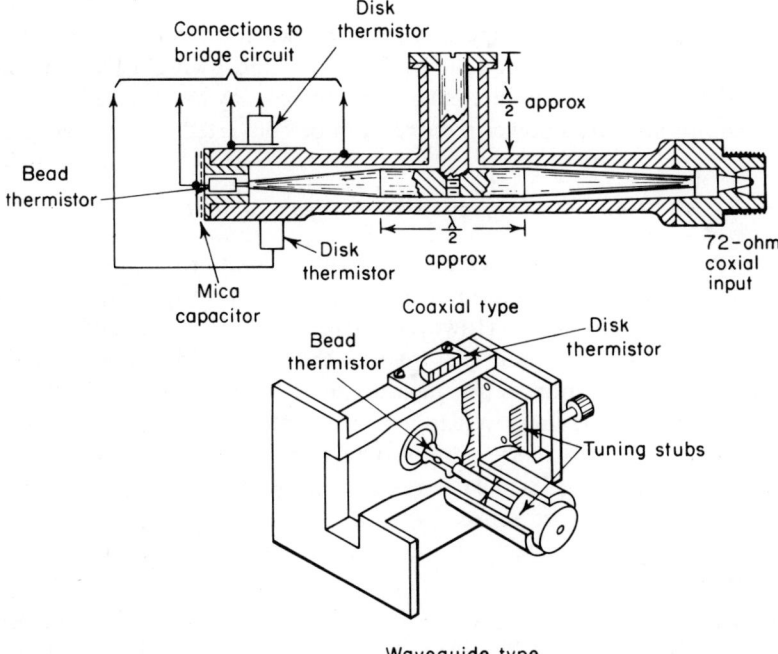

Fig. 11-18. Typical thermistor (bolometer) mounts.

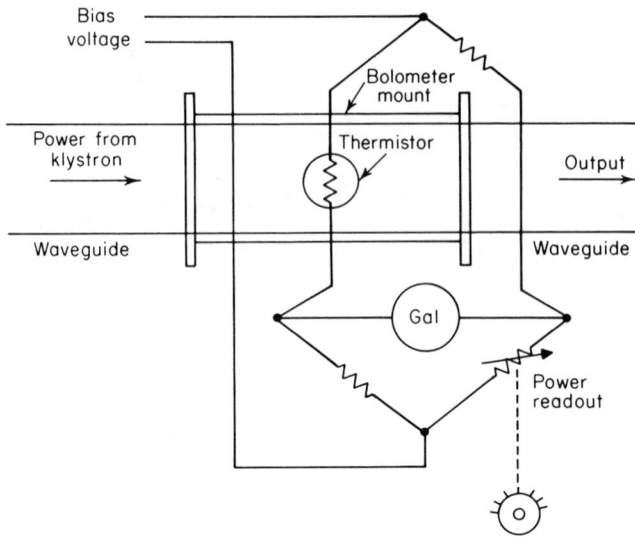

Fig. 11-19. Bolometer power measurement bridge.

the bolometer's resistance, so that the bridge is essentially balanced. When the unknown r-f energy (supplied by a Klystron in this case) is applied to the bolometer, the resulting temperature rise causes the element's resistance to change, tending to unbalance the bridge. By withdrawing a like amount of d-c or a-c bias power from the element, the bridge may be returned to balance, and the amount of bias power removed can be converted to a power readout.

11-6.2. Calorimetric Power Measurements

When the power to be measured is approximately 10 mW or higher, bolometer elements are not used directly, since the barretters or thermistor could be damaged. (However, there are thermistors capable of dissipating 50 mW without damage.) Instead, the basic bolometer circuit is used with some form of calorimetric measurement system. The basic difference between the two systems is that in a noncalorimetric meter the temperature sensor is placed directly in the path of the r-f power, whereas in a calorimetric meter the power is used to heat a resistance, and the *resultant heat is transferred* to the temperature sensor.

Calorimetric power meters fall into two classes: dry and fluid. Dry calorimeters depend on a static thermal path between the dissipative load or resistor and the temperature sensor (such as mounting both dissipating resistor and sensor in the same heat sink). This arrangement often requires several minutes for both resistor and sensor to stabilize and is time consuming. Fluid calorimeters, such as those shown in Figure 11-20, use a moving stream

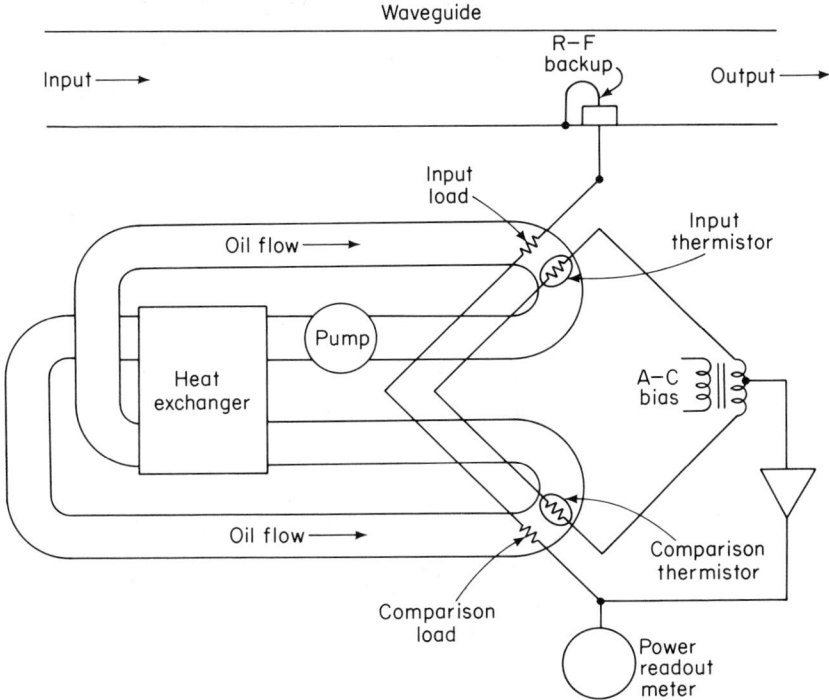

Fig. 11-20. Fluid calorimetric power meter (Hewlett-Packard).

of oil to transfer heat quickly to the sensing element. An amplifier-feedback arrangement, in conjunction with the series oil-flow system, reduces measurement time to a few seconds for full-scale response.

Basically, the fluid calorimetric power meter of Figure 11-20 consists of a self-balancing bridge that has identical temperature-sensitive resistors in two legs, an indicating meter, and two load resistors, one for the unknown input power and one for the comparison power. The input load resistor and input thermistor are in close thermal proximity so that heat generated in the input load resistor (by dc from the r-f pickup, proportional to microwave power) heats the input thermistor and unbalances the bridge. The unbalanced signal is amplified and applied to the comparison load resistor, which is in close thermal proximity to the comparison thermistor. The heat generated in the comparison load resistor is transferred to the comparison thermistor and nearly rebalances the bridge.

The meter measures the power supplied to the comparison load to rebalance the bridge. The characteristics of the thermistors are the same, and the heat transfer characteristics from each load are the same, so the power dissipated in each load is the same, and the meter may be calibrated directly in input power.

11-6.3. Peak Power Measurements

It is often necessary to meaure peak power of pulses used in microwave work. This can be done by using some form of bolometer. However, it is usually more practical to use some form of *video-comparator* circuit. In such circuits, a known d-c voltage (across a known impedance) is adjusted to a level equal to the pulse being measured.

In a typical video-comparator type of peak power test instrument, such as shown in Figure 11-21, the r-f signal is picked up by an r-f probe and ap-

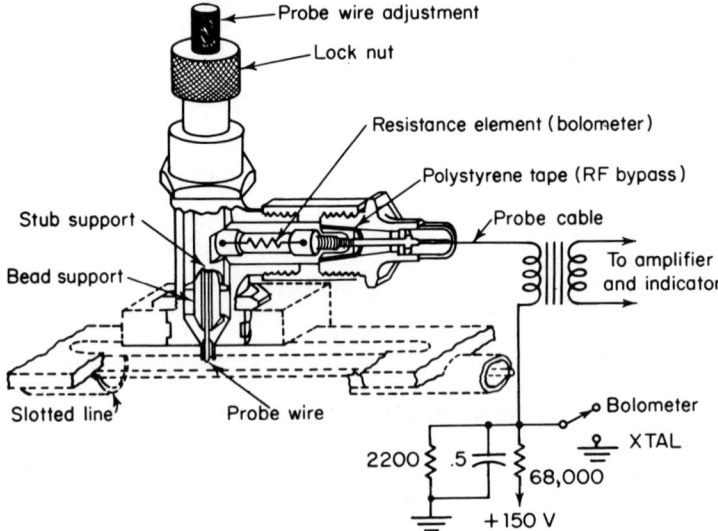

Fig. 11-21. Peak power measurement circuit (video comparator) (Hewlett-Packard).

plied to a diode detector. The demodulated diode output and the output of a d-c reference supply are simultaneously fed to the input of an oscilloscope through a chopper. The d-c reference volgage is adjusted so that it is exactly equal to the peak value of the demodulated pulse. The level of the required d-c reference voltage is then indicated on the panel meter, calibrated to read r-f power.

11-7. MICROWAVE MEASUREMENT DEVICES

Although there are many devices used in microwave measurements, those most commonly used are the r-f probe (which may include a bolometer or a diode detector), a slotted line, a directional coupler, a dummy load, and a waveguide short. The following paragraphs provide a brief description of these basic microwave test devices.

11-7.1. Radio-Frequency Probe

Figure 11-22 shows the construction of a typical r-f probe, mounted on a slotted line. The probe wire is adjustable for depth of penetration (coupling), so that the amount of r-f pickup can be controlled. In practice, the coupling is kept to a minimum to reduce distortion of the fields inside the transmission medium.

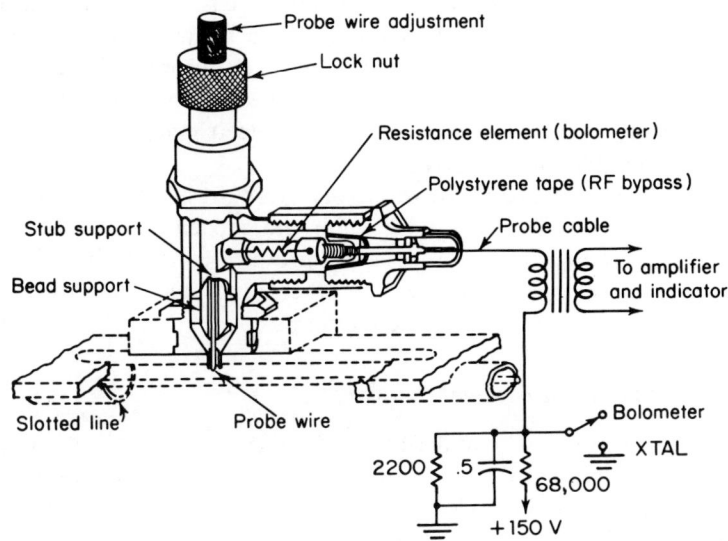

Fig. 11-22. Typical RF probe and slotted line.

The bolometer element shown in Figure 11-22 may be replaced by a diode detector. When the switch is set to BOLOMETER, a bias current is applied. The bias current is removed when the switch is set to CRYSTAL. Either way, the r-f probe output is coupled to an indicating circuit through a transformer (omitted with some probes). The indicating meter may be a microammeter located directly in the probe cable circuit, or it may be an ammeter preceded by a d-c amplifier to which the probe voltage is fed. Usually, if amplification (greater sensitivity) is desired, the microwave signal source is amplitude-modulated with an audio-frequency square wave, so that the detector output can, in turn, be amplified by normal audio amplifier circuits (rather than a d-c amplifier).

11-7.2. Slotted Line

As shown in Figure 11-22, the slotted line is a coaxial or waveguide section of transmission line, with a longitudinal slot cut into its outer shell to permit the insertion of a probe. The slot is generally at least one wavelength long and is narrow enough to cause very little loss from radiation leakage. Through this slot a probe can be placed in the electromagnetic field inside the section and moved up and down to explore the voltage field. The most common use for a slotted line is to measure standing-wave ratios (SWR) in microwave equipment.

11-7.3. Diode Detector

The diode detectors used in microwave r-f probes are essentially the same as those used at lower frequencies. Their function is to convert r-f power into d-c voltage. This voltage can then be monitored with a d-c voltmeter and calibrated in terms of power or whatever other value is desired. Diode probes are often used with VTVMs or electronic voltmeters where there is some amplification involved.

When used in microwave work, diode detectors must be designed in conjunction with the r-f probe to provide minimum capacitance and maximum input resistance. A high capacitance could produce considerable capacitive reactance in a microwave circuit. On the other hand, maximum input resistance keeps the current at an absolute minimum.

11-7.4. Directional Coupler

A directional coupler functions to pass r-f signals in one direction only and to sample a portion of the r-f signals (Figure 11-23A). The detector in the directional coupler produces a d-c voltage in proportion to the r-f signal passing through the coupler.

As shown in Figure 11-23B, the basic directional coupler consists of a short section of waveguide coupled to the main-line waveguide section by means of two small holes. The main-line waveguide section is, in turn, coupled into the waveguide system under measurement. The short section of waveguide contains a match load in one end and a coaxial output probe in the other. The degree of coupling between the main-line waveguide and the short-section (or auxiliary) waveguide is determined by the size of the two holes.

Operation of the directional coupler is shown in Figures 11-23C and 11-23D, In Figure 11-23C, power is shown flowing from left to right, and two samples are extracted or coupled out at points *C* and *D*, with the probe connected to point *F* and the load connected to point *E*. Since the two paths (represented by *C-D-F* and *C-E-F*) to the coaxial probe are the same length, the two

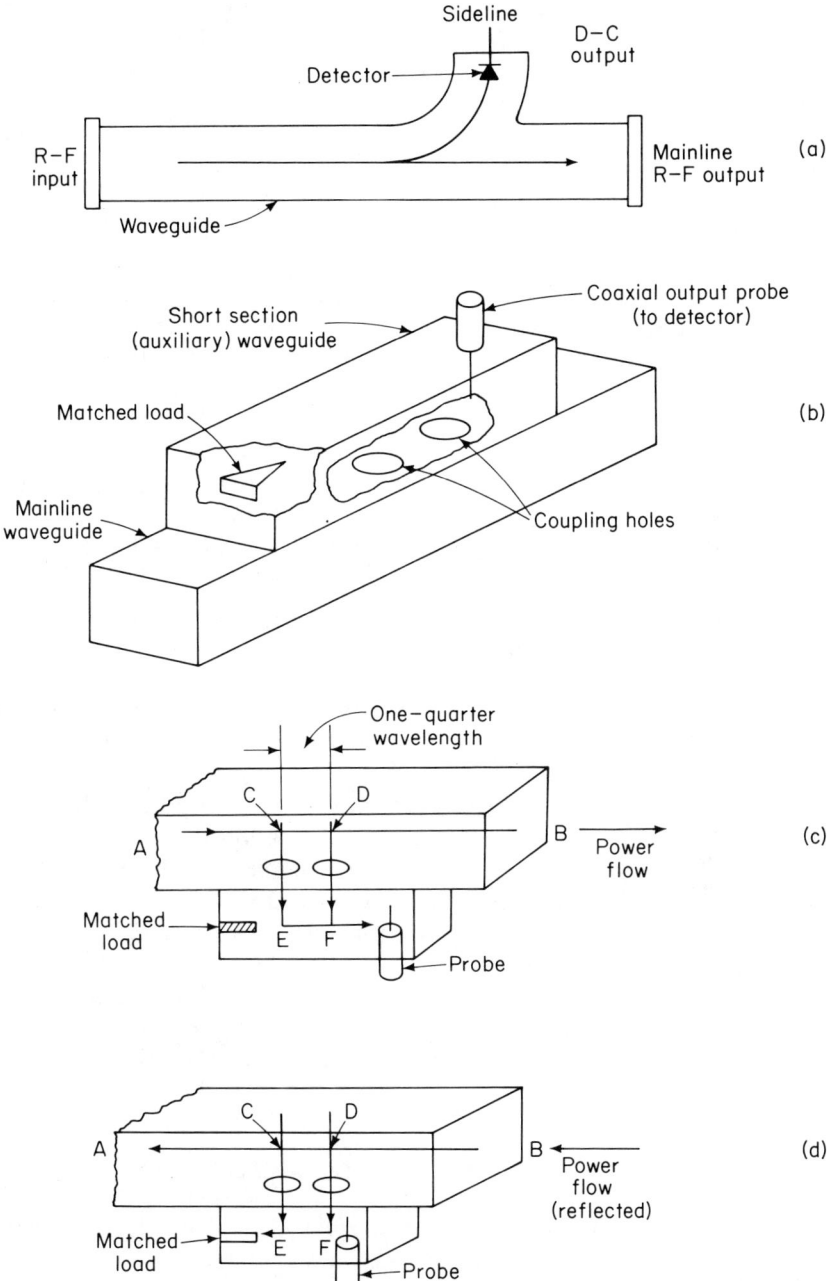

Fig. 11-23. Directional coupler used in microwave waveguides.

samples arrive at point *F* in phase and are picked up by the coaxial probe to give a power output reading. If, however, path *C-D-F-E* is one-half wavelength longer than path *C-E* (because the two holes are one-quarter wavelength apart), then the two samples will arrive at point *E* (the load) 180° out of phase, producing cancellation. Thus the load absorbs no directly transmitted power flowing in the left-to-right direction.

Figure 11-23D shows the same coupler with reflected power flowing in the reverse direction. Again, samples are removed at points *C* and *D*. The two paths *D-F-E* and *D-C-E* are the same length, and the two samples of reflected power arrive at point *E* in phase. Thus the load absorbs reverse power flowing in the right-to-left direction. However, path *D-C-E-F* to the probe is one-half wavelength longer than path *D-F*, and the resulting 180° phase shift causes cancellation at point *F*, so that no power indication results.

Therefore, the coaxial probe measures power only from a wave traveling from left to right in the main line, and any reflections causing power to flow from right to left have no effect on the output reading. Most reflected power (traveling from right to left) is absorbed in the load.

In practice, there is a nominal attenuation or coupling factor between the coaxial output and the main line for power flowing from left to right (forward power). Usually, this is in the order of 20 dB (100:1 power ratio). Therefore, a 100-W forward power would produce a 1-W forward power indication at the coaxial output.

The ability to reject power in the reverse direction is called the *directivity attenuation* or, more often, the *directivity*. Usually, this is at least 20 dB, and is often 40 dB.

One point is often overlooked when measuring reverse power with such a coupler. The directivity and the nominal attenuation (coupling factor) must be *added* for reverse power indications. For example, assume that a directional coupler has a 20-dB attenuation between forward power and the coaxial output. Also assume that the same directional coupler has a 20-dB directivity. A 10,000-W forward power would produce a 100-W forward power indication at the coaxial output. This same 10,000-W signal would be attenuated to 100 W of reverse power by the 20-dB directivity. The 100 W of reverse power would then be measured as 1 W by the 20-dB attenuation to the coaxial output.

11-7.5. Dummy Loads and Waveguide Shorts

A dummy load is used to prevent the radiation of power that could cause interference. Dummy loads are required for service and test procedures on microwave equipment, as they are for lower-frequency units. However, a microwave dummy load must be of special design in order not to create mismatches between load and waveguide. A conventional wire-wound resistor (which is a common dummy load at lower frequencies) will create a high inductive reactance at microwave frequencies. This reactance will cause a severe mismatch between the waveguide and the dummy load, resulting in power loss, and so on. If such a dummy load were connected directly to a microwave transmitter output, it might even affect transmitter tuning. At the normal operating frequency, a microwave antenna is almost purely resistive. The dummy load that replaces such an antenna must also present almost pure resistance with an absolute minimum of inductive reactance.

Many times, *sliding or variable* dummy loads are used when testing microwave components in the laboratory. This permits the dummy load to be matched to the *exact impedance* of the waveguide. *Sliding waveguide shorts* are also used to terminate some waveguide test setups. A nonadjustable short can produce standing waves at microwave frequencies.

11-8. FIXED-FREQUENCY VERSUS SWEPT-FREQUENCY MEASUREMENTS

Both fixed-frequency and swept-frequency techniques are used for the measurement of power, attenuation, frequency, impedance, and so on, in microwave test equipment.

Fixed-frequency techniques offer the highest precision attainable for individual measurements. This is because the small inherent mismatch errors that must be tolerated on a broad frequency-sweep basis may be individually tuned out. Consequently, fixed-frequency techniques are widely used in "standard" measurements and in applications in which the system under test is operating either at a single frequency or within a very narrow band.

Swept-frequency techniques are used to obtain measurements quickly and easily over a range of frequencies. Important microwave parameters such as standing-wave ratio (SWR), directivity, impedance, attenuation, and so on, can be determined on a swept-frequency basis, and the user can quickly determine whether there is a narrowband condition, such as resonance at a particular frequency, in the device being tested.

11-9. MICROWAVE IMPEDANCE MEASUREMENTS

The most important consideration in microwave impedance measurement is in *impedance-matching* a load to its source. This is usually more important than actual impedance value. If the load and source are mismatched, part of the power will be reflected back along the waveguide toward the source. This reflection not only prevents maximum power transfer, but also can be responsible for erroneneous measurements and can easily cause circuit damage in high-power applications. The power reflected from the load interferes with the incident (forward) power, causing standing waves of voltage and current along the line. The ratio of standing-wave maximums to minimums is directly related to the impedance mismatch of the load. The standing-wave ratio (SWR) therefore provides a valuable means of determining impedance and mismatch.

A classic example of standing-wave ratio measurements is that made when a microwave antenna is attached to a waveguide. When the antenna and waveguide are perfectly matched for impedance, all of the energy or signal will be transferred to or from the antenna, and there will be no loss. If there is a mismatch (as is the case in any practical application), some of the energy will be reflected back into the line. This energy will cancel part of the desired signal.

If the voltage (or current) is measured along the line, there will be voltage or current maximums (where the reflected signals are in phase with the outgoing signals), and volgage or current minimums (where the reflected signal is out of phase, partially canceling the outgoing signal). These maximums and minimums are called *standing waves*. The ratio of the maximum to the minimum is the standing-wave ratio, or SWR. This ratio can be related to either voltage or current. Since voltage is easier to measure, it is usually used, resulting in the common term *voltage standing-wave ratio*, or VSWR. The calculations for VSWR are shown in Figure 11-24.

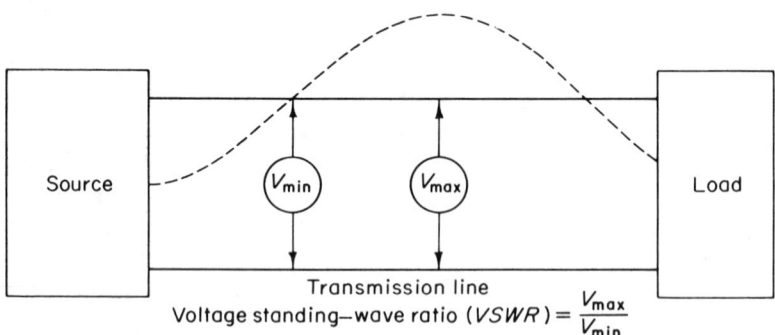

Voltage standing–wave ratio $(VSWR) = \dfrac{V_{max}}{V_{min}}$

Fig. 11-24. Calculations for voltages standing wave ratio (VSWR).

A standing-wave ratio of 1:1 means that there are no maximums or minimums (the voltage is constant at any point along the line) and that there is a perfect match between antenna and transmission line.

11-9.1. Fixed-Frequency Impedance (SWR) Measurements

Standing-wave ratio can be measured directly at fixed frequencies by using a slotted line. In use, the slotted line is placed immediately

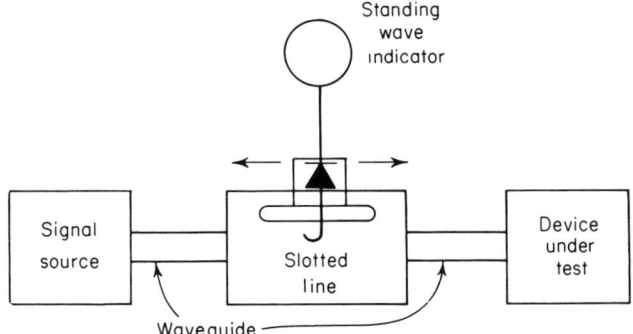

Fig. 11-25. Basic slotted line test connections.

ahead of the load, as shown in Figure 11-25, and the source is adjusted for some fixed amplitude-modulation frequency (usually 1 kHz) at the desired microwave frequency. The slotted line probe is loosely coupled to the r-f field in the line, thus sensing relative amplitudes in the standing-wave pattern as the probe is moved along the line. The ratio of maximum to minimum voltage is then read on the standing-wave indicator (usually an amplifier and meter).

For example, assume that the forward signal is 10 V and that the reflected signal is 2 V. This would produce standing waves with 12-V maximums (where the forward and reflected voltages are in phase and adding) and 8-V minimums (where the forward and reflected voltages are out of phase and subtracting). This would mean a voltage standing-wave ratio (VSWR) of $\frac{12}{8}$ or 1.5.

Because the probe must not be allowed to extract any appreciable power from the slotted line, high sensitivity and low noise are required in the detector and indicator. Therefore, the indicator is sharply tuned to the modulation frequency of the source (usually with a 1-kHz filter), thereby reducing noise and allowing the use of a high-gain audio amplifier and voltmeter circuit.

Other considerations relative to accurate slotted line measurements include elimination of harmonics from the source prior to entering the slotted line, a minimum of frequency modulation in the source (preferably no frequency modulation of the source), and a low residual standing-wave ratio in the slotted line itself.

11-9.2. Swept-Frequency Impedance Measurements

Figure 11-26 shows the basic circuit for swept-frequency impedance measurement of microwave equipment, using the *reflection coefficient* system. (This reflection system could also be used at a fixed frequency but is well suited to swept-frequency measurements.)

The basic idea of this method is to hold the forward voltage constant and measure *return loss* of a given load rather than the direct ratio of forward

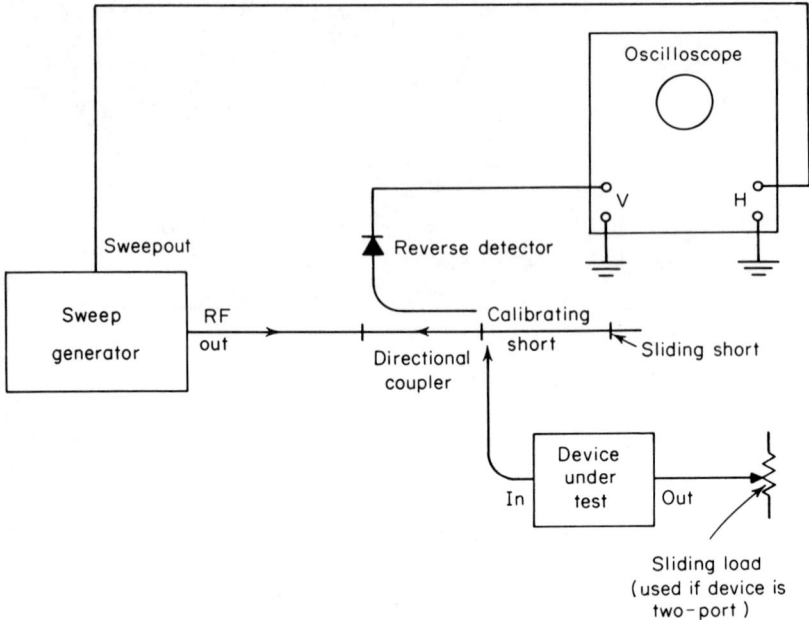

Fig. 11-26. Microwave sweep-frequency impedance measurement.

and reflected signal voltages. This technique eliminates the need for a slotted line and amplitude modulation of the signal source.

In operation, a calibrating short (sliding short) is placed on the waveguide output to set up a 100 per cent reflection to the directional coupler (mounted in the reverse direction) as the generator sweeps the desired operating-frequency range. The output from the reversed directional coupler is applied to the oscilloscope and is adjusted to some convenient reference point *representing the forward signal level*. When the reference is established, the short is replaced by the unknown load. The *amplitude decrease* on the oscilloscope indicates the *load return loss*.

For example, assume that the forward voltage is 10 V and the reflected voltage is 2 V (as in the previous example). With the sliding short in place, the forward (and reflected) signal would be 10 V. The oscilloscope is then adjusted to some convenient level, such as ten vertical scale divisions. After the short is replaced by the unknown load (say an antenna), the oscilloscope indication would drop to two divisions (indicated by a 2-V reflected signal). The difference between the forward and reflected voltages can then be converted to VSWR, if desired.

The term *reflection coefficient* is often used in microwave test equipment instead of VSWR.

$$\text{Reflection coefficient} = \frac{\text{Reflected voltage}}{\text{Forward voltage}}$$

For example, using 10-V forward and 2-V reflected, the reflection coefficient is $\frac{2}{10} = 0.2$.

11-9.2.1. Reflection Coefficient Versus VSWR

Reflection coefficient can be converted to VSWR by dividing (1 + reflection coefficient) by (1 − reflection coefficient). For example, using the 0.2 reflection coefficient, the VSWR is

$$\frac{1 + 0.2}{1 - 0.2} = \frac{1.2}{0.8} = 1.5 \text{ VSWR}$$

VSWR can be converted to reflection coefficient by dividing (VSWR − 1 by VSWR + 1). For example, using the 1.5 VSWR, the reflection coefficient is

$$\frac{1.5 - 1}{1.5 + 1} = \frac{0.5}{2.5} = 0.2 \text{ reflection coefficient}$$

11-9.3. Impedance Measurements with Bidirectional Coupler

Both fixed-frequency and swept-frequency impedance measurements can be made by using a bidirectional coupler. The bidirectional coupler is used to measure both outgoing and reflected power, simultaneously. As shown in Figure 11-27, the bidirectional coupler consists of two enclosed sections attached to each side of a straight section of waveguide and along its narrow dimensions. Each enclosed section countains an r-f pickup probe at one end and a matched load at the other end.

The sections are supplied with energy from within the main waveguide through three openings spaced one-quarter wavelength apart. Energy from the source (for example, a transmitter) going toward a load (for example, an antenna) enters the enclosed sections through the three openings on each side so that the r-f probe farthest away from the transmitter is used to measure direct power and the one nearest the transmitter is used to measure reflected power.

Because the openings in each section are spaced one-quarter wavelength apart, the energy travels a quarter wavelength between each of the three openings and is coupled into the enclosed sections by a predetermined attenuation below that within the waveguide.

The center opening in each of the enclosed sections is larger than each of the holes on either side, thus allowing twice as much energy to enter through

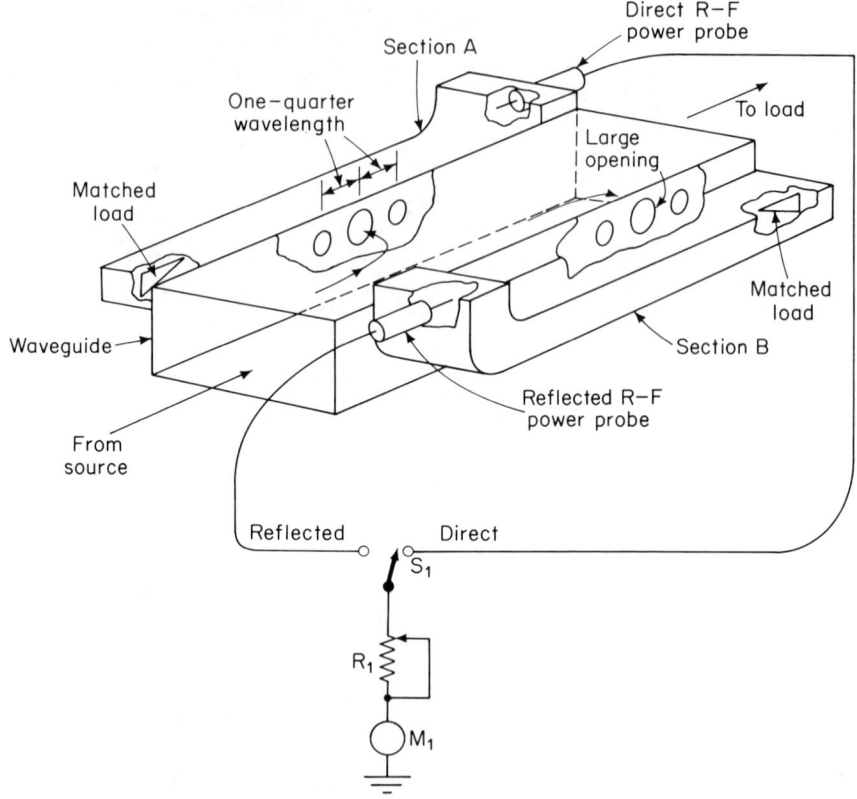

Fig. 11-27. Reflection coefficient (or SWR) measurement using bidirectional coupler.

that opening. The transmitted energy entering section A, because of the location and dimensions of the openings, combines properly in phase and amplitude to make it measurable by the indicating meter connected to the direct power probe. The transmitted energy that enters section B will produce zero power because of the phase displacement of the three openings. Energy passing the first opening will be out of phase with that of the center opening and in phase with the energy at the third opening. Because the center opening is sufficiently large to supply twice the magnitude of energy as that supplied by either of the other two, the reflected energy is cancelled.

The end of the enclosed section is terminated in a matched load. Section B thus measures reflected energy in exactly the same manner that section A measured direct energy.

Since the direction of energy flow is reversed, the energy will now appear at the reflected power probe in section B. Reflected energy, upon entering section A, will be cancelled out in the same manner that section B cancels the transmitter power.

The reflected voltage appears on meter M_1 when S_1 is in the reflected position. Likewise, M_1 indicates outgoing voltage when S_1 is in the outgoing position. Once the outgoing and reflected voltages have been found, the reflection coefficient (not SWR) is determined by using the equations of Section 11-9.2. If desired, the reflection coefficient can be converted to SWR, using the equations of the same Section 11-9.2.1.

This same basic technique is used in commercial SWR meters, at microwave frequencies as well as lower frequencies. (However, the bidirectional couplers for lower-frequency SWR meters are coaxial rather than waveguide.) In the SWR meters, the meter scale is calibrated in SWR, although the actual readout is in reflection coefficient. With such systems, switch S_1 is set to read outgoing voltage, and resistor R_1 is adjusted until the meter needle is aligned with some "set" or "calibrate" line (near the right-hand end of the meter scale). Switch S_1 is then set to read reflected voltage, and the meter needle moves to the corresponding SWR indication.

11-10. MICROWAVE ATTENUATION MEASUREMENTS

It is possible to measure the attenuation of microwave components at fixed frequencies. However, the swept-frequency techniques are becoming more popular for reasons discussed in Section 11-8. In addition, it is possible to use an X-Y recorder instead of an oscilloscope with swept-frequency measurements. An X-Y recorder (Chapter 5) requires a sweep voltage or trigger voltage (supplied by the sweep generator) and provides a permanent record on chart paper instead of an oscilloscope display. The X-Y recorder/sweep generator combination can also be used for other microwave measurements, such as SWR, impedance, power, and frequency, if desired.

The basic setup for making swept-frequency attenuation measurements is shown in Figure 11-28. A portion of the sweep generator output signal is removed by the directional coupler and fed back to the power output level-control circuits of the sweep generator. This closed loop provides a means of maintaining the sweep generator output constant (automatic-leveling control). The second directional coupler is arranged so that the reflected channel of the impedance test setup (as described in Section 11-9.2) now becomes the transmission channel.

The system is first calibrated by placing a length of loss-free waveguide (or a sample of the waveguide with which the device under test is to be used) between the sweep generator output and the X-Y recorder. A reading is obtained on the X-Y recorder with the section of waveguide in the circuit.

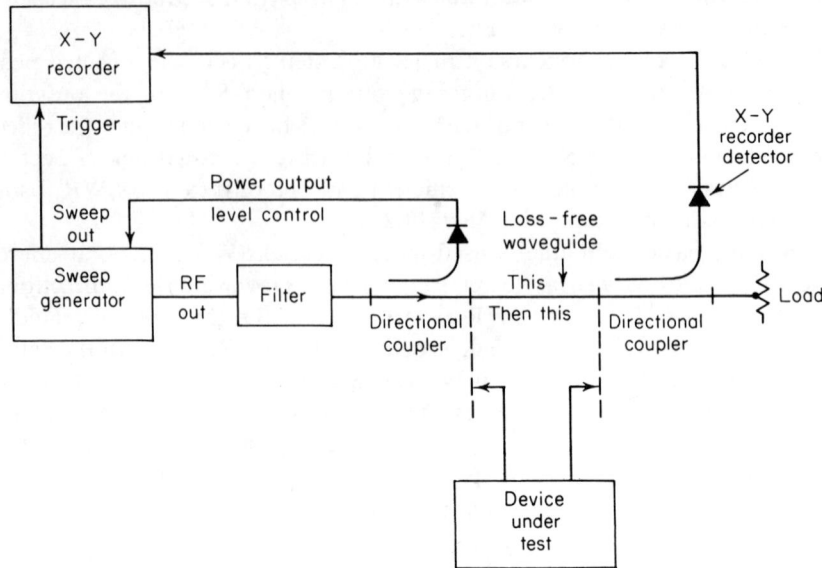

Fig. 11-28. Microwave swept-frequency attenuation measurement.

The X-Y recorder amplitude controls are adjusted to some specific reference level representing zero attenuation. The waveguide is then removed, and the device to be tested is connected in its place. The resulting signal will then be detected, and the value will be displayed on the X-Y recorder. Usually, X-Y recorder indications are related to voltage rather than power. Therefore, the ratio between the two voltages can be converted to a loss figure expressed in decibels, if desired. This type of measurement is known as *insertion loss*. For insertion loss measurements in microwave equipment, decibel is defined as

$$1 \text{ dB} = 10 \log \frac{P_1}{P_2}$$

where P_1 is the power absorbed at the load without the device in the line, and P_2 is the power absorbed with the device in the line.

11-11. MICROWAVE FREQUENCY MEASUREMENTS

There are four items of test equipment used in microwave frequency measurements; the lecher line, the L-C meter, the cavity meter, and some form of heterodyne or beat-frequency meter.

The *lecher line* is used primarily in experimental microwave work to measure wavelength. A typical lecher line is simply a two-wire line with a sliding shorting bar (or meter with sliding contacts) across the two lines, as shown in Figure 11-29.

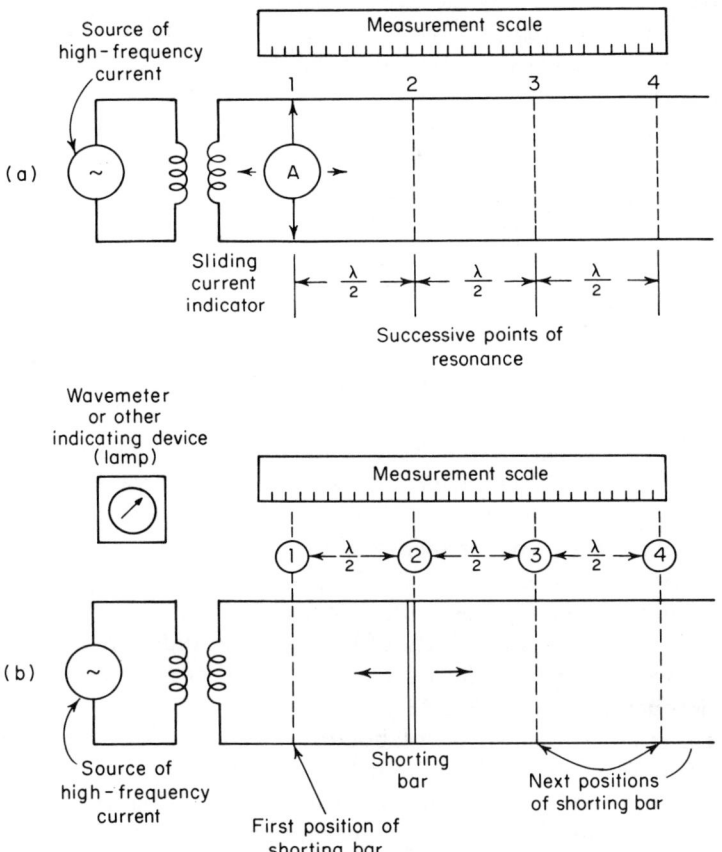

Fig. 11-29. Lecher lines for microwave frequency measurement.

In Figure 11-29A, the current-indicating device (probably a thermocouple ammeter) is moved along the line to measure points of maximum and minimum current. The distance between successive maximums or minimums is equal to a half wavelength.

In Figure 11-29B, the shorting bar is moved until the line becomes resonant at the generator frequency, as indicated by the wavemeter. The actual wavelength can be measured and the frequency calculated from this measurement. Lecher lines are used at frequencies up to about 3 GHz.

The L-C meter is simply a resonant circuit with variable inductance or capacitance. L-C meters are usually not accurate above about 1.5 GHz and can cause several problems in use. For example, the L-C meter must be very lightly coupled to the circuit under test. Otherwise, inaccuracies could occur, and the meter indicating device could burn out. For these reasons, the L-C meter has been replaced by the cavity wavemeter.

The *cavity wavemeter* is typical of passive microwave frequency test instruments. Figure 11-30 is a typical cavity wavemeter. As shown, the cavity

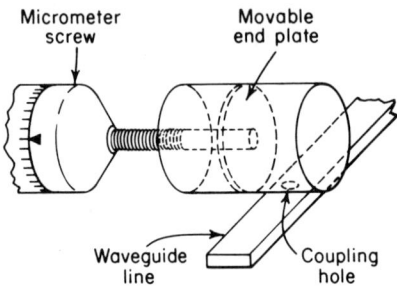

Fig. 11-30. Typical cavity wavemeter for microwave frequency measurement.

wavemeter is a two-part device that absorbs part of the input power in a tunable resonant cavity. When the cavity is tuned to resonate at the input frequency, a dip occurs in the output power level that is measured by a meter or oscilloscope. Frequency is then read directly on a calibrated dial driven by the cavity tuning mechanism.

The accuracy of cavity wavemeters depends on the cavity Q, tuning-dial calibration, backlash, and effects of temperature and humidity variations. Present-day cavity wavemeters are available with accuracies of a few parts in 10^6, at frequencies up to 40 GHz.

Wavemeters are often used in laboratories to check signal generator accuracy. The wavemeters are calibrated against some form of heterodyne or beat-frequency standard at routine intervals. (Refer to Chapter 8.) The wavemeters are then used to check signal generator frequency just prior ot use or are incorporated into the test circuit to provide continuous monitoring of the signal generator frequency.

The *heterodyne frequency* meters used for microwave work are similar to those of lower frequency (Chapter 8). In basic operation, a signal of known frequency is mixed with a signal of unknown frequency. The unknown signal frequency is then adjusted until there is an indication that it is at the same frequency as the known signal (or multiple thereof).

In microwave work, the heterodyne frequency measurement system most often involves a sweep generator, transfer oscillator, and electronic counter. Such systems provide frequency measurements up to about 40 GHz, with accuracies up to a few parts in 10^8.

Microwave Frequency Measurements 397

Figure 11-31 shows a typical heterodyne measurement setup used to calibrate cavity wavemeters. In operation, the sweep generator is set for a leveled, rapidly sweeping output across a small segment of the cavity wavemeter's band, with the main power being fed through the wavemeter to a diode detector. The detected wavemeter dip is fed to the vertical input of the oscil-

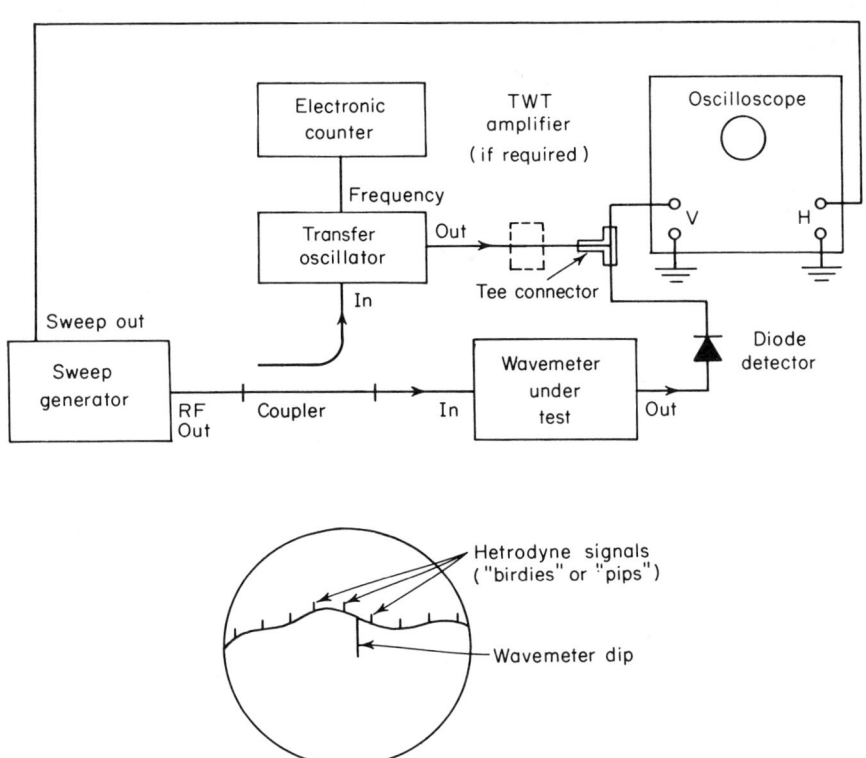

Fig. 11-31. Microwave sweep-frequency measurement of wavemeters.

loscope. A portion of the sweep generator output is applied to the input of the transfer oscillator (also known as a frequency converter). The transfer oscillator is essentially a crystal-controlled oscillator of high accuracy and a mixer circuit. (Usually, some form of crystal mixer is used.) Harmonics of the transfer oscillator mix with the sweeping input signal, producing heterodyne signals (pips) at intervals equal to the transfer oscillator fundamental frequency. The actual transfer oscillator frequency is read out by the electronic counter.

For example, if the transfer oscillator frequency is 200 MHz and the input frequency is sweeping from 10 to 20 GHz (or any other band), pips will appear every 200 MHz throughout the swept band, as shown in Figure 11-

31. With such a display on the oscilloscope, the wavemeter is adjusted for coincidence of its dip and that of the reference pips, noting the accuracy of the cavity wavemeter dial calibration at each point.

As shown in Figure 11-31, the output of a typical transfer oscillator can be used directly up to about 18 or 20 GHz. At higher frequencies, it is usually necessary to use some form of amplification (such as a traveling-wave tube) for transfer oscillator output.

APPENDIX

APPENDIX: DIGITAL LOGIC CIRCUITS

It is necessary to have a knowledge of digital circuits to understand operation of laboratory test equipment. For some time, digital meters and electronic counters have used logic circuits. Today, logic techniques are used with many types of test equipment. For example, such devices as bridges, meters, and other measuring instruments have incorporated circuits that convert the value being measured into binary-coded-decimal (BCD) form for output to another system.

The information in this Appendix is not intended as a basic course in digital circuits or logic symbology but as a summary and refresher. If the readers understand the data in this Appendix, they should have no difficulty in working their way through the logic diagrams and schematics found in the instruction manuals of test equipment manufacturers.

A-1. LOGIC SYMBOLOGY

Logic symbols and diagrams are a form of language used in digital circuits. One of the major problems encountered in digital work is the fact that each manufacturer speaks a different dialect of the same language. No two major manufacturers use identical symbols in the instruction manuals to represent the same circuit element.

This problem was supposedly eliminated when MIL-STD-806 was introduced. However, many manufacturers still do not follow this government standard exactly, some with good reason. For example, the logic symbols used by Hewlett-Packard in their instruction manuals clarify many points left unsaid by MIL-STD-806. Throughout the following discussions, the military standard symbols will be used (where an MIL-STD-806 symbol exists).

Manufacturer's modifications will also be given (as an alternate) where they clarify operation of the circuit or symbol function.

A-2. BINARY LOGIC

In the binary system, information is broken down into elementary bits. For example, all numbers can be made up using only zeros and ones, rather than zero through nine, as in the familiar decimal system. Consequently, instead of requiring ten different values to represent 1 digit, digital equipment using the binary method needs only *two values* (thus the term binary) for each digit.

Each bit in the binary system can exist in only one of two possible states. For logic, these states are designated "1" and "0." A 1 means *yes, assertion, enable,* or *true*. Consequently, 0 means *no, negation, disable* (or *inhibit*), or *false*. Use of the words *true* and *false* does not imply that one state is more important than the other. The states are conditions, and both states are equally significant and are used equally in a two-state system.

A single bit is always used to represent a function in binary logic. For example, suppose the information to be conveyed is the presence (or absence) of a count. If the count has been received (or perhaps stored), the bit representing the count would be in the 1 state for presence or truth, otherwise the bit would be in the 0 state, denoting absence or falsity.

The word bit is used to denote

1. An assertion or negation of a variable
2. Storage of a variable

In electronic equipment, true and false are generally represented by voltage or current levels and must be defined. For example, true could be $+12$ V, false, -12 V; true could be presence of current, false the absence of current. Generally, voltage levels, rather than current levels, are used to define the true and false states.

In some (but not all) logic diagrams, a plus or minus sign may be used within any logic symbol to further define the true state for that element. A plus sign within a logic symbol means that the *relatively positive level* of the two logic voltages at which that circuit operates is said to be true. Note that the true voltage level does not have to be absolutely positive (that is, above ground or above 0-V reference). The two voltage levels at which a logic circuit operates could be -24 and -12 V. A plus sign within the logic symbol representing such a circuit would indicate that the -12-V level is true and the -24-V level is false. This is because the -12 V is closer to positive than -24 V. A minus sign within the symbol would indicate that the -24 V level is true (since it is more negative) and that the -12-V level is therefore false.

The sign should be used for all logic elements in which true and false levels are meaningful. As an alternate, the logic diagram can state in a note that all logic is positive true or all is negative true. (MIL-STD-806 does not distinguish positive true from negative true.)

A-3. BINARY NUMBERS

In digital equipment, numbers are usually represented in binary form. The binary bits representing the number are each assigned a *weight* or value; thus the number of bits required depends on the magnitude of the number to be represented. Because each bit can exist in only two states (1 or 0), the weights assigned to each successive bit can increase by a maximum factor of 2. Thus, in a pure binary form number, the weights assigned to successive bits are

32	16	8	4	2	1	.	$\frac{1}{2}$	$\frac{1}{4}$	$\frac{1}{8}$	$\frac{1}{16}$
2^5	2^4	2^3	2^2	2^1	2^0	.	2^{-1}	2^{-2}	2^{-3}	2^{-4}

Since each bit is either true or false, the *weights of all true bits are added* to obtain the number. For example, the decimal number 22.5 in pure binary is 10110.1000:

Binary weight	2^4	2^3	2^2	2^1	2^0	.	2^{-1}	2^{-2}	2^{-3}	2^{-4}
Decimal weight	16	8	4	2	1	.	$\frac{1}{2}$	$\frac{1}{4}$	$\frac{1}{8}$	$\frac{1}{16}$
Binary number	1	0	1	1	0	.	1	0	0	0

By adding all of the true bits ($16 + 4 + 2 + \frac{1}{2}$) the total number (22.5) may be found.

If all of the possible bits were true (giving a binary number 11111.1111), the decimal number represented would be $16 + 8 + 4 + 2 + 1 + \frac{1}{2} + \frac{1}{4} + \frac{1}{8} + \frac{1}{16}$, or $31 - \frac{15}{16}$. If the number 32 were needed, the 2^5 bit would have to be added to the left of the decimal point. (In the pure binary system, the decimal point is called the *binary point*.) Note that bits can also be added to the right of the binary point, thus increasing resolution.

A-4. BINARY-CODED-DECIMAL SYSTEM

The binary-coded-decimal (BCD) system combines the advantages of the binary system (the need for only two states, 1 or 0) and the convenience of the familar decimal representation. In the BCD system, a number is expressed in normal decimal coding, but each digit in the number is expressed in binary form.

For example, the number 37 in pure binary BCD form would appear as

	Tens Digit	Units Digit
BCD	0011	0111
(Pure binary weight)	(8421)	(8421)
Decimal	3	7

Four bits are needed for each digit. In general, four bits yields 2^4 or sixteen possible combinations, as shown in the following Table A-1.

TABLE A-1. TYPICAL BCD CODES USED IN TEST EQUIPMENT

Code Bits	Decimal Equivalents			
	8421	42*21	XS-3	2421
0000	0	0		0
0001	1	1		1
0010	2	2		2
0011	3	3	0	3
0100	4		1	4
0101	5		2	5
0110	6	4	3	6
0111	7	5	4	7
1000	8		5	
1001	9		6	
1010			7	
1011			8	
1100		6	9	
1101		7		
1110		8		8
1111		9		9

Note that there are many BCD codes used other than the *pure binary* 8421 code. Three typical codes (42*21, XS-3, and 2421) are shown in Table A-1. In every case, four bits are required for each digit. Likewise, six combinations of four bits are unused in each code. The six unused combinations are often referred to as *forbidden codes*.

Besides the pure binary and other BCD forms, it is possible to express numbers in test equipment applications by other means. One such system is the *10-line code*, where each digit is represented by ten bits, with each bit weighted zero through nine. For a given digit, only one bit of the 10-line code can be true at one time. This 10-line code is sometimes known as *multiple-line* code. A variation of this system is used in electronic counter readouts, as described in Chapter 6.

The *negation* or "not" function is also used in BCD systems. The not condition is indicated by a bar above the bit identification. For example, a "not eight" is $\bar{8}$. For all numbers where the 8-bit is true, the $\bar{8}$-bit is, by definition, false. Note that the $\bar{8}$-bit is a separate bit but is related by definition to the 8-bit. Conversely, for all numbers where the 8-bit is false (not true), the $\bar{8}$-bit is true.

A-5. BASIC DIGITAL LOGIC ELEMENTS AND SYMBOLS

Although there are many variations of digital logic elements and their symbols, there are only four basic classes or groups. These are *gates, amplifiers, switching elements,* and *delay elements*.

A-5.1. Gates

A gate is a circuit that produces an output on condition of certain rules governing input combinations. (The specific rules for various gates are discussed in later paragraphs of this Appendix.)

As shown in Figure A-1, the basic gate symbol has input lines connecting

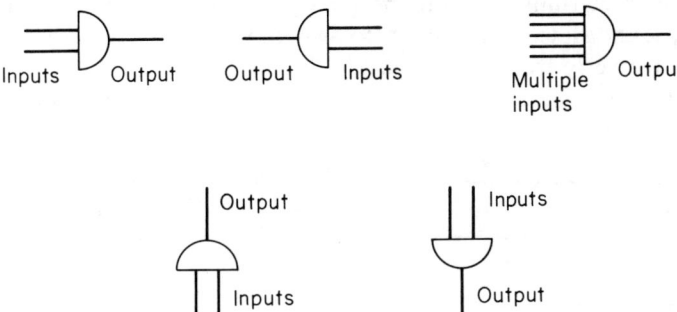

Fig. A-1. Basic gate symbols.

to the flat side of the symbol and output lines connecting to the curved side. Since inputs and outputs are thus easily identifiable, the symbol can be shown facing left or facing right (or facing up or down), as necessary.

A-5.2. Amplifiers

Amplifiers are not necessarily limited to digital work. However, when so used, the driving or input signals will normally be digital (pulsed, binary, and so on). Consequently, the output of the amplifier will be an amplified and otherwise modified form of the input.

As shown in Figure A-2, the amplifier symbol is an equilateral triangle,

Appendix: Digital Logic Circuits

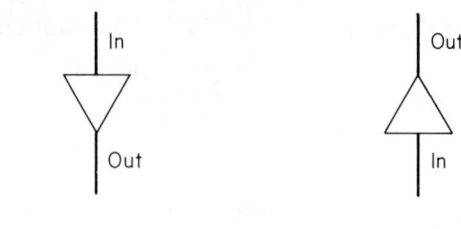

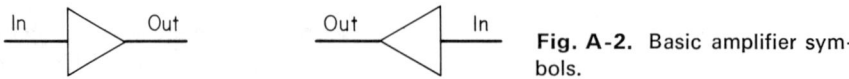

Fig. A-2. Basic amplifier symbols.

with the input applied to the center of one side and the output connected to the opposite point of the triangle. Like gates, the amplifier may be shown in any of four positions.

A-5.3. Switching Elements

Switching elements used in digital work are some form of *multivibrator:* bistable (flip-flop, Schmitt trigger), monostable (one-shot), and astable (free-running multivibrator). According to the type of circuit, inputs cause the state of the circuit to switch, reversing the outputs; that is, an output formerly true will switch to false, and vice versa.

As shown in Figure A-3, the most common basic symbol for switching

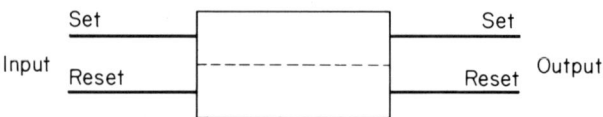

Fig. A-3. Basic switching element symbol.

circuits is a horizontal rectangle, divided horizontally, with the upper portion representing the set side and the lower portion representing the reset side.

A switching element is said to be set when the *output from the set side is true*. The element is reset when the output from the reset side is true.

Inputs are on the left; outputs are on the right. To avoid confusion, switching elements are always drawn facing the same way (or should be so drawn).

A-5.4. Delay Elements

A delay element provides a *time delay* between input and output signals. As shown in Figure A-4, the symbol accepts the input on the left

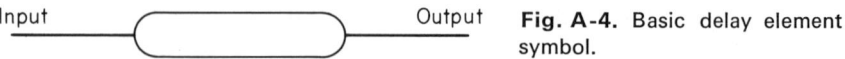

Fig. A-4. Basic delay element symbol.

and provides output on the right. Like switching elements, delay elements are always drawn facing the same way.

A-6. MODIFICATION OF LOGIC SYMBOLS

The basic logic symbols are usually modified to express circuit conditions. Although each manufacturer may have its own set of modifiers, together with those of MIL-STD-806, the following modifiers are in general use.

A-6.1. Truth Polarity

As previously discussed, positive (+) or negative (−) indicators may be placed inside a symbol to designate whether the true state for that circuit is positive or negative, *relative* to the false state. This is frequently done with gates and switching elements, as shown in Figure A-5. (The FF designation in Figure A-5 indicates that the particular switching element is a flip-flop.)

Where all symbols on a particular diagram have the same polarity, a note to the effect that all logic is positive-true or all is negative-true may be used instead of having individual polarity signs in each symbol.

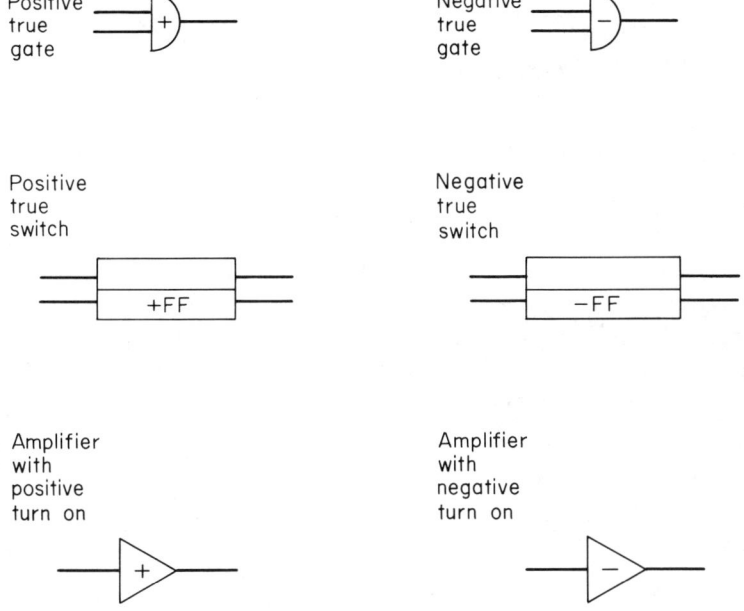

Fig. A-5. Truth polarity in symbols.

Polarity signs used in amplifier symbols do not have any direct logic significance. Rather, the polarity signs are a troubleshooting aid, indicating the *polarity required to turn the amplifier on*.

As shown in Figure A-5, the positive-true gate and positive FF operate with true levels positive with respect to the false levels. Similarly, the negative-true gate and FF operate with true levels negative with respect to the false levels.

A-6.2. Inversion

Generally, logic inversion is indicated by an inversion dot at inputs or outputs. (In some cases, inverted pulses are shown at the inputs and outputs.)

When the inversion dot appears on an input (generally only on gates and switching elements), the input will be effective when the input signal is of opposite polarity to that *normally required*. For example, if the switching element in Figure A-6 is normally positive-true or is used on a diagram where

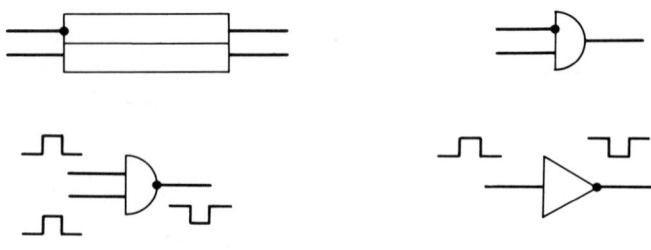

Dots may be solid or open

Fig. A-6. Methods of indicating inversion in symbols.

all logic is positive-true, a negative input at the inversion dot will set the circuit.

When the inversion dot appears at an output (generally only on gates and amplifiers), the output will be of opposite polarity to that *normally delivered*. For example, if the gate of Figure A-6 is used in a positive-true logic circuit, the output will be negative. Likewise, the amplifier in Figure A-6 will produce a positive output if the input is negative, and vice versa.

A-6.3. Alternating-Current Coupling

Capacitor inputs to logic elements are indicated by an arrow, as shown in Figure A-7. In the case of gates and switching elements, the element responds only to a change of the a-c-coupled input in the true-going direction.

AND Gates 409

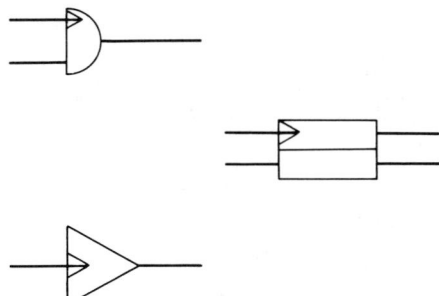

Fig. A-7. Methods of indicating a-c coupling in symbols.

An inversion dot used in conjunction with the coupling arrow indicates that the element responds to a change in the false-going direction.

In the case of an amplifier, a pulse edge of the same polarity as given in the symbol turns the amplifier on briefly, then off as the capacitor discharges. The output is then a pulse of the same width as the amplifier on-time. With an inversion dot at the amplifier output, the output pulse is inverted.

A-7. AND GATES

The symbol, basic circuit, and truth table for the AND gate is shown in Figure A-8.

By definition, for output C to be true, both inputs A and B must be true, hence the term AND gate. This can be illustrated by means of the truth table in Figure A-8. (Similar truth tables are found in many logic diagrams.) In this table, 1 represents true, and 0 represents false. Note that C is true (1) only when both A and B are true, and that the truth table does not define true and false; that is, the table does not spell out positive-true, negative-true, the voltage level for true or false, and so on.

If the AND gate is positive-true, a plus sign should be given in the symbol. A negative-true AND gate should have a minus sign in the symbol. (Note that MIL-STD-806 does not distinguish positive-true from negative-true.)

Assume that the respective true-false levels for the two gates in Figure A-8 are $+5$ V/0 V and -5 V/0 V, as shown. These two values can be substituted for 1 and 0, as in the following tables:

AND Gate Truth Tables					
Positive Gate			Negative Gate		
A	B	C	A	B	C
0 V	0 V	0 V	0 V	0 V	0 V
0 V	+5 V	0 V	0 V	−5 V	0 V
+5 V	0 V	0 V	−5 V	0 V	0 V
+5 V	+5 V	+5 V	−5 V	−5 V	−5 V

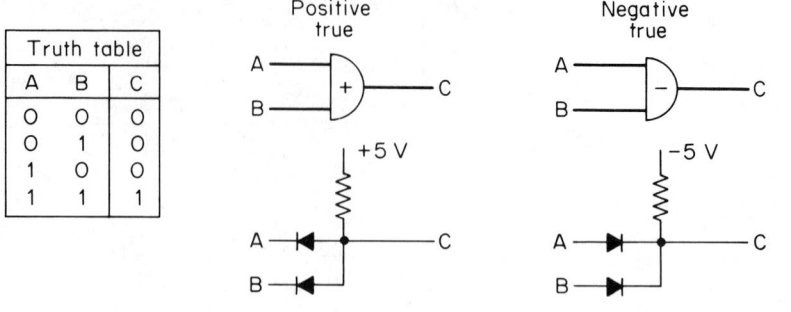

Truth table		
A	B	C
0	0	0
0	1	0
1	0	0
1	1	1

Multiple input

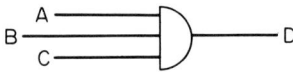

Truth table			
A	B	C	D
0	0	0	0
0	0	1	0
0	1	0	0
0	1	1	0
1	0	0	0
1	1	0	0
1	1	1	1

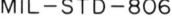

MIL–STD–806

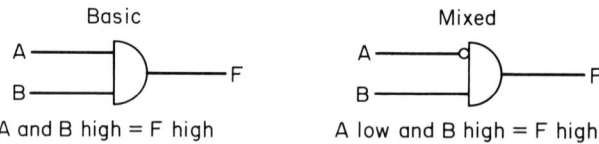

Basic — A and B high = F high

Mixed — A low and B high = F high

Hewlett–Packard

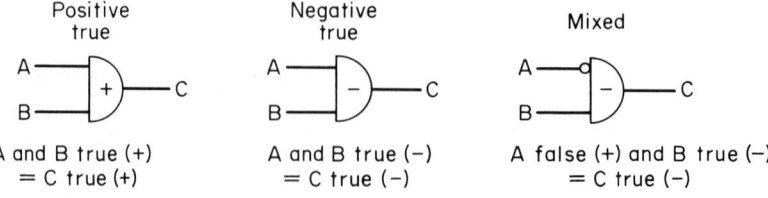

Positive true — A and B true (+) = C true (+)

Negative true — A and B true (−) = C true (−)

Mixed — A false (+) and B true (−) = C true (−)

Fig. A-8. AND gate symbols.

In both cases, the true voltage level appears at output C when the A and B input diodes are reverse-biased by true inputs. At all other times, the false level (approximately 0 V) exists at output C. For example, in the positive-true gate, if A is at 0 V, current will flow through the corresponding diode, lowering the voltage at C to approximately 0 V. The same will occur if B (or both A and B) is at 0 V. If both A and B are at $+5$ V, no current will flow through either diode, and C will rise to approximately $+5$ V.

AND gates are not restricted to two inputs but may have any number of inputs, including one input in special cases. (Single-input AND gates require that the input be of a given polarity and/or level to produce a true output. Single-input gates are not truly AND gates but are often so described.)

For multiple-input AND gates, the rule is that *all inputs must be true for the output to be true*, as shown in Figure A-8.

A-8. OR GATES

The symbol, basic circuit, and truth table for the OR gate are shown in Figure A-9.

By definition, for C to be true, *either* input A *or* input B must be true, hence the term OR gate. This is illustrated by the truth table of Figure A-9. Note that C is true (1) whenever any of the inputs is true and that the truth table does not define true and false. If a plus sign or a minus sign is given in the symbol, the OR gate is defined as positive-true or negative-true, respectively.

Note that the circuit for a negative-true OR gate is the same as for a positive-true AND gate, and vice versa. This produces inversion of the two circuit functions. For example the 0-V level is now defined as true, while the $+5$ V and -5-V levels are false for the two gates in Figure A-9. These two values can be substituted for 1 and 0, as in the following tables:

OR Gate Truth Tables					
Positive Gate			Negative Gate		
A	B	C	A	B	C
-5 V	-5 V	-5 V	$+5$ V	$+5$ V	$+5$ V
-5 V	0 V	0 V	$+5$ V	0 V	0 V
0 V	-5 V	0 V	0 V	$+5$ V	0 V
0 V	0 V	0 V	0 V	0 V	0 V

As with AND gates, OR gates may have more than two inputs. For multiple-input OR gates, the rule is that *any true input will produce a true output*, as shown in Figure A-9.

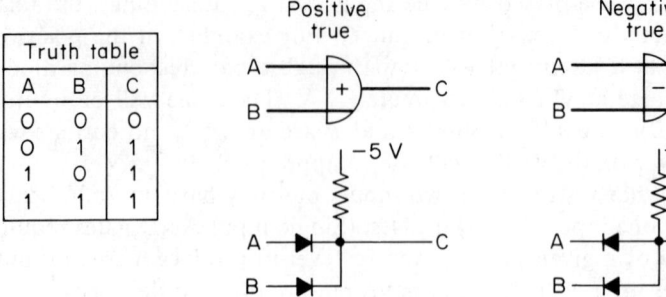

Multiple input

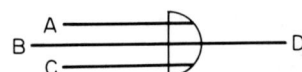

MIL-STD-806

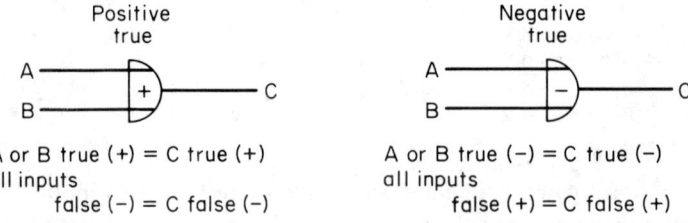

Fig. A-9. OR gate symbols.

A-9. NAND GATES

The symbol, basic circuit, and truth table for the NAND gate are shown in Figure A-10.

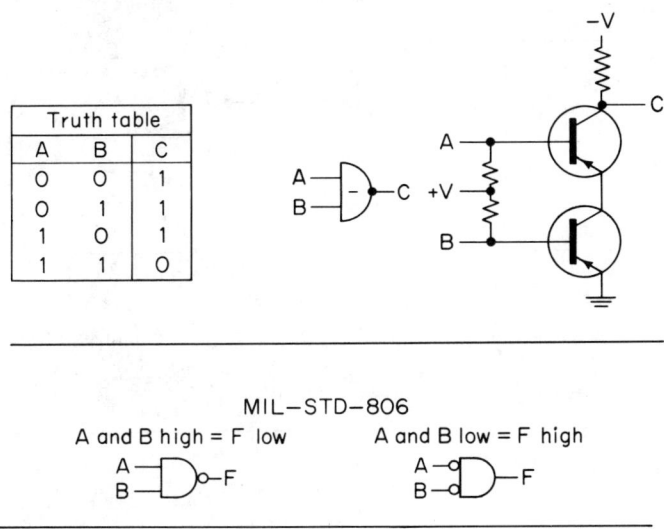

Fig. A-10. NAND gate symbol.

The NAND gate is a variation of the conventional AND gate, delivering an *inverted* (false) output when *all inputs are true*. When either or both inputs are false, the output is true. The term NAND is a contraction of "not and."

A NAND gate uses an active inverting element in the gate circuitry and may have any number of inputs.

Operation of the NAND gate in Figure A-10 is as follows. For output C to be positive (at or near ground level), both Q_1 and Q_2 must be conducting. This requires that *both A and B* inputs be negative. If either A or B is positive, then either Q_1 or Q_2 is biased off, and the output is negative (C will assume the level of $-V$).

A-10. NOR GATES

The symbol, basic circuit, and truth table for the NOR gate are shown in Figure A-11.

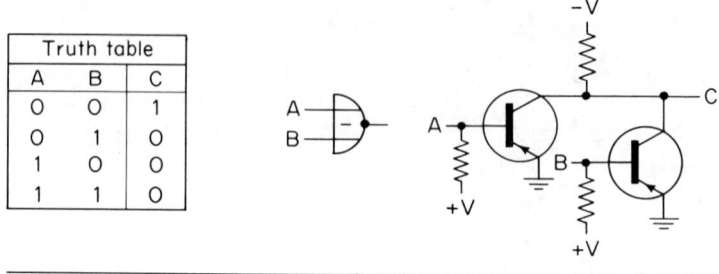

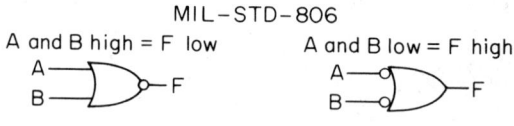

MIL-STD-806
A and B high = F low   A and B low = F high

Hewlett–Packard
(Positive true shown)

A ──┐
    │+ ├── C
B ──┘

Either A or B true (+) = C false (−)
All inputs false (−) = C true (+)

Fig. A-11. NOR gate symbol.

A NOR gate is a variation of the conventional OR gate, delivering an *inverted* (false) output when *any or all or its inputs are true*. When all inputs are false, the output is true. The term NOR is a contraction of "not or."

A NOR gate uses an active inverting element in the gate circuitry and may have any number of inputs.

Operation of the NOR gate in Figure A-11 is as follows. For output C to be positive (at or near ground level), either or both Q_1 and Q_2 must be conducting. This requires that either or both inputs A and B are negative. If both A and B are positive, Q_1 and Q_2 are biased off, and the output is negative (C will assume the level of $-V$).

A-11. EXCLUSIVE OR GATES

The symbol, basic circuit, and truth table for the EXCLUSIVE OR gate are shown in Figure A-12.

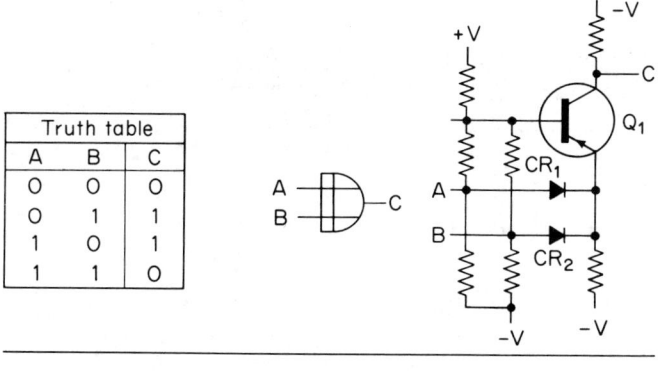

MIL-STD-806

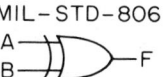

F is high if (and only if) any one input is high, and all other inputs are low.

Hewlett-Packard

A ─⫤D─ C
B ─⫤

One input (A or B) true makes C true. Both inputs (A and B) true or false makes C false

Fig. A-12. EXCLUSIVE OR gate symbol.

An EXCLUSIVE OR gate is a special type of OR gate. It has two inputs, and the output will be true if one, *but not both*, of the inputs is true. The converse statement is equally accurate: the output will be false if the inputs are *both true* or *both false*.

The EXCLUSIVE OR gate is independent of polarity and is not generally spoken of as being either positive-true or negative-true.

Operation of the EXCLUSIVE OR gate in Figure A-12 is as follows. If either A or B is positive, Q_1 will be biased on via CR_1 or CR_2, and the C output is positive. If A and B are both negative, Q_1 is biased on, but the voltage at the Q_1 emitter is negative, and the C output is negative. With A and B both positive, Q_1 is biased off, and the C output remains negative. (These results require proper selection of resistor values and operating voltages.)

A-12. ENCODE GATES

The symbol, basic circuit, and truth table for the ENCODE gate are shown in Figure A1-13.

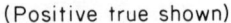
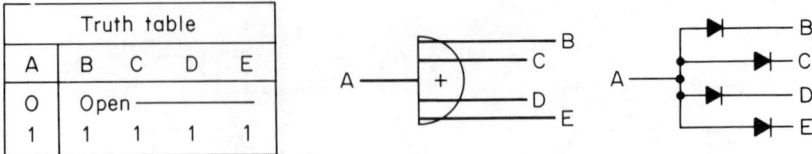

Input A true (+) makes all
outputs (B, C, D, E) true (+)

Fig. A-13. ENCODE gate symbol (Hewlett-Packard).

The ENCODE gate has one input and multiple outputs. When the input is true, all outputs are true. When the input is false, the output may be true or false, depending on the state of the logic element to which they are connected.

A plus or minus sign can be placed in the symbol to indicate the true state. In the circuit of Figure A-13, with *A* positive, all diodes conduct and all outputs are clamped positive. This would be true or 1 in a positive-true system. With *A* negative, in the same system, all diodes would be open (non-conducting), producing a false or 0.

A-13. AMPLIFIERS, INVERTERS, AND PHASE SPLITTERS

Amplifiers (see Figure A-14), both operational and differential, are discussed in Chapter 7. When an amplifier is used in digital work, it is assumed that the output will be essentially the same as the input but in amplified form; that is, a true input will produce a true output, and vice versa. When inversion occurs, an inversion dot (or possibly an inverted pulse symbol) is placed at the output. Usually, the element is then termed an *inverter* rather than an amplifier, even though amplification may occur.

If a plus or minus sign is used in the symbol, this indicates the *input polarity* required to turn the amplifier on.

One amplifier or inverter symbol may represent any number of amplification stages or, optionally, separate symbols may be shown for each stage. Logic symbols, by themselves, do not necessarily imply a specific number of components but rather relate to overall logic effect.

Sometimes an amplifier will be used as a *phase splitter*, that is, one input with dual outputs. One of the dual outputs is in phase with the input, while the other output is out of phase with the input.

A similar case exists with *differential amplifiers*, which have dual inputs and dual outputs (although some differential amplifiers have dual inputs and a single output).

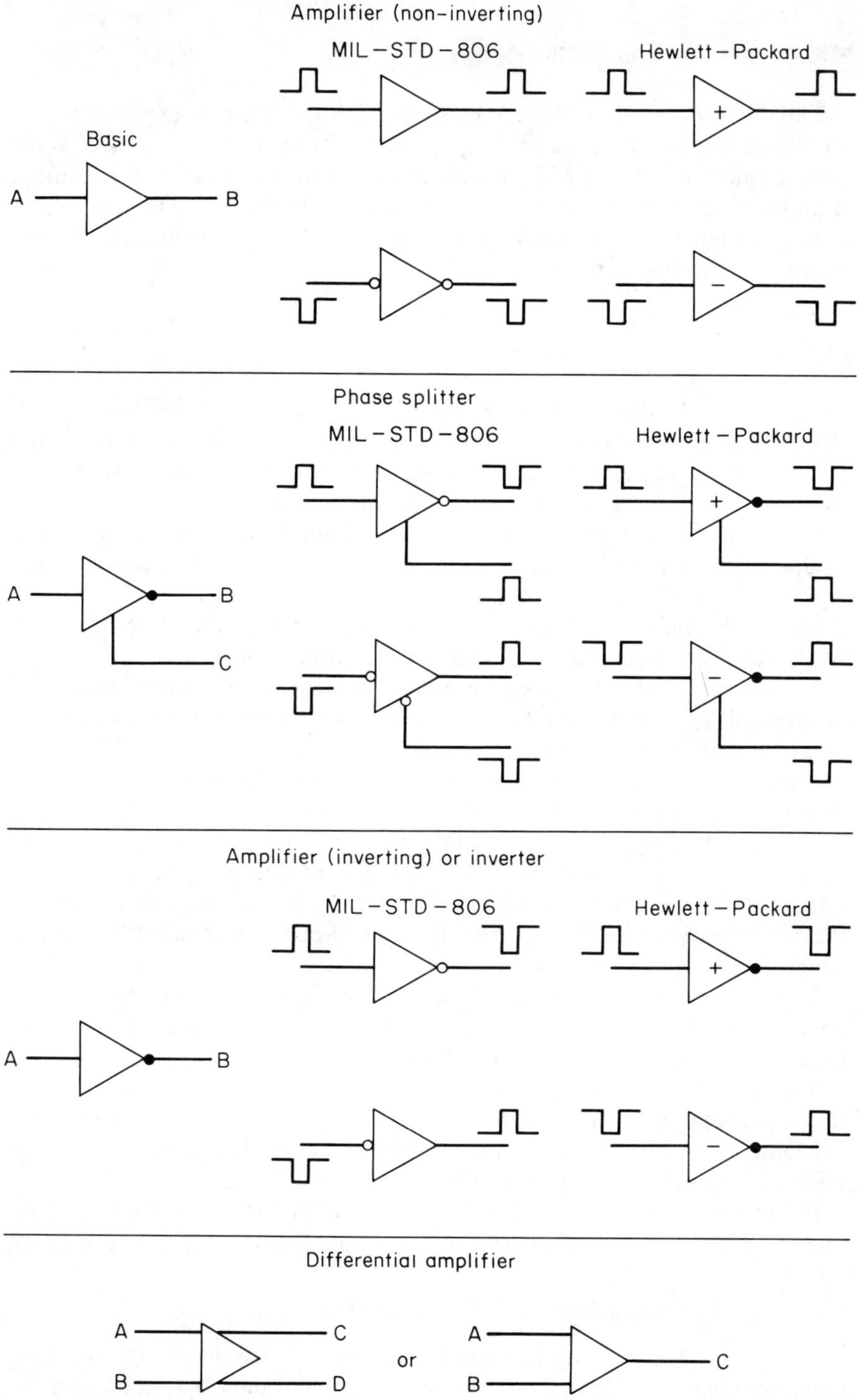

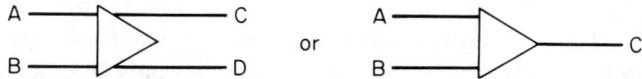

Fig. A-14. Amplifier, inverter, and phase-splitter symbols.

The rule for inversion dots on phase splitters and (particularly) differential amplifiers is that the dot indicates inversion of an output *with respect to the corresponding input* (not with respect to the opposite side of the amplifier). If an inversion dot were placed on the C output of the differential amplifier shown in Figure A-14, this would mean that the C output would be inverted with respect to the A input.

A-14. FLIP-FLOPS

As discussed in Chapter 6, a flip-flop is a bistable switching element, meaning that it takes an external signal to set the element and another signal to reset it. The flip-flop will remain in a given state until switched to the opposite state by the appropriate external signal.

There are many types of flip-flops. Those used most frequently in test equipment circuits are the reset-set (R-S), reset with clock, J-K, toggle, latching, and delay flip-flops.

The basic symbol and rules for application of flip-flops are shown in Figure A-15. The following notes provide an explanation of the rules.

The letters FF should appear in either the upper or the lower portion of the symbol, thus identifying the element as a flip-flop (rather than a one-shot multivibrator, Schmitt trigger, and so on).

A flip-flop is assumed to be the simple R-S type if no other identification is made. When a *clock* input is added, the identifying letter C is placed inside the symbol. A clock input is usually a repetitive pulse (for example, from a time base or programmer) and is *parallel-connected to both* the set and reset side (as discussed in Chapter 6). Clock pulses are transient operated; that is, they are effective on leading or trailing edges of pulses, somewhat like an a-c-coupled input.

If the clock input has no inversion dot, the input is effective on the true-going edge of the clock pulse. If an inversion dot is shown at the clock input, the clock input is effective on the false-going edge.

Multiple inputs on the *same side* of the FF symbol require the logical AND function (both inputs must be true to set or reset).

Multiple inputs diagonally on the corner of the symbol require the logical OR function (either input true will set or reset).

In some cases, a gate symbol (AND or OR) will be shown at the set or reset inputs. (This is particularly true where multiple inputs are required.)

A-14.1. *Reset-Set (R-S) Flip-Flop*

The R-S flip-flop has a minimum of two inputs, set and reset (A and B), and usually two outputs, set output and reset output (D and $\bar{D}$), as shown in Figure A-16.

The letter $\bar{D}$ indicates that the reset output, whether a 1 or a 0, is always

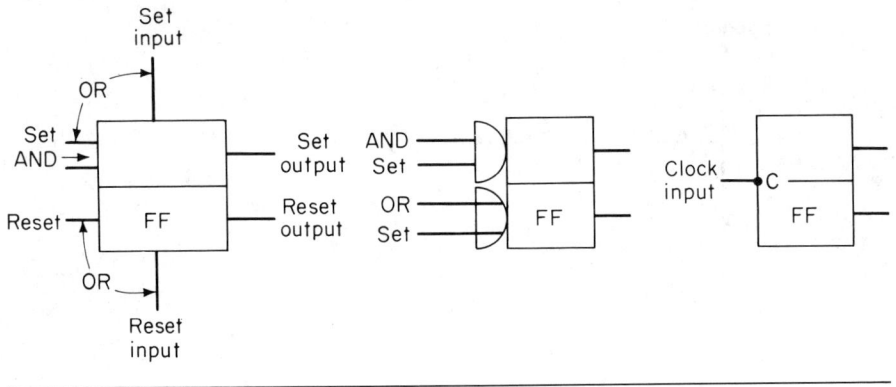
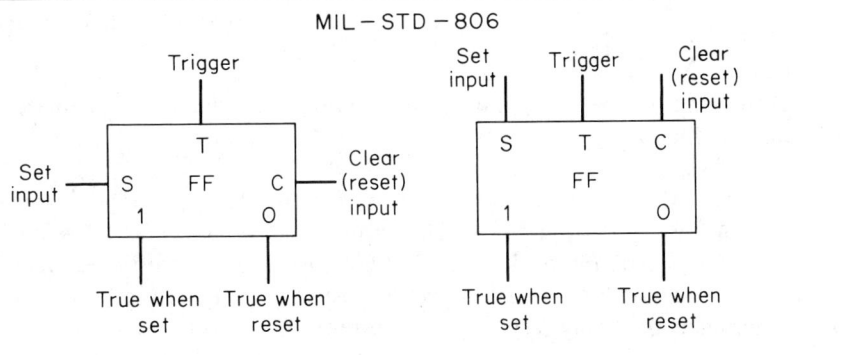
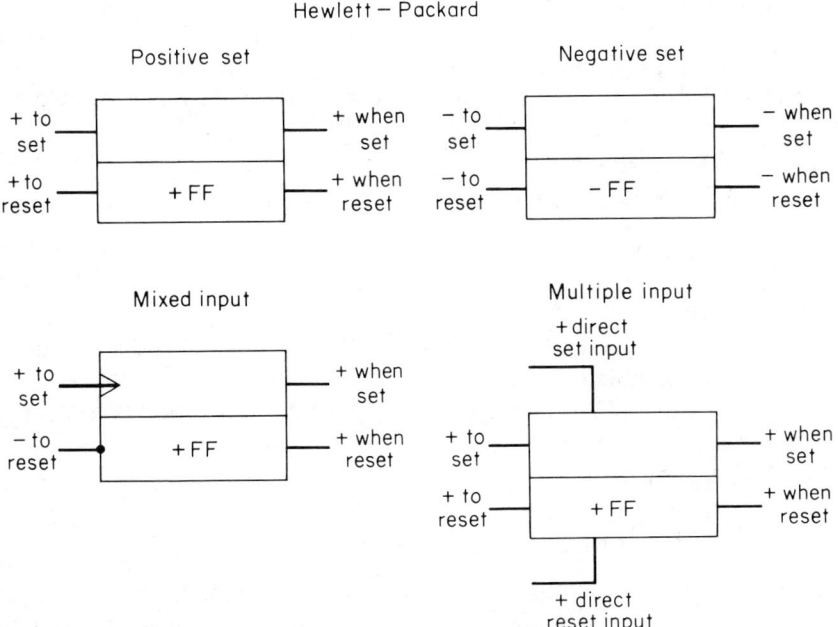

Fig. A-15. Basic flip-flop symbols.

	Truth table		
A	B	D	$\bar{D}$
1	0	1	0
0	1	0	1
0	0	No change	

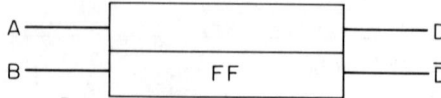

Fig. A-16. R-S flip-flop symbol.

the complement of the set output; that is, when D is true and $\bar{D}$ is false, the FF is defined as being in the set state. With D false and $\bar{D}$ true, the FF is in the reset state.

The FF is set by a true input to A (assuming no inversion dot on the symbol) and is reset by a true input to B. False inputs have no effect on a basic R-S.

Simultaneous true inputs to A and B are forbidden, since some intermediate output state would result. In practice, the FF would try to set and reset simultaneously, probably resulting in an undesired set or reset.

The truth table of Figure A-16 shows the three allowable input combinations for a basic R-S flip-flop.

If a-c-coupled inputs are used, the flip-flop would be set or reset by true-going transitions at A and B, respectively. If, in addition, input inversion dots are also used, false-going transitions at A or B would set or reset the flip-flop.

A-14.2. Reset-Set Flip-Flop with Clock Input

An R-S FF with clock (See Figure A-17) is similar to the basic R-S, except for the addition of the clock input (parallel-connected to both set and reset inputs).

A true input is required to both A and C to set the FF, and a true input to B and C is required for reset.

Since the clock input operates on a pulse edge, the setting or resetting must be present at A or B *before the clock pulse* transition occurs. This time relationship is shown (in positive-true form) in Figure A-17.

A-14.3. (J-K) Flip-Flops

A J-K flip-flop is used (instead of an R-S flip-flop) where there is a possibility of two simultaneous true inputs (which would result in an unpredictable output from an R-S FF).

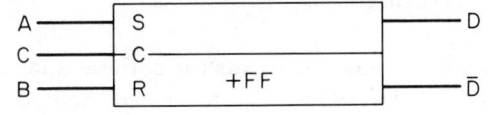

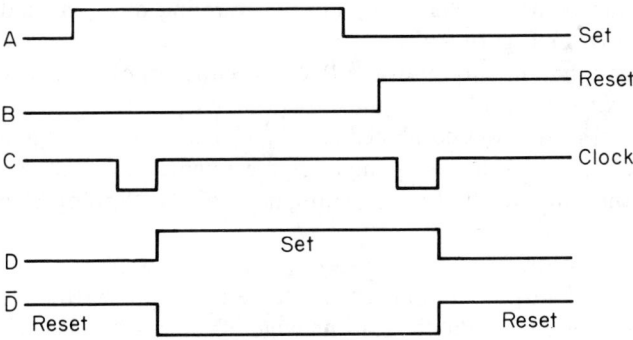

Fig. A-17. R-S flip-flop (with clock input) symbol.

Truth table					
A	B	Initial state		Resulting state	
		D	$\bar{D}$	D	$\bar{D}$
1	0			1	0
0	1			0	1
1	1	0	1	1	0
1	1	1	0	0	1
0	0			No change	

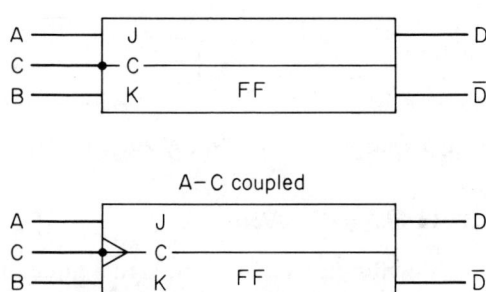

Fig. A-18. J-K flip-flop symbol.

With a J-K FF, simultaneous true inputs for both set and reset will *reverse the existing state* of the FF. This requires some method of *storing* two information conditions (the existing output state and the new input state) until the clock pulse time.

Storage can be accomplished by (1) a-c coupling or (2) a dual-rank flip-flop, as shown in Figure A-18.

The a-c-coupling method uses the RC time constant of the capacitive input for short-term storage of the input information.

The dual-rank method combines two flip-flops (input storage and output storage) and several gates as a single logic element. For simplicity of representation, the internal dual-rank arrangment of the flip-flop is not usually shown.

Overall operation of a J-K flip-flop (either a-c-coupled or dual-rank) is summarized as follows (and in the truth table of Figure A-18).

With true input at A only, the leading edge of the clock pulse acknowledges (stores) the input information at A; the trailing edge of the clock pulse sets the FF.

With true input at B only, the leading edge of the clock pulse acknowledges (stores) the input information at B; the trailing edge of the clock pulse resets the FF.

With true inputs at both A and B, the leading edge of the clock pulse acknowledges the input information at A and B; the trailing edge of the clock pulse *switches the existing state* of the FF.

A-14.4. Toggle Flip-Flop

The toggle flip-flop has only one input, as shown in Figure A-19. Each time input A goes true, outputs D and $\bar{D}$ switch states. Since two input pulses or cycles are required to produce one complete cycle of the output, the toggle flip-flop acts as a divide-by-two element and is commonly used in counting circuits (Chapter 6). The letter T inside the symbol identifies the toggle flip-flop.

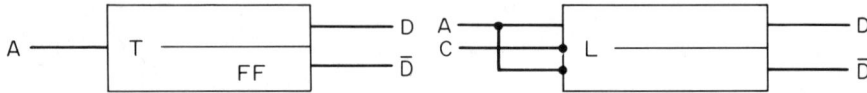

Fig. A-19. Toggle flip-flop symbol.   **Fig. A-20.** Latching flip-flop symbol.

A-14.5. Latching Flip-Flop

The latching flip-flop has a single signal input and a clock input. The symbol is identified by the letter L inside the symbol, as shown in Figure A-20.

The set input responds to a true condition at A, while the reset input responds to a false condition at A. The FF will "latch" into the state existing at A (true-set or false-reset) when the clock pulse at C goes from true to false (usually at the trailing edge of the clock pulse). One unusual feature (not always desirable) is that during the clock pulse duration, the FF is "unlatched," so that if A switches true and false several times during this period, outputs D and $\bar{D}$ will free-switch accordingly.

A-14.6. Delay Flip-Flop

The delay flip-flop is similar to the latching flip-flop, and its symbol is identified by the letter D, as shown in Figure A-21. The delay FF

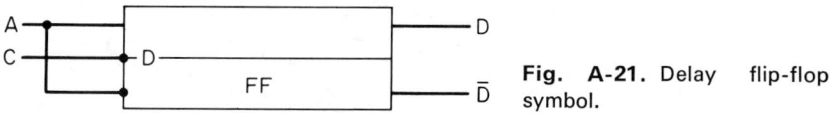

Fig. A-21. Delay flip-flop symbol.

will latch into the state existing at A when C goes from true to false. However, unlike the latching FF, the delay FF "delays" any switching of outputs until the trailing edge of the clock pulse occurs. The delay FF does not have an unlatched condition.

A-15. MULTIVIBRATORS

There are three types of multivibrators in general use with digital equipment: astable, one-shot (or monostable), and Schmitt trigger (bistable).

A-15.1. One Shot

The one-shot is a *monostable* switching element, using a multivibrator-type circuit. The one-shot element (sometimes known as a mono) is commonly used as an active *delay* device and is triggered into its unstable state by an external signal (see Figure A-22). After an interval determined by circuit constants, the one-shot returns automatically to the stable state. Thus a known, fixed delay time is provided.

One-shot inputs are frequently a-c-coupled, and triggering is accomplished when input A goes through a false-to-true transition. The abbreviation OS within the symbol identifies a one-shot, and a plus or minus sign may be used to indicate the true state of the D output during the on time.

A-15.2. Schmitt Trigger

The Schmitt trigger is a two-state (bistable) element, using a multivibrator-type circuit. The Schmitt trigger is commonly used for level

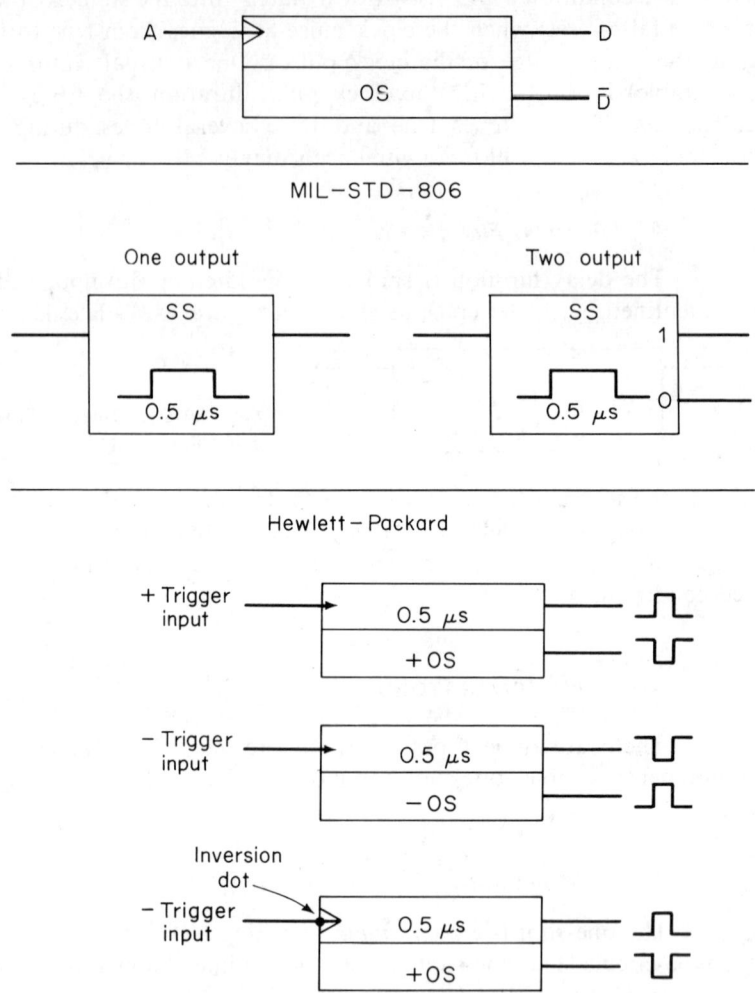

Fig. A-22. One-shot (monostable) multivibrator symbol.

sensing and signal squaring or shaping. When the input voltage is below a reference level, the element is in one state (see Figure A-23). When the input level goes above the reference level, the element switches to the other state. Switching between states takes place rapidly, making the element useful for squaring signals with poor rise times (converting sine waves to square waves) and for voltage level restoration.

With input A below the reference level (or false), D is false and $\bar{D}$ is true. When the input is above the reference level, D switches to true and $\bar{D}$ switches to false. (The reference level is established by circuit constants.)

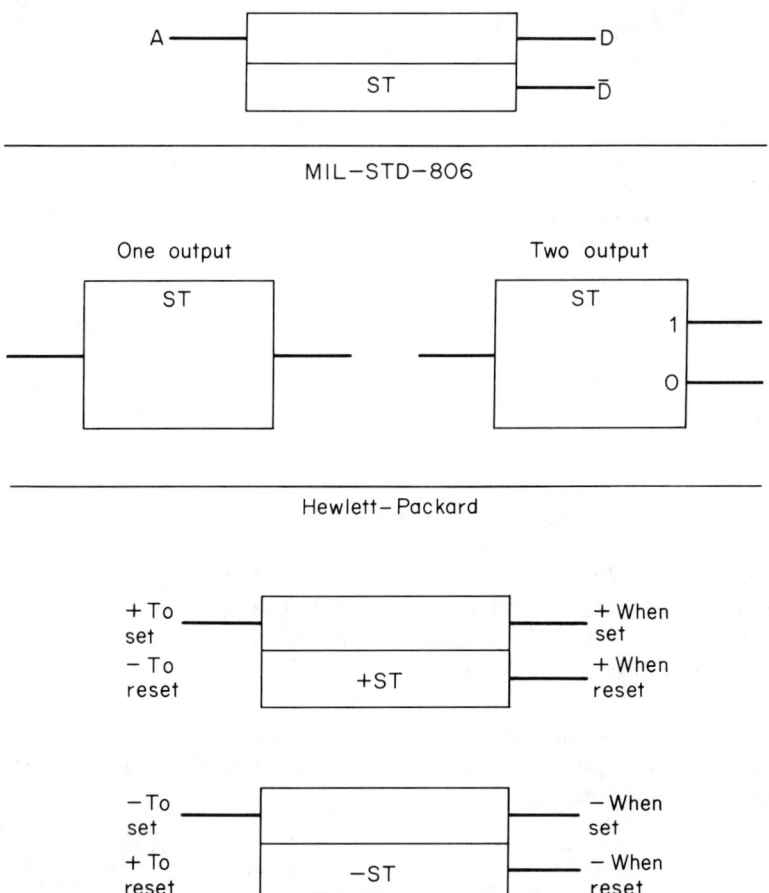

Fig. A-23. Schmitt trigger (bistable) multivibrator symbol.

A-15.3. Multivibrator

Although all logic switching elements are forms of multivibrators, an *astable* type is assumed when the term multivibrator is used without a modifier (such as one-shot, Schmitt trigger, and so on). An astable multivibrator will start free-running operation when input A goes true and will continue to generate complementary pulse trains at outputs D and $\bar{D}$ until A goes false, as shown in Figure A-24. Typical negative-true timing wave-forms are shown with the symbol. Note that the minus sign (in front of the letters MV) indicates both the relative level required to start operation and the direction of the first output pulse at D. (The wave-forms do not necessarily have to be symmetrical as shown.)

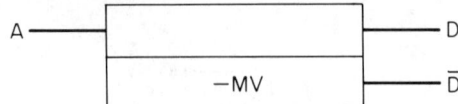

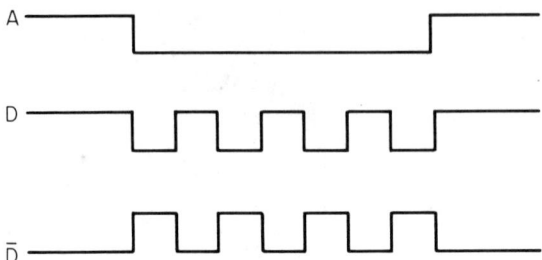

Fig. A-24. Multivibrator (astable) symbol.

A-16. DELAY ELEMENTS

A delay element provides a finite time delay between the input and output signals. The delay symbols, with examples of actual delay time, are shown in Figure A-25. Typical theoretical wave-forms for the elements are shown adjacent to the symbols.

Many types of delay elements are used in digital work. Two frequently used delays are the *tapped delay* and delays effective only on the leading or trailling edges of pulses. Such delay elements, together with the theoretical wave-forms, are shown in Figure A-25.

A-17. LOGIC SYMBOL IDENTIFICATION

Logic symbol identification is one area where manufacturers take off in several directions at once. The following is a summary of the methods of logic symbol identification in general use.

A-17.1. Reference Designations

Most logic diagrams show logic elements as a complete component rather than the many components that make up the element. For example, an amplifier is shown as a triangle, rather than as several dozen resistors, capacitors, transistors, and so on. Therefore, the amplifier has a reference designation of its own. On some logic diagrams, the logic elements symbols are mixed with symbols of individual resistors, capacitors, and so on.

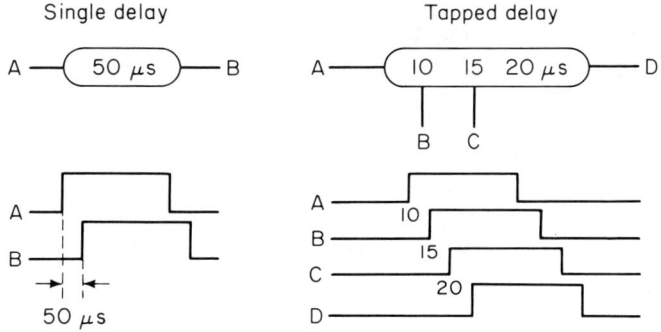

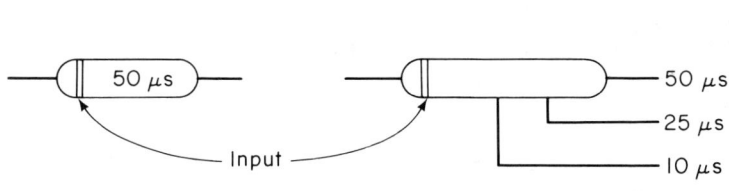

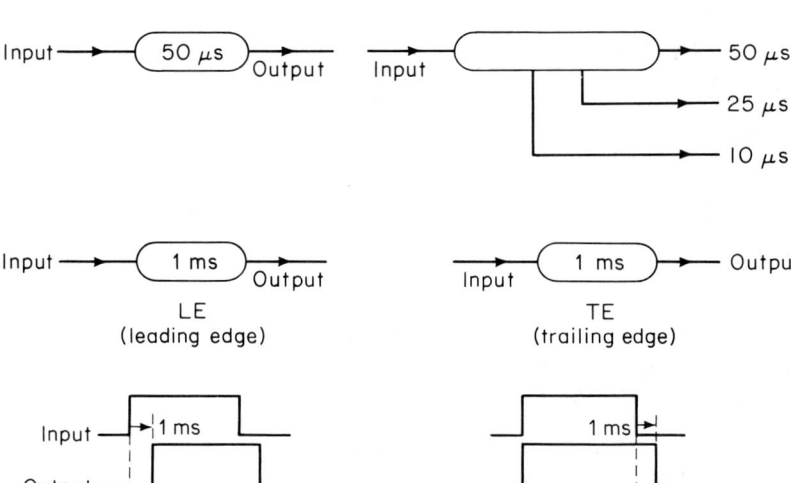

Fig. A-25. Delay element symbols.

Appendix: Digital Logic Circuits

Either way, the logic element symbol must be identified by a reference designation (to match descriptions in text or as a basis for parts listing).

The following logic symbol reference designations are in general use:

Gates *G*
Amplifiers and inverters *A*
Flip-Flops FF
One-Shot OS (sometimes SS for single-shot)
Multivibrator MV
Schmitt Trigger ST
Delay *D*

Figure A-26 shows some examples of how the reference designations are used. Note that the reference designations for switching elements are placed within the symbol. All other designations are (usually) placed beside the sym-

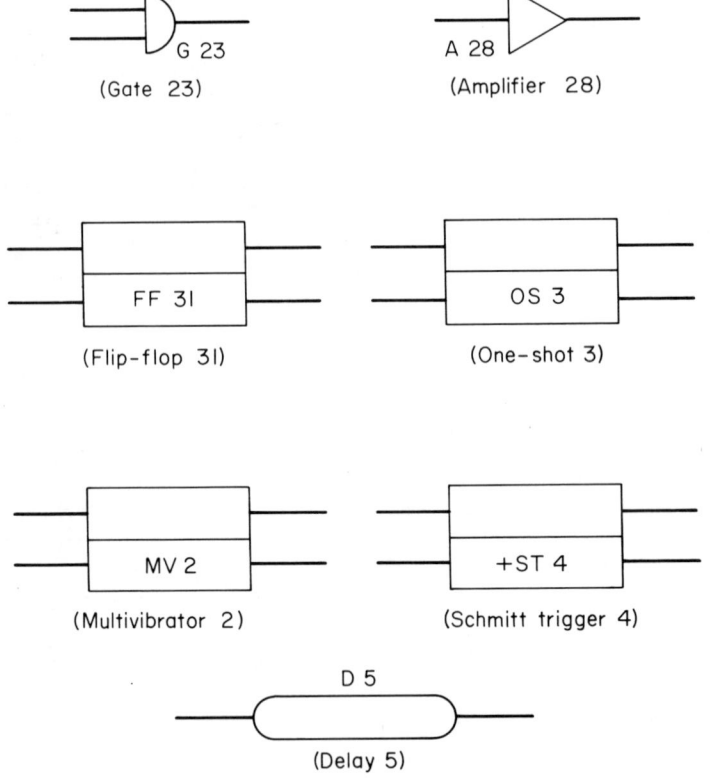

Fig. A-26. Reference designations for logic symbols.

bol. In the case of switching elements, the true-state sign can be used as a prefix to the designation.

A-17.2. Integrated Circuit Reference Designations

When logic elements appear in integrated circuit (or *microcircuit*) form, the system of reference designations is usually changed. Typical examples of microcircuit logic reference designations are shown in Figure A-27.

When a microcircuit is used to form one complete logic element, such as an amplifier or gate, the element can assume the microcircuit reference designator of MC rather than G or A.

When a microcircuit is used to form a complete switching circuit, the element can assume the microcircuit designator of MC, but should also include the appropriate abbreviation (such as FF, OS, and so on.) to identify its function.

When a switching element is composed of portions of different microcircuit packages, the designator MC of both packages should be included inside the symbol, and the appropriate identifying abbreviation should be located outside the symbol.

When more than one logic element is included in the microcircuit package, each logic element is identified with an A, B, C, and so on, suffix.

On some diagrams, the logic element is shown enclosed in dotted lines with the terminals identified. An example of this is shown in Figure A-27, where an amplifier symbol is enclosed in dotted lines and identified as IC45. This indicates that the amplifier is part of IC45 (integrated circuit 45), that the input is available at terminal 3, and that the output is available at terminal 7.

On some logic diagrams that correlate to a schematic diagram of the same circuit, the *active components* of the element may also be designated. For example, in Figure A-27, Q_6 is the active transistor of amplifier A_5, and Q_1 and Q_2 are the active transistors of FF_3. This is primarily an aid in troubleshooting.

A-17.3. Reference Names

Identification of logic elements by functional name in a diagram is normally done only with switching elements, as shown in Figure A-28. The upper portion of the symbol can be reserved for this purpose. If the type of switching element (FF, OS, and so on) is not part of the designation (in the lower portion of the symbol) then the reference name should include the appropriate abbreviation. If designations appear in both the upper and lower portions of the symbol, the name can be placed outside the symbol. For one-shots, it is often convenient to give the *time duration* as the reference name.

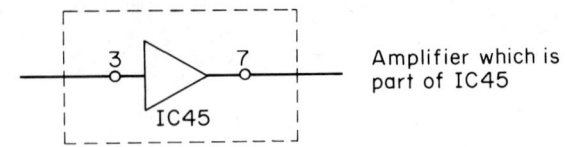

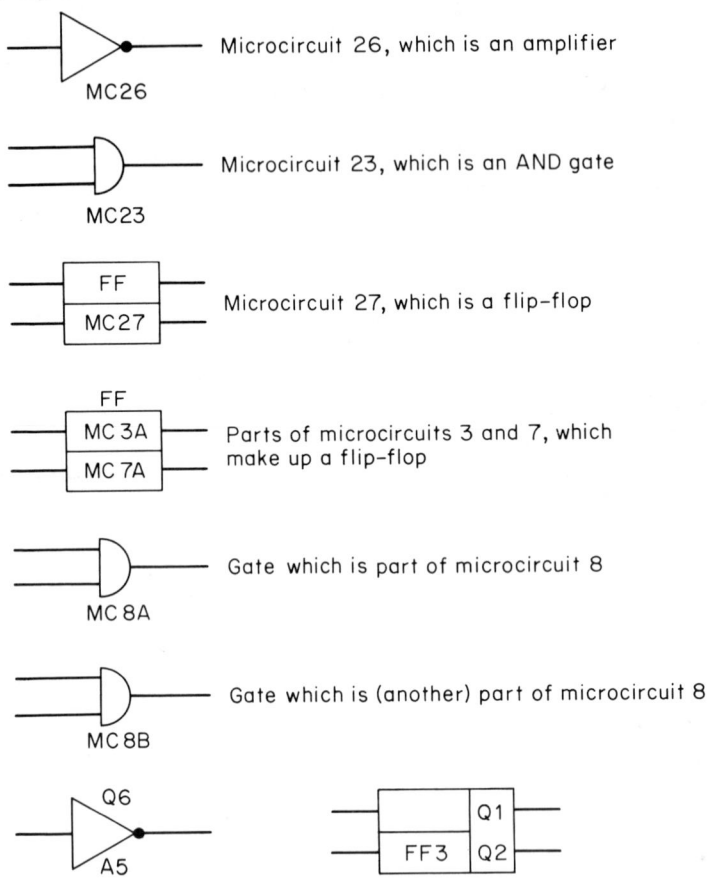

Fig. A-27. Reference designations for microcircuit (integrated circuit) logic symbols.

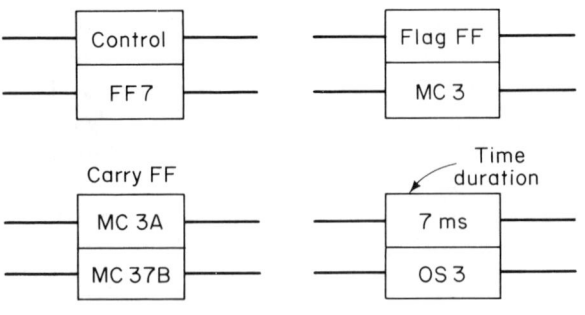

Fig. A-28. Reference names for logic symbols.

A-17.4. Location Information

Logic diagrams are intended to show the combination of logic elements that together form an instrument, a part of an instrument, or a system of instruments. To aid in correlating the diagram with physical locations in instruments, additional information is given (by some manufacturers) with logic symbols, as shown in Figure A-29. The number in the

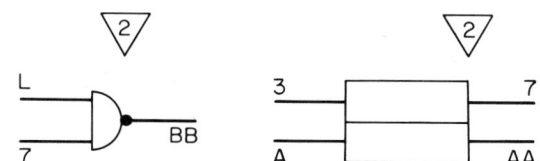

Fig. A-29. Location information on logic diagrams.

small triangle is sometimes used to indicate the circuit board on which the element appears. In the example of Figure A-29, the elements appear on circuit board No. 2.

If all elements are on the same board, individual identification in this way is unnecessary. Letters and numbers adjacent to inputs and outputs indicate pin numbers of the board (or microcircuit package) where inputs and/or outputs appear (as shown in Figure A-27).

A-18. LOGIC EQUATIONS

Logic equations are sometimes used to aid in explanation of logic circuits. At one time some manufacturers presented all logic information (in their instruction manuals) in equation form. This practice has generally been discontinued. Logic diagrams are used in instruction manuals. These diagrams may, or may not, be supplemented by the corresponding equations.

When logic equations are used, the familiar algebraic symbols have the following meanings:

$+$ means or
$\cdot$ means and
$=$ means equals (or is the result of)
$-$ means not (or complement of)

For example, the logic equation for a simple two-input AND gate is

$$C = B \cdot A$$

The logic equation for the combination of elements shown in Figure A-30 is

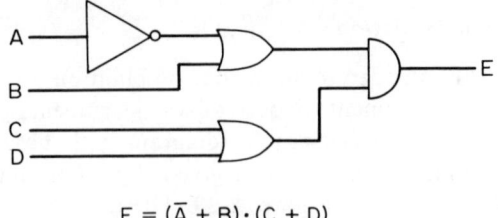

$$E = (\overline{A} + B) \cdot (C + D)$$
alternate
$$(\overline{A} + B) \cdot (C + D) = E$$

Fig. A-30. Logic diagram and corresponding logic equations.

$$E = (\bar{A} + B) \cdot (C + D)$$

Sometimes the same equation will be written

$$(\bar{A} + B) \cdot (C + D) = E$$

Generally, logic equations are better suited to *design* of digital equipment rather than as a tool for troubleshooting.

A-19. DIGITAL COMPONENT APPLICATIONS

The individual components described thus far can be arranged in various combinations to perform specific functions in test equipment. For example, as discussed in Chapter 6, *decade counters* using FFs can convert a series of pulses to BCD form (such as conversion of a pulse count to an 8421 code. Also, a *decoder* can be used to convert the BCD data to a 10-line or decade readout.

Some other examples using combinations of basic digital components to perform specific functions include analog-to-digital conversion, digital-to-analog conversion, storage registers, shift registers, level comparators, sign or polarity comparators, adders, encoders, and decoders. Operation of those circuits that appear frequently in digital test equipment are discussed as follows.

A-20. CONVERSION BETWEEN ANALOG AND DIGITAL INFORMATION

Chapters 2 and 6 discussed methods for converting a voltage (or current) to digital form. This was done by first converting the voltage to a frequency or a series of pulses, then converting the pulses to BCD form, then converting the BCD information to a decade readout.

It is also possible to convert a voltage directly to BCD form by means of an analog-to-digital (a/d) converter and to convert BCD data back to voltage by means of a digital-to-analog (d/a) converter.

Before going into the operation of these conversion circuits, let us discuss the signal formats for BCD data, as well as the four-bit system.

A-20.1. Typical Binary-Coded-Decimal Signal Formats

Although there are many ways in which pulse wave-forms can be used to represent the 1 and 0 digits of a BCD code, there are only three ways in common use. These are the NRZL (non-return-to-zero-level), the NRZM (non-return-to-zero-mark), and the RZ (return-to-zero) formats. Figure A-31 shows the relation of the three formats.

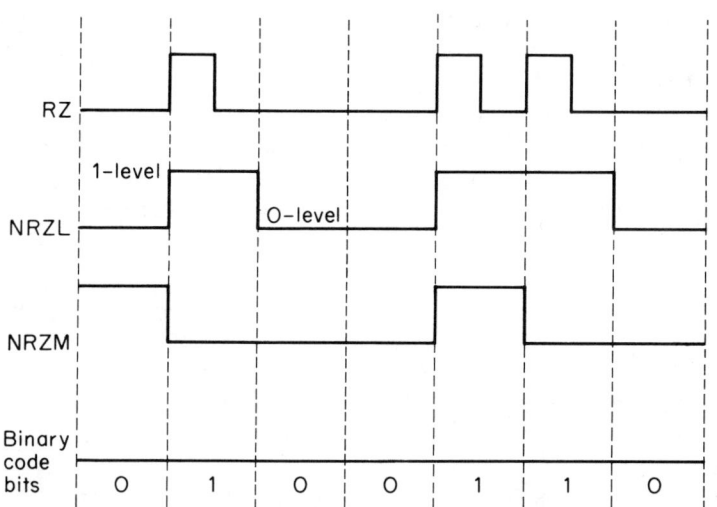

Fig. A-31. Typical BCD signal formats.

In NRZL, a 1-bit is one signal level, while a 0-bit is another signal level. These levels could be 5, 10, 0, or −5 V, or any other selected values, provided the 1 and 0 levels were *entirely different* and predetermined.

In NRZM, the level of the wave-form has no meaning. A 1-bit is represented by a *change* in level (either higher or lower), while a 0-bit is represented by *no change* in level. (Generally, NRZM is not used in test equipment work, except on special order where the test instrument must feed into an existing NRZM system.)

In RZ, a 1-bit is represented by a pulse of some definite width (usually a half-bit width) that returns to zero-signal level, while a 0-bit is represented by the absence of a signal (or a zero-level signal).

A-20.2. The Four-Bit System

As the name implies, a four-bit system is one that is capable of handling four information bits. Any fractional part *in sixteenths* can be stated by using only four binary digits. As discussed in Sections A-3 and A-4, the number 15 is represented by 1111 in the binary system. (Zero is then represented by 0000.) Therefore, any number that is between 0 and 15 requires only four binary digits. For example, the number 1101 is 13, the number 0011 is 3, and so on.

Although not all BCD codes use the four-bit system, it is quite common and does provide a high degree of accuracy for conversion between analog and digital data. In actual practice, a four-bit analog-to-digital converter (sometimes known as a binary encoder) samples the voltage level to be converted and compares the voltage to half-scale, quarter-scale, eighth-scale, and sixteenth-scale (in that order) of some given full-scale voltage. The encoder then produces bits in sequence, with the decision made on the most significant (or half-scale) first.

Figure A-32 shows the relation between three voltage levels to be encoded (or converted) and the corresponding binary code (in NRZL form).

As shown, each of three voltage levels is divided into four equal time increments. The first time increment is used to represent the half-scale bit, the

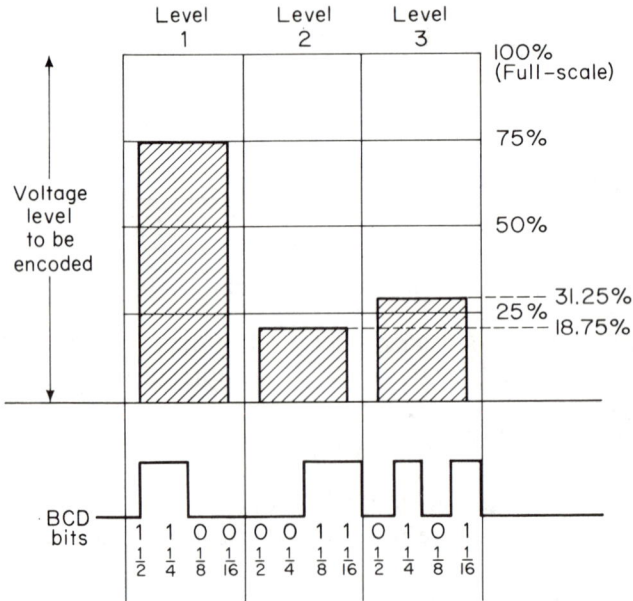

Fig. A-32. Relationship between three voltage levels to be encoded and corresponding BCD code (using four-bit system).

second increment is used for the one-quarter bit, the third increment is used for the one-eighth-scale bit, and the fourth increment is for the one-sixteenth bit.

In level 1, the first two time increments are at a binary 1, while the second two increments are at a binary 0. This would be represented as 1100, or 12. Twelve is three-fourths of 16. Therefore, level 1 is 75 per cent of full scale. For example, if full scale were 100 V, level 1 would be at 75 V.

In level 2, the first two time increments are at a binary 0, while the second two increments are at a binary 1. This would be represented as 0011, or 3. Therefore, level 3 is three-sixteenths of full scale (or 18.75 V).

This could be expressed in another way. In the first or half-scale increment, the encoder produces a binary 0. This is because the voltage (18.75) is less than half scale (50). The same is true of the second, or quarter-scale, increment (18.75 V are less than 25 V). In the third or one-eighth-scale increment the encoder produces a binary 1, as it does in the fourth or one-sixteenth-scale increment. This is because the voltage being compared is greater than one-eighth of full scale (18.75 is greater than 12.5), and greater than one-sixteenth of full scale (18.75 is greater than 6.25).

Therefore, the half and quarter increments are at zero or off, while the one-eighth- and one-sixteenth-scale increments are on. Also, $\frac{1}{8} + \frac{1}{16} = \frac{3}{16}$, or 18.75 per cent.

In level 3, the first time increment is 0, the second time increment is 1, the third is 0, while the fourth is 1. This is a binary 0101, or 5, and represents $\frac{5}{16}$ of full scale (31.25 V).

A-20.3. Analog-to-Digital Conversion

One of the most common methods of *direct* analog-to-digital conversion involves the use of an encoder that uses a sequence of half-split, trial-and-error steps. This produces code bits in serial (all on one line, one after another) form.

The heart of an encoder is a *conversion ladder*. Such a ladder (which is a form of digital-to-analog converter) is shown in Figure A-33. The ladder provides a means of implementing a four-bit binary-coding system and produces an output voltage that is equivalent to the switch positions. The switches can be moved to either a 1 or a 0 position, which corresponds to a four-place binary number. The output voltage will describe a percentage of the full-scale reference voltage, depending on the binary switch positions.

For example, if all of the switches were in the 0 position, there would be no output voltage. This would produce a binary 0000 (conventional zero), and it should be represented by zero voltage.

If switch *A* were in the 1 position and the remaining switches were in the 0 position, this would produce a binary 1000 (conventional 8). Since the total

Appendix: Digital Logic Circuits

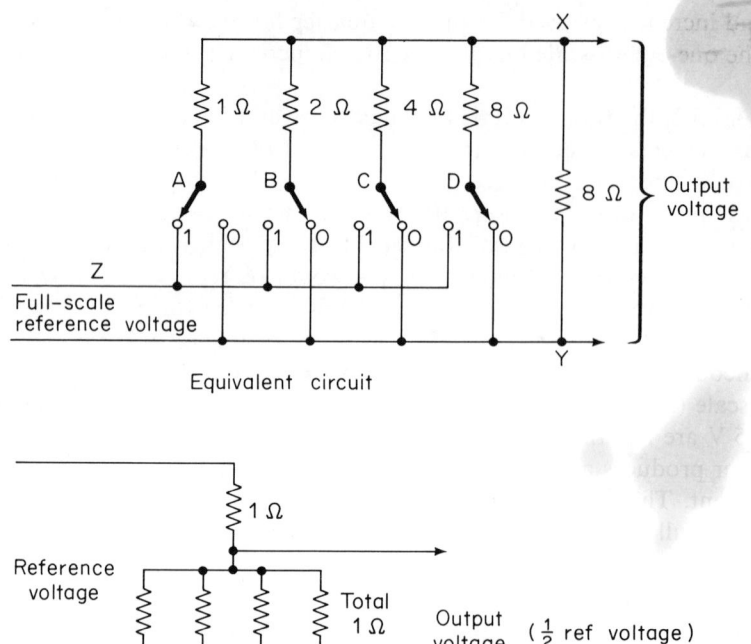

Fig. A-33. Binary conversion ladder used in four-bit system.

in a four-bit system is 16 (as previously described), this should represent one half (8 = one-half of 16). Therefore, the output voltage should be one-half of the full-scale reference voltage. This is accomplished as follows.

The 2-, 4-, and 8-Ω switch resistors and the 8-Ω output resistor are all connected in parallel. This produces a value of 1 Ω across points X and Y. The reference voltage is applied across the 1-Ω switch resistor (across points Z and X) and the 1-Ω combination of resistors (across points X and Y). In effect, this is the same as two 1-Ω resistors in series. Since the full-scale reference voltage is applied across both resistors in series and the output is measured across only one of the resistors, the output voltage will be one-half of the reference voltage.

In a practical encoder circuit, the same basic ladder is used to supply a comparison voltage to a *comparison circuit*. This circuit compares the voltage to be encoded against the binary-coded voltage from the ladder. The resultant output of the comparison circuit is a binary code representing the voltage to be encoded.

The mechanical switches shown in Figure A-33 are replaced by electronic switches, usually flip-flops. When the FF is in one state (the ON state), the corresponding ladder resistor is connected to the reference voltage. In the OFF state, the resistor is disconnected from the reference voltage. The switches are triggered by four pulses (representing each of the four binary bits) from a time base. (The time base is sometimes known as a clock, or programmer; the time base pulses are also known as command pulses.) A fifth pulse is used to turn the comparison circuit on and off so that as each switch is operated, a comparison can be made for each of the four bits.

Figure A-34 is a simplified block diagram of such an encoder. Here the reference voltage (equivalent to a full-scale input voltage, or voltage to be encoded) is applied to the ladder through the electronic switches. The ladder

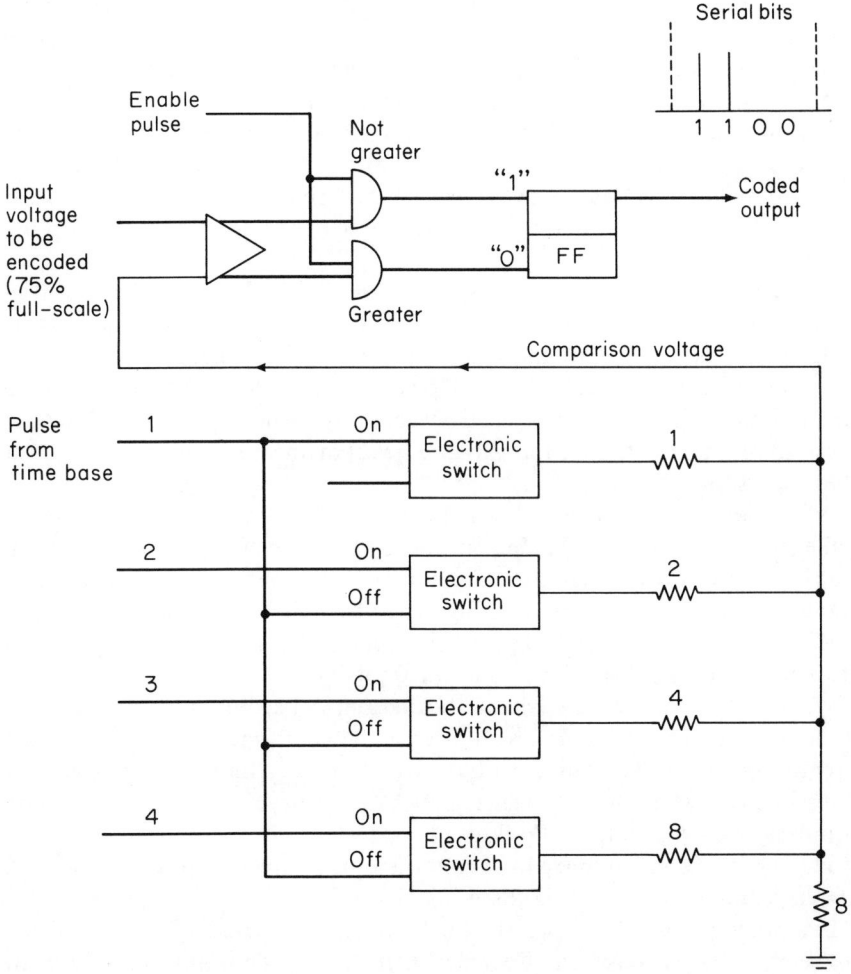

Fig. A-34. Analog-to-digital converter using four-bit system.

output (comparison voltage) is controlled by switch positions that, in turn, are controlled by pulses from the time base. The sequence of these pulses is also shown in Figure A-34.

The following describes the sequence of events necessary to produce a series of four binary bits that describe the input voltage as a percentage of full scale (in one-sixteenth increments). Assume that the input voltage is three-fourths full scale.

When pulse 1 arrives, switch 1 is turned on and the remaining switches are turned off. The ladder output is a half-scale voltage that is applied to the differential amplifier. The balance of this amplifier is set so that its output will be sufficient to turn on one AND gate and turn off the other AND gate if the ladder voltage is greater than the input voltage. Likewise, the differential amplifier will reverse the AND gates if the ladder voltage is not greater than the input voltage. Both AND gates are placed in a condition to turn on (the gates are *enabled*) by the fifth pulse from the time base.

In our sample (three-fourths full scale), the ladder output would be less than the input voltage when pulse 1 was applied to the ladder. As a result, the not-greater AND gate would turn on, and the output flip-flop would be set to the 1 position. Therefore, for the first of the four bits, the flip-flop output is a 1.

When pulse 2 arrives, switch 2 is turned on and switch 1 remains on. Both switches 3 and 4 remain off. The ladder output is now a three-fourths-scale voltage that is applied to the differential amplifier. Therefore, the ladder voltage equals the input voltage. However, the ladder output is still not greater than the input voltage. Consequently, when the AND gates are enabled by the fifth pulse, the AND gates remain in the same condition as does the output flip-flop.

When pulse 3 arrives, switch 3 is turned on. Switches 1 and 2 remain on, while switch 4 is off. The ladder output is now a seven-eighths-scale voltage that is applied to the differential amplifier. The ladder output is now greater than the input voltage. Consequently, when the AND gates are enabled, they reverse. The not-greater AND gate turns off and the greater AND gate turns on. The output flip-flop then sets to the 0 position.

When pulse 4 arrives, switch 4 is turned on. All switches are now on. The ladder output is now maximum (full-scale). This output is applied to the differential amplifier. The ladder voltage is still greater than the input voltage. Consequently, when the AND gates are enabled, they remain in the same condition as does the output flip-flop.

Therefore, the four binary bits from the output are 1, 1, 0, and 0, or 1100. This is a binary 12, which is three-fourths of 16.

In a practical encoder, when the fourth pulse has passed, all of the switches are reset to the off position. This places them in a condition to spell out the next four-bit binary word.

A-20.4. Digital-to-Analog Conversion

A digital-to-analog converter performs the opposite function to that of the a/d converter just described. A d/a converter produces an output voltage (usually direct current) that corresponds to the binary code. Information to be converted is usually applied to the d/a input in serial form. However, some d/a converters receive four-line data.

As shown in Figure A-35, a conversion ladder is also used in the d/a con-

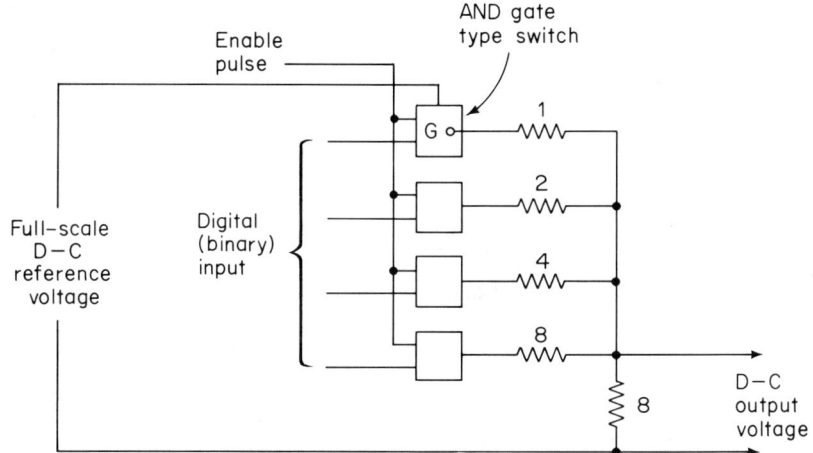

Fig. A-35. Digital-to-analog converter using four-bit system.

verter. The output of the conversion ladder is a d-c voltage that represents a percentage of the full-scale reference voltage. The output percentage is dependent on the switch positions. In turn, the switches are set to ON or OFF by corresponding binary pulses. If the information is applied to the switches in four-line form, each line can be connected to the corresponding switch. If the data are applied in serial form, the data must be converted to parallel (four-line) form by a shift register and/or storage register (both of which are described in later sections of this chapter.)

The switches in a d/a converter are essentially a form of AND gate. Each gate completes the circuit from the reference voltage to the corresponding ladder resistor when both the enable pulse and the binary pulse coincide.

Assume that the digital number to be converted is 1000 (a binary 8). When the first pulse is applied to the ladder switches, switch A is enabled, and the reference voltage is applied to the 1-Ω resistor. When switches B, C, and D receive their enable pulses, there will be no binary pulses (or they will be in their low-level 0 condition, depending on the binary pulse format used). Therefore, switches B, C, and D will not complete the circuits to the 2-, 4-, and 8-Ω ladder resistors. These resistors combine with the 8-Ω output resistor to produce a 1-Ω resistance, in series with the 1-Ω ladder resistance. This divides the reference voltage in half to produce a half-scale output.

A-21. STORAGE REGISTERS

The storage register is a combination of gates and flip-flops used to store binary information. A typical storage register is illustrated in Figure A-36. This register can store four bits of binary information, applied

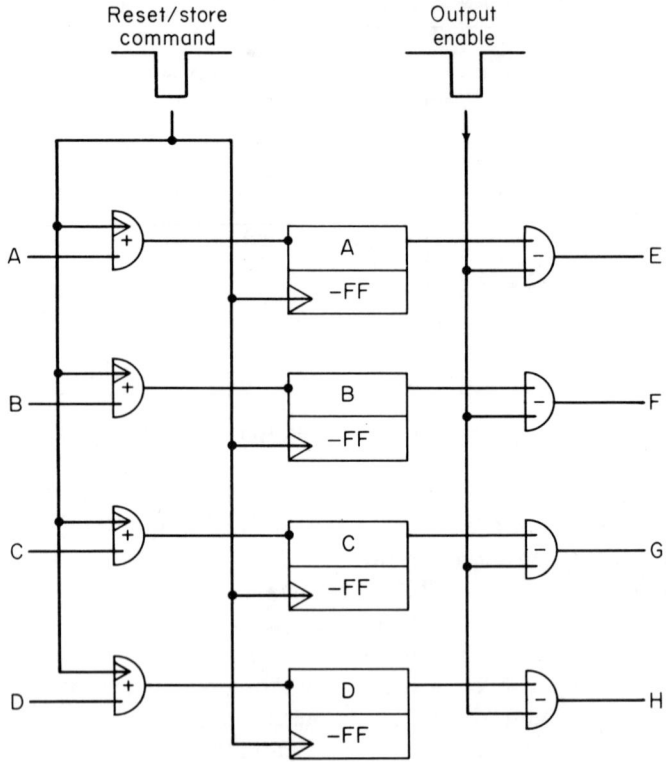

Fig. A-36. Typical storage register.

to inputs *A*, *B*, *C*, and *D*. The inputs must be positive-true. Assume that *A* and *C* are at positive levels (true), and that *B* and *C* are negative (false).

Before data are stored, all four flip-flops are reset by the *leading edge* of the reset/store command pulse (from the time base, clock, or programmer). This clears the register of any previously stored data.

The trailing (positive-going) edge of the reset/store command pulse is coupled into the four positive-true input AND gates. Inputs that are true (A and C) will enable a true output from the corresponding gates, which in turn will set flip-flops *A* and *C* (positive input required). The other two FFs (*B* and *D*) remain in the reset state. The information at *A*, *B*, *C*, and *D* is now stored in the four flip-flops.

When the stored data are required, the output enable lines are made negative (by a pulse from the time base), and output lines *E* and *G* from the set flip-flops are true (negative-true, now). Outputs *F* and *H* will be false (posi-

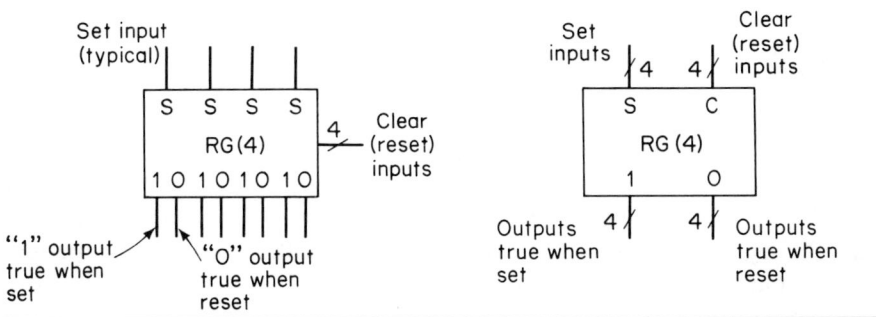

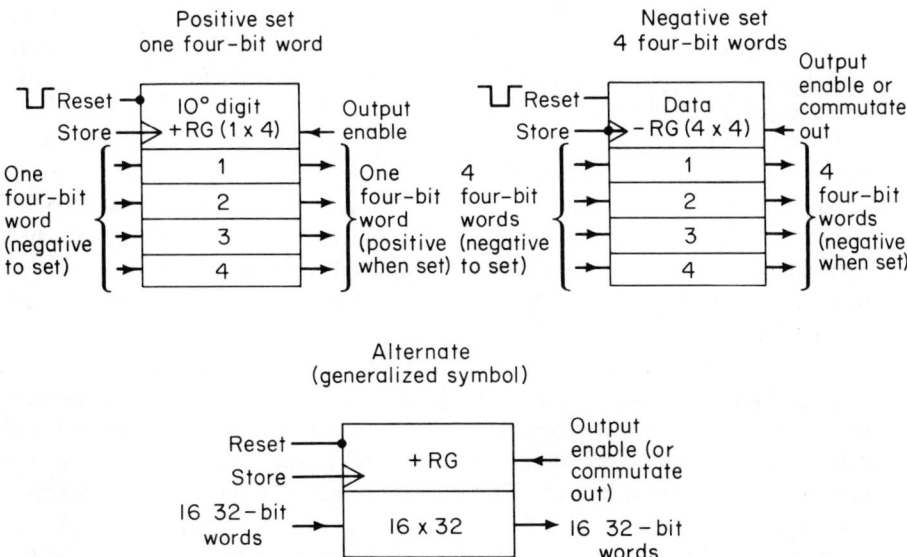

Fig. A-37. Typical storage register symbols (Hewlett-Packard).

tive). (If positive-true outputs were required from the FFs, then the reset output of the FFs could be used.)

The storage register of Figure A-36 is called a 4 × 1 register. The 4 denotes the number of bits per word or character. The 1 indicates the number of characters or words. Storage capacity is expanded by adding additional 4 × 1 registers, one for each added character. The expanded register is then called a 4 × 2, 4 × 3, and so on, register. If the characters require more than four bits, the register is increased to 5 × 1, 5 × 2, and so on, by the addition of storage FFs and associated gates. (See Figure A-37.)

By using one store pulse, parallel data (all four binary pulses occuring simultaneously) are stored. Then by successively enabling the register for each character, parallel-to-serial conversion is obtained (corresponding bits of each register are OR-gated together). The process of successively enabling character storage register is sometimes called *commutation*.

A-22. SHIFT REGISTERS

Shift registers have many functions. In test equipment circuits, a shift register is sometimes used to perform arithmetic functions of multiplication and division. When a number is multiplied by 10, it is shifted to the left, and a zero is added. For example, the number 377 × 10 is 3770 (377 shifted to the left and 0 added). Division, the opposite of multiplication, requires a right shift. The use of shift registers to perform arithmetic operations is primarily a computer function and will not be discussed here. Instead, we will concentrate on the use of shift registers for series-to-parallel conversion of binary pulses (which is the most common use of a shift register in test equipment applications).

As shown in Figure A-38, a shift register is made up of FFs and delay elements. The FFs are connected together so that each one transfers its existing data to the next FF when the reset input is applied. (The reset input pulse applied to all FFs simultaneously is often known as the *advance* input, when referring to a shift register). One FF is required for each bit to be converted, thus four FFs are shown for four binary bits. This is similar (but not identical) to the decade discussed in Chapter 6.

When the advance (reset) line is pulsed, as it is between every binary pulse bit, the FFs in the set (1) state switch to 0, and those in the reset (0) state remain in the 0 state. The output signal from a switching FF is shaped and delayed in the delay circuits and then directed to the 1 input of the next FF to complete the shift operation. Data can be inserted into the first FF between each shift.

Assume that four binary bits in serial form are applied to the input of the register in Figure A-38 and that the pulses appear as 0101 (a binary 5), as

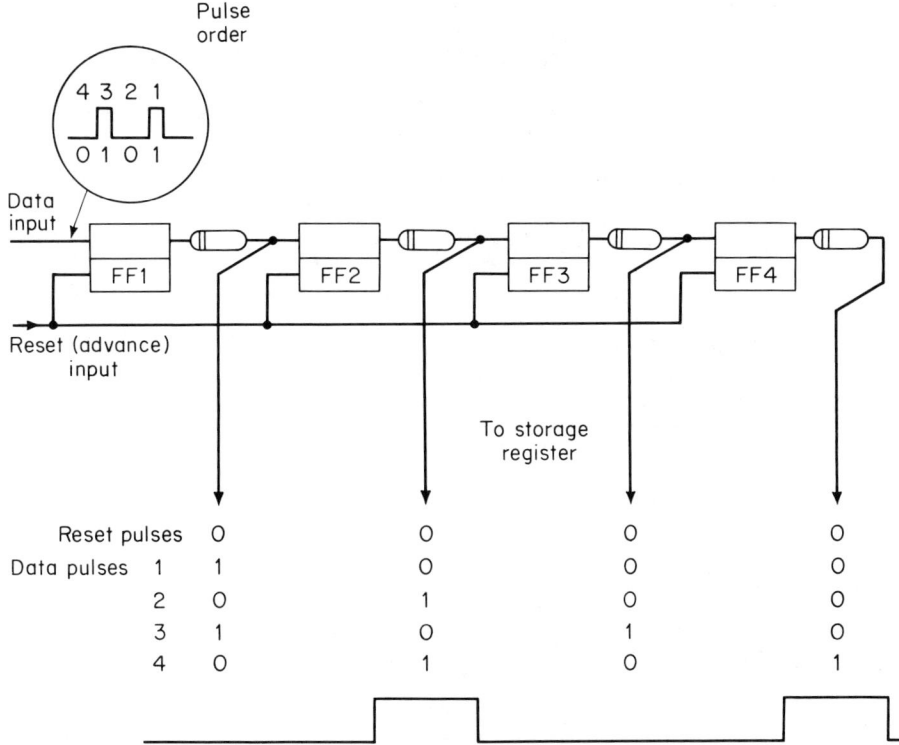

Fig. A-38. Typical shift register.

shown. All of the FFs are in the 0 state before each binary pulse. At pulse 1, FF_1 goes to the 1 state, and the remaining FFs stay in the 0 state. Between pulses 1 and 2, all FFs go to 0, but the 1 condition of FF_1 is delayed and applied to FF_2.

At pulse 2, FF_1 stays in the 0 condition (since the second bit is 0), but FF_2 moves to a 1 state because it received the delayed pulse from FF_1.

At pulse 3, FF_1 goes to the 1 state, FF_2 goes to 0, and FF_3 now goes to 1, since it received the delayed pulse from FF_2.

At pulse 4, FF_1 remains in the 0 state, FF_2 moves to the 1 state (pulse from FF_1), FF_3 goes to 0 state (since it received a 0 from FF_2), and FF_4 (thus far unaffected) moves to a 1 state (received from FF_3).

At the end of the four bits, before all FFs are reset to 0, the existing states (0101) are fed into a storage register (Section A-20). The output of the storage register can then be used as needed—for example, to operate the electronic switches of a digital-to-analog converter.

As shown in Figure A-39, the shift register is usually presented in symbol form on logic diagrams.

MIL-STD-806

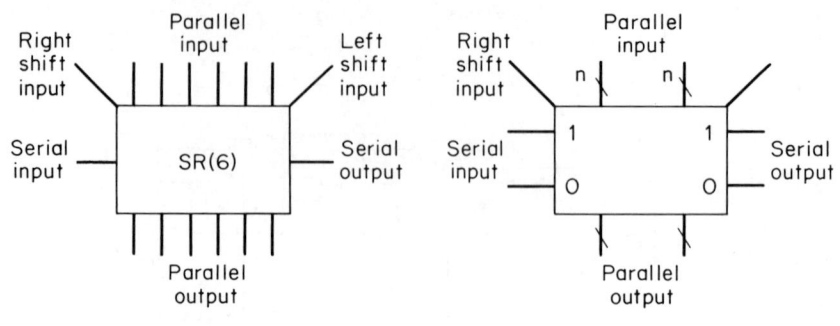

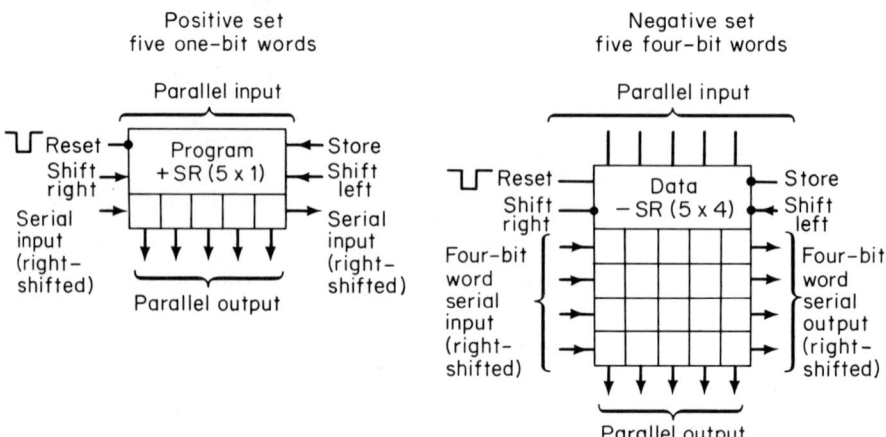

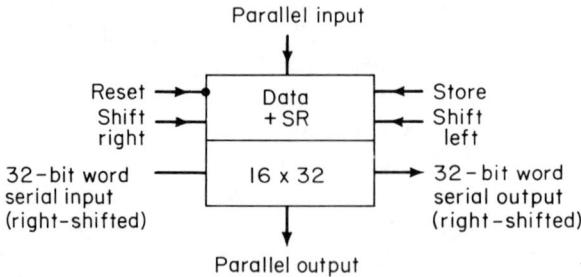

Fig. A-39. Typical shift register symbols (Hewlett-Packard).

A-23. SIGN COMPARATORS

Sign comparators are used in both computer and test equipment digital applications. In computers, a sign comparator is used to determine the sign of numbers (plus or minus). In test equipment, a sign comparator is used to determine (or compare) the polarity of d-c voltages. Sign comparators are used for this purpose in digital voltmeters.

As shown in Figure A-40, a sign comparator consists of two FFs, four AND

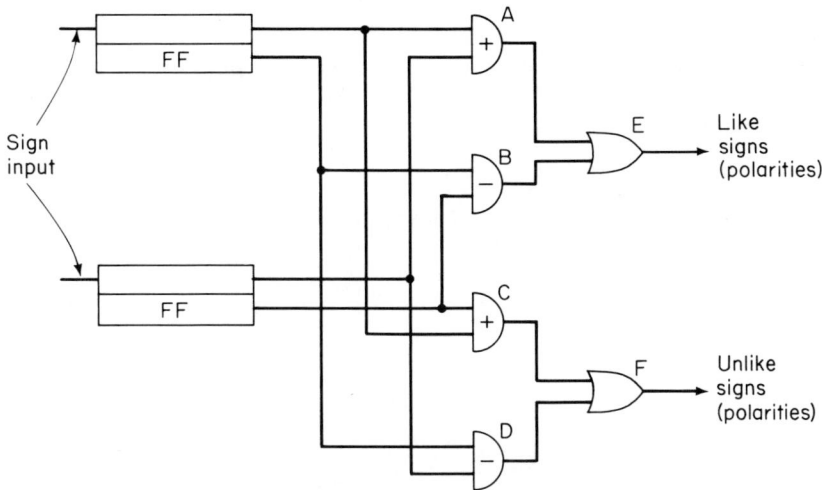

Fig. A-40. Typical sign (polarity) comparator.

gates, and two OR gates. The two FFs, each indicating the polarity of an applied voltage, can produce four possible combinations, two for like polarities and two for unlike polarities.

If the FFs have the same output, one of the AND gates (*A* for positive or *B* for negative) will have an output, and OR gate *E* will provide an output signal; that is, there will be an output indicating *like* polarities, whether both FFs have a positive or negative output. One of the two AND gates will provide an output signal and, since an OR gate requires only one input to produce an output, OR gate *E* will provide a *like* polarity output.

If the two polarities are opposites (one positive and one negative) there are only two possible combinations. For either combination, one of two AND gates (*C* or *D*) will provide an output to OR gate *F*, indicating an *unlike* polarity output.

A-24. ADDERS

Adders (half adders and full adders) are used primarily in computer circuits to perform mathematical functions. However, adders can also appear in test equipment circuits, usually as some form of comparison circuit. It is therefore necessary to understand adders as a basis for understanding the comparison circuit.

A-24.1. Half-Adder

When only two binary digits are to be added, a half-adder can be used. The output will be a sum ($1 + 0 = 1$, $0 + 1 = 1$) or a carry ($1 + 1 = 10$, which is a 0 sum and carry 1). The two digits to be added have only four possibilities, as shown in Figure A-41. The digit to be added (A) is the addend and the other digit (B) is the augend.

Addend (digit A)	Augend (digit B)	Sum	Carry
0	0	0	0
0	1	1	0
1	0	1	0
1	1	0	1

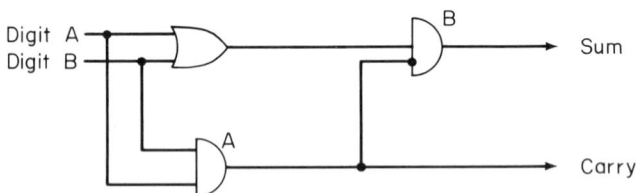

Fig. A-41. Typical half-adder.

If both A and B digits are 0 (or false) there will be no output from the OR gate. Likewise, there will be no output from AND gate A, hence, no carry; that is, the output from AND gate A will be 0 (false). Note that this same carry output is also applied to AND gate B through an inversion. Therefore, if the carry output is 0, the inverted input to AND gate B will be 1. With a 0 from the OR gate and a 1 from the inverted input, there can be no output from AND gate B. Therefore, we have 0 sum and 0 carry.

If either A or B digit equals 1, there will be an output from the OR gate to AND gate B, but the AND gate A output will remain 0 (to carry output).

This 0 output is inverted to a 1 and applied to AND gate *B*. Now, a 1 from the OR gate and a 1 from the inversion result in a sum output of 1 and a 0 carry output.

If both *A* and *B* digits equal 1, they will pass through the OR gate to produce a 1 and also through the AND gate *A* to produce a carry output of 1. This carry 1 is inverted to AND gate *B*, so a 1 and a 0 are applied to AND gate *B*, and there is no sum output. Therefore, the sum is 0, and the carry is 1.

A-24.2. Full-Adder

A full-adder circuit is shown in Figure A-42. This circuit makes it possible to add two three-digit binary numbers. As shown, there are eight possible combinations. However, aside from all 0s, there are only three dif-

	A Addend	B Augend	C Carry (in)	Sum	Carry (out)
1	1	0	0	1	0
2	0	1	0	1	0
3	0	0	1	1	0
4	1	1	0	0	1
5	1	0	1	0	1
6	0	1	1	0	1
7	1	1	1	1	1
8	0	0	0	0	0

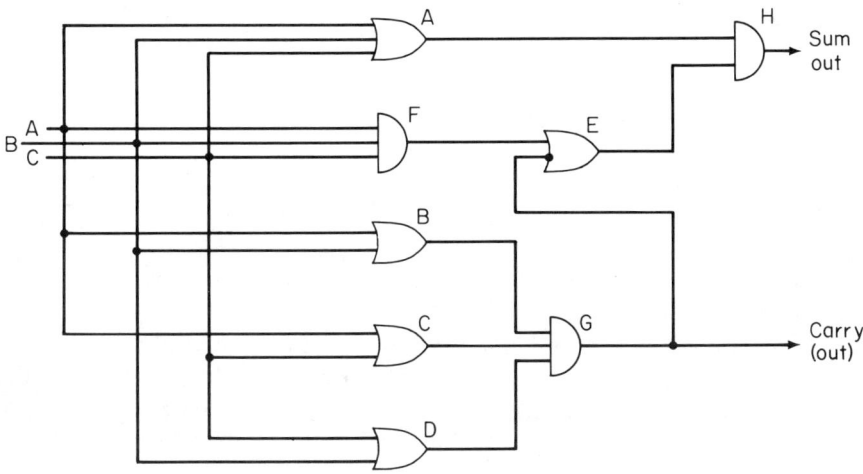

Fig. A-42. Typical full-adder.

ferent cases, depending on the number of 1s (trues—either one, two, or three 1s in any combination.

For a single 1, there are three possibilities: $1 + 0 + 0$, $0 + 1 + 0$, and $0 + 0 + 1$. Thus there is always a sum but never a carry.

For any two 1s ($1 + 1 + 0$; $1 + 0 + 1$; $0 + 1 + 1$), there is a carry, but never a sum.

For three 1s ($1 + 1 + 1$), there are both a carry and a sum.

Where every digit is 0 ($0 + 0 + 0$) there is neither a sum nor a carry.

If all digits are 1s, the three OR gates B, C, and D each have two 1s. So there is a 1 from each, resulting in three 1s in AND gate G. This provides a carry, which is inverted and appears as a 0 at OR gate E. But the three 1s into AND gate F provide a 1 input to OR gate E; hence OR gate F has an output. OR gate A has three 1s, so there are two 1s input to AND gate H. This produces both a sum and a carry output ($1 + 1 + 1 = 11$).

If two digits are 1s, the three OR gates B, C, and D will each have at least one 1. AND gate G will have three 1s, providing a carry output. AND gate F will have two 1s and no output (0 output). AND gate G will have an output (carry), but it will be inverted. Therefore, both inputs to OR gate E will be 0, as will its output to AND gate H. This produces a carry but no sum ($1 + 1 + 0 = 10$; $1 + 0 + 1 = 10$; $0 + 1 + 1 = 10$).

If only one digit is a 1, there cannot be three inputs to AND gate G, hence no carry. The inversion will change this carry 0 to a 1, which OR gate E will pass to AND gate H. Another input from OR gate A is applied to AND gage H, producing a sum output. This results in a sum but no carry ($1 + 0 + 0 = 1$; $0 + 1 + 0 = 1$; $0 + 0 + 1 = 1$).

If all digits are 0, there will be no input to AND gate G, hence no carry. There will be no output for OR gate A, hence no output from AND gate H, and no sum ($0 + 0 + 0$).

A-25. COMPARISON CIRCUITS

The half-adder can be used as a comparison circuit for detecting errors. If two unlike digits are fed into the half-adder, the output will be a sum. If the digits are alike (0,0 or 1,1), there will be no sum output (even though there may be a carry output). Therefore, the half-adder is a detector of equality or inequality.

Figure A-43 shows how a half-adder is used to compare the contents of two registers. The binary count in register A is transferred to register B, and the half-adder must make sure that both registers contain the same count. The half-adder compares the contents of A and B registers, digit by digit. As long as the half-adder has no sum output, the digit in A is the same as its respective digit in B. If there is a sum output, the digit in A is unlike the digit in B.

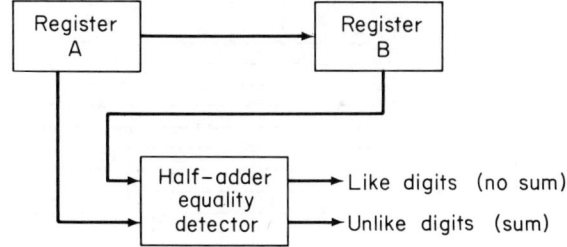

Fig. A-43. Half-adder used as comparison circuit (equality detector).

A-26. MULTIPLE-LINE ENCODERS AND DECODERS

In test equipment applications, it is often necessary to convert BCD data to multiple-line form, and vice versa. When multiple-line data is converted to BCD, the function is known as encoding, requiring the use of an *encoder* (sometimes known as a matrix). A decoder is the opposite of an encoder; that is, the decoder converts BCD data to multiple-line form. (An example of such a decoder is discussed in Chapter 6.)

A-26.1. Multiple-Line Encoder

The encoder shown in Figure A-44 converts 10-line decimal information to 42*21 BCD code. For example, assume that input line 3 is true but all other input lines are false. This would make OR gates 1 and 2 true, while the 4 and 2* OR gates would be false. The resultant output would be 0011, or a binary 3.

An encoder may have any number of inputs (only one of which is true at one time) and any number of binary bits on the output (various combinations of which may be made true). Encoders are often shown simplified on logic diagrams as rectangular boxes with the type of matrix (10-line to 42*21, and so on) indicated within the symbol.

A-26.2. Multiple-Line Decoder

The decoder shown in Figure A-45 converts a 42*21 BCD code to 10-line decimal form. Note that with this type of decoder, the $\overline{42}*\overline{21}$ inputs are also required.

Assume that the BCD input is 7 (1101 for a 42*21 code). Under these conditions, gate *A* would be true, but gates *B* and *C* would be false (the $\overline{4}$ input line would be false). Therefore, the 6, 7, 8, and 9 output gates could be true, but the 0, 1, 2, 3, 4, and 5 output gates must be false. The 7 output gate has all three inputs true (from gate *A*, the 1 line and the $\overline{2}$ line) and therefore produces

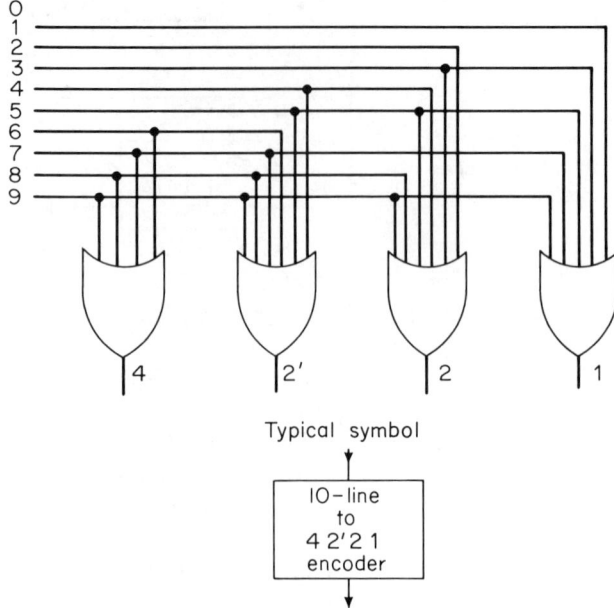

Fig. A-44. Typical multiple-line to BCD encoder.

a true output. Gates 6, 8, and 9 have at least one false input and thus produce false outputs.

Decoders are sometimes shown simplified on logic diagrams as rectangular boxes with the logic arrangement (42*21 to 10-line, and so on) indicated within the symbol.

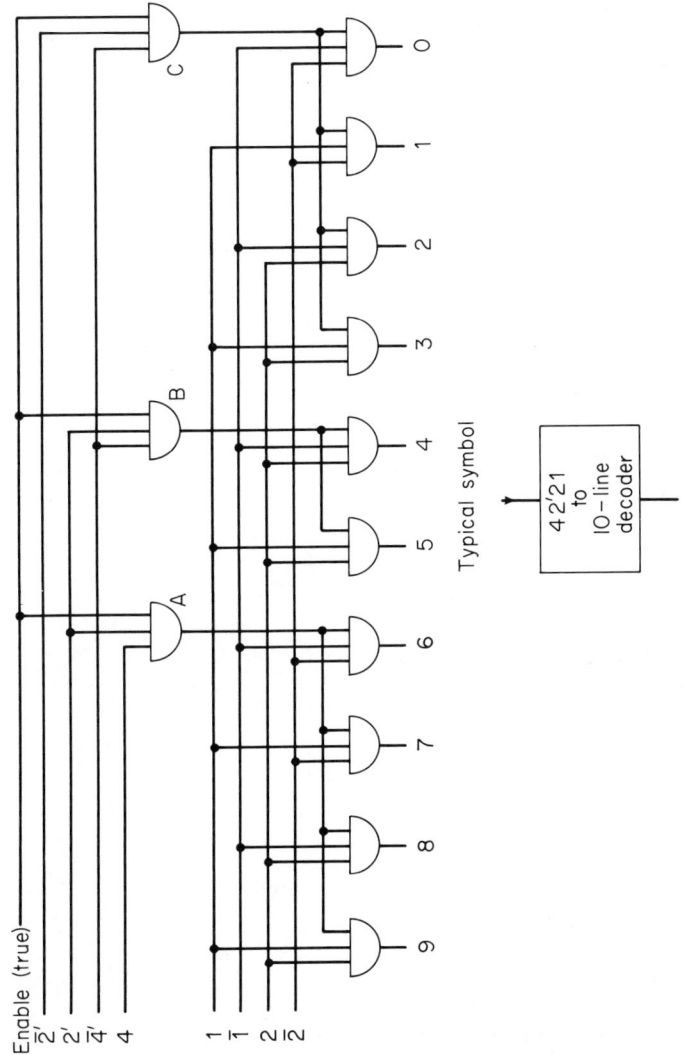

Fig. A-45. Typical BCD to multiple-line decoder.

INDEX

A

AC amplifiers 251
AC bridges 77
AC coupling 408
AC measurements:
  dual-slope meter 61
  electronic meters 20
AC meters 10
AC scales 26
AC voltage measurements, differential meter 66
Accuracy, meter 33
Adders 446
Admittance meter 89
Ammeter, basic 3
  measurements 41
  test and calibration 49
Amplification:
  by a constant 267
  high impedance input 272
Amplifiers 250
  characteristics 261–266
  differential 255
  logic 416
  logic symbols 405
  operational 258
  precautions 266
  test 261–265
Analog-to-digital (A/D) conversion 432, 435
Analog meters 1
AND gate 409

Attenuation measurements, microwave 393
Audio generators 122–130
Automatic bridge 86
Average value converter 61
  average voltage 12

B

Bandpass amplifier 274
Barretter 378
Beam deflection, oscilloscope 165
Binary-coded decimal (BCD) 403
  signal formats, 433
Binary logic 402
Binary numbers 403
Bolometer 378
Bridge rectifier, meter 12
Bridge-type test equipment 74

C

Calorimetric power measurement 380
Capacitance bridge 83
Capacitance-changing transducers 335
Capacitance (low) probes 318–320
Capacitance measurements, bridge 80
Cathode follower probes 323
Cathode-ray tube (oscilloscope) 164
Cavity wavemeter 396
Cesium beam (frequency standard) 292
Chopper 17

Clip-on meter 13
Clock, electronic 308
Color television generator 150–162
Common mode 69
Comparison circuits, logic 448
Conductance measurement, bridge 81
Counter:
   accuracy 247
   basics 232
   display 246
   electronic 232
   readout 239
Coupler, directional 384
Coupling, AC 408
Crystal frequency standard 294
Currents:
   ground 69
   measurements 41
   measurements (oscilloscope) 219
   probe 13
Curve tracers 195

D

D'Arsonval movement 2
DC amplifiers 252
DC scales 24
DC voltage measurements, differential meter 67
Decade box 46
Decade circuits (counter) 240–246
Decibel measurements 42
Decibel scales:
   correction factor 29
   using 30
Decibel-to-RMS conversion 30
Decoders, logic 449
Delay element 406
   logic 426
Delay flip-flop 423
Demodulator probes 49, 322
Detector, diode 384
Difference amplification 272
Differential amplifier 255
Differential meter 51
   basic 64
   circuits 66

Differentiation of signals 269
Digital-to-analog (D/A) conversion 432, 439
Digital circuits, introduction 51
Digital delay generator 139–144
Digital logic 405
Digital logic circuits 401
Digital meter 51
   basic 52
Diode detector 384
Directional coupler 384
Dissipation factor (D) 76
Distortion measurement 337
Divider circuit (counter) 239
Drift (in meters) 16
Dual-slope meter 59
Dummy load 386

E

Effective voltage 12, 13
Electronic clock 308
Electronic counter 232
Electronic meter, basic 15
Electronic switch (dual-trace) 171
Encode gates 415
Equations, logic 431
Exclusive OR gate 414

F

FET meter 15
Flip-flop, logic 418
Flowmeter 331
FM stereo generator 144–150
Four-bit system 434
Frequency:
   broadcast 281
   calibration (oscilloscope) 303
   calibration (standard) 300
   comparison, direct 305
   compensation, meters 48
   measurement 386
   measurement (counter) 235
   measurement (oscilloscope) 166, 220–227

Index 457

Frequency (cont.):
  meter, heterodyne 396
  standard equipment 289
  standard terms (definitions) 312
  standards 278
  synthesizer 130–139
  test and calibration, meters 47
Frequency-amplitude displays 351
Frequency-to-voltage conversion 276
Full adder 447
Function generator 122–132
Fuse, meter 31

G

Galvanometer:
  basic 10
  moving-coil 2
Gas detector 334
Gates 405
Generating transducers 320
Generators, signal 95
Ground currents 69
Guard circuit 58, 70

H

Half adder 446
Half-wave probe 321
Harmonic distortion 339
Heterodyne frequency meter 396
Hydrogen maser (frequency standard) 290

I

Impedance measurements, microwave 387
Impedance (vector) meter 91
Inductance bridge 83
Inductance-changing transducer 335
Inductance measurement, bridge 82
Injection, signal 121
Integrated circuit (IC) reference designations 429
Integrating potentiometric meter 58
Integration of signals 268

Intermodulation distortion 342
Inversion 408
Inverter, logic 416

J

J-K flip-flop 420

K

Kelvin-Varley divider 67
Klystron 374–378
  tube 372

L

Ladder, binary 435
Latching flip-flop 422
LC meter 92, 396
Lecher line 395
Linear transformer 335
Liquid-level measuring 335
Lissajous figures 225, 303
Logic:
  digital 401
  elements and symbols 405
  equations 431
  symbol identification 426
  symbology 401
  symbols 407

M

Magnetic-induction transducers 333
Marker adder 120
Marker generator 116–122
Mercury cell 44
Meter(s):
  AC 10
  accuracy 33
  analog 1
  basics 1
  clip-on 13
  digital and differential 51
  drift 17
  frequency response 47

Meters (cont.):
　internal resistance 5
　laboratory 15
　movement 34
　operating precautions 36
　operating procedures 36
　protection 31
　scales and ranges 22
　thermocouple 13
Microwave(s):
　attenuation measurements 393
　circuits, transfer of energy 370
　frequency measurements 394
　impedance measurements 387
　measurement 382
　power measurement 378
　resonant circuit 366
　signal sources 372, 374
　test equipment 364
Mirrored scales 33
Modulation measurement 343
Movements, meter 34
Moving-coil galvanometer 2
Multiple-line encoders and decoders 449
Multiplier, calculating values 7
Multivibrators, logic 423–425

N

NAND gate 413
NOR gate 414

O

Ohmmeter:
　basic 8
　measurements 37
　scales 22
　test and calibration 42
Ohms-per-volt 7
One-shot multivibrator (logic) 423
Operational amplifier 258
OR gate 411
Oscilloscope 163
　basic circuits 167–173
　current measurements 219

Oscilloscope (cont.):
　operating controls and indicators 173–182
　operating precautions 200–202
　operation 202
　phase measurements 227
　sampling 184
　storage 182
　time and frequency measurements 220–270
　trace recording 204
　voltage and current measurements 205–220

P

Parallax 33
Pattern drift method (time and frequency) 304
Peak-to-peak voltage 12
Peak power measurement 382
Peak-reading amplifier 276
Peak-responding meter 21
Peak voltage 12
Period measurement (counter) 236
Phase comparison 311
Phase measurements, oscilloscope 227
Phase splitters, logic 416
Piezoelectric-type transducers 333
Potentiometric method of voltage measurement 65
Power measurement, microwave 378
Primary standard, voltage 43
Probe 317
　with amplifier 267
　compensation 325
　current 13
　demodulator 49
　operation 326
　RF 11, 383
Protection, meters 31
Pulse generators 139–144

Q

Q meter 67
Quartz crystal (frequency standard) 294

Index   459

R

Radiation detector 334
Ramp-type digital meter 53
Ratio measurement (counter) 239
Recorders 163, 198
Rectifier, meter 11
Reference designations, logic 429
Register:
  shift 442
  storage 440
Relay, meter 32
Resistance, internal, meter 5
Resistance-changing transducer 328
Resistance measurement 37
  bridge 81
  differential voltmeter 71
  dual-slope meter 63
  electronic meters 18
  external power source 19
Resistance-type probes 319
Resonant circuits, microwave 366
RF probe 11, 320, 383
RF signal generator 96–106
RMS-to-dB conversion 30
RMS-responding meter 21
RMS voltage 12
R-S flip-flop 418
Rubidium vapor (frequency standard) 293
RX meter 88

S

Sampling oscilloscope 184
Schmitt trigger, logic 423
Self-generating transducers 330
Shift register 442
Shunt, ammeter 3
Shunts, calculating values 4
Sign comparators 445
Signal:
  conditioning 330
  generators 95
  injection 121
  sources, microwave 372
  tracing proves 324

Simpson Instruments Model 262, 270
Sines, table of 229
Sinewave analysis (distortion) 338
Sinewave relationships 12
Skin effect, microwave test equipment 365
Slotted line 384
Solar-cell transducers 333
Spectrum analyzer 349
Squarewave analysis (distortion) 339
Staircase ramp meter 55
Standard cell 44
Stereo (FM) generator 144–150
Storage factor (Q) 76
Storage oscilloscope 182
Storage register 440
Strip-chart recorder 198
Subtraction amplification 272
Summation of signals 271
Sweep and swept-frequency
  generators 106–116
Switching elements 406
Synthesizers, frequency 130–139

T

Tachometer transducers 332
Television generators (color) 150–162
Thermistor 378
Thermocouple:
  meter 13
  transducer 331
Time-amplitude displays 351
Time-base generator 116–122
Time broadcasts 281
Time calibration 307
Time comparison 309
Time-domain reflectometer 358
Time-interval measurements (counter) 238
Time measurements, oscilloscope 220–227
Time period measurement 53
Time standards 278
  equipment 289
  terms (definitions) 312

Toggle flip-flops 422
Totalizing (counter) 234
Transadmittance amplifier 273
Transducers 317, 327
Transformer-ratio bridge 84
Transistor curve tracers 195
Transmission lines, microwave 365
Traveling wave tube (TWT) 373
Triplett Model 850 Electronic
  Voltmeter 24
True-RMS converter 62
Truth polarity 407
Turnover (in meters) 27

U

Unity gain amplifier 271
Universal bridge 78

V

Varistor diode 32
Vector impedance meters 91
Vectorscope 189–194
Voltage-to-current converter 273
Voltage divider probe 318–320
Voltage-to-frequency conversion 55
Voltage gain amplifier 274
Voltage measurements 39
  oscilloscope 167, 215–218

Voltage-to-time conversion 54
Voltage-to-voltage amplifier 274
Voltmeter:
  basic 6
  measurements 39
  test and calibration 43
VOM (volt-ohmmeter) 1
VTVM (vacuum-tube voltmeter) 1
  basic 15

W

Wave analysis instruments 337
Wave analyzers 344
Waveguide resonant circuits 366
Waveguides 365
Wavemeter, cavity 396
Weston cell 44
Wheatstone bridge 74
WWV 283

X

X-Y recorder (plotter) 199

Z

Zero-beat method (frequency
  measurement) 301